ANTENNAS FOR RADAR AND COMMUNICATIONS: A POLARIMETRIC APPROACH • *Harold Mott*

INTEGRATED ACTIVE ANTENNAS AND SPATIAL POWER COMBINING • *Julio A. Navarro and Kai Chang*

FREQUENCY CONTROL OF SEMICONDUCTOR LASERS • *Motoichi Ohtsu (ed.)*

SOLAR CELLS AND THEIR APPLICATIONS • *Larry D. Partain (ed.)*

ANALYSIS OF MULTICONDUCTOR TRANSMISSION LINES • *Clayton R. Paul*

INTRODUCTION TO ELECTROMAGNETIC COMPATIBILITY • *Clayton R. Paul*

INTRODUCTION TO HIGH-SPEED ELECTRONICS AND OPTOELECTRONICS • *Leonard M. Riaziat*

NEW FRONTIERS IN MEDICAL DEVICE TECHNOLOGY • *Arye Rosen and Harel Rosen (eds.)*

ELECTROMAGNETIC PROPAGATION IN MULTI-MODE RANDOM MEDIA • *Harrison E. Rowe*

NONLINEAR OPTICS • *E. G. Sauter*

InP-BASED MATERIALS AND DEVICES: PHYSICS AND TECHNOLOGY • *Osamu Wada and Hideki Hasegawa (eds.)*

DESIGN OF NONPLANAR MICROSTRIP ANTENNAS AND TRANSMISSION LINES • *Kin-Lu Wong*

FREQUENCY SELECTIVE SURFACE AND GRID ARRAY • *T. K. Wu (ed.)*

ACTIVE AND QUASI-OPTICAL ARRAYS FOR SOLID-STATE POWER COMBINING • *Robert A. York and Zoya B. Popović (eds.)*

OPTICAL SIGNAL PROCESSING, COMPUTING AND NEURAL NETWORKS • *Francis T. S. Yu and Suganda Jutamulia*

InP-Based Materials and Devices

InP-Based Materials and Devices
Physics and Technology

Edited by

OSAMU WADA
FESTA Laboratories

HIDEKI HASEGAWA
Hokkaido University

A WILEY-INTERSCIENCE PUBLICATION
JOHN WILEY & SONS, INC.
NEW YORK / CHICHESTER / WEINHEIM / BRISBANE / SINGAPORE / TORONTO

This text is printed on acid-free paper. ♾

Copyright © 1999 by John Wiley & Sons, Inc. All rights reserved.

Published simultaneously in Canada.

No part of this publication may be reproduced, stored in a retrieval system or transmitted in any form or by any means, electronic, mechanical, photocopying, recording, scanning or otherwise, except as permitted under Sections 107 or 108 of the 1976 United States Copyright Act, without either the prior written permission of the Publisher, or authorization through payment of the appropriate per-copy fee to the Copyright Clearance Center, 222 Rosewood Drive, Danvers, MA 01923, (978) 750-8400, fax (978) 750-4744. Requests to the Publisher for permission should be addressed to the Permissions Department, John Wiley & Sons, Inc., 605 Third Avenue, New York, NY 10158-0012, (212) 850-6011, fax (212) 850-6008, E-Mail: PERMREQ @ WILEY.COM.

Library of Congress Cataloging-in-Publication Data:

InP-based materials and devices : physics and technology / edited by
   Osamu Wada, Hideki Hasegawa.
       p.   cm.
    "A Wiley-Interscience publication."
    Includes index.
    ISBN 0-471-18191-9 (cloth : alk. paper)
    1. Optoelectronics—Materials.   2. Indium phosphide.
 3. Optoelectronic devices.   I. Wada, O. (Osamu), 1946–
 II. Hasegawa, H. (Hideki), 1941–   .   III. Series.
 TA1750.I568   1999
 621.381′045—dc21   98-38547

Printed in the United States of America

10 9 8 7 6 5 4 3 2 1

Contents

Contributors		vii
Preface		ix
1	**Introduction** *Osamu Wada and Hideki Hasegawa*	1
2	**Demand for InP-Based Optoelectronic Devices and Systems** *Junichi Yoshida*	11
3	**Applications of InP-Based Transistors for Microwave and Millimeter-Wave Systems** *Mehran Matloubian*	37
4	**Material Physics of InP-Based Compound Semiconductors** *Yoshikazu Takeda*	71
5	**InP Bulk Crystal Growth and Characterization** *David Francis Bliss*	109
6	**Metal-Organic Chemical Vapor Deposition of InP-Based Materials** *Takashi Fukui*	165
7	**InP and Related Compound Growth Based on MBE Technologies with Gaseous Sources** *Harald Heinecke*	187

8	Physics and Technological Control of Surfaces and Interfaces of InP-Based Materials *Hideki Hasegawa*	247
9	Dry Process Technique for InP-Based Materials *Kiyoshi Asakawa*	289
10	Heterostructure Field Effect Transistors and Circuit Applications *Juergen Dickmann*	339
11	Heterojunction Bipolar Transistors and Circuit Applications *Hin-Fai Frank Chau and William Liu*	391
12	Lasers, Amplifiers, and Modulators Based on InP-Based Materials *Niloy K. Dutta*	449
13	Photodiodes and Receivers Based on InP Materials *Kenko Taguchi*	501
14	Hybrid Integration and Packaging of InP-Based Optoelectronic Devices *Werner Hunziker*	537
Index		583

Contributors

KIYOSHI ASAKAWA, Femtosecond Technology Research Association (FESTA), Tsukuba, Japan

DAVID FRANCIS BLISS, U.S. Air Force Research Laboratory, Hanscom Air Force Base, Massachusetts

HIN-FAI FRANK CHAU, EiC Corporation, Fremont, California

JUERGEN DICKMANN, DaimlerChrysler AG, Ulm, Germany

NILOY K. DUTTA, Photonics Research Center, University of Connecticut, Storrs, Connecticut

TAKASHI FUKUI, Research Center for Interface Quantum Electronics, Hokkaido University, Sapporo, Japan

HIDEKI HASEGAWA, Research Center for Interface Quantum Electronics, Hokkaido University, Sapporo, Japan

HARALD HEINECKE, BMW AG, Munich, Germany

WERNER HUNZIKER, Institute of Quantum Electronics, Swiss Federal Institute of Technology, Zurich, Switzerland

WILLIAM LIU, EiC Corporation, Fremont, California

MEHRAN MATLOUBIAN, Hughes Research Laboratories, Malibu, California

KENKO TAGUCHI, Opto-Electronics Research Laboratories, NEC Corporation, Tsukuba, Japan

YOSHIKAZU TAKEDA, Department of Materials Science and Engineering, Nagoya University, Nagoya, Japan

OSAMU WADA, Femtosecond Technology Research Association (FESTA), Tsukuba, Japan

JUNICHI YOSHIDA, NTT Opto-Electronics Laboratories, Atsugi, Japan

Preface

Indium phosphide (InP)-based materials and devices have made remarkable progress in recent years in the fields of optoelectronics and of microwave and millimeter-wave electronics. Despite the presence of a competing, even more mature, technology based on GaAs and related compounds, InP-based technology has grown and indeed, has established a unique area of semiconductor science and technology in which key optoelectronic devices and high-speed ICs with unsurpassed performance and high reliability are produced for use in various optical and microwave communication systems. With the recent very rapid evolution of information technology toward the development of a heavily multimedia society based in the twenty-first century, as evidenced by the recent worldwide explosion in Internet and cellular phone usage, it is expected that the roles of InP-based materials and devices will continue to expand very rapidly, thus becoming ever more important in the future.

Reflecting this rapid growth in InP-based technology, an international conference dedicated solely to InP-based materials and devices was held in 1988. This annual conference has been evolving ever since. The plan to publish this book was developed during a discussion that occurred while organizing IPRM'95, which was held in May 1995 in Sapporo, Japan. IPRM had a successful tenth anniversary in Tsukuba, Japan in May 1998. We are very pleased that we are able to publish this book celebrating the commemorative year for the international community involved in research on InP-based materials and devices.

The aim of the book is to review the basic physics and technologies that underlie InP-based materials and devices as well as their applications. Primary emphasis is placed on providing suitable material on present-day physics and standard technologies, but an effort has also been made to describe various ongoing technological challenges involved in overcoming the limitations of conventional technologies, so that readers can obtain insights into the future of this remarkable technology. The book starts with system needs calling for InP-based technology, and continues with basic physics and materials, processing, and electronic/optoelectronic device technology, together with information on circuit applications and packaging.

We wish to express our sincere gratitude to all of the distinguished contributors to this book who have written such excellent chapters in a timely fashion. In organizing the book and reviewing the manuscripts, we have also received help from a number of people, including the members of the IPRM'95 Program Committee. We would like to thank them, in particular H. Asahi (Osaka University), M. Abe, D. Iwai, T. Kikkawa, H. Kuwatsuka, and M. Sugawara (Fujitsu), T. Enoki and E. Sano (NTT), T. Kamijoh (Oki), A. Kasukawa (Furukawa), F. Koyama (Tokyo Institute of Technology), and O. Oda (Japan Energy) for their useful suggestions and comments.

OSAMU WADA
HIDEKI HASEGAWA

CHAPTER ONE

Introduction

OSAMU WADA
Femtosecond Technology Research Association (FESTA)

HIDEKI HASEGAWA
Hokkaido University

1.1	Applications of InP-Based Devices	1
1.2	Basic Technologies and Challenges	4
1.3	Book Outline	7
	References	9

1.1 APPLICATIONS OF InP-BASED DEVICES

The evolution of information technology has been very rapid, as can be seen from the recent worldwide explosive expansion of use of the Internet and cellular phones. The speed of technological development is becoming ever faster. As an example, Fig. 1.1 depicts the increase in the total volume of domestic information transmission in Japan for the last two decades. Changes in the throughput of trunk-line optical communication systems and microprocessor large-scale-integration (LSI) performance are also shown. The total information transmission volume consists of contributions by various media, including printed matter, such as newspapers and journals, and the contribution from electrocommunication technology, including

---

InP-Based Materials and Devices: Physics and Technology, Edited by Osamu Wada and Hideki Hasegawa.
ISBN 0-471-18191-9 © 1999 John Wiley & Sons, Inc.

television and radio broadcasting, telecommunications, and data communications [1]. However, since the development of optical communication technology in the early 1980s [2], where InP-based lasers and photodiodes played a key role, the information volume transmitted by electrocommunication technology has been increasing rapidly, and now occupies almost one-third of the overall information volume, as shown in Fig. 1.1.

The most recent commercial optical communication systems run at a bit rate of 10 Gb/s, and the feasibility of a transmission throughput exceeding 1 Tb/s has been demonstrated experimentally by utilizing both time-division multiplexing (TDM) and wavelength-division multiplexing (WDM) techniques [3–6]. This continuous progress in optical communication systems has been brought about by a variety of related new technologies that have realized low-transmission-loss silica optical fibers, high-optical-gain erubium-doped fiber amplifiers (EDFAs) in the 1.55-μm band, and a variety of InP-based LEDs, lasers, and photodiodes in the long-wavelength region, ranging from 1.3 to 1.55-μm. The progress of electronic signal-processing technology is another key factor that has contributed to this information volume expansion, and indeed progress has been remarkable, particularly in the area of Si very large scale integrated circuits (VLSIs), GaAs high-speed integrated circuits (GaAs-ICs), and system technologies. Along with such developments, production of InP-based optoelectronic devices has also been increasing lately, as illustrated by the trend in Japanese production of InP-based light source devices shown in Fig. 1.2. The production increase is most pronounced in 1.3-μm lasers, reflecting the fact that the application of optoelectronics is growing rapidly in medium-distance transmission systems, including local area networks (LANs). This technical

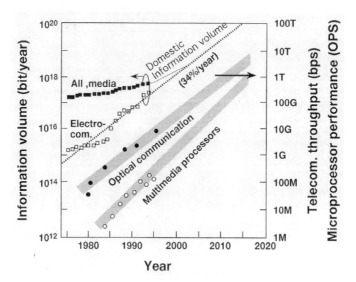

FIGURE 1.1 Increasing trends in Japanese domestic information transmission volume, throughput of trunk-line optical transmission systems, and performance of multimedia processor LSIs as functions of the calendar year.

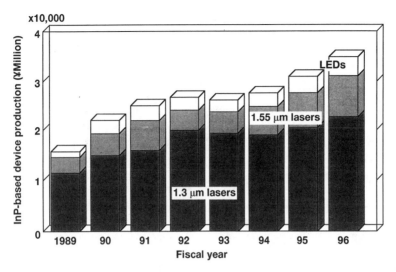

FIGURE 1.2 Trends in Japanese production of InP-based light source devices versus fiscal year. (From Ref. 7.)

evolution will continue and the social demand for a robust information infrastructure will increase. There is no indication that the volume of information to be transmitted and processed will slacken; on the contrary, it will definitely increase continuously in the future. In fact, it is anticipated that the communications and signal-processing systems will have to fulfill, circa 2010, a throughput of at least 1 Tb/s or even 5 Tb/s [8] and a performance of 100 GOPS (gigaoperations per second). Such an improvement in system performance by more than 100-fold can no longer be achieved simply by modifications in existing systems. Substantial improvements in device and system technologies are crucial for success.

Among many devices that are required for wider applications of optoelectronic systems in the future, the most important include ultrafast optoelectronic devices for very high speed TDM systems and those for even faster OTDM (optical TDM) systems as well as WDM devices for highly flexible network systems. Low-cost optical devices and modules that are applicable to optical access networks will also certainly become indispensable for the future penetration of optical communication technology into premises [fiber to the home (FTTH)]. In parallel with the application of InP-based devices in telecommunications, other application fields, such as optical interconnections within electronic systems and equipment, will become increasingly important in solving the problem of interconnect-speed limitation in conventional electronic systems [9]. Development of uniform array devices and their packaging techniques will become critical in this application. Such an application is also unique in the sense that it cannot be realized by conventional optical devices only, but will lead to a merger of electronics and optoelectronics.

In the electronic device field, the real benefit of using InP-based material systems has been widely recognized, due to recent demonstrations of high-performance transistors and system demonstrations, including 40-Gb/s optical transmission systems [10]. Microwave to millimeter-wave devices and monolithic microwave integrated circuits (MMICs) are expected to show high-performance, low-cost advantages for widening the market in wireless LANs, car radar, satellite communication systems, and even microwave-photonics transmission systems, in which the microwave signal is transmitted by light through modulation at the microwave to millimeter-wave frequency. The last example again suggests a merger of electronics and optoelectronics.

1.2 BASIC TECHNOLOGIES AND CHALLENGES

InP-based materials have various advantages, such as a bandgap energy suitable for light emitters and receivers in the long-wavelength region, an extremely high saturation velocity of electrons suitable for the active channel in high-power and high-speed electronic devices, as well as high thermal conductivity and high threshold of optical catastrophic degradation, both of which have merit for producing reliable devices. Due to these advantages, this series of materials has been used in many electronic and optoelectronic devices. On the other hand, these materials are mechanically brittle and still less mature in wafer- and device-processing technologies such as surface passivation, MIS (metal–insulator–semiconductor) and MS (metal–semiconductor) gate technology, and ohmic contact preparation [11]. Thus more studies and developments are still required in various areas of technology to fully utilize the intrinsic advantages of these materials. Since each of the technical areas of InP-based materials and devices is discussed in detail in the following chapters, only the recent technological trends are outlined in the remainder of this section.

Figure 1.3 shows the relationship between the energy gap and the corresponding photoemission wavelength at 300 K as functions of the lattice constant for a variety of binary and ternary alloy compound semiconductor material systems with a focus on InP-based systems [12]. An InGaAsP quaternary alloy can vary its energy gap to cover 1.3- and 1.55-μm wavelengths while keeping the lattice-matching condition to the InP substrate. Since the first room-temperature continuous-wave (CW) operation of InGsAsP/InP lasers in 1976 [13], InGsAsP/InP heterostructures have been the standard material system most often used in light emitters, including light-emitting diodes (LEDs) and a variety of Fabry–Perot (FP) and distributed-feedback (DFB) lasers. InGaAs ternary alloy lattice matched to InP at the indium fraction of $x = 0.53$ has a bandgap of 0.73 eV at room temperature. It has been used as the photoabsorption layer in pin photodiodes (PIN-PDs) and avalanche photodiodes (APDs) for wavelengths of up to 1.6 μm. The development of highly controllable, uniform growth techniques, such as metal-organic chemical vapor deposition (MOCVD), has contributed to the industrial application of these devices.

Due to the narrow gap and the resultant small effective mass, electrons in the InGaAs ternary material exhibit a mobility in excess of 10,000 cm^2/V·s and a saturation velocity of over 2.4×10^7 cm/s [14]. Additionally, InAlAs/InGaAs lattice-matched heterostructures on InP offer a conduction band discontinuity as large as 0.5 eV. Combination of these features has led to very high performance of high-electron-mobility transistors (HEMTs), or more generally, heterostructure field-effect transistors (HFETs) exceeding that of their GaAs-based counterparts. The small bandgap of InGaAs is also advantageous for low-turn-on-voltage heterojunction bipolar transistors (HBTs), and is very attractive for application to ultrahigh-speed, low-power-consumption circuits.

Figure 1.3 also indicates that there exist strict limitations between the lattice-matching condition and the energy band lineup, particularly the conduction band discontinuity, which determines the efficiency of electron confinement within quantum-well and modulation-doped heterostructures. Attempts have been made to resolve these constraints by introducing a variety of new material systems with new growth techniques. The material systems, which target primarily temperature-insensitive 1.3-μm lasers, are indicated by various black and shaded vertical bars in Fig. 1.3. They include InGaAlAs/InAlAs/InP strained-layer quantum-well structures [15], InGaP/InAsP/InP strained-layer quantum-well structures [16], InGaAs/InGaP/InGaAs strained-layer quantum-well structures grown by introducing a composition-graded InGaAs buffer layer on a GaAs substrate [17], InAlGaAs/InGaAs/InGaP structures grown on a substrate cut from an InGaAs ternary bulk crystal [18], and GaInNAs alloy systems grown on GaAs substrates [19].

Similarly, advancement of technologies is being pursued in a variety of basic technological areas required for InP-based material preparation and device fabrica-

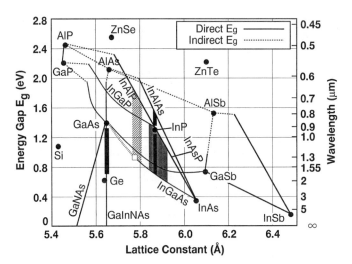

FIGURE 1.3 Variation of energy gap and corresponding photoemission wavelength with lattice constant at room temperature for various InP-based semiconductor alloys. Areas marked by shaded and black vertical bars indicate the alloy combination and composition variations introduced by recent novel heterostructure growth techniques, as explained in the text. (From Ref. 12.)

6 INTRODUCTION

tion. Taking optoelectronic device fabrication as an example, Fig. 1.4 illustrates various steps that are essential for device fabrication, together with the technology under investigation. As shown in the left half of Fig. 1.4, a variety of technical refinements are being carried out within each of the conventional fabrication steps, such as an increase in bulk and epitaxial wafer size, an improvement in uniformity and the development of more cost-effective, reliable, and highly uniform device processing, packaging, and so on.

It is important to note that new technical trends are appearing in many examples in order of linking or combining neighboring fabrication steps to overcome the existing technical barriers. Some of them are included in Fig. 1.4 as "link" technology. For example, the use of lattice-mismatched heteroepitaxy and the use of ternary substrates, as mentioned above, as well as the wafer-fusion technique, in which two different wafers are joined at an elevated temperature for reconstruction of interface bonding [20], are regarded as link technologies to solve problems originating from lattice-matching requirements. Selective-area epitaxy using growth masks in MOCVD [21] and in situ processing using a system consisting of growth and dry process chambers [22] are attempts to combine the growth–process sequence and may become future fabrication techniques that are particularly useful for the fabrication of nanostructure devices. Replacement of conventional lateral device structures by vertical device structures, such as in vertical cavity surface-emitting lasers (VCSELs), as well as the use of either hybrid or monolithic integration in the form

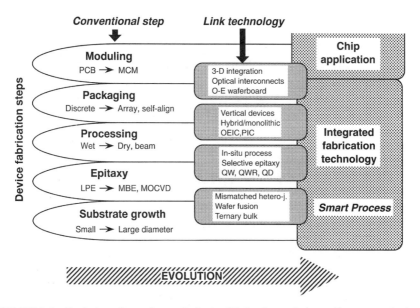

FIGURE 1.4 Evolution of optoelectronic device fabrication techniques. Recent trends show the development of a variety of new techniques to link two adjacent fabrication steps to overcome constraints existing within each of the conventional steps, suggesting a trend toward the overall integration of device fabrication technology.

of optoelectronic integrated circuits (OEICs) and photonic integrated circuits (PICs), are link technology that would solve the complexity and yield problems in the processing and packaging steps [23]. Even at a higher level of device implementation, the introduction of optical interconnection technique would simplify module design and fabrication [23].

Thus the trend toward integration is present not only in device structures but also in various steps in material growth and device fabrication. All of the trends of linking conventional fabrication steps to breaking through the problems that exist in the conventional method may eventually merge all the necessary fabrication technologies into a single technology, which one might call a "smart" processing technology, through integration. Such trends for technological merger between electronics and optoelectronics as well as among various fabrication process steps are expected to give an answer to the important question of how to produce high-performance, highly functional, low-cost devices that are necessary to realize future Tb/s systems.

1.3 BOOK OUTLINE

The purpose of this book is to cover the basic physics and technologies of InP-based materials and devices as well as their applications to systems. A primary emphasis is placed on the description of basic and well-defined technologies of the present day. However, an effort was also made to include ongoing technical challenges to overcome the limitations in each of the conventional technological areas, so that readers can have an insight into the future on the basis of sound knowledge of the present technology. The book begins with the present chapter, an overview of the present status and future challenges of InP-based material and device technology. In the following two chapters the roles and required performance of InP-based devices in optoelectonic and high-speed electronic systems are discussed in more detail. Chapter 4 begins a description of each of a variety of basic technological areas of InP-based material and device technology. Starting with materials physics, these descriptions include processing, then electronic devices and their applications, and finally, optoelectronic devices and their applications.

In Chapter 2, optoelectronic devices and their application to optical communication networks are described, with an emphasis on the evolution of integration technology. High-frequency electronic devices, including HEMTs, HBTs, and their integrated circuits, and the evolution of microwave systems are discussed in Chapter 3. Both chapters represent the state-of-the-art technologies based on InP-based devices and point out keys to future technical challenges.

Chapters 4 to 7 deal with InP-based materials physics and growth techniques. The basic physics of various alloy systems based on InP are covered in Chapter 4. The thermodynamical, crystallographic, electronic and optical properties of InP-based materials are summarized in tables useful for frequent reference; basic heterostructure materials important in practical device applications are included. InP bulk crystal growth and characterization are described in Chapter 5. A variety of crystal growth techniques are explained and improvements in crystal quality, including defect control, are discussed. Also, bulk crystal characterization techniques and results on recent crystals are demonstrated. Recent progress in metal-organic chemical vapor deposition techniques for InP-based materials and heterostructures are described in Chapter 6. Chapter 7 covers molecular beam epitaxy of InP-based materials, focusing on recent advances in MOMBE and CBE. In these chapters, not only the growth of conventional heterostructures and quantum-well structures, but also more recent progress in InP-based material growth, such as selective-area growth for waveguide device applications and self-organization growth of quantum dots, is included.

Chapters 8 and 9 deal with processing-related topics. The physics and technology of surfaces and interfaces of InP-based materials are discussed in Chapter 8. The basic physics on band lineup at heterointerfaces and Schottky barrier formation as well as Fermi-level pinning at surfaces and interfaces of InP-based materials are reviewed and techniques for their control are described. Chapter 9 deals with dry processing of InP-based materials, with an emphasis on the dry etching of nanostructures. Etching mechanisms, profile control, and damage characterization are described and the potential for their application to nanostructure devices is pointed out.

Chapters 10 and 11 deal with electronic devices, describing the heterostructure field-effect transistor (HFET) and hetrojunction bipolar transistor (HBT), respectively, together with their integrated circuits. Chapter 10 focuses on InP-based HFETs, and the basic principles, design, performance of devices, and state-of-the-art circuit demonstrations of operation at frequencies exceeding 40 GHz are presented. HBT technology, including basic structures, fabrication processes, characteristics and modeling, and state-of-the-art demonstrations of analog, digital, and mixed-signal circuits is described in Chapter 11.

Optoelectronic device technology is discussed in Chapters 12 to 14. Chapter 12 covers InP-based lasers, amplifiers, modulators, and laser-based integrated devices for high-speed and WDM optical communications and interconnections. The basics of lasers, recent progress in strained quantum-well lasers and amplifiers, modulators, laser arrays, and photonic integrated circuits are discussed. These devices show an operation bit rate of a few tens of Gb/s. Chapter 13 deals with photodetectors, with a focus on PIN-PDs and APDs, which are the most important elements in current optical communication systems. Not only basic operation and design of conventional structures, but also various recent improvements, such as superlattice APDs and integrated circuits, are reviewed. The integration and hybrid packaging technology for optoelectronic devices are discussed in Chapter 14. To fulfill increasing needs for low-cost-device fabrication, the establishment of simple, reliable hybrid integration techniques becomes as important as reduction in the cost of the device itself. This chapter focuses on this issue by introducing a packaging technique using waferboards and self-alignment for multichannel device arrays.

In editing the book, an attempt has been made to retain consistent terminology and notation throughout the volume. However, the structure of each chapter has been organized independently so that the author of each chapter can describe a topic most effectively. It is recommended that readers refer to other chapters by using the cross-references in the text and the subject index at the end of the book.

REFERENCES

1. *Communication White Paper*, 1992, 1996, Ministry of International Trade and Industry, Tokyo.
2. T. Miki, "Optical transport networks," *Proc. IEEE*, **81**, 1594–1609 (1993).
3. H. Onaka, H. Miyata, G. Ishikawa, K. Otsuka, H. Ooi, Y. Kai, S. Kinoshita, H. Nishimoto, and T. Chikama, "1.1 Tb/s WDM transmission over a 150 km 1.3 μm zero-dispersion single-mode fiber," presented at OFC'96, San Jose, Calif., 1996, *Tech. Dig. OFC'96*, postdeadline paper PD19.
4. A. H. Gnauck, A. R. Charplyvy, R. W. Tkach, J. L. Zyskind, J. W. Tkach, A. L. Lucero, Y. Sun, R. M. Jopson, F. Forghieri, R. M. Derosier, C. Wolf, and A. R. McCormick, "One terabit/s transmission experiment," presented at OFC'96, San Jose, Calif., 1996, *Tech. Dig. OFC'96*, postdeadline paper PD20.
5. T. Morioka, H. Tanaka, S. Kawanishi, O. Kamatani, K. Takiguchi, K. Uchiyama, M. Saruwatari, H. Takahashi, M. Yamada, T. Kanamori, and H. Ono, "100 Gbit/s × 10 channel OTDM/WDM transmission using a single supercontinuum WDM source," presented at OFC'96, San Jose, Calif., 1996, *Tech. Dig. OFC'96*, postdeadline paper PD21.
6. Y. Yano, T. Ono, K. Fukuchi, T. Ito, H. Yamazaki, M. Yamaguchi, and K. Emura, "2.6 Terabit/s WDM transmission experiment using optical optical duobinary coding," *Proc. ECOC'96*, Oslo, 1996, Vol. 5, paper ThB.3.1, pp. 3–6.
7. Optoelectronics Industry and Technology Development Association, Reports of Optoelectronics Industry Trend, OITDA, Tokyo, 1989–1997.
8. *Optoelectronic Technology Road Map for Optical Communications*, Version 1, OITDA, Tokyo, 1996.
9. J. W. Goodman, F. I. Leonberger, S. Y. Kung, and R. A. Athale, "Optical interconnections for VLSI systems," *Proc. IEEE*, **72**, 850–866 (1981).
10. S. Kuwano, N. Takachio, K. Iwashita, T. Ohtsuji, Y. Imai, T. Enoki, K. Yoshino, and K. Wakita, "160 Gbit/s (4-ch × 40 Gbit/s electrically multiplexed data) WDM transmission over 320-km dispersion-shifted fiber," presented at OFC'96, San Jose, Calif., 1996, *Tech. Dig. OFC'96*, postdeadline paper PD25.
11. A. Katz, *Indium Phosphide and Related Materials: Processing, Technology, and Devices*, Artech House, Norwood, Mass., 1992.
12. *Properties of Indium Phosphide*, EMIS Data Rev. Ser., No. 6, INSPEC/IEE, London, 1991.
13. J. J. Hsieh, J. A. Rossi, and J. P. Donnelly, "Room temperature CW operation of GaInAsP/InP double heterostructure diode lasers emitting at 1.1 μm," *Appl. Phys. Lett.*, **28**, 709–710 (1976).
14. R. F. Leheny, "High-field transport measurements," in *GaInAsP Alloy Semiconductors*, ed. T. P. Peasall, Wiley, Chichester, West Sussex, England, 1982, pp. 275–294.
15. C. E. Zah, R. Bhat, B. Pathak, F. Favire, W. Lin, M. C. Wang, N. C. Andreadakis, D. M. Hwang, M. A. Koza, T. P. Lee, Z. Wang, D. Darby, D. Flanders, and J. J. Hsieh, "High performance uncooled 1.3 μm $Al_xGa_yIn_{1-x-y}As$/InP strained-layer quantum well lasers for subscriber loop applications," *IEEE J. Quantum Electron.*, **30**, 511–523 (1994).

16. N. Yokouchi, N. Yamanaka, N. Iwai, T. Matsuda, and A. Kasukawa, "InAsP/InGaP all ternary strain-compensated multiple quantum wells and its application to long wavelength lasers," *Proc. 7th Int. Conf. Indium Phosphide and Related Materials,* Sapporo, Japan, 1995, pp. 57–60.
17. T. Uchida, H. Kurakake, H. Soda, and S. Yamazaki, "CW operation of a 1.3-μm strained quantum well laser on a graded InGaAs buffer with a GaAs substrate," *Proc. 7th Int. Conf. Indium Phosphide and Related Materials,* Sapporo, Japan, 1995, pp. 22–25.
18. K. Otsubo, H. Shoji, T. Kusunoki, T. Suzuki, T. Uchida, Y. Nishijima, K. Nakajima, and H. Ishikawa, "Low threshold and record high T0 (140 K) long wavelength strained quantum well lasers on InGaAs ternary substrates," *Extended Abst. 1997 Int. Conf. Solid State Devices and Materials,* Hamamatsu, Japan, 1997, late news paper C-4-6, pp. 556–557.
19. M. Kondow, K. Uomi, A. Niwa, T. Kitatani, S. Watahiki, and Y. Yazawa, *Extended Abstr. 1995 Int. Conf. Solid State Devices and Materials,* Osaka, Japan, 1995, p. 1016.
20. A. R. Hawkins, W. Wu, P. Abraham, K. Streubel, and J. Bowers, "High gain-bandwidth-product silicon heterointerface photodetector," *Appl. Phys. Lett.,* **70**, 303–305 (1997).
21. M. Aoki, H. Sano, M. Suzuki, M. Takahashi, K. Uomi, and A. Takai, "Novel structure MQW electroabsorption modulator/DFB-laser integrated device fabricated by selective area MOCVD growth," *Electron. Lett.,* **27**, 2138–2140 (1991).
22. T. Ishikawa and Y. Katayama, "Advanced semiconductor processing technology," in *Optoelectronic Integration: Physics, Technology, and Applications,* ed. O. Wada, Kluwer Academic, Norwell, Mass., 1994, Chap. 4, pp. 107–142.
23. I. Hayashi, "Future OEICs: the basis for photoelectronic integrated systems," in *Integrated Optoelectronics,* ed. M. Dagenais, R. F. Leheny, and J. Crow, Academic Press, San Diego, Calif., 1995, Chap. 17, pp. 645–675.

CHAPTER TWO

Demand for InP-Based Optoelectronic Devices and Systems

JUNICHI YOSHIDA
NTT Opto-Electronics Laboratories

2.1	Introduction	12
2.2	Evolution of Optical Fiber Communication Systems	12
	2.2.1 Trunk Networks: Toward the Multimedia Backbone Network	13
	2.2.2 Access Networks: B-ISDN Services to the Subscriber	16
2.3	Evolution of InP-Based Optoelectronic Devices	17
	2.3.1 High-Performance Lasers and Modulators: Devices for Systems Operating Beyond 10 Gb/s	17
	2.3.2 Spot-Size-Converter Waveguides: Toward Cost-Effective, High-Performance Structures	18
	2.3.3 Optoelectronic and Photonic Integrated Circuits: Devices for Photonic Networks	20
2.4	Challenges of InP-Based Optoelectronic Devices	22
	2.4.1 Exploring Much Higher Performance: Devices for High-Speed Systems and WDM Networks	22
	2.4.2 Exploring Reduction in Device Costs for Key Components of Optical Systems	25
2.5	Conclusions	26
	References	27

InP-Based Materials and Devices: Physics and Technology, Edited by Osamu Wada and Hideki Hasegawa.
ISBN 0-471-18191-9 © 1999 John Wiley & Sons, Inc.

2.1 INTRODUCTION

Indium phosphide optoelectronic devices play key roles in optical fiber communication systems. The recent rapid growth of multimedia services is driving a demand for optical communication systems with much more bandwidth, so the optoelectronic devices used in these systems must be improved. In this chapter we describe the progress of optical fiber communication systems and InP-based optoelectronic devices and discuss the global demand for their evolution. Also outlined here is the direction of their development as key devices for the next generation of optical communication networks: photonic networks.

2.2 EVOLUTION OF OPTICAL FIBER COMMUNICATION SYSTEMS

The transmission capacity and data rate of optical fiber transmission systems have been increasing rapidly over the past two decades. This reflects social needs for various kinds of convenient, user-friendly communication networks and systems. The demand for high-speed computer communication networks and systems such as the Internet and LANs (local area networks) has shown tremendous growth, which in turn has stimulated various service deployments, trials, and experiments around the world. The volume of multimedia data files, especially files containing video or picture data, is growing larger and larger. Consequently, the required transmission speed for these multimedia services should be higher than that of today's networks. For example, data files of a movie program contain several gigabytes of information; therefore, the required tranmission speed should be 10 Mb/s or more for transferring the program within a tolerable delay. Not only movie program distribution but also various kinds of software or journal article downloading services, which are inevitable for multimedia services, require a similar transmission speed. These circumstances lead to the conclusion that large-capacity data transmission systems are inevitable for both trunk and access networks in the multimedia era.

In Europe and the United States, nationwide projects such as RACE (Research and Development of Advanced Communication Technologies in Europe) and ACTS (Advanced Communication Technologies and Services), and those organized by DARPA (Defence Advanced Research Projects Agency), have been demonstrating the potential and feasibility of various optical communication systems, networks, and services through a wide range of trials [1–4]. ACTS projects conducted by the EC (European Community) have been developing strong activities in optical fiber transmision systems, such as high-speed optical communication systems for transmission speeds beyond 10 Gb/s, WDM (wavelength-division multiplex) systems for both trunk and access networks, and PON (passive optical network), with various kinds of multimedia services, such as video distribution, the Internet, and computer communication [5]. In the United States, 1.5-μm-wavelength eight-channel D-WDM (dense wavelength-division multiplex) ring and transport network experiments with 200-GHz channel spacing have been conducted between the New Jersey and Washington, DC areas by the MONET (multiwavelength optical networking)

2.2 EVOLUTION OF OPTICAL FIBER COMMUNICATION SYSTEMS

program. This is a consortium of several U.S. operating companies and laboratories [6,7]. In Japan, field experiments for multimedia services have been conducted by NTT since 1994 [8]. In the Japanese multimedia field experiment, optical fiber transmission systems operating at 2.5 and 10 Gb/s have been used as backbone networks, and 156-Mb/s optical fiber transmission systems have been introduced for high-speed applications (Fig. 2.1). Regarding access networks, an optical fiber subscriber system with video distribution capability overlaid on digital communication services has been investigated and is a strong candidate for the FTTH (fiber to the home) system [9,10]. These experiments with new services and FTTH systems are exploring the demand for multimedia services and are testing hardware and software under actual operating conditions. Even before the results of these experiments are final, however, the shift in demand from N-ISDN (narrowband ISDN) to B-ISDN (broadband ISDN) has become clear, and the demand for more bandwidth has become obvious. Optical fiber transmission systems can thus play an essential role not only in trunk networks but also in access networks.

2.2.1 Trunk Networks: Toward the Multimedia Backbone Network

Figure 2.2 shows the evolution of optical fiber transmission systems in Japan [11,12]. A 10-Gb/s system is already in commercial service in Japan using 1.55-μm dispersion-shifted single-mode fibers [13,14], while 2.5-Gb/s systems have been

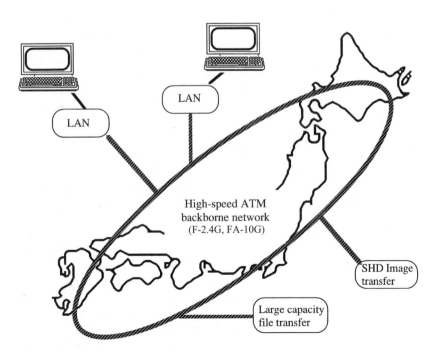

FIGURE 2.1 High-speed computer communication experiments in Japan.

14 DEMAND FOR InP-BASED OPTOELECTRONIC DEVICES AND SYSTEMS

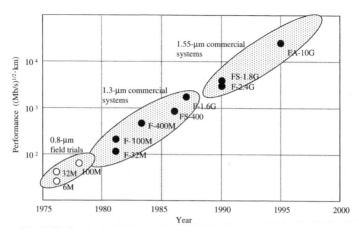

FIGURE 2.2 Evolution of optical fiber transmission systems in Japan. Performance = (bit rate)$^{1/2}$ × (repeater span).

widely used in Europe and the United States using standard 1.3-μm zero-dispersion fibers. All these recent systems have utilized the 1.5-μm wavelength region as the transmission wavelength, but the difference in the type of fibers reflected the difference in the approach to system upgrading in terms of dispersion management. Because of the dispersion, it is difficult to attain 10-Gb/s transmission systems with the standard 1.3-μm zero-dispersion fiber, and dispersion compensation has been a big issue.

For a larger transmission capacity, D-WDM systems have been studied for several years [15,16]. A system with eight 2.5-Gb/s WDM channels has been deployed in the United States [17], and similar systems have also been studied in Europe and Japan. Table 2.1 summarizes major projects on D-WDM systems throughout the world. These projects utilizes 4- to 8- or 16-channel WDM; it is desirable to have as many channels as possible to achieve a higher transmission capacity. However, about 25 nm of the bandwidth of conventional fiber amplifiers (EDFA: erbium-doped fiber amplifier) and nonlinear effects such as four-wave mixing are the major factors that limit the number of channels of D-WDM systems [18]. To overcome these problems, many works, such as broadband EDFAs with over 80 nm of bandwidth [19], flat-gain EDFAs [20,21], non-equal-channel-spacing arrayed-waveguide grating filters [22], dispersion compensation fibers [23,24], and variable group-delay programmable optical filters [25], have been studied intensively. All-optical networks using D-WDM systems have also been studied intensively [26] throughout the world in efforts to develop the next generation of optical communication networks: photonic networks. System experiments with a total transmission capacity of over 1 Tb/s have already been reported by several laboratories [27–30]. Although the technology needed for simple point-to-point D-WDM transmission systems has already been deployed, it is necessary to investigate technological issues of D-WDM systems thoroughly to exploit the advantages of optical wavelengths and

TABLE 2.1 Various Projects on D-WDM in the World

	Europe			United States		Japan	
Project[a]:	OPEN	PHOTON	METON	MONET	NTON	WP/VWP	FRONTIER
Network configuration	4/8/16 ch.–2.5/10	8 ch.–2.5/5/10	4 ch.–622M/2.5	8 ch.–2.5	4/8 ch.–2.5	8/16 ch.–2.5	16/32 ch.–2.5/10
Field trials	Paris–Brussels, Oslo–Hjorring/ Thisted	Munich–Vienna	Stockholm	NY, NJ– Washington, DC	San Franciso	Yokusuka– Atsugi– Masashino	
Participants	Alcatel, Belgacom, CSELT, CNET, etc.	Siemens, DT, HHI, TU-Wien, Philips, etc.	Ericson, CSELT, CT, CNET, Thomson, etc.	AT&T, Lucent, Bellcore, Bell Atlantic, Bell South, etc.	HP, Hughes, Nortel, etc.	NTT	NTT

[a]OPEN, Optical Pan-European Network; PHOTON, Pan-European Photonic Transport Overlay Network; METON, Metropolitan Optical Network; MONET, Multiwavelength Optical Networking; NTON, National Transparent Optical Network Technology; WP/VWP, Wavelength Path/Virtual Wavelength Path; FRONTIER, Frequency-Routing Time-Division Interconnection Network.

frequencies throughout the entire range of network hierarchies. Wavelength routing, reconfiguration, and wavelength-interchangeable optical cross-connect technologies are the issues that need to be studied more deeply from the viewpoint of both system and device technologies and of cost [31,32].

2.2.2 Access Networks: B-ISDN Services to the Subscriber

For access networks, optical fiber subscriber systems utilizing the PDS (passive double star) configuration [33] have attracted much attention among telecommunication operators all over the world because of their potential for reducing system costs. ATM (asynchronous transfer mode)-based PDS systems have been discussed intensively within the FSAN Initiative (Full Service Access Network Initiative) [34], which includes representatives from France (CNET), Germany (Deutsche Telekom), Italy (CSELT and Telecom Italia), Japan (NTT), the Netherlands (PTT Telekom), the United Kingdom (British Telecom), and Spain (Telefonica). Nine equipment manufacturers are also included in the initiative. Its activities include finding common architecture and standard components for the full-service access networks that will be introduced within the next few years. Part of the discussion within the initiative was presented at a conference in June 1996 in London [35]. Access networks that the FSAN Initiative has been discussing include FTTx (fiber to the x, where x = cabinet, curb, building, or home) topologies based on ATM-PDS systems (Fig. 2.3). Two system configurations, a one-fiber WDM (1.3-μm/1.5-μm) system and a 1.3-μm two-fiber system, have been under discussion. The target system will supply broadband services at rates up to 10 Mb/s (including multimedia services) not only to business users but also to home users. Most of the telecommunication operators expect to deploy services for business users first, but their final target is to provide FTTH [34,35] so that each home user can at any time enjoy real-time bidirectional communication with full motion pictures on his or her own terminal.

Basic concepts of the D-WDM subscriber network have been investigated, to-

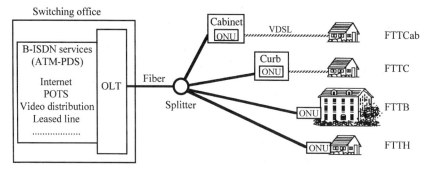

FIGURE 2.3 Access network topologies: FTTx. FTTCab, Fiber-to-the-Cabinet; FTTC, Fiber-to-the-Curb; FTTB, Fiber-to-the-Building; FTTH, Fiber-to-the-Home.

ward configurations of access networks in the future, by using a reflective modulator receiver [36] or a DFB (distributed feedback) one-chip transceiver [37] in the ONU (optical network unit). Figure 2.4 shows an outline of a system utilizing a reflection-type receiver. Downstream optical signals are divided into two portions in the ONU. One is received as downstream signals, and the other is modulated by the upstream electrical input signals and directed back to the central office. Since neither laser diodes nor wavelength-control devices are necessary at the customer ONU, this configuration will enable the cost of the subscriber terminal equipment (the ONU) to be kept sufficiently low. In the central office, however, D-WDM systems require sophisticated multiple-wavelength generators, optical frequency controllers, or switching devices. Mature technologies for these devices have not yet been developed, nor has sufficient performance been obtained from devices whose costs are reasonably low. Despite the extensive work that has already been reported [38–40], further research and development on WDM devices will be necessary to construct actual systems.

2.3 EVOLUTION OF InP-BASED OPTOELECTRONIC DEVICES

Research on InP-based optoelectronic devices such as semiconductor laser diodes, photodiodes, modulators, switches, and amplifiers has been accomplished together with the evolution of optical transmission systems. InP-based optoelectronics devices for future transmission systems must evolve in many directions: toward higher performance, more functionality, and lower cost. The utilization of hybrid and monolithic integration technologies will make it possible to produce high-performance devices that are compact, reliable, and inexpensive.

2.3.1 High-Performance Lasers and Modulators: Devices for Systems Operating Beyond 10 Gb/s

Increasing the transmission speed into the gigabit-per-second range without shortening the repeater span requires the use of single-longitudinal-mode laser diodes

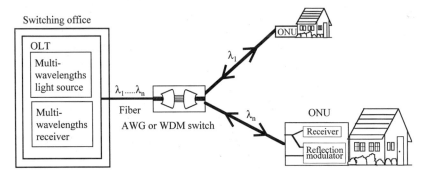

FIGURE 2.4 Concept of a D-WDM subscriber system utilizing a reflection modulator.

such as DFB (distributed feedback) or DBR (distributed Bragg reflector) lasers in place of Fabry–Perot laser diodes. Furthermore, systems operating at these higher bit rates require the low-chirp characteristics of the laser oscillation frequency to minimize the dispersion effect of the optical fiber. To simplify the structure of the light source, DFB lasers with an integrated MQW (multiple-quantum-well) electroabsorption modulator [41–44] have been quite promising and have been widely utilized in transmission experiments [45–47]. A DFB laser with an integrated modulator [48] capable of 40-Gb/s operation at a bias voltage of 3 V has already been developed (Fig. 2.5).

The chirping characteristics and driving voltages of semiconductor optical modulators have been improved remarkably by introducing strained-MQW structures, and MQW InP-based electroabsorption modulators have exhibited a bandwidth greater than 40 GHz [49–51] and an operating voltage below 1 V [52]. A 160-Gb/s (4 × 40 Gb/s) transmission experiment using these InGaAs/InAlAs MQW modulators has been reported [53]. Electroabsorption semiconductor optical modulators have been limited by chirping characteristics that are relatively large compared with those of LN (lithium niobate) modulators [54,55]. Analysis of the chirping characteristics of MQW modulators under both tensile and compressive strain conditions, however, has recently revealed that there is an optimum strain condition at which the chirp parameter α is negative under an applied field lower than those conventionally applied [56,57]. This kind of study has been performed on simulation software by changing quantum-well parameters to improve the device characteristics. Optical CAD (computer-aided design) systems [58] will thus become powerful tools not only for device design and analysis but also for evaluating device characteristics.

2.3.2 Spot-Size-Converter Waveguides: Toward Cost-Effective High-Performance Structures

InP-based optical devices have been used mainly in the trunk network, where the cost or price of optical devices is not of great concern because such devices are shared by a number of customers. The recent rapid growth of new services, howev-

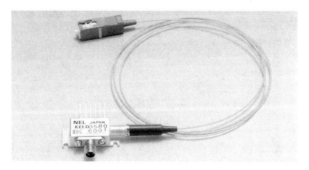

FIGURE 2.5 High-speed electroabsorption-modulator-integrated DFB laser module.

er, is encouraging telecommunication operators to consider opticalization of the access networks, for which costs and prices are critical. Reducing the cost of the optical devices has thus become a big issue. Most of the work that has been done to reduce the cost of the optical devices used for access networks has been concentrated on reducing coupling losses between lasers and fibers or waveguides by increasing the spot size of the output light beam from the laser diode [59–65]. To form laser diodes with a large spot size, two types of semiconductor waveguide structure have generally been utilized (Fig. 2.6). The principle of spot-size conversion or mode-field matching, which has been widely adopted in glass waveguide devices, is also used in semiconductor devices. These types of laser diodes are called SS-LDs (spot-size-converter-integrated laser diodes) [66]. By using SS-LDs, lensless optical coupling with the fiber or a waveguide can be achieved with large coupling tolerance without reducing the coupling efficiency (Fig. 2.7). Consequently, the sophisticated module-assembling requirements needed for conventional laser modules, such as precise optical alignment with submicron tolerance using coupling lenses, can be eliminated. This leads to less assembling time and fewer components and thus contributes to cost reduction [67].

The SS-LD is one of the monolithically integrated InP devices that will be used in low-cost devices. One example of a practical application of SS-LDs is a 1.3-/1.5-μm WDM optical module for a FTTH system [68]. Although the optical module itself is an optically hybrid integrated structure, the SS-LD is mounted directly on the glass waveguide (PLC: planar lightwave circuit) platform [69] without coupling lenses (Fig. 2.8). The monolithically integrated spot-size-converter waveguide has thus become widely recognized as a convenient structure not only for reducing the cost of laser modules but also for matching the propagation mode in semiconductor

Vertical Taper Structure

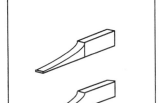

Horizontal Taper Structure

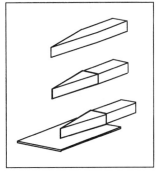

1990 AT&T
1992 ETH (Waveguide)
1993 Gent (Waveguide)
1994 Fujitsu, Gent
1995 NTT, NEC, Mitsubishi

1992 FTZ (Waveguide)
1993 NTT
1994 Alcatel, BT, AT&T, CNET(AMP, Waveguide)
1995 NTT, Furukawa, Hitachi

FIGURE 2.6 Typical structures of monolithically integrated tapered waveguide.

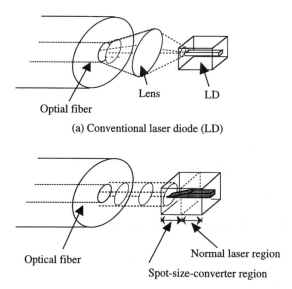

FIGURE 2.7 Comparison of optical coupling between (*a*) a conventional laser diode (LD) and (*b*) a spot-size-converter-integrated laser diode (SS-LD).

waveguides to that of optical fibers or that of other types of waveguides. This structure has therefore been used in other semiconductor optical devices, such as amplifiers [70,71] and switches [72], to improve their coupling characteristics.

2.3.3 Optoelectronic and Photonic Integrated Circuits: Devices for Photonic Networks

Monolithic integration with increased functionality onto a single chip has attracted much attention for more than 25 years, since the first proposal of integrated optics [73] and the OEIC (optoelectronic integrated circuit) [74]. The concept of PIC (photonic integrated circuits) [75] has also been proposed for components on which all devices are interconnected by monolithically integrated guided-wave optoelectronic devices [76,77]. There is a great potential for improving InP-based optoelectronic devices remarkably by using hybrid and monolithic integration technologies. The monolithic integration technology, in particular, has unlimited potential for providing novel functions and revolutionary performance. Telecommunication systems today employ 1.3- or 1.5-μm wavelengths, and hence InP-based PICs have many advantages over other materials, such as GaAs or Si, due to material compatibility with devices for telecommunication systems, such as lasers, photodiodes, or optical modulators. The development of monolithically integrated devices has already begun with the integration of a small number of devices and is going to expand (Fig. 2.9).

2.3 EVOLUTION OF InP-BASED OPTOELECTRONIC DEVICES

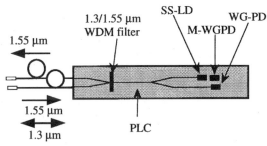

FIGURE 2.8 Hybrid integrated WDM optical module. A SS-LD and two WG-PDs are mounted directly on a PLC platform.

Most of the progress in OEIC technology has been made in photoreceivers rather than phototransmitters, and 20-Gb/s-class monolithically integrated photoreceivers with transimpedance preamplifiers have been reported [78–82]. Photoreceiver arrays [83,84] have also been a research subject among devices for WDM systems or for optical interconnection applications. A high-speed optical signal multiplexer has been made by integrating five semiconductor integrated delay-line waveguides with five MQW modulators on each waveguide [85]. The delay-line waveguide has optical paths differing by 20 ns. By using this multiplexer and 20-Gb/s mode-locked-laser pulse inputs, coded 100-Gb/s optical signal outputs were obtained.

Semiconductor-based MMI (multimode interference) coupler integrated devices [86] can be used in structures ranging from simple couplers to waveguide circuits with large numbers of input and outputs. MMI is the principle of multimode waveguides in which an input field profile is reproduced along the waveguide direction by self-imaging. The structure is essentially low in insertion loss, polarization insensitive, and fabrication-process tolerant [87]. A novel structure introducing an active waveguide into the MMI structure for improving the quantum efficiency of laser diodes was reported recently [88]. Because of their wide applicability and ease

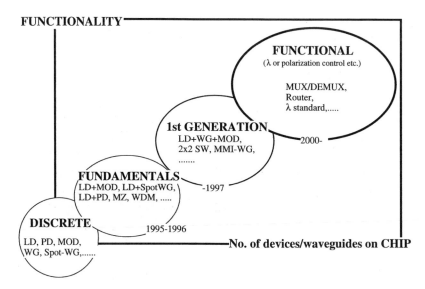

FIGURE 2.9 Evolution of monolithic integration technologies and devices.

of fabrication, MMI devices will be one of the promising technologies for innovative devices and functions in the photonic network.

2.4 CHALLENGES OF InP-BASED OPTOELECTRONIC DEVICES

2.4.1 Exploring Much Higher Performance: Devices for High-Speed Systems and WDM Networks

High-Speed Optoelectronic Devices The transmission capacity of trunk networks and of access networks needs to be increased continually to keep pace with the growing demand for bandwidth. For high-speed trunk networks, laboratory experiments obtained with 40-Gb/s electrically multiplexed [53] and 200-Gb/s optically multiplexed [89] TDM optical transmission systems have been reported. D-WDM optical transmission systems are also becoming solutions to the problems posed by future telecommunication networks [15–17,26]. As described in Section 3.1, semiconductor optical modulators and modulator-integrated DFB lasers have been capable of 40-Gb/s operation at quite low driving voltages (1 to 2 V). Furthermore, an optical pulse generator with a 50-GHz repetition frequency [90] has been achieved by using an electroabsorption-modulator-integrated mode-locked laser.

For advanced systems in the future, an OTDM (optical TDM) system [91] is quite attractive for a high-speed system transmission at more than 100 Gb/s. This is because it is intended to provide the key functions by using optical devices without the help of high-speed electronic circuits such as 100-Gb/s driver and receiver circuits. Mode-lock technologies have been quite useful for generating such high-speed optical pulses because of their capability of producing transform-limit optical pulses with low timing jitter. The passive mode-lock technique is one of the promising methods for this purpose, since it does not need any electrical modulation. In 1992 [92], by introducing the MQW structure and a colliding-pulse mode-lock (CPM) laser configuration, 350-GHz optical-pulse-train generation was demonstated as a monolithic CPM laser on an InP substrate. Recently, 1.54-THz optical pulse generation at a 1.5-μm wavelength by using passively mode-locked DBR laser with harmonic mode-locking technique has been reported [93]. For photodetectors, pioneering work on a high-coupling-efficiency waveguide photodetector with a 110-GHz bandwidth [94] and on a traveling-wave photodetector with a 172-GHz bandwidth [95] has been reported. The operating voltages for such high-speed optical devices must be as low as possible. Around 1 to 2 V and about 1 V should be desirable for devices operating above 40 and at 100 Gb/s, respectively [96].

In addition to work on light sources and detectors, however, further work on a high-speed optical MUX/DEMUX (multiplexer/demultiplexer) for OTDM systems is necessary. The monolithically integrated 100-Gb/s optical signal multiplexer [78] described in Section 2.3.3 is a typical example of this kind of device, and a picosecond optical switching device [97] using a surface-reflection structure has also been reported. This device showed its potential capability of providing the low-power (2 pJ) polarization-independent operation [98,99] required for ultrafast OTDM systems.

InP-Based Integrated Waveguide Devices InP-based waveguide devices or PICs for the all-optical network or the photonic network have been limited by the polarization dependence of their characteristics. This dependence is due to the rectangular shape of the cross section of semiconductor waveguides, which in turn is a result of the high refractive index of InP and of the difficulty of controlling the submicron-order etching process. Based on the recognition that polarization-insensitive operation is essential for the integrated waveguide devices to be used in the photonic network, efforts have been made to reduce the polarization dependence of the InP-based waveguide devices. Polarization-insensitive devices have been reported by introducing strained quantum wells [100–102], a quasisymmetric submicron-wide bulk layer [103], a combination of a window structure and a tapered waveguide at the end of the submicron-wide bulk active layer [104], or operation detuned from the exciton absorption peak wavelength [105,106].

Another potential drawback of integrated waveguide devices is their high-temperature sensitivity, and this is a weakness shared by all InP semiconductor optoelectronic devices. The temperature coefficient of the refractive index of InP is 2.7×10^{-5} deg^{-1} [107], while that of typical silica is about 1×10^{-5} deg^{-1} [108]. Temperature controllers have therefore been essential for controlling the wavelength characteristics of InP-based waveguide devices such as wavelength filters and arrayed waveguide gratings. A waveguide structure compensating the temperature dependence by using polymers with refractive indices whose temperature coefficients are negative [109] has already been made. However, it is almost impossible to find conventional group III-V semiconductors whose refractive indices have negative temperature coefficients.

Other temperature-insensitive optical components for fiber gratings [110] and dielectric wavelength filters [111] have been reported. A novel concept stating that the temperature dependence of the refractive indices of semiconductor waveguides can be compensated by combining two waveguides with different refractive indices has been proposed [112,113]. An InP-based arrayed waveguide grating [114] has been made by utilizing this novel concept. It is unique in that the compensation is effected by combining materials whose temperature coefficients have the same sign (positive) but different values. Figure 2.10 shows the principle of its temperature-insensitive operation. The temperature dependence of the center wavelength for the interference waveguide filter can be controlled by choosing appropriate refractive indices for the waveguide. Furthermore, this principle is not limited to InP-based waveguides but is also applicable to materials such as glass and organic materials. The concept should therefore be useful for controlling the temperature dependence of all kinds of waveguide devices.

InP-based integrated waveguide devices are obviously heading for higher performance and richer functionality by combining the technologies of semiconductor waveguides, O/E–E/O conversion devices, and modules (Fig. 2.11). Although much work must be done to solve the remaining problems, such as the polarization sensitivity, propagation loss, and dynamic range [115], the potential of the devices and promising results for future high-speed multifunctional systems as described above have been obtained.

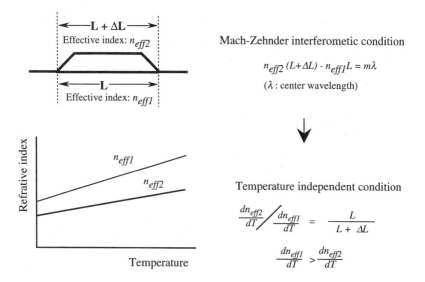

FIGURE 2.10 Principle of temperature independent operation of a Mach–Zehnder filter.

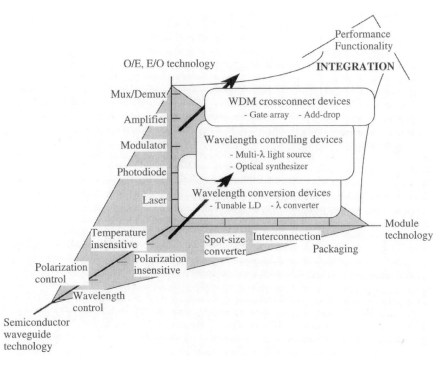

FIGURE 2.11 Direction of InP-based integrated waveguide devices toward photonic networks.

2.4.2 Exploring Reduction in Device Costs for Key Components of Optical Systems

Opticalization of access networks requires that the cost of optoelectronic devices be reduced so that optical fibers can be deployed in these networks. As described in Section 2.3.2, an SS-LD is one example of a device that will be used to lower the cost of access networks. Module assembling or packaging processes, which are the most critical issues in reducing the cost of optoelectronic devices, will be drastically improved by the implementation of SS-LD technology [67,116]. An SS-LDs fabrication technology using 2-in. InP wafers [117] has been reported to provide good uniformity and reproducibility. Because more than 5000 chips can be fabricated on a 2-in. wafer, the fabrication cost of the laser chip should be almost the same as that of LDs for the consumer market. However, the characteristics and reliability required for devices for use in telecommunication networks (even in access networks) are quite different from those required for devices for the consumer market. The operating temperature range, for example, should be 0 to 60°C for indoor use [35] and −40 to +85°C for outdoor use. The laser diode should therefore be designed to have both high output power and high quantum efficiency, even at high temperature [118–123]. Recently, a 1.3-μm laser fabricated on an InGaAs substrate was reported with excellent characteristics, and its high-temperature performance was predicted [124,125]. A novel active-layer compound (GaInNAs, whose lattice constants can be matched to a GaAs substrate) was proposed for telecommunication wavelengths, and an excellent temperature coefficient of 107 K at the oscillation wavelength of 1.2 μm was reported [126].

To reduce the complexity of the driver electronics and its power dissipation, the laser diode should operate at the zero-bias condition. The lower the bias, however, the bigger the turn-on delay and jitters. Therefore, low-threshold laser diodes are obviously essential. To develop this kind of low-threshold laser diode, it is necessary to further investigate overall output characteristics, taking into account required output power, operating bit rates, and operating temperature [127]. Moreover, the photodetectors have to operate at a bias voltage as low as 1 V [128] because the operation voltage for CMOS receiver circuits has been reduced to 3.3 V. Figure 2.12 is an overview of the approach to reducing the cost of optoelectronic devices. One of the key technologies is nonhermetically sealed packaging. Although preliminary experimental results and promising data have been reported [129–132], further intensive study will be needed to establish a technology similar to that widely used in the volume manufacturing of electronic devices.

2.5 CONCLUSIONS

The evolution of InP-based optoelectronic devices and systems has been described with regard to both high performance and cost reduction. Rapid growth of multimedia service demands has been driving the opticalization of the entire telecommunication network. These service demands and opticalization of the network lead existing optical communication networks to larger capacity, more flexibility, and bit-rate- and media-transparent next-generation communication networks: photonic

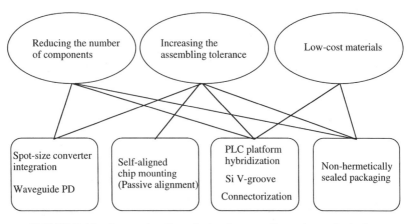

FIGURE 2.12 Approaches for reducing the cost of optoelectronic devices.

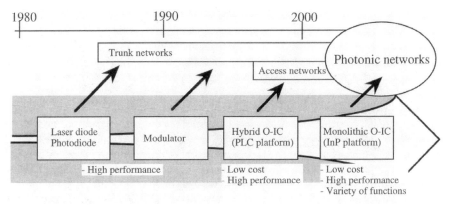

FIGURE 2.13 Role of InP-based optoelectronic devices in future systems.

networks. As illustrated in Fig. 2.13, InP-based optoelectronic devices are expected to play a key role in photonic networks.

REFERENCES

1. European Commission, *Research and Technology Development in Advanced Communications Technologies in Europe: RACE 1995,* EC, Brussels, February 1995.
2. A. J. N. Houghton, "Progress in the ACTS programme towards pan-European photonic networks," *Proc. ECOC'96,* Oslo, 1996, Vol. 2, paper TuB.1.3, pp. 19–26.
3. G.-K. Chang, G. Ellinas, J. K. Gamelin, M. Z. Iqbal, and C. A. Brackett, "Multiwavelength reconfigurable WDM/ATM/SONET network testbeds," *IEEE J. Lightwave Technol.,* **14,** 1320–1340 (1996).
4. P. E. Green, Jr., "Optical networking update," *IEEE J. Sel. Areas Commun.,* **14,** 764–779 (1996).
5. European Commission, *ACTS'96 Project Summaries,* EC, Brussels, September 1996.
6. R. E. Wagner, R. C. Alferness, A. A. M. Saleh, and M. S. Goodman, "MONET: multiwavelength optical networking," *IEEE J. Lightwave Technol.,* **14,** 1349–1355 (1996).
7. R. S. Vodhanel, L. D. Garret, S. H. Patel, and W. Kraeft, "National-scale WDM networking demonstration by the MONET consortium," presented at OFC'97, Dallas, Texas, 1997, postdeadline paper PD27.
8. NTT Multimedia Planning and Promotion Office, "Joint utilization tests of multimedia communications," *NTT Rev.,* **6,** 8–10 (1994).
9. I. Yamashita, F. Mano, E. Yoneda, and K. M. Kawase, "Fiber-optic subscriber network systems field trial," *NTT Rev.,* **4,** 16–22 (1992).
10. N. Shibata, "Evolution of fiber-optic access networks toward the multimedia era," *Tech. Dig. First Optoelectronics and Communications Conf.,* Makuhari, Japan, 1996, pp. 8–9.
11. I. Yamashita, "Optical networks towards the 21st century and the role of InP-based devices," *Proc. 7th Int. Conf. Indium Phosphide and Related Materials,* Sapporo, Japan, 1995, pp. 3–6.

12. H. Ishio, "Evolution of next generation optical transmission technologies," *NTT J. Technol.*, **4**, 8–12 (1992).
13. K. Hagimoto, "Experimental 10 Gbit/s transmission system and its IC technology," *Tech. Dig. GaAs IC Symp.*, 1993, paper A2, pp. 7–10.
14. K. Nakagawa and K. Hagimoto, "Very high speed FA–10G optical fiber transmission system," *NTT Res. Dev.*, **44**, 241–246 (1995).
15. K. Fichew and A. Hauton, "Optical network research and development in Europe," *Tech. Dig. First Optoelectronics and Communications Conf.*, Makuhari, Japan, 1996, paper 18B2-3, pp. 324–325
16. I. P. Kaminow, C. R. Doerr, C. Dragone, T. Koch, U. Koren, A. A. M. Saleh, A. J. Kirby, C. M. Ozveren, B. Schofield, R. E. Thomas. R. A. Barry, D. M. Catagnozzi, V. W. S. Chan, B. R. Hemenway, Jr., D. Marquis, S. A. Parish, M. L. Stevens, E. A. Swanson, S. G. Finn, and R. G. Gallager, "A wideband all-optical WDM network," *IEEE J. Sel. Areas Commun.*, **14**, 780–799 (1996).
17. H. Kogelnik, "WDM networks: a U.S. perspective," *Proc. ECOC'96*, Oslo, 1996, Vol. 5, paper MoA.2.2, pp. 81–85.
18. F. Forghieri, R. W. Tkach, and A. R. Chraplyvy, "WDM systems with unequal spaced channels," *IEEE J. Lightwave Technol.*, **13**, 889–897 (1995).
19. A. Mori, Y. Oishi, M. Yamada, H. Ono, Y. Nishida, K. Oikawa, and S. Sudo, "1.5 μm broadband amplification by tellurite-based EDFAs," presented at OFC'97, Dallas, Texas, 1997, postdeadline paper PD1.
20. P. Wysocki, J. Judkins, R. Espindola, M. Andrejco, A. Vengsarkar, and K. Walker, "Erbium-doped fiber amplifier flattened beyond 40 nm using long-period gratings," presented at OFC'97, Dallas, Texas, 1997, postdeadline paper PD2.
21. M. Fukushima, Y. Tashiro, and H. Ogoshi, "Flat gain erbium-doped fiber amplifier in 1570nm–1600nm region for dense WDM transmission systems," presented at OFC'97, Dallas, Texas, 1997, postdeadline paper PD3.
22. K. Okamoto, M. Ishii, Y. Hibino, Y. Ohmori, and H. Toba, "Fabrication of unequal channel spacing arrayed-waveguide grating multiplexer modules," *Electron. Lett.*, **31**, 1464–1465 (1995).
23. K. O. Hill, F. Bilodeau, B. Malo, T. Kitagawa, S. Thereault, D. C. Johnson, J. Albert, and K. Takiguchi, "Aperiodic in-fiber Bragg gratings for optical fiber dispersion compensation," presented at OFC'94, San Jose, Calif., 1994, postdeadline paper PD2.
24. C. D. Chen, J. M. P. Delavaux, B. W. Hakki, O. Mizuhara, T. V. Nguyen, R. J. Nuyts, K. Ogawa, Y. K. Park, C. S. Skolnick, R. E. Tench, J. Thomas, L. D. Tzeng, and P. D. Yeates, "A field demonstration of 10 Gb/s–360 km transmission through embedded standard (non-DSF) fiber cables," presented at OFC'94, San Jose, Calif., 1994, postdeadline paper PD27.
25. K. Takiguchi, K. Jinguji, and Y. Ohmori, "Variable group-delay dispersion equalizer based on a lattice-form programmable optical filter," *Electron. Lett.*, **31**, 1240–1241 (1995).
26. V. O. K. Li, A. E. Willner, P. D. Dapkus, K.-C. Lee, A. D. Norte, E. Park, W. Shieh, and A. Mathur, "Wavelength conversion and wavelength routing for high-efficiency all-optical networks: a proposal for research on all-optical networks," *J. High Speed Networks,* **4**, 5–25 (1995).
27. H. Onaka, H. Miyata, G. Ishikawa, K. Otsuka, H. Ooi, Y. Kai, S. Kinoshita, M. Seino,

H. Nishimoto, and T. Chikama, "1.1 Tb/s WDM transmission over a 150 km 1.3 μm zero-dispersion single-mode fiber," presented at OFC'96, San Jose, Calif., 1996, postdeadline paper PD19.
28. A. H. Gnauck, A. R. Charplyvy, R.W. Tkach, J. L. Zyskind, J. W. Sulhoff, A. L. Lucero, Y. Sun, R. M. Jopson, F. Forghieri, R. M. Derosier, C. Wolf, and A. R. McCormick, "One Terabit/s transmission experiment," presented at OFC'96, San Jose, Calif., 1996, postdeadline paper PD20.
29. T. Morioka, H. Takara, S. Kawanishi, O. Kamatani, K. Takiguchi, K. Uchiyama, M. Saruwatari, H. Takahashi, M. Yamada, T. Kanamori, and H. Ono, "100 Gbit/s × 10 channel OTDM/WDM transmission using a single supercontinuum WDM source," presented at OFC'96, San Jose, Calif., 1996, postdeadline paper PD21.
30. Y. Yano, T. Ono, K. Fukuchi, T. Ito, H. Yamazaki, M. Yamaguchi, and K. Emura, "2.6 terabit/s WDM transmission experiment using optical duobinary coding," *Proc. ECOC'96*, Oslo, 1996, Vol. 5, paper ThB.3.1, pp. 3–6.
31. T. Miki, "The potential of photonic networks," *IEEE Commun.*, December 1994, pp. 23–27.
32. A. S. Acampora, "The scalable lightwave network," *IEEE Commun.*, December 1994, pp. 36–42.
33. T. Miki, "Fiber optic subscriber networks and system development," *IEICE Trans. E*, **74**, 93–100 (1991).
34. T. Aoyama, "FSAN (full service access networks) international conference," *NTT Rev.*, **8**, 42–43 (1996).
35. W. Warzanskyj, G. Adams, R. Cadarella, U. Ferrero, H. Hofmeister, M. Knuckey, N. Gieschen, R. Grabenhorst, M. de Grandis, K. Murano, K. Okada, J. V. Schijndel, E. Soleres, J. Stern, W. Verbiest, and M. Yamashita, "Services, architectures, topologies and economic issues," *Proc. Full Services Access Networks Conf.*, London, 1995, paper 1.
36. T. Wood, E. Carr, B. Kasper, R. Linke, and C. Burrus, "Bidirectional fiber-optical transmission using a multiple-quantum-well (MQW) modulator/detector," *Electron. Lett.*, **22**, 528–529 (1988).
37. H. Nakajima, J. Charil, D. Robein, A. Gloukhian, B. Pierre, J. Landreau, S. Grosmaire, and A. Leroy, "Full-duplex operation of an in-line transceiver emitting at 1.3 μm and receiving at 1.55 μm," *Electron. Lett.*, **32**, 473–474 (1996).
38. N. J. Frigo, P. P. Iannone, P. D. Magli, T. E. Darcie, M. M. Downs, B. N. Desai, U. Koren, T. L. Koch, C. Dragone, H. M. Presby, and G. E. Bodeep, "A wavelength-division multiplexed passive optical network with cost-shared components," *IEEE Photon. Technol. Lett.*, **6**, 1365–1367 (1994).
39. M. Zirngibl, C. H. Joyner, L. W. Stulz, C. Dragone, H. M. Presby, and P. Kaminow, "LARNet, a local access router network," *IEEE Photon. Technol. Lett.*, **7**, 215–217 (1996).
40. U. Hilbk, G. Bader, H. Bunning, M. Burmeister, E. Grossmann, Th. Hermes, B. Hoen, K. Peters, F. Raub, J. Saniter, and F. J. Westphal, "Experimental WDM upgrade of a PON using an arrayed waveguide grating," *Proc. ECOC'96*, Oslo, 1996, Vol. 3, paper WeB.1.5, pp. 31–34.
41. Y. Kawamura, K. Wakita, Y. Yoshikuni, Y. Itaya, and H. Asahi, "Monolithic integration of InGaAsP/InP DFB lasers and InGaAs/InAlAs MQW optical modulators," *Electron. Lett.*, **22**, 242–243 (1986).

42. H. Soda, M. Furuse, K. Sato, N. Okazaki, S. Yamazaki, and H. Ishikawa, "Low drive voltage semi-insulating BH structure monolithic electroabsorption modulator/DFB laser light source," *Tech. Dig. OFC'90*, San Jose, Calif., 1990, paper TH12.

43. M. Aoki, H. Sano, M. Suzuki, M. Takahashi, K. Uomi, and A. Takai, "Novel MQW electro-absorption modulator/DFB-laser integrated device grown by selective area MOCVD," *Tech. Dig. OFC'92,* San Jose, Calif., 1992, paper FB6.

44. T. Kato, T. Sasaki, N. Kida, K. Komatsu, and I. Mito, "Novel MQW DFB laser diode/modulator integrated light source using band gap energy control epitaxial growth technique," *Proc. ECOC'91,* Paris, 1991, paper WeB7–1.

45. H. Haisch, W. Baumert, C. Hache, E. Kuhn, K. Satzke, M. Schilling, J. Weber, and E. Zielinski, "10 Gbit/s standard fiber TDM transmission at 1.55 μm with low chirp monolithically integrated MQW electroabsorption modulator/DFB-laser realized by selective area MOVPE," *Proc. ECOC'94*, Florence, Italy, 1994, pp. 801–804.

46. T. Kataoka, Y. Miyamoto, K. Hagimoto, K. Sato, I. Kotaka, and K. Wakita, "20 Gbt/s transmission experiments using an integrated MQW modulator/DFB laser module," *Electron. Lett.*, **30**, 872–873 (1994).

47. A. Ramdane, D. Delpart, F. Devaux, D. Mathoorasing, H. Nakajima, A. Ougazzaden, J. Landreau, and A. Gloukhian, "Integrated MQW laser-modulator with 36 GHz bandwidth and negative chirp," *Proc. ECOC'95*, Brussels, 1995, paper ThB.2.2, pp. 893–896.

48. H. Takeuchi, K. Tsuzuki, K. Sato, S. Matsumoto, M. Yamamoto, Y. Itaya, A. Sano, M. Yoneyama, and T. Otsuji, "NRZ operation at 40 Gb/s of a compact module with an MQW electroabsorption modulator integrated DFB laser," *Proc. ECOC'96*, Oslo, 1996, *IEEE Commun.*, December Vol. 5, paper ThD.3.1, pp. 55–58.

49. K. Wakita, I. Kotaka, O. Mitomi, H. Asai, Y. Kawamura, and M. Naganuma, "High speed InGaAs/InAlAs multiple quantum well optical modulators with bandwidths in excess of 40 GHz at 1.55 μm," *Tech. Dig. CLEO,* Anaheim, Calif., 1990, paper CTu C6.

50. F. Devaux, P. Bordes, A. Ougazzaden, M. Carre, and F. Huet, "Experimental optimization of MQW electroabsorption modulators with up to 40 GHz bandwidths," *Electron. Lett.*, **30**, 1347–1348 (1994).

51. T. Ido, S. Tanaka, M. Suzuki, and H. Inoue, "An ultra-high-speed (50 GHz) MQW electro-absorption modulator with waveguides for 40 Gb/s optical modulation," presented at IOOC'95, Hong Kong, 1995, postdeadline paper PD1–1.

52. K. Yoshino, K. Wakita, I. Kotaka, S. Kondo, Y. Noguchi, S. Kuwano, N. Takachio, T. Otsuji, Y. Imai, and T. Enoki, "40-Gbit/s operation of InGaAs/InAlAs MQW electroabsorption modulator module with very low driving-voltage," *Proc. ECOC'96*, Oslo, 1996, Vol. 3, paper WeD.3.5, pp. 203–206.

53. S. Kuwano, N. Takachio, K. Iwashita, T. Otsuji, Y. Imai, T. Enoki, K. Yoshino, and K. Wakita, "160-Gbit/s (4-ch × 40-Gbit/s electrically multiplexed data) WDM transmission over 320-km dispersion-shifted fiber," presented at OFC'96, San Jose, Calif., 1996, postdeadline paper PD25

54. F. Koyama and K. Iga, "Frequency chirping in external modulators," *IEEE J. Lightwave Technol.*, **6**, 87–93 (1988).

55. G. Ishikawa, H. Ooi, K. Morito, K. Kamite, Y. Kotaki, and H. Nishimoto, "10-Gb/s optical transmission systems using modulator-integrated DFB lasers with chirp optimization," *Proc. ECOC'96*, Oslo, 1996, Vol. 3, paper WeP.09, pp. 245–248.

56. T. Yamanaka, K. Wakita, and K. Yokoyama, "Potential chirp-free characteristics (negative chirp parameter) in electroabsorption modulation using a wide tensile-strained quantum well structure," *Appl. Phys. Lett.*, **68**, 3114–3116 (1996).
57. T. Yamanaka, K. Wakita, and K. Yokoyama, "Pure strain effect on reducing the chirp parameter in InGaAsP/InP quantum well electroabsorption modulators," *Appl. Phys. Lett.*, **70**, 87–89 (1997).
58. K. Yokoyama, "High performance design for InP-based strained-layer quantum well laser diodes," *Proc. Int. Workshop on Physics and Computer Modeling of Devices Based on Low-Dimensional Structures*, Aizuwakamatsu, Japan, 1995, IEEE Computer Society Press, New York, pp. 112–121.
59. T. Koch, U. Koren, G. Eisenstein, M. G. Young, M. Oron, C. R. Gilis, and B. I. Miller, "Tapered waveguide InGaAs/InGaAsP multiple-quantum-well lasers," *IEEE Photon. Technol. Lett.*, **2**, 88–90 (1990).
60. K. Kasaya, Y. Kondo, M. Okamoto, O. Mitomi, and M. Naganuma, "Monolithically integrated DBR lasers with simple tapered waveguide for low-loss fiber coupling," *Electron. Lett.*, **29**, 2067–2068 (1993).
61. P. Doussiere, P. Garabedian, C. Graver, E. Derouin, E. G. Goarin, G. Michaud, and R. Meilleur, "Tapered active stripe for 1.5 μm InGaAsP/InP strained multiple quantum lasers with reduced beam divergence," *Appl. Phys. Lett.*, **64**, 539–541 (1994).
62. I. Moerman, M. D'Hondt, W. Vanderbauwhere, G. Coudenys, J. Haes, P. De Dobbelaere, R. Baets, P. Van Daele, and P. Demeester, "Monolithic integration of a spot size transformer with a planar buried heterostructure in InGaAsP/InP laser using the shadow masked growth technique," *IEEE Photon. Technol. Lett.*, **6**, 888–890 (1994).
63. I. F. Leelman, L. J. Rivers, M. J. Harlow, S. D. Perrin, and M. J. Robertson, "1.56 μm InGaAsP/InP tapered active layer multiple quantum well laser with improved coupling to cleaved singlemode fibre," *Electron. Lett.*, **30**, 857–859 (1994).
64. H. Kobayashi, M. Ekawa, N. Okazaki, O. Aoki, S. Ogita, and H. Soda, "Tapered thickness MQW waveguide BH MQW lasers," *IEEE Photon. Technol. Lett.*, **6**, 1080–1081 (1994).
65. Y. Tohmori, Y. Suzaki, H. Fukano, M. Okamoto, Y. Sakai, O. Mitomi, S. Matsumoto, M. Yamamoto, M. Fukuda, M. Wada, Y. Itaya, and T. Sugie, "Spot-size converted 1.3 μm laser with butt-jointed selectively grown vertically tapered waveguide," *Electron. Lett.*, **31**, 1069–1070 (1995).
66. Y. Itaya, Y. Tohmori, H. Okamoto, O. Mitomi, M. Wada, K. Kawano, H. Fukano, K. Yokoyama, Y. Suzaki, M. Okamoto, Y. Kondo, I. Kotaka, M. Yamamoto, M. Kohtoku, Y. Kadota, K. Kishi, Y. Sakai, H. Ohashi, and M. Nakao, "Spot-size converter integrated laser diodes (SS-LDs)," *IEICE Trans. Electron.*, **E80-C**, 30–37 (1997).
67. J. Yoshida, M. Kawachi, T. Sugie, M. Horiguchi, Y. Itaya, and M. Fukuda, "Strategy for developing low-cost optical modules for the ONU of optical subscriber systems," *Tech. Dig. IEEE Workshop on Access Networks*, Nuremberg, Germany, 1995, pp. 6.1.1–6.1.8.
68. N. Uchida, Y. Yamada, Y. Hibino, Y. Suzuki, T. Kurosaki, N. Ishihara, M. Nakamura, T. Hashimoto, Y. Akahori, Y. Inoue, K. Moriwaki, K. Kato, Y. Tohmori, M. Wada, and T. Sugie, "Low-cost and high-performance hybrid WDM module integrated on a PLC platform for fiber-to-the-home," *Proc. ECOC'96*, Oslo, 1996, Vol. 2, paper TuC.3.1, pp. 107–114.
69. Y. Yamada, S. Suzuki, K. Moriwaki, Y. Hibino, Y. Tohmari, Y. Akatu, Y. Nakasuga, T. Hashimoto, H. Terui, M. Yanagisawa, Y. Inoue, Y. Akahori, and R. Nagase, "Application

of planar lightwave circuit platform to hybrid integrated optical WDM transmitter/receiver module," *Electron. Lett.*, **31**, 16–17 (1995).

70. G. Glastre, D. Rondi, A. Enard, and R. Blondeau, "Polarization insensitive 1.55 μm semiconductor integrated optical amplifier with access waveguides grown by LP-MOCVD," *Electron. Lett.*, **27**, 899–900 (1991).

71. H. J. Bruckner, B. Mersali, S. Sainson, M. Feuillade, A. Ougazzaden, Ph. Krauz, and A. Carenco, "Taper-waveguide integration for polarization insensitive InP/InGaAsP based optical amplifiers," *Electron. Lett.*, **30**, 1290–1291 (1994).

72. K. Kawano, S. Sekine, H. Takeuchi, M. Wada, M. Kohtoku, N. Yoshimoto, T. Ito, M. Yanagibashi, S. Kondo, and Y. Noguchi, "4 × 4 InGaAlAs/InAlAs MQW directional coupler waveguide switch modules integrated with spot-size converters and their 10 Gbit/s operation," *Electron. Lett.*, **31**, 96–97 (1995).

73. S. E. Miller, "Integrated optics: an introduction," *Bell Syst. Tech. J.*, **48**, 2059–2069 (1969).

74. S. Somekh and A. Yariv, "Fiber optic communications," *Proc. Conf. International Telemetry*, Los Angeles, 1972, p. 407.

75. T. Koch and U. Koren, "Semiconductor photonic integrated circuits," *IEEE J. Quantum Electron.*, **27**, 641–653 (1991).

76. M. Degenais, R. F. Leheny, and J. Crow, eds., *Integrated Optoelectronics*, Academic Press, San Diego, Calif., 1994.

77. O. Wada ed., *Optoelectronic Integration: Physics, Technologies and Applications*, Kluwer Academic Publishers, Norwell, Mass., 1994.

78. L. M. Lunardi, S. Chandrasekhar, A. H. Gnauck, C. A. Burrus, and R. A. Hamm, "20 Gb/s monolithic p-i-n/HBT photoreceiver module for 1.55 μm applications," *Proc. ECOC'95*, Brussels, 1995, paper We.L.2.1, pp. 657–660.

79. A. L. Gutierrez-Aitken, K. Yang, X. Zhang, G. I. Haddad, and P. Bhattacharya, "16 GHz bandwidth InAlAs/InGaAs monolithically integrated PIN-HBT photoreceiver," *Proc. ECOC'95*, Brussels, 1995, paper We.L.2.2, pp. 661–664.

80. J. Spicher, B.-U. H. Klepser, M. Beck, A. Rudra, R. Sachot, and M. Ilegems, "A 20-Gbit/s monolithic photoreceiver using InAlAs/InGaAs HEMT's and regrown p-i-n photodiode," *Proc. 8th Int. Conf. Indium Phosphide and Related Materials*, Schwäbisch-Gmünd, Germany, 1996, paper WA1–4, pp. 439–442.

81. P. Fay, W. Wohlmuth, C. Caneau, and I. Adesida, "18.5-GHz bandwidth monolithic MSM/MODFET photoreceiver for 1.55-mm wavelength communication systems," *IEEE Photon. Technol. Lett.*, **8**, 679–681 (1996).

82. A. Umbach, S. van Waasen, H.-G. Bach, R. M. Bertenburg, G. Janssen, G. G. Mekonnen, W. Passenberg, W. Schlaak, C. Schramm, G. Unterbörsch, P. Wolfram, and F.-J. Tegude, "High-speed photoreceiver by monolithic integration of a waveguide fed photodiode and a GaInAs/AlInAs-HEMT based distributed amplifier," *Proc. 8th Int. Conf. Indium Phosphide and Related Materials*, Part 2, Schwäbisch-Gmünd, Germany, 1996, postdeadline paper ThB. 2.6, pp. 26–27.

83. K. Takahata, Y. Muramoto, Y. Akatsu, Y. Akahori, A. Kozen, and Y. Itaya, "10-Gb/s two-channel monolithic photoreceiver array using waveguide p-i-n PD's and HEMT's," *IEEE Photon. Technol. Lett.*, **8**, 563–565 (1996).

84. L. D. Garrett, S. Chandrasekhar, A. G. Dentai, J. Zyskind, J. Shulhoff, C. A. Burrus, and E. C. Burrows, "Performance of an eight-channel p-i-n/HBT OEIC photoreceiver array

module in an optically preamplified WDM system experiment," *IEEE Photon. Technol. Lett.*, **8**, 1689–1691 (1996).

85. F. Zamkostian, K. Sato, H. Okamoto, K. Kishi, I. Kotaka, M. Yamamoto, Y. Kondo, H. Yasaka, Y. Yoshikuni, and K. Oe, "An InP optical multiplexer integrated with modulators for 100 Gb/s transmission," *Proc. ECOC'94*, Florence, Italy, 1994, pp. 105–108.

86. R. Ulrich, "Image formation by phase coincidences in optical waveguides," *Opt. Commun.*, **6**, 259–264 (1975).

87. L. B. Soldano and E. C. M. Penning, "Optical multi-mode interference devices based on self-imaging: principles and applications," *IEEE J. Lightwave Technol.*, **13**, 615–627 (1995).

88. K. Hamamoto, E. Gini, C. Holtsmann, and H. Melchior, "Single-transverse-mode active-MMI 1.5 µm InGaAsP buried-hetero laser diode," presented at European Conf. Integrated Optics, Stockholm, 1997, postdeadline paper PD5.

89. S. Kawanishi, H. Takara, T. Morioka, O. Kamatani, and M. Saruwatari, "200 Gb/s, 100 km TDM transmission using supercontinuum pulses with prescaled PLL timing extraction and all-optical demultiplexing," presented at OFC'95, San Diego, Calif., 1995, postdeadline paper PD28.

90. K. Sato, I. Kotaka, Y. Kondo, and M. Yamamoto, "Active mode locking at 50 GHz repetition frequency by half-frequency modulation of monolithic semiconductor lasers integrated with electroabsorption modulators," *Appl. Phys. Lett.*, **69**, 2626–2628 (1996).

91. R. A. Barry, V. W. S. Chan, K. L. Hall, E. S. Kintzer, J. D. Moores, K. A. Rauchenbach, E. A. Swanson, L. E. Adams, C. R. Doerr, S. G. Finn, H. A. Haus, E. P. Ippen, W. S. Wong, and M. Hander, "All-optical network consortium: ultrafast TDM networks," *IEEE J. Sel. Areas Commun.*, **14**, 999–1013 (1996).

92. Y.-K. Chen and M. C. Wu, "Monolithic colliding-pulse mode-locked quantum-well lasers," *IEEE J. Quantum Electron.*, **28**, 2176–2185 (1992).

93. S. Arahira, Y. Matsui, and Y. Ogawa, "Mode-locking at very high repetition rates more than terahertz in passively mode-locked distributed-Bragg-reflector laser diodes," *IEEE J. Quantum Electron.*, 32, 1211–1224 (1996).

94. K. Kato, A. Kozen, Y. Muramoto, Y. Itaya, T. Nagatsuma, and M. Yaita, "110-GHz, 50%-efficiency mushroom-mesa waveguide pin photodiode for a 1.55 µm wavelength," *IEEE Photon. Technol. Lett.*, **6**, 719–721 (1994).

95. K. S. Giboney, R. L. Nagarajan, T. E. Reynolds, S. T. Allen, R. P. Mirin, M. J. W. Rodwell, and J. E. Bowers, "Traveling-wave photodetectors with 172-GHz bandwidth and 76-GHz bandwidth-efficiency product," *IEEE Photon. Technol. Lett.*, **7**, 412–414 (1995).

96. H. Abe, T. Mimura, and N. Chinone, *Optical and Microwave Semiconductor Technology*, Science Forum, Tokyo, 1996, p. 325 (in Japanese).

97. R. Takahashi, Y. Kawamura, T. Kagawa, and H. Iwamura, "Ultrafast 1.55-µm photoresponses in low-temperature-grown InGaAs/InAlAs quantum wells," *Appl. Phys. Lett.*, **65**, 1790–1792 (1994).

98. R. Takahashi, W.-Y. Choi, Y. Kawamura, and H. Iwamura, "Femtosecond all-optical AND gates based on low-temperature grown Be-doped strained InGaAs/InAlAs multiple quantum wells," *Proc. Annual Meeting of IEEE-LEOS*, Boston, 1995, pp. 343–345.

99. R. Takahashi, Y. Kawamura, and H. Iwamura, "Ultrafast 1.55-µm all-optical switching

using low-temperature-grown multiple quantum wells," *Appl. Phys. Lett.*, **68**, 153–155 (1996).

100. K. Magari, M. Okamoto, H. Yasaka, K. Sato, Y. Noguchi, and O. Mikami, "Polarization insensitive traveling wave type amplifier using strained multiple quantum well structure," *IEEE Photon. Technol. Lett.*, **4**, 556–558 (1990).

101. L. F. Tiemeijer, P. J. A. Thijs, T. van Dongen, R. W. M. Slootweg, J. M. M. van der Heijden, J. J. M. Binsma, and M. P. C. M. Krijn, "High performance 1300nm polarization insensitive laser amplifiers employing both tensile and compressive strained quantum wells in a single active layer," presented at ECOC'92, Berlin, 1992, postdeadline paper ThPD. II.6.

102. M. A. Newkirk, B. I. Miller, U. Koren, M. G. Young, M. Chien, R. M. Jopson, and C. A. Burrus, "1.5 μm multiquantum-well semiconductor optical amplifier with tensile and compressive strain wells for polarization-independent gain," *IEEE Photon. Technol. Lett.*, **4**, 406–408 (1993).

103. S. Kitamura, K. Komatsu, and M. Kitamura, "Polarization-insensitive semiconductor optical amplifier array grown by selective MOVPE," *IEEE Photon. Technol. Lett.*, **6**, 173–175 (1994).

104. P. Doussiere, P. Garabedian, C. Graver, D. Bonnevie, T. Fillion, E. Derouin, M. Monnot, J. G. Provost, D. Leclerc, and M. Klenk, "1.55 μm polarization independent semiconductor optical amplifier with 25 dB fiber to fiber gain," *IEEE Photon. Technol. Lett.*, **6**, 1170–1172 (1994).

105. N. Yoshimoto, Y. Shibata, S. Oku, S. Kondo, Y. Noguchi, K. Wakita, K. Kawano, and M. Naganuma, "Fully polarization insensitive Mach–Zehnder optical switches using a wide-well InGaAlAs/InAlAs MQW structure," *Tech. Dig. Intern. Topical Meeting on Photonics in Switching*, Sendai, Japan, 1996, paper PWB3, pp. 78–79.

106. N. Yoshimoto, S. Kondo, Y. Noguchi, T. Yamanaka, and K. Wakita, "Polarization-insensitive field-induced refractive index change using a lattice-matched InGaAlAs/InAlAs multiple quantum well structure," *Appl. Phys. Lett.*, **69**, 4239–4241 (1996).

107. M. Cardona, "Temperature dependence of the refractive index and the polarizability of free carriers in some III-V semiconductors," *Proc. Int. Conf. Semiconductor Physics*, Prague, 1960, pp. 388–394.

108. M. Kawachi, "Silica waveguides on silicon and their application to integrated-optic components," *Opt. Quantum Electron.*, **22**, 391–416 (1990).

109. Y. Kokubun, M. Takizawa, and S. Taga, "Three-dimensional athermal waveguides for temperature independent lightwave devices," *Electron. Lett.*, **30**, 1223–1225 (1994).

110. J. B. Judkins, J. R. Pedrazzani, D. J. DiGiovanni, and A. M. Vengsarkar, "Temperature-insensitive long-period fiber gratings," presented at OFC'96, San Jose, Calif., 1996, postdeadline paper PD1.

111. M. Shirasaki, "Temperature independent interferometer for WDM filters," *Proc. ECOC'96*, Oslo, 1996, Vol. 3, paper WeD.1.6, pp. 147–150.

112. H. Tanobe, Y. Kondo, Y, Kadota, H. Yasaka, and Y. Yoshikuni, "A temperature insensitive InGaAsP/InP wavelength filter," *Tech. Dig. Integrated Photonics Research 1996*, Boston, 1996, paper IME3.1.

113. H. Tanobe, Y. Kondo, Y. Kadota, H. Yasaka, and Y. Yoshikuni, "A temperature insensitive InGaAsP–InP optical filter," *IEEE Photon. Technol. Lett.*, **8**, 1489–1491 (1996).

114. H. Tanobe, Y. Kondo, Y. Kadota, K. Okamoto, and Y. Yoshikuni, "Temperature insensi-

tive arrayed waveguide grating on InP substrates," *Tech. Dig. OFC'97*, Dallas, Texas, 1997, paper ThM4, pp. 298–299.

115. M. Erman, "What technology is required for the pan-European network, what is available and what is not," *Proc. ECOC'96*, Oslo, 1996, Vol. 5, paper TuB.2.2, pp. 87–94.

116. J. Yoshida, "Technologies for cost reduction of LD modules," *Tech. Dig. OECC'96*, Makuhari, Japan, 1996, paper 18D3-1, pp. 402–403.

117. M. Wada, M. Kohtoku, K. Kawano, S. Kondo, Y. Tohmori, Y. Kondo, K. Kishi, Y. Sakai, I. Kotaka, Y. Noguchi, and Y. Itaya, "Laser diodes monolithically integrated with spot-size converters fabricated on 2-inch InP substrates," *Electron. Lett.*, **31**, 1252–1253 (1995).

118. P. J. Thijs, T. van Dongen, L. F. Tiemeijer, and J. J. M. Binsma, "High-performance λ = 1.3 μm InGaAsP–InP strained quantum well lasers," *J. Lightwave Technol.*, **12**, 28–37 (1994).

119. C. E. Zah, R. Bhat, B. Pathak. F. Favire, M. C. Wang, W. Lin, N. C. Andreadakis, D. M. Hwang, M. A. Koza, T. P. Lee, Z. Wang, D. Darby, D. Flanders, and J. J. Hsieh, "High-performance uncooled 1.3-μm AlGaInAs/InP strained-layer quantum-well lasers for fiber-in-the-loop applications," *Tech. Dig. OFC'94*, San Jose, Calif., 1994, paper ThG1.

120. H. Oohashi, T. Hirono, S. Seki, H. Sugiura, J. Nakano, M. Yamamoto, Y. Tohmori, and K. Yokoyama, "1.3?μm InAsP compressively strained multiple-quantum-well lasers for high-temperature operation," *J. Appl. Phys.*, **77**, 4119–4121 (1995).

121. R. F. Kazarinov and G.L. Belenky, "Novel design of AlGaInAs–InP lasers operating at 1.3 μm," *IEEE J. Quantum Electron.*, **31**, 423–426 (1995).

122. S. Seki, H. Oohashi, H. Sugiura, T. Hirono, and K. Yokoyama, "Design criteria for high-efficiency operation of 1.3 μm InP-based strained-layer multiple-quantum-well lasers at elevated temperatures," *IEEE Photon. Technol. Lett.*, **7**, 839–841 (1995).

123. T. Nakamura, T. Tsuruoka, K. Fukushima, A. Uda, Y. Hosono, K. Kurata, and T. Torikai, "High efficiency 1.3 μm strained multi-quantum well lasers entirely grown by MOVPE for passive optical network use," *Proc. Annual Meeting of IEEE LEOS'96*, Boston, 1996, paper MA.2, pp. 8–9.

124. H. Shoji, T. Uchida, T. Kusunoki, M. Matsuda, H. Kurokane, S. Yamazaki, K. Nakajima, and H. Ishikawa, "Fabrication of $In_{0.25}Ga_{0.75}As/InGaAsP$ strained SQW lasers on $In_{0.05}Ga_{0.95}As$ ternary substrate," *IEEE Photon. Technol. Lett.*, **6**, 1170–1172 (1994).

125. H. Ishikawa, "Theoretical gain of strained quantum well grown on a InGaAs ternary substrate," *Appl. Phys. Lett.*, **63**, 712–713 (1993).

126. S. Nakatsuka, M. Kondow, K. Kitatani, Y. Yazawa, and M. Okai, "Low threshold index-guide GaInNAs laser diode for optical communications," *Tech. Dig. Spring Annual Meeting of Japan Society of Applied Physics*, paper 30p-NG-5, 1997 (in Japanese).

127. Y. Suzuki, T. Kurosaki, H. Oohashi, N. Uchida, Y. Itaya, and H. Toba, "Criterion of high-speed zero-biased laser diodes for passive optical networks (PON)," *Proc. 8th IEEE Workshop on Optical Access Networks*, Atlanta, Ga., 1997, paper IV.4.3.

128. K. Kato, M. Yuda, A. Kozen, Y. Muramoto, K. Noguchi, and O. Nakajima, "Selective-area impurity-doped planar edge-coupled waveguide photodiode (SIMPLE-WGPD) for low-cost, low-power consumption optical hybrid modules," *Electron. Lett.*, **32**, 2078–2079 (1996).

129. J. V. Collins, R. A. Payne, C. W. Ford, A. R. Thurlow, I. F. Lealman, and P. J. Fiddyment,

"Technology developments for low-cost laser packaging," *Tech. Dig. OFC'95*, San Diego, Calif., 1995, paper WS11, pp. 222–223.

130. J. W. Osenbach, R. B. Comizzoni, T. L. Evanosky, and N. Chand, "Degradation of InGaAs–InP lasers in hot-humid ambients: determination of temperature-humidity-bias acceleration factors," *IEEE Photon. Technol. Lett.*, **7,** 1252–1254 (1995).

131. M. Fukuda, F. Ichikawa, H. Sato, S. Tohno, and T. Sugie, "Pigtail type laser modules entirely molded in plastic," *Electron. Lett.*, **31,** 1745–1747 (1995).

132. M. Fukuda, F. Ichikawa, H. Sato, Y. Hibino, K. Moriwaki, S. Tohno, T. Sugie, and J. Yoshida, "Plastic packaging of semiconductor laser diodes," *Proc. 46th Electronic Components and Technology Conf.*, Orlando, Fla., 1996, pp. 1101–1108.

CHAPTER THREE

Applications of InP-Based Transistors for Microwave and Millimeter-Wave Systems

MEHRAN MATLOUBIAN
Hughes Research Laboratories

3.1	Introduction	38
3.2	Applications	38
3.3	Communication and Radar Systems	40
3.4	InP-Based HEMTs	43
	3.4.1 Low-Noise Amplifiers	46
	3.4.2 Power HEMT Amplifiers	48
	3.4.3 Mixers and Frequency Multipliers	50
	3.4.4 InP HEMT Digital Circuits	51
3.5	InP-Based HBTs	53
	3.5.1 InP-Based Digital ICs	56
	3.5.2 InP-Based Analog ICs	57
	3.5.3 InP-Based Power HBTs	57
3.6	System Realization and Device Comparison	58
3.7	Summary and Future Trends	61
	References	62

InP-Based Materials and Devices: Physics and Technology, Edited by Osamu Wada and Hideki Hasegawa.
ISBN 0-471-18191-9 © 1999 John Wiley & Sons, Inc.

3.1 INTRODUCTION

There has been remarkable progress in the development of InP-based high electron mobility transistors (HEMTs) and heterojunction bipolar transistors (HBTs) over the last eight years. The unique performance requirements for military systems such as radars and communication systems have been the driving force in the development of high-performance InP-based device technologies. The necessity for superior military system performance has required continuous development of increasingly complex circuits with higher operating frequencies, higher gains, lower noise figures, higher output powers, higher power-added efficiencies, and wider bandwidths. At millimeter-wave frequencies, the availability of large spectrums of bandwidth and the allocation of frequency bands for commercial use has stimulated great interest for the development of low-cost circuits for commercial products at these frequencies. The utilization of InP-based device technologies for commercial applications potentially offers an attractive return on the large investment that companies have made in developing the technology for military applications.

In this chapter, applications of InP-based HEMT and HBT technologies for microwave and millimeter-wave systems are reviewed. Detailed physics and fabrication of HEMTs and HBTs are discussed in Chapters 10 and 11, respectively. The issues that are covered in this chapter in addition to some of the system applications of InP HEMT and HBT technologies are the present state-of-the-art performance of devices and circuits in these two technologies, comparison with competing III-V or Si-based technologies, requirements for various circuit components of a system, when it makes sense to use InP-based device technologies for system applications, and some cost/performance trade-offs.

3.2 APPLICATIONS

There are a wide variety of both military and potential commercial applications for InP-based HEMTs and HBTs. Some of these applications are shown in Fig. 3.1. In

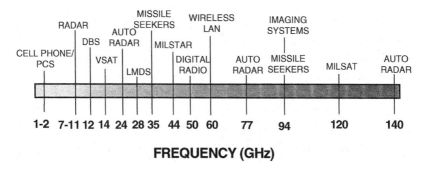

FIGURE 3.1 Applications of microwave and millimeter-wave systems and frequencies of interest.

addition to overcrowding of the frequency spectrum at microwave frequencies, there are several other factors in pushing communications and radar systems to higher frequencies and in selecting frequencies for particular applications. Operating at millimeter-wave frequencies allows a radar system to achieve the same resolution at microwave frequencies using a much smaller antenna, or alternatively, to achieve much higher resolution using the same antenna size used at microwave frequencies. For example, one particular application that potentially has a very high volume commercial market is collision warning radar or adaptive cruise control for automotive applications [1,2]. In this application, the radar is mounted in front of the car, so small antenna size is needed so as not to interfere with the styling of the car. Most present work is concentrated in developing automotive radar for operation at 76 GHz, with a desire to go up to 140 GHz to further reduce antenna size.

Another factor in selecting frequencies for a particular application deals with absorption of microwaves and millimeter-waves by the atmosphere (Fig. 3.2). As can be seen from Fig. 3.2, there is peak absorption at around 60 GHz, whereas absorption at 35 and 94 GHz is low (called the *atmospheric window*). Due to the high atmospheric attenuation at 60 GHz, the frequency range from 59 to 64 GHz, for example, has been allocated in the United States for unlicensed commercial applications such as local area networks (LANs) [3]. The high absorption allows reuse of this frequency in adjacent buildings with minimum interference. In addition, this frequency range is used for secure satellite-to-satellite communications to prevent interception of transmission. On the contrary, 35 and 94 GHz are used in

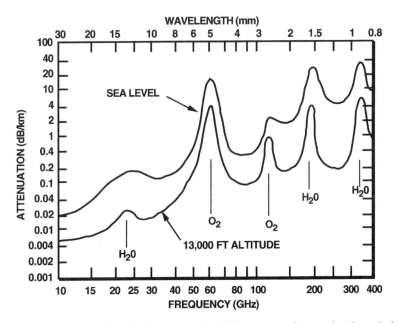

FIGURE 3.2 Transmission of microwave and millimeter-wave frequencies through the atmosphere at sea level and at 13,000 ft altitude.

systems where long-range or atmospheric penetration is highly desirable, such as missile seeker applications and millimeter-wave imaging systems and radar to aide in landing planes in fog [4].

Other potential applications of InP-based device technologies include local multipoint distribution services (LMDS) [5], intended to provide wireless cable distribution and other interactive digital services (Internet access); millimeter-wave digital radios for communication links between cellular phone base stations, and direct broadcast satellite (DBS) receivers. These and many other applications could potentially provide the markets needed to exploit InP-based device technologies.

3.3 COMMUNICATION AND RADAR SYSTEMS

InP-based HEMTs and HBTs have their greatest impact in system performance in the radio-frequency (RF) front end (transmit and receive functions) of radar and communication systems. A simplified block diagram of a generic radar or communication system is shown in Fig. 3.3. Such a transmit/receive (T/R) system is used in a number of commercial and military applications, including automotive radars, missile seekers, and communication links. A typical system consists of low-noise and power amplifiers, mixers and frequency multipliers, oscillators, digital-to-analog converters (DACs), and analog-to-digital converters (ADCs). The signal-to-noise ratio (S/N) at the input to the receiver of a radar is given by the equation

$$\frac{S}{N} = \frac{G^2 \lambda^2 \sigma P_T}{(4\pi)^3 R^4 k T_0 B_n L F} \tag{1}$$

where G is the gain of the antenna, λ the wavelength, σ the radar target cross section, P_T the transmitted power, R the distance to target, k is Boltzmann's constant, T_0 is the room temperature in kelvin, B_n the noise bandwidth, L is other losses (such as atmospheric attenuation), and F is the noise figure of the receiver. From this equation it can be seen that the signal-to-noise ratio of a radar is directly proportional to the power transmitted and inversely proportional to the receiver noise figure. To in-

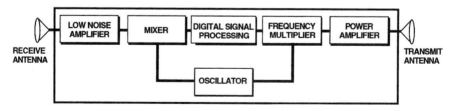

FIGURE 3.3 Simplified block diagram of a generic microwave/millimeter-wave radar or communication system.

crease the range of a radar given a limited transmitted power, the noise figure of the receiver has to be improved.

For a low-noise amplifier (LNA) the parameters of interest are the noise figure (NF), gain, bandwidth, linearity and dynamic range, and power consumption. Depending on the application, some of these parameters become more important than others. For multistage amplifiers, the NF of the amplifier is dominated by the NF of the first stage if the first stage has a high gain (about 10 dB) and low NF. For a three-stage amplifier, the overall NF of the amplifier F is given by

$$F = F_1 + \frac{F_2 - 1}{G_1} + \frac{F_3 - 1}{G_1 G_2} \quad (2)$$

where F_1, F_2, F_3, and G_1, G_2, G_3 are the noise figure and gain of the first, second, and third stages respectively. As can be seen from Eq. (2), it is desirable to have the lowest-noise device at the front end of the amplifier or system to get the best performance. Better low-noise amplifiers will enable more efficient detection of the faint signals from satellites. This means that smaller antennas can be used [lower G, antenna gain, in Eq. (1)]. For military applications, having a better low-noise amplifier for a radar receiver gives longer range and tactical advantage, so it is critical in military applications to have the best possible receiver performance. For commercial applications, performance that meets the system specifications while at the same time meet the cost goals is most desirable.

Power amplifiers are typically one of the most critical components of radar and communication systems. The parameters of interest for power amplifiers are power-added efficiency (PAE), gain, output power, linearity and bandwidth, and power consumption. For some applications, such as hand-held personal communications systems (PCSs), low operating voltage (<3 V) is highly desirable to allow battery operation. The higher output power of power amplifiers gives radars more detection range and allows communication systems to link over longer distances. The efficiency of power amplifiers is extremely important in reducing dc power consumption (extending battery life for portable communications and satellite applications), increasing device reliability, and reducing the need for cooling large array of devices. For a multistage power amplifier, the efficiency of the last stage (output stage) will dominate the overall efficiency of the amplifier if the drivers have high gain (about 10 dB). For a three-stage power amplifier, the overall PAE η_T is given by:

$$\frac{1}{\eta_T} \simeq \frac{1}{\eta_{d1} G_2 G_3} + \frac{1}{\eta_{d2} G_2} + \frac{1}{\eta_{d3}} \quad (3)$$

where η_{d1}, η_{d2}, η_{d3} and G_1, G_2, G_3 are the drain efficiency and gain of the first, second, and third stages, respectively. As can be seen from Eq. (3), it is desirable to have the device with the best efficiency at the output stage of the multistage amplifier. For most commercial applications, such as automotive radars and LANs, the output power requirement is low (about 10 mW). For communication applications such as PCS, the output power is somewhat higher and is in the range 300 mW to 1 W. For military T/R systems, the output power requirement ranges from about 5 W for millimeter-wave systems to more than 50 W for microwave systems.

For mixers and frequency multipliers the nonlinearity of the device (FET, HBT, or diode) is used to downconvert or upconvert the signal. For mixers, the RF (radio-frequency, high-frequency input signal)-to-IF (intermediate-frequency, low-frequency output signal) conversion loss is one of the critical parameters. The IF frequency used for a system depends on a number of factors, including the bandwidth of information in the carrier frequency. In addition, other important mixer parameters include the LO (local oscillator) power and intermodulation distortion (IMD), which is a measure of the power of unwanted mixer signals generated near the desired IF signal. Frequency multipliers are used to upconvert the information signal onto a carrier signal. For frequency multipliers, the upconversion efficiency and the LO power required for upconversion are the critical parameters.

Oscillators play a critical role in systems and often are one of the most difficult circuits to realize. The oscillator sets the frequency of the system operation and since for most applications (in particular commercial applications) the frequency allowed for system operation is limited to a particular frequency and bandwidth, it is important to achieve the required performance reproducibly. In designing amplifiers the bandwidth is intentionally designed to be larger than system specifications to account for circuit fabrication tolerance. For oscillators it is difficult to increase bandwidth without significantly reducing the output power and phase noise of the oscillator. For radar applications, the phase noise of the oscillator is critical in achieving the required performance. To achieve reproducible oscillator performance, highly reproducible device performance is needed.

The IF signal is digitized using analog-to-digital converters (ADCs) to allow digital signal processing or digital storage of information. For military radar applications, high-resolution (10 to 16 bits) ADCs with a 20- to 200-MHz bandwidth, and medium-resolution (6 to 8 bits) ADCs with 500-MHz to 5-GHz bandwidth are needed. The general trend in systems has been toward direct digital synthesis (DDS) [6,7]. In DDS, the mixers and oscillators used in conventional systems are eliminated and replaced by ADCs. This reduces the total number of circuits in the system, but it requires ADCs with wider bandwidths of operation and provides a potential market for high-performance III-V-based ICs. In addition to radar applications, high-performance ADCs are used in communication links as well as commercial instrumentation (oscilloscopes). Digital circuits can consume a large percentage of power in a system. The power consumption of digital circuits can be lowered by using devices with lower threshold voltage swing and scaling devices to smaller geometries to minimize current consumption.

All the circuit components described in this section have been realized using InP-based HEMT and HBT technologies. In the following sections, the state of the art in performance for these InP-based circuits is reviewed. Other important factors to consider are the competing technologies and when it makes sense to use InP-based technologies in system applications.

3.4 InP-BASED HEMTs

Field-effect transistors (FETs) have progressed remarkably in performance over the last 25 years, from the early developments of GaAs MESFETs with about 1-μm gate lengths [8] to the present advanced InP-based HEMTs with 0.1-μm gate lengths and f_{max} values of 600 GHz [9]. The evolution of InP-based HEMT technology is reviewed in this section. The detailed physics of operation of HEMTs and their fabrication are discussed in Chapter 10. The bandgap versus lattice constant for common III-V semiconductors is shown in Fig. 3.4. Figure 3.5 compares several FET technologies. In the GaAs MESFET structure shown in Fig. 3.5a, all the layers are made using GaAs with different doping concentrations. Typically, the same doped GaAs layer acts as the channel as well as the Schottky layer. The mobility of the electrons in a MESFET channel is lower than in undoped GaAs due to scattering of the electrons by ionized donors. To reduce the gate length of MESFETs, the thickness of the channel also has to be reduced to allow good control of the current flow through the channel by the gate. But as the channel thickness is reduced, the doping has to be increased to keep constant channel resistance and drain current. This criteria limit the cutoff frequency of GaAs MESFETs to approximately 100 GHz [10–12]. MESFETs have demonstrated very good power performance, but their frequency of operation is typically limited to less than 30 GHz. GaAs MESFET technology is very mature and MESFET MMICs are

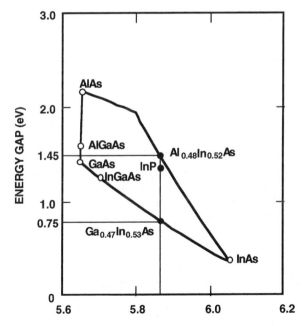

FIGURE 3.4 Bandgap versus lattice constant for common III-V compound semiconductors.

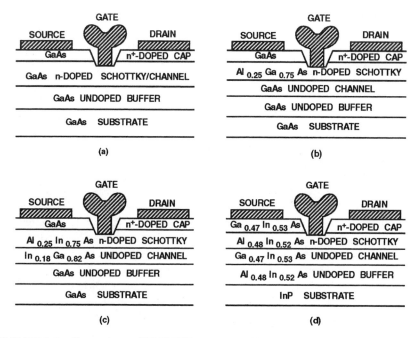

FIGURE 3.5 Comparison of III-V FET structures: (*a*) GaAs MESFET; (*b*) AlGaAs/GaAs HEMT; (*c*) AlGaAs/InGaAs pHEMT; (*d*) AlInAs/GaInAs HEMT.

presently being manufactured in large volumes for low-frequency (<5 GHz) commercial applications.

In AlGaAs/GaAs HEMT structure (Fig. 3.5*b*), performance improvement is achieved over conventional GaAs MESFETs by using a wider-bandgap (about 1.7 eV) AlGaAs layer as the Schottky layer and a lower-bandgap (1.42 eV) GaAs layer as the channel. The Schottky layer is doped, whereas the channel is undoped. The electrons transfer from the wider-bandgap AlGaAs layer into the lower-bandgap GaAs channel, forming a two-dimensional electron gas (2DEG) near the AlGaAs/GaAs interface. In effect, by incorporating an AlGaAs/GaAs heterojunction a conduction band edge discontinuity is created that physically separates the donors in the wider-bandgap AlGaAs from the mobile electrons in the lower-bandgap GaAs channel. This separation significantly reduces ionized impurity scattering of the electrons in the channel and leads to higher carrier mobility and higher f_T values [13].

By adding indium to the GaAs channel to form GaInAs channels (Fig. 3.5*c*), the device characteristics are further enhanced. The electron mobility, conduction band discontinuity (ΔE_c), and 2DEG carrier density all increase with increasing indium content of the channel. However, increasing the indium content of the channel is limited by the fact that GaInAs is not lattice matched to the GaAs substrate, and the layer is increasingly strained as the indium content is increased. Due to the lattice mismatch between the GaInAs channel and the GaAs substrate, the AlGaAs/GaInAs HEMTs are commonly referred to as GaAs pseudomorphic HEMTs (GaAs pHEMTs). To avoid relaxation of the crystal structure and the creation of dislocations in the channel, the indium content of AlGaAs/GaInAs/GaAs HEMT structures is typically limited to a maximum of approximately 30%. In addition, the channel thickness of GaAs pHEMTs (depending on Indium content) is kept to less than 150 Å. GaAs-based pHEMTs have demonstrated excellent low-noise performance [14–16] as well as state-of-the-art power performance at microwave and millimeter-wave frequencies [17–21]. GaAs-based pHEMTs are gradually replacing MESFETs for some microwave applications and have been the primary device of choice for millimeter-wave power applications.

The InP-based HEMT technology (Fig. 5d) has the advantage over GaAs pHEMTs that GaInAs channels with higher indium content can be grown on InP substrates [22]. GaInAs with 53% indium ($Ga_{0.47}In_{0.53}As$ with a bandgap of 0.76 eV) is lattice matched to InP substrates and instead of an AlGaAs Schottky layer $Al_{0.48}In_{0.52}As$ (1.48 eV), which is also lattice matched to InP, is used. Compared with GaAs-based pHEMTs, the $Al_{0.48}In_{0.52}As/Ga_{0.47}In_{0.53}As/InP$ materials system exhibits higher electron mobility and a higher electron peak velocity, higher conduction-band discontinuity, leading to higher 2DEG density; and higher transconductance [23]. InP-based HEMTs exhibit higher f_T and f_{max} value than those of GaAs pHEMTs of the same gate length. To further increase the mobility of the channel, the indium content of the channel can be increased beyond 53% [24]. This results in a pseudomorphic InP-based HEMT. The unique properties of the InP-based HEMT materials system have established this system as the leading transistor for millimeter-wave low-noise applications [25–27]. They have set record performances for the fastest transistor operating at room temperature with the highest reported f_T value of 343 GHz [24] as well as the highest reported f_{max} value of 600 GHz [9]. They have also set the world record for the fastest monolithic integrated circuit operating at a frequency of 213 GHz [28]. The photograph of this oscillator is shown in Fig. 3.6. A common-gate oscillator configuration with coplanar waveguide (CPW) transmission lines is used in the design of this oscillator. The oscillator uses an InP-based HEMT fabricated using a self-aligned gate process [24]. These HEMTs have an intrinsic f_{max} value of over 400 GHz, and the fact that these oscillators operate at a frequency of 213 GHz demonstrates that these devices have gain at frequencies of over 200 GHz.

All the RF circuit components of a typical communication or a radar system (shown in Fig. 3.3) can be fabricated using InP HEMTs. In addition to low-noise amplifiers, InP-based HEMTs have been used for power amplification, mixers, frequency multipliers, oscillators, switches, and digital circuits. Even though all the different circuits can be fabricated using InP HEMT technology, the choice of optimum device for each application depends on a number of factors that will be discussed later.

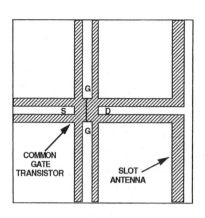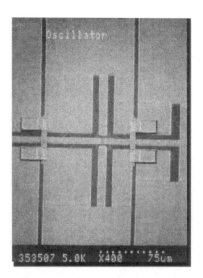

FIGURE 3.6 SEM photograph of a 213-GHz monolithic InP HEMT oscillator. (From Ref. 28.)

3.4.1 Low-Noise Amplifiers

Qualitatively, to achieve better low-noise performance from an FET, the device requires high transconductance, high f_T and f_{max} values, and low source resistance. InP-based HEMTs, with their intrinsic material properties, are ideally suited for low-noise amplifiers and have demonstrated state-of-the-art low-noise performance at microwave and millimeter-wave frequencies. Transconductance values as high as 1500 to 1700 mS/mm have been reported for InP-based HEMTs, and typical 0.1-μm InP low-noise HEMTs exhibit values of 800 to 1000 mS/mm, compared with 600 mS/mm for comparable GaAs pHEMTs. Figure 3.7 shows some of the best reported noise figures for discrete GaAs-based pHEMTs and InP-based HEMTs. The lowest noise figure achieved for GaAs pHEMTs is 1.5 to 1.6 dB at 60 GHz and 2.1 dB at 94 GHz. For InP-based HEMTs the lowest noise figure at 60 GHz is 0.8 to 0.9 dB at 60 GHz and 1.2 to 1.3 dB at 94 GHz. In addition to the lower noise figure offered by InP HEMTs, they offer higher gain than do GaAs-based HEMTs, which can help further to reduce the noise figure for a multi-stage amplifier. InP low-noise HEMTs have also been used in cryogenic receivers for radio astronomy applications. A Q-band LNA cooled to 18 K has demonstrated a noise temperature of 12 K, which is comparable to superconductor–insulator–superconductor (SIS) receivers [29].

Using InP HEMT technology, low-noise hybrid (MIC) amplifiers and MMICs have been developed ranging in frequency from 2 to 160 GHz [30]. MMIC LNAs at microwave frequencies have been reported with NF values of 0.4 dB and 35 dB gain at 2 GHz [31] and 1 dB NF and 21 dB gain at X-band [32]. At millimeter-wave fre-

3.4 InP-BASED HEMTs

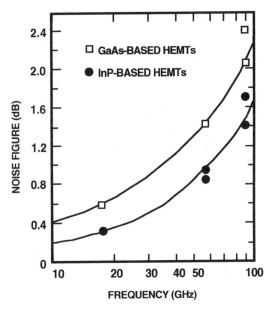

FIGURE 3.7 Comparison of noise figure of GaAs-based pHEMTs and InP-based HEMTs.

quencies, MMIC LNAs with minimum NF values of 1.9 dB and 22 dB gain at Q-band [33], 2 and 16 dB gain at V-band [33], and 3.3 and 20 dB gain at W-band have been reported [30]. Figure 3.8 shows the photograph and performance of a three-stage W-band LNA. The amplifier has an average noise figure of 4 dB and gain of 21 dB from 88 to 96 GHz [27]. Recently, low-noise MMICs with a gain of 12.5 dB at a frequency of 155 GHz have been reported [34].

Low-noise InP HEMTs are typically biased at a drain-to-source bias of 1 V and a drain current of 100 to 140 mA/mm. This operating point results in very low dc power consumption, which is highly desirable for battery operation as well as for satellite communication. However, the low operating voltage results in low 1-dB gain compression point [and low third-order intercept (IP3)]. By optimization of the layer design and using higher operating bias voltage for the output stage, it is possible to achieve highly linear characteristics using InP-based HEMTs [35]. InP HEMTs have also been incorporated in wideband amplifier designs. Amplifiers with 17 dB gain and 0.1 to 70 GHz bandwidth [36] and 10 dB gain with 92 GHz bandwidth [37] have been reported. These broadband amplifiers have applications in wideband receivers and instrumentation.

Extensive work on InP-based low-noise HEMTs from material and device design to optimization of metallization for ohmic and Schottky contacts has led to the development of highly reliable HEMTs. Mean time to failure (MTTF) of more than 10^8 h for operation at 45°C has been demonstrated [38]. This MTTF is more than adequate for using InP low-noise HEMTs for satellite applications.

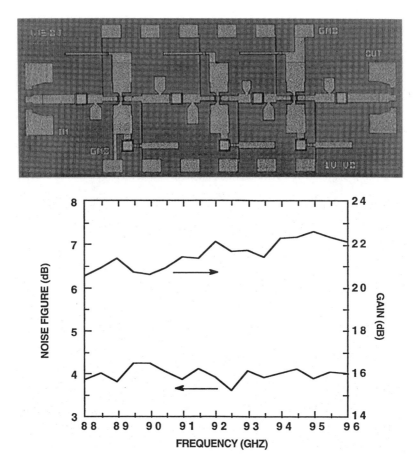

FIGURE 3.8 Three-stage W-band InP-based HEMT LNA with an average noise figure of 4 dB and 21 dB of gain. (After Ref. 27.)

3.4.2 Power HEMT Amplifiers

For power applications, InP-based HEMTs offer a number of advantages over GaAs-based HEMTs. The thermal conductivity of InP is 40% higher than GaAs (Fig. 3.9), allowing a lower operating channel temperature for the same power dissipation. In addition, the higher electron densities in the channel, coupled with the higher electron velocity, can lead to higher current densities. Despite these advantages, until recently very little work had been done on InP-based HEMTs for power applications [39–42]. This has been due primarily to the low gate-to-drain breakdown voltage and low Schottky barrier height of low-noise InP-based HEMTs. But by proper device layer design it is possible to overcome the drawbacks of InP-based HEMTs for power applications [43]. Optimization of the InP HEMT layer structure has led to the development of power HEMTs with state-of-the-art power performance at microwave and millimeter-wave frequencies. Output power densities of

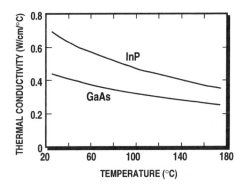

FIGURE 3.9 Comparison of thermal conductivity of GaAs and InP substrates.

more than 1 W/mm with power-added efficiencies of about 60% have been reported at 4 and 12 GHz [44,45]. By using a composite channel-layer design and optimizing the Schottky layer, gate-to-drain breakdown voltages of more than 20 V have been reported for 0.25-μm-gate-length high-performance InP-based HEMTs [44]. By using double-recess gate process, 0.15-μm-gate-length HEMTs with breakdown voltages as high as 19 V have been reported [46].

At millimeter-wave frequencies, state-of-the-art power results have been reported using InP-based HEMTs. At 60 GHz, single-device output powers as high as 192 mW with a power-added efficiency (PAE) of 30% and a maximum PAE as high as about 50% with 15 mW of output power have been reported [47]. By combining two discrete transistors, output powers of about 300 mW have been achieved at 60 GHz [48]. At 94 GHz, PAE values of more than 30% have been reported, which is more than double the PAE values reported for GaAs pHEMTs at the same frequency. In general, InP-based HEMTs have demonstrated higher gains and higher PAEs than those of GaAs pHEMTs at millimeter-wave frequencies [9]. At frequencies of around 60 GHz and below, GaAs pHEMTs have typically higher power densities, but the power density of around 0.5 W/mm achieved for InP HEMTs at 60 GHz is adequate for most applications. Presently, at frequencies below 60 GHz, where the operating voltages are high, the reliability of InP-based power HEMTs with a low bandgap channel has not been demonstrated.

In addition to millimeter-wave power applications where high gains and PAEs are desirable, InP HEMTs are also ideal for applications requiring low-voltage (battery operated) and high-efficiency operation. For low-voltage operation of power transistors, low knee voltage, high drain current density, and high power gain are essential for achieving high PAE values [49]. The efficiency of transmitters is critical in extending the life of batteries for millimeter-wave portable wireless applications. Using InP-based power HEMT technology, one can easily obtain electron sheet charge densities (2DEG) of more than 5×10^{12} cm^{-2}, which has led to transistors with current densities of 1 A/mm. The high electron density combined with the high electron mobility in the channel also leads to a low source resistance and low knee voltage. The combination of high current density, low knee voltage, and high gain make the InP-based power HEMT technology ideal for low-voltage applications.

Development of high-performance InP-based power HEMTs has also led to the development of MMICs with state-of-the-art performance. At 44 GHz an MMIC using a single 600-μm-wide device has demonstrated an output power of 250 mW (420 mW/mm) with 33% PAE [50]. To achieve higher output powers at 44 GHz, four 450-μm-wide HEMTs (1800-μm total gate periphery) have been combined on a single InP chip [51]. Figure 3.10a shows this chip, which measures only 1.4 × 1.2 mm^2 and has a thickness of 50 μm. Using a thinner chip allows a more compact circuit layout and lower thermal resistance for the chip. This amplifier has demonstrated state-of-the-art power performance with an output power of 800 mW with a PAE value of 30% and 6 dB of gain (Fig. 3.10b). Comparison of state-of-the-art power performance for discrete as well as power MMICs at 44 GHz is shown in Fig. 3.11.

Single-stage 94-GHz power amplifiers have been reported with an output power of 58 mW, PAE of 33%, and a power gain of 6.4 dB [9]. These amplifiers have a power gain that is 2.5 to 3 dB higher than that of GaAs pHEMT MMICs and a PAE value that is more than twice as high as that of pHEMT MMICs [52]. In addition, devices from the same wafer have demonstrated a noise figure of 1.4 dB with 7 dB of gain, which is comparable to that of InP HEMTs optimized for low-noise operation. Using these HEMTs, low-noise and power amplifiers can be integrated on the same chip to develop T/R modules at W-band.

3.4.3 Mixers and Frequency Multipliers

As discussed in Section 3.2, mixers and frequency multipliers are an integral part of communication and radar systems. Mixers that can easily be integrated with InP low-noise HEMTs are desirable for development of receivers. Several types of mil-

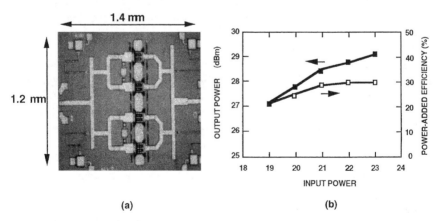

FIGURE 3.10 (a) A 44-GHz InP HEMT MMIC amplifier combining four 450-μm-wide HEMTs; (b) measured performance of the amplifier with an output power of 800 mW and a PAE of 30%. (From Ref. 43.)

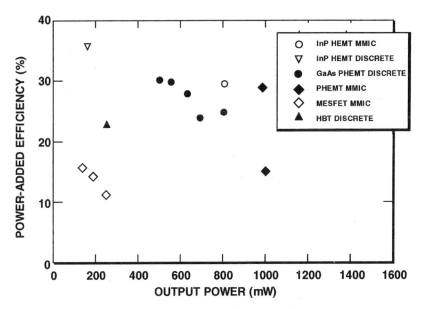

FIGURE 3.11 Comparison of state-of-the-art performance of power transistors at 44 GHz.

limeter-wave mixers with excellent performance have been fabricated in a standard low-noise InP HEMT process [27,53,54]. These include resistive HEMT mixers, active mixers, and rat-race diode mixers. A W-band (94-GHz) active InP HEMT mixer, along with its performance, is shown in Fig. 3.12. The mixer has a conversion loss of 3.3 dB at IF of 2 GHz with an LO power of 9.6 dBm at 94 GHz [54]. A W-band resistive HEMT mixer has been demonstrated with a conversion loss of 9.5 dB, requiring an LO power of only 3.4 dBm at 94 GHz [54]. Frequency multipliers have also been fabricated using InP low-noise HEMTs. A V-band (54 GHz) frequency quadrupler has been demonstrated with 14.25 dB of conversion gain and 3.25 dBm of output power [53].

3.4.4 InP HEMT Digital Circuits

The high-frequency performance of InP-based HEMTs should lead to high-performance digital circuits using this technology. In addition, it is possible to use longer gate lengths with InP-based HEMTs and achieve a performance comparable to that of GaAs MESFETs and pHEMTs with shorter gate lengths. Using longer gate lengths should improve device yield and therefore circuit yield. But compared to low-noise applications, very little work has been done using InP HEMTs for digital circuits. Some early work using InP HEMT technology resulted in 25-GHz divide-by-2 digital frequency dividers in a 0.2-μm gate-length emitter-coupled logic (ECL) design [55]. More recently, dynamic frequency dividers as high as 75 GHz have been demonstrated using 0.1-μm InP HEMT technology [56]. One of the critical

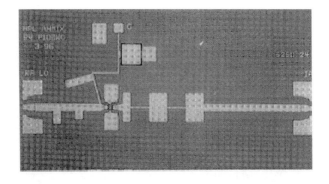
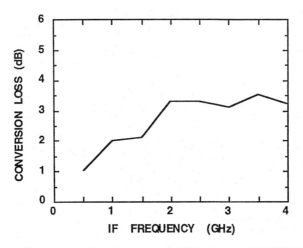

FIGURE 3.12 Photograph and performance of a W-band active InP HEMT mixer. The LO power is 9.6 dBm at 94 GHz. The chip is 1.3×2.9 mm^2. (From Ref. 54.)

parameters for using InP HEMTs for digital applications is the control of threshold voltage across the wafer. Using InP layers as gate recess etch-stop layers has resulted in very high uniformity HEMTs, and 40-GHz source-coupled-FET-logic (SCFL) static frequency dividers have been demonstrated using this process [57]. Another motivation for development of InP HEMT digital circuits is the ability to integrate analog and digital circuits into one chip. Fabrication of a 37-GHz SCFL HEMT static frequency divider and a 62-GHz two-stage MMIC HEMT LNA in the same process has been demonstrated [58].

In addition to the circuits discussed, InP-based HEMTs can also be used for development of low-insertion-loss high-frequency switches. The high mobility and high sheet charge in the channel results in low sheet resistance and therefore low on-resistance for the switch. The low input capacitance results in high-speed switch response, and low feedback capacitance in high input-to-output isolation in the off-state.

3.5 InP-BASED HBTs

The idea of heterojunction bipolar transistors (HBTs) is almost as old as the bipolar transistor itself [59,60]. In conventional silicon bipolar transistors, the emitter injection efficiency and the current gain of the device depend on the ratio of emitter doping to base doping of the transistor. To achieve current gain, the emitter has to be doped heavier than the base, therefore limiting the base doping and resulting in high base resistance. In HBTs, the emitter consists of a wider-bandgap material than the base. This difference in bandgap between the emitter and base is used to engineer the device such that the current gain of the device is not dependent on an upper limit on base doping. Base can be doped heavier to decrease the base resistance and improve high-frequency performance of the device. In effect, HBTs give the device designer more flexibility to engineer a device for high-frequency performance [61].

There are a number of different HBT structures, but for comparison only four of these structures are shown in Fig. 3.13. All of these structures consist basically of a wide-bandgap emitter and a narrower bandgap base. The AlGaAs/GaAs HBT (Fig. 3.13a) is a mesa structure and is the most mature HBT technology. A number of different circuits have been demonstrated using this technology, and presently AlGaAs/GaAs HBTs are being fabricated in high volume for commercial applications. A slight variation of this technology is to replace the AlGaAs emitter with GaInP [62–65]. This results in better etch selectivity between the emitter and the base layers and simplifies device fabrication.

The first high-performance InP-based HBT was reported in 1983 [66]. This device consisted of an $Al_{0.48}In_{0.52}As$ emitter layer and $Ga_{0.47}In_{0.53}As$ base and collector layers (Fig. 3.13b). A number of variations of this HBT structure have been reported, including HBTs with InP emitters [67] and HBTs with InP collectors [68] (Fig. 3.13b), referred to as double-heterojunction bipolar transistors (DHBTs). The HBTs with GaInAs collector [single-heterojunction bipolar transistors (SHBTs)] has lower base–collector breakdown voltage due to the low bandgap of the GaInAs collector, but typically it has a higher-frequency performance than that of an InP DHBT. Using InP as the collector increases the base–collector breakdown voltage (as well as V_{CEO}) and improves the output conductance of the device. The DHBT has been used primarily for power applications, but it also has applications for high-resolution ADCs [69].

InP-based HBTs have several advantages over GaAs-based HBTs. The lower bandgap of GaInAs results in a lower turn-on voltage and lower power dissipation. The lower surface recombination velocity of the InP-based HBTs (three orders of magnitude lower than that of GaAs) provides gain at low current densities, improving the ability to scale devices to smaller emitter sizes for LSI implementation and further reduction of power consumption. Due to increasing demand for portable communication systems, power consumption or power dissipation is becoming increasingly important for microwave and millimeter-wave systems. A comparison of the Gummel plots (I_c and I_b versus V_{be}) for GaAs-based and InP-based HBTs is shown in Fig. 3.14 and demonstrates some of the advantages of InP-based technology. As can be seen from this figure, the GaAs HBT does not have current gain at

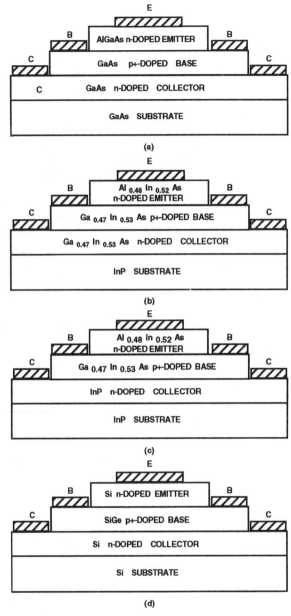

FIGURE 3.13 Comparison of HBT structures: (*a*) AlGaAs/GaAs HBT; (*b*) AlInAs/GaInAs HBT; (*c*) AlGaAs/GaInAs/InP DHBT; (*d*) Si/SiGe HBT.

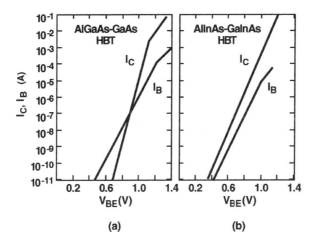

FIGURE 3.14 Comparison of the Gummel plot of (a) AlGaAs/GaAs HBT and (b) AlInAs/GaInAs HBT.

low currents ($I_c < I_b$). The InP HBT has a lower turn-on voltage due to the lower bandgap of GaInAs. In addition, shorter electron transit time through the base and collector allows for higher frequency of operation while maintaining reasonable breakdown voltage. InP-based HBTs have a significantly lower $1/f$ noise than GaAs HBTs, making them ideal for use in low-phase noise oscillators. In addition, InP HBTs are also radiation hard, making them suitable for satellite as well as military applications. A wide variety of HBT ICs have been demonstrated. The reliability of the InP-based HBT IC technology has been demonstrated [70]. At current densities of 70,000 A/cm², a MTTF of greater than 10^7 h has been projected for operating temperature of 125°C.

Si/SiGe HBT technology (Fig. 3.13d) also appears very promising for high-frequency circuit applications [71,72]. f_T and f_{max} values of over 100 GHz have been demonstrated using this technology. Si/SiGe HBTs are very attractive for realization of low-cost microwave circuits due to the availability of large-diameter silicon substrates. One limitation to the realization of MMICs using SiGe technology has been the high losses of transmission lines on silicon substrates. Several approaches have been taken to overcome this problem. In one approach high-resistivity silicon substrates are used [73]. But large-diameter high-resistivity silicon substrates are not available and are presently too expensive. In another approach, a top low-loss spin-on dielectric is added for fabrication of transmission lines [74]. This approach has the advantage that the SiGe HBT can be fabricated using existing mass-production silicon fabrication lines. One problem with the present Si/SiGe HBT technology has been the low value of V_{CEO}. SiGe HBTs with an output power of about 100 mW have been reported at microwave frequencies [75,76]. This power is adequate for these amplifiers to be used in a number of commercial applications. Also some recent reports have shown the potential to improve the breakdown voltage of SiGe HBTs for higher output powers [77].

III-V-based HBTs have demonstrated f_T and f_{max} values of over 200 GHz [78]. But these HBTs have very low operating voltages (typically less than 2 V). Recent work on bandgap engineering of the base–collector junction has resulted in record combination of breakdown voltage and cutoff frequency for HBTs [79]. By proper scaling of the material structure and minimizing device parasitics, it should be possible to achieve breakdown voltages of more than 10 V and an f_{max} value of over 150 GHz in the near future.

3.5.1 InP-Based Digital ICs

One of the primary applications of InP-based HBTs has been the development of high-frequency digital circuits. The high uniformity and control of threshold voltage, low threshold voltage and low power dissipation make this device technology ideal for high-frequency digital applications. A wide variety of high- speed digital ICs using InP-based SHBTs have been demonstrated [80]. The typical f_T of npn SHBTs used for development of digital circuits is 75 GHz with an f_{max} of 80 GHz. By slight modification of material layer structure, f_T can be increased to 170 GHz for different applications [81]. InP-based HBTs have demonstrated world record performance for digital ICs such as a static divide-by-4 frequency divider operating at 39.5 GHz [82] (Fig. 3.15), a 3-bit ADC converter operating at 8 GHz [83], a 12-bit DAC with a –60-dB spur-free dynamic range [84], and a second-order delta-sigma modulator with 12-bit dynamic range at 3.2 GHz [85]. The largest InP HBT IC fabricated is a 15-bit accumulator with 1500 transistors [81].

One of the advantages of InP-based HBTs is the ability to scale devices to much smaller dimensions because of the lower surface recombination velocity in this technology compared to GaAs HBTs. This allows further reduction of the power as

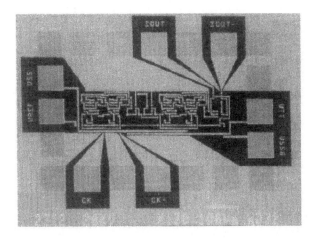

FIGURE 3.15 A 39.5-GHz InP HBT divide-by-4 circuit. (After Ref. 82.)

well as the area of the circuit. Scaled HBTs with emitter areas as small as 0.3 μm^2 have been demonstrated [86]. Small-area HBTs with peak f_T and f_{max} values of 170 and 130 GHz, respectively, have been developed. This technology combined with a reduction of interconnect metal pitch has led to the development of compact, high-performance, low-power ICs such as a 40-GHz comparator [87].

3.5.2 InP-Based Analog ICs

InP-based HBTs have demonstrated lower $1/f$ noise and lower phase noise oscillators than those of $Al_{0.30}Ga_{0.70}As$/GaAs HBTs [88,89]. A number of InP-based HBT oscillators at microwave and millimeter-wave frequencies have been demonstrated [90,91]. A fundamental 62-GHz MMIC InP HBT VCO with a 300-MHz tuning range and an output power of 4 dBm has demonstrated a phase noise of –78 dBc/Hz at 100 kHz offset [91]. Recently, fundamental InP-HBT oscillators as high as 94 GHz have been demonstrated [92]. The oscillator used an InP HBT with an f_T value of 70 GHz and an f_{max} value of 185 GHz.

Mixers can also be fabricated easily in the InP HBT process. Both InP HBT active mixers as well as Schottky diode mixers have been demonstrated. A double-balanced active HBT mixer operating from dc to 16 GHz has been developed with a conversion loss of 6 dB at IF of 1 MHz [93]. Schottky diodes fabricated in the InP HBT process have a lower threshold voltage than that of GaAs-based diodes, which results in reduction of the LO power for InP-based mixers [89]. A 39-GHz diode mixer using the InP HBT process has demonstrated a conversion loss of 8 dB with a noise figure of 11.5 dB at an IF of 1 MHz. A conversion loss of 9.5 dB was obtained from a 94-GHz diode mixer using an LO power of 5 dBm [94]. This mixer had a noise figure of 10 dB at an IF of 100 MHz.

A wide variety of amplifiers have also been demonstrated using InP HBTs. A feedback amplifier with dc to 33-GHz bandwidth was developed with 8.6 dB gain [95]. The widest-bandwidth InP HBT amplifier reported is a 2- to 50-GHz distributed amplifier with a peak gain of 6.3 dB [96]. Low-noise amplifiers based on GaAs HBTs have been reported with a noise figure of 1.6 dB at 2 GHz [97] and 6 dB at 30 GHz [98]. Even though this performance is not state of the art, it is adequate for some applications. Using InP-based HBTs, it is also possible to develop low-noise amplifiers. For InP-based HBTs, minimum device noise figures of 0.46, 1.09, and 3.33 dB were obtained at 2, 6, and 18 GHz, respectively [99]. This performance is better than that of GaAs-based HBTs but not as good as that of GaAs- or InP-based HEMTs.

3.5.3 InP-Based Power HBTs

The InP-based power HBT technology is less mature than that of GaAs-based power HBTs, but state-of-the-art power results have been obtained using InP-based materials system [68,100]. The larger thermal conductivity of the InP substrate (Fig. 3.9) will help in reducing the operating temperature of the device, which is critical for HBTs to achieve stable operation [101]. As mentioned previously, for power applications a DHBT structure (Fig. 3.13c) with an InP collector is used to achieve high breakdown voltages. Optimization of the grading of the base–collector junction has resulted in obtaining a record combination of breakdown voltage (35 V) and f_T (71 GHz) for a bipolar transistor [102]. Figure 3.16 shows the power performance of a 12-finger (each finger 2×30 μm^2) InP-based power HBT at 9 GHz measured using an on-wafer load-pull system. Output powers as high as 2 W (5.6 W/mm) with a PAE value of 70% have been demonstrated. At 2 GHz an InP DHBT operating in class C bias conditions has demonstrated an output power of 1.2 W with a PAE value of 90% [103].

Another aspect of InP-based HBTs not covered in this chapter is their compatibility with long-wavelength photonic devices operating at 1.3 and 1.55 μm, leading to optoelectronic integrated circuits (OEICs). This allows integration of analog, digital, and optical circuits using basically the same device technology. The optical properties of InP-based devices is covered in detail in Chapter 2.

3.6 SYSTEM REALIZATION AND DEVICE COMPARISON

A system designer often has the task of deciding what device technology to use for realization of a system. Choosing the right device technology for an application depends on several factors, including system specifications, system architecture, and system cost. The choice of the device technology can sometimes be limited simply

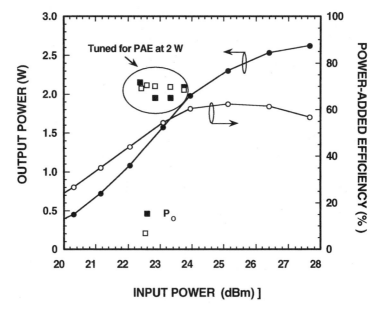

FIGURE 3.16 Power performance of an InP HBT with an InP collector measured on-wafer at 9 GHz. (After Ref. 79.)

3.6 SYSTEM REALIZATION AND DEVICE COMPARISON 59

by not having access to a particular technology. For military applications, the best possible system performance is desirable while the cost of the system is secondary. For commercial applications, the system designer uses the device technology that is necessary to meet the system specifications, while the system cost is typically the primary factor. In fact, for commercial applications the system performance sometimes has to be sacrificed (lower system specifications) to meet the cost goals. In this section different device technologies are compared to see when it makes sense to use InP-based device technologies.

For low-noise applications, InP-based HEMTs have the best performance of any existing transistor technology. But at microwave frequencies, the noise figures for low-noise GaAs-based pHEMTs are adequate for most applications (Fig. 3.17a). In fact, low-noise GaAs-based pHEMTs have been inserted in a number of commercial applications, including receivers for direct broadcast satellite (DBS) systems. The performance gap increases, as the frequency increases but the primary factor in selecting a device technology for commercial applications (other than availability) is still the system specifications. For 60-GHz LANs where InP-based low-noise HEMTs have a clear advantage over GaAs-based pHEMTs, InP-based HEMTs can potentially be inserted in a high-volume commercial market.

Choosing the device technology for power applications (for amplifiers or oscillators) is very complicated and depends on the frequency range of interest, output power, gain, efficiency, bandwidth, linearity, and phase noise. Figure 3.17b compares various power transistor technologies, depending on the frequency range of interest. At microwave frequencies, there are a number of competing technologies (in addition, SiC and GaN FETs, not shown in Fig. 3.17b, are also potential contenders at microwave frequencies). For microwave power applications, both FETs and HBTs have demonstrated very good power performance. GaAs-based pHEMTs have become the device of choice among FETs for microwave power applications and are gradually replacing GaAs MESFETs for most applications. GaAs HBTs are being used for some applications where lower phase noise or single power supply operation is desirable. Although GaAs HBTs are more mature than InP HBTs, as mentioned earlier, InP-based HBTs have also demonstrated state-of-the-art power performance. In addition, InP-based HBTs have the potential of achieving very high performance at millimeter-wave frequencies and might be able to compete with GaAs pHEMTs up to 60 GHz. Even though at microwave frequencies good power performance have been demonstrated using InP-based HEMTs, for most applications this technology cannot compete with other technologies, due to low drain-to-source breakdown voltage of these devices (one exception is low-voltage high-efficiency operation, discussed earlier). At W-band frequencies and beyond, InP-based HEMT technology has the best performance of any competing transistor technology (Gunn diodes and IMPATT diodes can also generate high powers at W-band frequencies but typically with low efficiencies), and InP-based HEMT technology has the potential of operation into the submillimeter-wave frequency range.

In comparing InP HEMTs and HBTs for system applications, InP-based HEMTs offer the best performance for low-noise applications, while InP HBTs are more commonly used for low-phase-noise oscillators, due to their lower $1/f$ noise. HBTs

60 APPLICATIONS OF InP-BASED TRANSISTORS

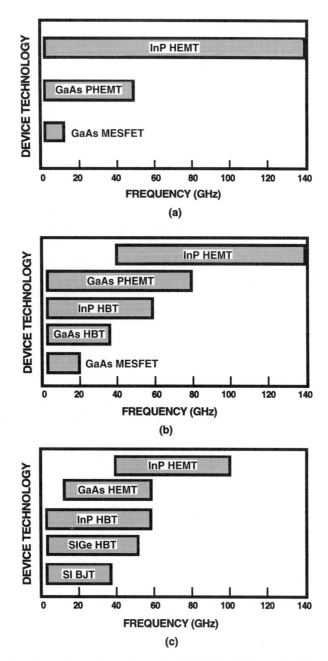

FIGURE 3.17 Comparison of transistor technologies for (*a*) low-noise, (*b*) power, and (*c*) digital circuit applications.

require more fabrication steps than HEMTs, but they do not require electron beam lithography to define sub–0.2 μm gates, which is needed for millimeter-wave HEMTs. HBTs can be fabricated with much higher uniformity and yield than HEMTs and have been the primary device of choice used for high-performance digital circuits. However, the rapid development of SiGe ICs, with their potential for lower-cost and higher-complexity circuits, can become competitive in performance with InP-based HBT ICs (Fig. 3.17c).

Even though InP-based ICs are higher in cost than GaAs- or Si-based ICs, in some cases InP-based technology can have an overall cost saving in a system. The higher gain of InP-based HEMTs compared to GaAs-based pHEMTs can reduce the number of amplifier stages necessary to achieve a particular gain. This will reduce the size of MMIC and increase MMIC yield (by reducing the number of devices) and therefore reduce the cost of the chip. Also, higher-performance InP HBT ICs and use of DDS can reduce the overall number of components in a system and effectively reduce the system cost even though the InP HBT IC will cost more than Si ICs. Another example is the use of fundamental InP HEMT or HBT oscillators at millimeter-wave frequencies rather than using low-frequency oscillators, which then requires power amplifiers and frequency doublers to generate power at millimeter-wave frequencies.

3.7 SUMMARY AND FUTURE TRENDS

Over the last few years, InP-based HEMT and HBT circuits have demonstrated state-of-the-art performance. They have been inserted in a number of military systems as well as used in some niche applications (such as low-noise InP HEMTs used for radio astronomy [29]). As mentioned earlier, in some applications the use of InP technology can lower the overall system cost; however, for most applications the cost of InP-based MMICs is too high for insertion into commercial systems. The cost of InP-based MMICs has to be reduced by an order of magnitude to allow insertion into commercial systems.

The single most important factor to reduce the cost of InP-based ICs is the availability of larger-diameter, low-defect-density, low-cost substrates. Presently, semi-insulating InP substrates are available with a maximum diameter of 3 in. and cost approximately $750 compared to 4 in. GaAs wafers (6 in. GaAs is also presently available) that cost less than $250. Clearly, for InP-based MMICs to be cost competitive, the cost per area of the substrate has to be reduced since the fabrication cost of GaAs- and InP-based devices is comparable. One approach being pursued is to grow InP-based type of devices on GaAs substrates [104]. This will allow development of larger-diameter, lower-cost starting material for fabrication. In addition, using growth techniques other than MBE (such as MOCVD, Chapter 6) can significantly lower the cost of the InP starting material.

Reducing the size of the circuits could also reduce cost significantly. The objective is to minimize the wasted real estate of the expensive InP substrate. New electromagnetic simulation tools that are now commercially available can be used to look at interaction between different circuit elements, which is critical in allowing the designer to make the circuits more compact without sacrificing performance. Use of CPW design rather than microstrip design for development of circuits is another factor that can reduce cost dramatically [105]. The back-side processing, wafer thinning, and etching of vias necessary for development of microstrip circuits account for approximately 25% of the fabrication cost of MMIC wafer. In addition, back-side processing is one of the major yield-limiting factors in MMIC fabrication. Use of CPW designs can eliminate the back-side process and reduce cost. The primary reason for use of microstrip by circuit designers over CPW has been the lack of availability of circuit models for CPW elements. Once again, the development of new electromagnetic simulation tools and development of in-house model libraries for CPW elements has allowed circuit designers to move more toward CPW designs in recent years.

For fabrication of HEMTs with sub–0.15-μm T-gates, electron beam lithography is one of the major cost factors. Electron beam lithography requires the use of very expensive machines that can write one wafer at a time and can take several hours to write each wafer. Each wafer is processed individually with no ability to do batch processing, and it requires careful SEM inspection during developing of electron beam resist to achieve the desired gate length. Use of I-line steppers and phase-shift lithography is a new trend for fabrication of short-gate-length devices that can significantly reduce the cost of MMICs [106].

Use of etch-stop layers in the design of HEMT and HBT layer structure as well as using selective etch processing to simplify batch processing of wafers will also reduce fabrication cost. Processing of high-performance InP-based transistors requires careful monitoring of currents and etch depths during the gate recess etch for HEMTs and mesa etching for HBTs. Use of etch-stop layers allows batch processing for simplified and more uniform processing, which should result in higher circuit yield and lower cost.

As system need is moved to even higher frequencies (76, 94, and 140 GHz), InP-based device technologies will be needed for insertion into systems. Potentially it will take 5 to 10 years before InP-based ICs find their way into commercial systems. One alternative approach for low-cost system insertion is to use discrete InP-based devices or very compact simple ICs (less than 1 mm^2 area) and flip-chip mounting them onto a low-cost substrate (Duroid, alumina) that contains all the passive components [107]. This approach will allow low-cost system development and should allow technology insertion into commercial applications in the near future.

REFERENCES

1. N. P. Morenc, "MMICs for automotive radar applications," *IEEE MTT-S Dig.*, 1996, pp. 39–41.
2. L. Raffaelli, "Millimeter-wave automotive radars and related technology," *IEEE MTT-S Dig.*, 1996, pp. 35–38.

3. R. L. Van Tuyl, "Unlicensed millimeter-wave communications, a new opportunity for MMIC technology at 60 GHz," *Tech. Dig. GaAs IC Symp.*, 1996, pp. 3–5.
4. D. C. W. Lo, G. S. Dow, B. R. Allen, L. Yujiri, M. Mussetto, T. W. Huang, H. Wang, and M. Biedenbender, "A monolithic W-band high-gain LNA/detector for millimeter-wave radiometric imaging applications," *IEEE MTT-S Dig.*, 1995, pp. 1117–1120.
5. N. J. Kolias, M. J. Vaughan, and R. C. Compton, "High power quasi-optical sources for millimeter-wave base-station applications," *Tech. Dig. GaAs IC Symp.*, 1996, pp. 221–224.
6. D. M. Akos, and J. B. Y. Tsui, "Design and implementation of a direct digitization GPS receiver front end," *IEEE Trans. Microwave Theory Tech.*, **44**(12), 2334–2339 (1996).
7. G. Van Andrews, J. B. Delaney, M. A. Vernon, M. P. Harris, C. T. M. Chang, T. C. Eiland, C. E. Hastings, V. L. DiPerna, M. C. Brown, and W. A. White, "Recent progress in wideband monolithic direct digital synthesizers," *IEEE MTT-S Dig.*, 1996, pp. 1347–1350.
8. J. V. DiLorenzo, and D. D. Khandelwal, *GaAs FET Principles and Technology*, Artech House, Norwood, Mass., 1982.
9. P. M. Smith, S.-M. J. Liu, M.-Y. Kao, P. Ho, S. C. Wang, K. H. G. Duh, S. T. Fu, and P. C. Chao, "W-band high efficiency InP-based power HEMT with 600 GHz f_{max}," *IEEE Microwave Guided Wave Lett.*, **5**(7), 230–232 (1995).
10. J. V. DiLorenzo and W. R. Wisseman, "GaAs power MESFET: design, fabrication, and performance," *IEEE Trans. Microwave Theory Tech.*, **27**(5), 367–378 (1979).
11. J. M. Golio, "Ultimate scaling limits for high-frequency GaAs MESFET's," *IEEE Trans. Electron Devices*, **35**(7), 839–848 (1988).
12. M. B. Das, "Millimeter-wave performance of ultrasubmicrometer-gate field-effect transistors: a comparison of MODFET, MESFET, and PBT structures," *IEEE Trans. Electron Devices*, **34**(7), 1429–1440 (1987).
13. L. D. Nguyen, L. E. Larson, and U. K. Mishra, "Ultrahigh-speed modulation-doped field-effect transistors: a tutorial review," *Proc. IEEE*, **80**(4), 494–518 (1992).
14. K. L. Tan, R. M. Dia, D. C. Streit, L. K. Shaw, A. C. Han, M. D. Sholley, P. H. Liu, T. Q. Trinh, T. Lin, and H. C. Yen, "60-GHz pseudomorphic $Al_{0.25}Ga_{0.75}As/In_{0.28}Ga_{0.72}As$ low-noise HEMT's" *IEEE Electron Device Lett.*, **12**(1), 23–25 (1991).
15. T. Katoh, N. Yoshida, H. Minami, T. Kashiwa, and S. Orisaka, "A 60 GHz-band ultra low noise planar-doped HEMT," *IEEE MTT-S Dig.*, 1993, pp. 337–340.
16. K. L. Tan, R. M. Dia, D. C. Streit, T. Lin, T. Q. Trinh, A. C. Han, P. H. Liu, P. D. Chow, and H. C. Yen, "94-GHz 0. 1 μm T-gate low-noise pseudomorphic InGaAs HEMT's," *IEEE Electron Device Lett.*, **11**(12), 585–587 (1990).
17. S. Bouthillette, A. Platzker, and L. Aucoin, "High-efficiency 40 watt PsHEMT S-band MIC power amplifiers," *IEEE MTT-S Dig.*, 1994, pp. 667–670.
18. B. Kraemer, R. Basset, P. Chye, D. Day, and J. Wei, "Power pHEMT module delivers 12 watts, 40% PAE over the 8.5 to 10.5 GHz band," *IEEE MTT-S Dig.*, 1994, pp. 683–686.
19. P. M. Smith, C. T. Creamer, W. F. Kopp, D. W. Ferguson, P. Ho, and J. R. Willhite, "A high-power Q-band pHEMT for communication terminal applications," *IEEE MTT-S Dig.*, 1994, pp. 809–812.
20. R. Lai, M. Biedenbender, J. Lee, K. Tan, D. Streit, P. H. Liu, M. Hoppe, and B. Allen, "0. 15 μm InGaAs/AlGaAs/GaAs HEMT production process for high performance and high yield V-band power MMICs," *Tech. Dig. GaAs IC Symp.*, 1995, pp. 105–108.

21. H. Wang, Y. Hwang, T. H. Chen, M. Biedenbender, D. C. Streit, D. C. W. Lo, G. S. Dow, and B. R. Allen, "A W-band monolithic 175-mW power amplifier," *IEEE MTT-S Dig.*, 1995, pp. 419–422.

22. L. F. Eastman, "Use of molecular beam epitaxy in research and development of selected high speed compound semiconductor devices" *J. Vac. Sci. Technol B*, **1,** 131–134 (1983).

23. U. K. Mishra, A. S. Brown, M. J. Delaney, P. T. Greiling, and C. F. Krumm, "The AlInAs–GaInAs HEMT for microwave and millimeter-wave applications," *IEEE Trans. Microwave Theory Tech.*, **37**(9), 1279–1285 (1989).

24. L. D. Nguyen, A. S. Brown, M. A. Thompson, and L. M. Jelloian, "50-nm self-aligned-gate pseudomorphic AlInAs/GaInAs high electron mobility transistors," *IEEE Trans. Electron Devices*, **39**(9), 2007–2014 (1992).

25. K. H. G. Duh, P. C. Chao, P. Ho, M. Y. Kao, P. M. Smith, J. M. Ballingall, and A. A. Jabra, "High performance InP-based HEMT millimeter-wave low-noise amplifiers," *IEEE MTT-S Int. Microwave Symp. Dig.*, 1989, pp. 805–808.

26. K. L. Tan, D. C. Streit, P. D. Chow, R. M. Dia, A. C. Han, P. H. Liu, D. Garske, and R. Lai, "140 GHz 0.1 μm gate length pseudomorphic $In_{0.52}Al_{0.48}As/In_{0.60}Ga_{0.40}As/InP$ HEMT," *Tech. Dig. IEDM*, 1991, pp. 239–242.

27. R. Isobe, C. Wong, A. Potter, L. Tran, M. Delaney, R. Rhodes, D. Jang, L. Nguyen, and M. Le, "Q- and V-band MMIC chip set using 0. 1 μm millimeter-wave low noise InP HEMTs," *IEEE MTT-S Dig.*, 1995, pp. 1133–1136.

28. S. E. Rosenbaum, L. M. Jelloian, A. S. Brown, M. A. Thompson, M. Matloubian, L. E. Larson, R. F. Lohr, B. K. Kormanyos, G. M. Reibez, and L. P. B. Katehi, "A 213 GHz AlInAs/GaInAs/InP HEMT MMIC oscillator," *Tech. Dig. IEDM*, 1993, pp. 924–926.

29. M. W. Pospieszalski, V. J. Lakatosh, L. D. Nguyen, M. Lui, T. Liu, M. Le, M. A. Thompson, and M. J. Delaney, "Q- and E-band cryogenically-coolable amplifiers using AlInAs/GaInAs/InP HEMTs," *IEEE MTT-S Dig.*, 1995, pp. 1121–1124.

30. P. M. Smith, "Status of InP HEMT technology for microwave receiver applications," *IEEE MTT-S Dig.*, 1996, pp. 5–8.

31. S. E. Rosenbaum, L. M. Jelloian, L. E. Larson, U. K. Mishra, D. A. Pierson, M. A. Thompson, T. Liu, and A. S. Brown, "A 2 GHz three-stage AlInAs/GaInAs/InP HEMT MMIC low-noise amplifier," *IEEE Microwave and Guided Wave Lett.*, **3**(8), 265–267 (1993).

32. S. E. Rosenbaum, C. S. Chou, C. M. Ngo, L. E. Larson, T. Liu, and M. A. Thompson, "A 7 to 11 GHz AlInAs/GaInAs/InP MMIC low noise amplifier," *IEEE MTT-S Dig.*, 1993, pp. 1103–1104.

33. L. Tran, R. Isobe, M. Delaney, R. Rhodes, D. Jang, J. Brown, L. Nguyen, M. Le, M. Thompson, and T. Liu, "High performance, high yield, millimeter-wave MMIC LNAs using InP HEMTs," *IEEE MTT-S Dig.*, 1996, pp. 9–12.

34. R. Lai, H. Wang, Y. C. Chen, T. Block, P. H. Liu, D. C Streit, D. Tran, P. Siegel, M. Barsky, W. Jone, and T. Gaier, "D-band MMIC LNAs with 12 dB gain at 155 GHz fabricated on a high yield InP HEMT MMIC production process," *Proc. 9th Int. Conf. Indium Phosphide and Related Materials*, Hyannis, Mass., 1997, pp. 241–244.

35. K. Y. Hur, K. T. Hetzler, R. A. McTaggart, D. W. Vye, P. J. Lemonias, and W. E. Hoke, "Ultralinear double pulse doped AlInAs/GaInAs/InP HEMTs," *Electron. Lett.*, **32**(16), 1516–1518 (1996).

36. C. J. Madden, R. L. Van Tuyl, M. V. Le, and L. D. Nguyen, "A 17 dB gain, 0.1–70 GHz InP HEMT amplifier," *Tech. Dig. ISSCC*, 1994, pp. 178–179.
37. J. Pusl, B. Agarwal, R. Pullela, L. D. Nguyen, M. V. Le, M. J. W. Rodwell, L. Larson, J. F. Jensen, R. Y. Yu, and M. G. Case, "Capacitive-division traveling-wave amplifier with 340 GHz gain-bandwidth product," *IEEE MTT-S Dig.*, 1995, pp. 1661–1664.
38. M. Hafizi, and M. J. Delaney, "Reliability of InP-based HBT's and HEMT's: experiments, failure mechanisms, and statistics," *Proc. 6th Int. Conf. Indium Phosphide and Related Materials*, Santa Barbara, Calif., 1994, pp. 299–302.
39. M. Matloubian, A. S. Brown, L. D. Nguyen, M. A. Melendes, L. E. Larson, M. J. Delaney, M. A. Thompson, R. A. Rhodes, and J. E. Pence, "20-GHz high-efficiency AlInAs–GaInAs on InP power HEMT," *IEEE Microwave and Guided Wave Lett.*, 3(5), 142–144 (1993).
40. M. Matloubian, L. M. Jelloian, A. S. Brown, L. D. Nguyen, L. E. Larson, M. J. Delaney, M. A. Thompson, R. A. Rhodes, and J. E. Pence, "V-band high-efficiency high-power AlInAs/GaInAs/InP HEMTs," *IEEE Trans. Microwave Theory Tech.*, 41(12), 2206–2210 (1993).
41. K. C. Hwang, P. Ho, M. Y. Kao, S. T. Fu, J. Liu, P. C. Chao, P. M. Smith, and A. W. Swanson, "W-band high power passivated 0.15 μm InAlAs/InGaAs HEMT device," *Proc. 6th Int. Conf. Indium Phosphide and Related Materials*, Santa Barbara, Calif., 1994, pp. 18–20.
42. K. Y. Hur, R. A. McTaggart, B. W. LeBlanc, W. E. Hoke, P. J. Lemonias, A. B. Miller, T. E. Kazior, and L. M. Aucoin, "Double recessed AlInAs/GaInAs/InP HEMTs with high breakdown voltages," *Tech. Dig. IEEE GaAs IC Symp.*, 1995, pp. 101–104.
43. W. Lam, M. Matloubian, A. Igawa, C. Chou, A. Kurdoghlian, C. Ngo, L. Jelloian, A. Brown, M. Thompson, and L. Larson, "44-GHz high-efficiency InP-HEMT MMIC power amplifier," *IEEE Microwave Guided Wave Lett.*, 4(8), 2777–2781 (1994).
44. M. Matloubian, L. M. Jelloian, M. Lui, T. Liu, and M. Thompson, "Ultra-high breakdown high-performance AlInAs/GaInAs/InP power HEMTs," *Tech. Dig. IEDM*, 1993, pp. 915–917.
45. M. Matloubian, L. D. Nguyen, A. S. Brown, L. E. Larson, M. A. Melendes, and M. A. Thompson, "High power and high Efficiency AlInAs/GaInAs on InP HEMTs," *IEEE MTT-S Dig.*, 1991, pp. 721–724.
46. K. Y. Hur, R. A. McTaggart, B. W. LeBlanc, W. E. Hoke, P. J. Lemonias, A. B. Miller, T. E. Kazior, and L. M. Aucoin, "Double recessed AlInAs/GaInAs/InP HEMTs with high breakdown voltages," *Tech. Dig. IEEE GaAs IC Symp.*, 1995, pp. 101–104.
47. P. Ho, P. M. Smith, K. C. Hwang, S. C. Wang, M. Y. Kao, P. C. Chao, and S. M. J. Liu, "60 GHz Power Performance of 0.1 μm Gate-Length InAlAs/InGaAs HEMTs," *Proc. 6th Int. Conf. Indium Phosphide and Related Materials*, Santa Barbara, Calif., 1994, pp. 411–414.
48. M. Matloubian, "InP-based HEMTs for millimeter-wave and submillimeter-wave power applications," *Int. Conf. Millimeter and Submillimeter Waves and Applications III*, Denver, Colo., August 1996, pp. 22–31.
49. L. E. Larson, M. Matloubian, J. J. Brown, A. S. Brown, R. Rhodes, D. Crampton, and M. Thompson, "AlInAs/GaInAs on InP HEMTs for low power supply voltage operation of high power-added efficiency microwave amplifiers," *Electron. Lett.*, 29(15), 1324–1326 (1993).

50. A. Kurdoghlian, W. Lam, C. Chou, L. Jelloian, A. Igawa, M. Matloubian, L. Larson, A. Brown, M. Thompson, and C. Ngo, "High-efficiency InP-based HEMT MMIC power amplifier," *Tech. Dig. IEEE GaAs IC Symp.*, 1993, pp. 375–377.

51. W. Lam, M. Matloubian, A. Igawa, C. Chou, A. Kurdoghlian, C. Ngo, L. Jelloian, A. Brown, M. Thompson, and L. Larson, "44-GHz high-efficiency InP-HEMT MMIC power amplifier," *IEEE Microwave Guided Wave Lett.*, **4**(8), 277–278 (1994).

52. D. C. Streit, K. L. Tan, R. M. Dia, J. K. Liu, A. C. Han, J. R. Velebir, S. K. Wang, T. Q. Trinh, P. D. Chow, P. H. Liu, and H. C. Yen, "High-gain W-band pseudomorphic InGaAs power HEMT's," *IEEE Electron Dev. Lett.*, **12**(4), 149–150 (1991).

53. L. Tran, M. Delaney, R. Isobe, D. Jang, and J. Brown, "Frequency translation MMICs using InP HEMT technology," *IEEE MTT-S Dig.*, 1996, pp. 261–264.

54. R. Virk, L. Tran, M. Matloubian, M. Le, M. Case, and C. Ngo, "Comparison of W-band MMIC mixers using InP HEMT technology," *IEEE MTT-S Dig.*, 1997, pp. 435–438.

55. J. F. Jensen, U. K. Mishra, A. S. Brown, R. S. Beaubian, M. A. Thompson, and L. M. Jelloian, "25 GHz static frequency dividers in AlInAs/GaInAs HEMT technology," *Tech. Dig. IEEE Int. Solid-State Circuits Conf.*, 1988, pp. 268–269.

56. C. Madden, D. R. Snook, R. L. Van Tuyl, M. V. Le, and L. D. Nguyen, "A novel 75 GHz InP HEMT dynamic divider," *Tech. Dig. GaAs IC Symp.*, 1996, pp. 137–140.

57. Y. Umeda, K. Osafune, T. Enoki, H. Ito, and Y. Ishii, "SCFL static frequency divider using InAlAs/InGaAs/InP HEMTs," *Proc. 25th European Microwave Conf.*, Bologna, Italy, 1995, pp. 222–228.

58. Y. Umeda, T. Enoki, K. Osafune, H. Ito, and Y. Ishii, "InAlAs/InGaAs/InP-HEMT technologies for high-yield analog/digital ICs," *Tech. Dig. IEEE Microwave and Monolithic Circuits Symp.*, 1996, pp. 115–118.

59. W. Shockley, U.S. patent 2,569,347 (September 25, 1951).

60. H. Kroemer, "Theory of a wide-gap emitter for transistors," *Proc. IRE*, **45**, 1535–1537 (1957).

61. H. Kroemer, "Heterostructure bipolar transistors and integrated circuits," *Proc. IEEE*, **70**(1), 13–25 (1982).

62. M. J. Mondry, and H. Kroemer, "Heterojunction bipolar transistor using a (Ga,In)P emitter on a GaAs base, grown by molecular beam epitaxy," *IEEE Electron Device Lett.*, **6**(4), 175–177 (1985).

63. F. Ren, C. R. Abernathy, S. J. Pearton, J. R. Lothian, P. W. Wisk, T. R. Fullowan, Y.-K. Chen, L. W. Yang, S. T. Fu, R. S. Brozovich, and H. H. Lin, "Self-aligned InGaP/GaAs heterojunction bipolar transistors for microwave power application," *IEEE Electron Device Lett.*, **14**(7), 332–334 (1993).

64. W. Liu, A. Khatibzadeh, T. Henderson, S.-K. Fan, and D. Davito, "X-band GaInP/GaAs power heterojunction bipolar transistor," *IEEE MTT-S Dig.*, 1993, pp. 1477–1480.

65. K. Riepe, H. Leier, U. Seiler, A. Marten, and H. Sledzik, "High-efficiency X-band GaInP/GaAs HBT MMIC power amplifier for stable long pulse and CW operation," *Tech. Dig. IEDM*, 1995, pp. 795–798.

66. R. J. Malik, J. R. Hayes, F. C. Capasso, K. Alavi, and A. Y. Cho, "High-gain $Al_{0.48}In_{0.52}As/Ga_{0.47}In_{0.53}As$ vertical n-p-n heterojunction bipolar transistors grown by molecular-beam epitaxy," *IEEE Electron Device Lett.*, **4**(10), 383–385 (1983).

67. J. Laskar, R. N. Nottenburg, J. A. Baquedano, A. F. J. Levi, and J. Kolodzey, "Forward

transit delay in $In_{0.53}Ga_{0.47}As$ heterojunction bipolar transistors with nonequilibrium electron transport," *IEEE Trans. Electron Devices*, **40**(11), 1942–1948 (1993).
68. M. Hafizi, P. A. Macdonald, T. Liu, and D. B. Rensch, "Microwave power performance of InP-based double heterojunction bipolar transistors for C- and X-band applications," *IEEE MTT-S Dig.*, 1994, pp. 671–674.
69. J. F. Jensen, A. E. Cosand, W. E. Stanchina, R. H. Walden, T. Liu, Y. K. Brown, M. Montes, K. Elliott, and C. G. Kirkpatrick, "Double heterostructure InP HBT technology for high resolution A/D converters," *Tech. Dig. GaAs IC Symp.*, 1994, pp. 224–227.
70. M. Hafizi, W. E. Stanchina, R. A. Metzger, J. F. Jensen, and F. Williams, "Reliability of AlInAs/GaInAs heterojunction bipolar transistors," *IEEE Trans. Electron Devices*, **40**(12), 2178–2185 (1993).
71. U. König, A. Gruhle, and A. Schüppen, "SiGe devices and circuits: where are advantages over III/V?" *Tech. Dig. IEEE GaAs IC Symp.*, 1995, pp. 14–17.
72. D. L. Harame, J. H. Comfort, J. D. Cressler, E. F. Crabbé, J. Y.-C. Sun, B. S. Meyerson, and T. Tice, "Si/SiGe Epitaxial-Base Transistors: I. Materials, physics, and circuits," *IEEE Trans. Electron Devices*, **42**(3), 455–468 (1995).
73. J.-F. Luy, K. M. Strohm, and E. Sasse, "Si/SiGe MMIC technology," *IEEE MTT-S Dig.*, 1994, pp. 1755–1757.
74. M. Case, P. Macdonald, M. Matloubian, M. Chen, L. Larson, and D. Rensch, "High-performance microwave elements for SiGe MMICs," *Proc. IEEE/Cornell Conf. Advanced Concepts in High Speed Semiconductor Devices and Circuits*, 1995, pp. 85–92.
75. U. Erben, M. Wahl, A. Schüppen, and H. Schumacher, "Class-A SiGe HBT power amplifiers at C-band frequencies," *IEEE Microwave Guided Wave Lett.*, **5**(12), 435–436 (1995).
76. L. Larson, M. Case, S. Rosenbaum, D. Rensch, M. Chen, P. Macdonald, M. Matloubian, D. Harame, J. Malinowski, B. Meyerson, M. Gilbert, and S. Maas, "Si/SiGe HBT technology for low-cost monolithic microwave integrated circuits," *Tech. Dig. IEEE Intern. Solid-State Circuits Conf.*, 1996, pp. 80–81.
77. K. D. Hobart, F. J. Kub, N. A. Papanicoloau, W. Kruppa, and P. E. Thompson, "Si/$Si_{1-x}Ge_x$ heterojunction bipolar transistors with high breakdown voltage," *IEEE Electron Device Lett.*, **16**(5), 205–207 (1995).
78. S. Yamahata, K. Kurishima, H. Ito, and Y. Matsuoka, "Over-220-GHz-f_T-and-f_{max} InP/InGaAs double-heterojunction bipolar transistors with a new hexagonal-shaped emitter," *Tech. Dig. GaAs IC Symp.*, 1995, pp. 163–166.
79. C. Nguyen, T. Liu, M. Chen, H.-C. Sun, and D. Rensch, "AlInAs/GaInAs/InP double heterojunction bipolar transistor with a novel base–collector design for power applications," *Tech. Dig. IEDM*, 1995, pp. 799–802.
80. J. F. Jensen, L. M. Burns, and W. E. Stanchina, "High-Speed InP HBT Circuits," in *InP HBTs: Growth, Processing, and Applications*, ed. B. Jalali and S. J. Pearton, Artech House, Norwood, Mass., 1995.
81. W. E. Stanchina, J. F. Jensen, R. H. Walden, M. Hafizi, H.-C. Sun, T. Liu, G. Raghavan, K. E. Elliot, M. Kardos, A. E. Schmitz, Y. K. Brown, M. E. Montes, and M. Yung, "An InP-based HBT Fab for high-speed digital, analog, mixed-signal, and optoelectronic ICs," *Tech. Dig. GaAs IC Symp.*, 1995, pp. 31–34.
82. J. F. Jensen, M. Hafizi, W. E. Stanchina, R. A. Metzger, and D. B. Rensch, "39.5 GHz

static frequency divider implemented in AlInAs/GaInAs HBT technology," *Tech. Dig. GaAs IC Symp.*, 1992, pp. 101–104.

83. C. Baringer, J. Jensen, L. Burns, and R. Walden, "3-bit, 8 GSPS flash ADC," *Proc. 8th Int. Conf. Indium Phosphide and Related Materials*, Schwäbisch-Gmünd, Germany, 1996, pp. 64–67.

84. T. A. Schaffer, H. P. Warren, M. J. Bustamante, and K. W. Kong, "A 2 GHz 12-bit digital-to-analog converter for direct digital synthesis applications," *Tech. Dig. GaAs IC Symp.*, 1996, pp. 61–64.

85. J. F. Jensen, G. Raghavan, A. E. Cosand, and R. H. Walden, "A 3.2 GHz second-order delta-sigma modulator implemented in InP HBT technology," *IEEE J. Solid State Circuits*, **30**(10), 1119–1127 (1995).

86. M. Hafizi, "Submicron, fully self-aligned HBT with an emitter geometry of 0.3 μm," *IEEE Electron Device Lett.*, **18**(7), 358–360 (1997).

87. M. Hafizi and J. Jensen, "Ultra-fast, low-power integrated circuits in a scaled submicron HBT IC technology," *IEEE RFIC Symp.*, 1997, pp. 87–90.

88. J. Cowles, L. Tran, T. Block, D. Streit, C. Grossman, G. Chao, and A. Oki, "A comparison of low frequency noise in GaAs and InP-based HBTs and VCOs," *IEEE MTT-S Dig.*, 1995, pp. 689–692.

89. L. Tran, J. Cowles, T. Block, H. Wang, J. Yonaki, D. Lo, S. Dow, B. Allen, D. Streit, A. Oki, and S. Loughran, "Monolithic VCO and mixer for Q-band transceiver using InP-based HBT process," *Tech. Dig. IEEE Microwave and Monolithic Circuits Symp.*, 1995, pp. 101–104.

90. K. W. Kobayashi, L. T. Tran, A. K. Oki, T. Block, and D. C. Streit, "A coplanar waveguide InAlAs/InGaAs HBT monolithic Ku-band VCO," *IEEE Microwave and Guided Wave Lett.*, **5**(9), 311–312 (1995).

91. H. Wang, K. W. Chang, L. Tran, J. Cowles, T. Block, D. C. W. Lo, G. S. Dow, A. Oki, D. Streit, and B. R. Allen, "Low phase noise millimeter-wave frequency sources using InP-based HBT technology," *Tech. Dig. GaAs IC Symp.*, 1995, pp. 263–266.

92. J. Cowles, L. Tran, H. Wang, E. Lin, T. Block, D. Streit, and A. Oki, "InP-based HBT technology for millimeter-wave MMIC VCOs," *Tech. Dig. IEDM*, 1996, pp. 199–202.

93. L. M. Burns, J. F. Jensen, W. E. Stanchina, R. A. Metzger, and Y. K. Allen, "DC-to-Ku band MMIC InP HBT double-balanced active mixer," *Tech. Dig. Int. Solid-State Circuits Conf.* 1991, pp. 124–125.

94. E. W. Lin, H. Wang, K. W. Chang, L. Tran, J. Cowles, T. Block, D. C. W. Lo, G. S. Dow, A. Oki, D. Streit, B. R. Allen, "Monolithic millimeter-wave Schottky-diode-based frequency converters with low drive requirements using an InP HBT-compatible process," *Tech. Dig. GaAs IC Symp.*, 1995, pp. 218–221.

95. M. Rodwell, J. F. Jensen, W. E. Stanchina, R. A. Metzger, D. B. Rensch, M. W. Pierce, T. V. Kargodorian, and Y. K. Allen, "33-GHz monolithic cascode AlInAs/GaInAs heterojunction bipolar transistor feedback amplifier," *IEEE J. Solid State Circuits*, **26**(10), 1378–1382 (1991).

96. K. W. Kobayashi, J. Cowles, L. T. Tran, T. R. Block, A. K. Oki, and D. C. Streit, "A 2-50 GHz InAlAs/InGaAs-InP HBT Distributed Amplifier," *Tech. Dig. GaAs IC Symp.*, 1996, pp. 207–210.

97. K. W. Kobayashi, L. T. Tran, A. K. Oki, and D. C. Streit, "Noise optimization of a GaAs HBT direct-coupled low noise amplifier," *IEEE MTT-S Dig.*, 1996, pp. 815–818.

98. A. P. Freundorfer, Y. Jamani, and C. Falt, "A Ka-band GaInP/GaAs HBT four-stage LNA," *IEEE MTT-S Dig.*, 1996, pp. 17–20.
99. Y. K. Chen, D. A. Humphrey, L. Fan, J. Lin, R. A. Hamm, D. Sivco, A. Y. Cho, and A. Tate, "Noise characteristics of InP-based HBTs," *Proc. 7th Int. Conf. Indium Phosphide and Related Materials*, Sapporo, Japan, 1995, pp. 851–856.
100. M. Chen, C. Nguyen, T. Liu, and D. Rensch, "High-performance AlInAs/GaInAs/InP DHBT X-band power cell with InP emitter ballast resistor," *Proc. IEEE/Cornell Conf. Advanced Concepts in High Speed Semiconductor Devices and Circuits,* 1995, pp. 573–582.
101. W. Liu, and A. Khatibzadeh, "The collapse of current gain in multi-finger heterojunction bipolar transistors: its substrate temperature dependence, instability criteria, and modeling," *IEEE Trans. Electron Devices*, **41**(10), 1698–1707 (1994).
102. C. Nguyen, T. Liu, M. Chen, H.-C. Sun, and D. Rensch, "AlInAs/GaInAs/InP double heterojunction bipolar transistor with a novel base–collector design for power applications," *IEEE Electron Devices Lett.*, **17**(3), 133–135 (1996).
103. T. Liu, M. Chen, C. Nguyen, and R. Virk, "InP-based DHBT with 90% power-added efficiency and 1 W output power at 2 GHz," presented at Topical Workshop for Heterostructure Microelectronics, Japan, August 1996.
104. A. Cappy, "Metamorphic InGaAs/AlInAs heterostructure field effect transistors: layer growth, device processing and performance," *Proc. 8th Int. Conf. Indium Phosphide and Related Materials*, Schwäbisch-Gmünd, Germany, 1996, pp. 3–6.
105. M. Berg, T. Hackbarth, B. E. Maile, and J. Dickmann, "W-band MMIC amplifiers based on quarter micron gate-length InP HEMTs and coplanar waveguides," *Proc. 8th Int. Conf. Indium Phosphide and Related Materials*, Schwäbisch-Gmünd, Germany, 1996, pp. 72–75.
106. J.-E. Müller, A. Bangert, T. Grave, M. Kärner, H. Riechert, A. Schäfer, H. Siweris, L. Schleicher, H. Tischer, L. Verweyen, W. Kellner, and T. Meier, "A GaAs HEMT MMIC chip set for automotive radar systems fabricated by optical stepper lithography," *Tech. Dig. GaAs IC Symp.*, 1996, pp. 189–192.
107. M. Matloubian, "Flip-chip components for realization of low-cost millimeter wave systems," *Proc. Asia Pacific Microwave Conf.*, New Delhi, 1996, pp. 1619–1622.

CHAPTER FOUR

Material Physics of InP-Based Compound Semiconductors

YOSHIKAZU TAKEDA
Nagoya University

4.1	Introduction	72
4.2	III-V Compound Semiconductors	72
	4.2.1 Crystal Structure and Bond	72
	4.2.2 {001} and {110} Surfaces	74
	4.2.3 Valence-Force-Field Model	76
	4.2.4 Energy Band Structure	79
	4.2.5 Material Parameters	81
4.3	InP	81
	4.3.1 InP and Alloys	81
	4.3.2 Velocity-Field Characteristics and Surface Properties	85
	4.3.3 Bulk and Epitaxial Growth	87
4.4	GaInAs and GaInAsP	88
	4.4.1 Energy Band Structure Versus Lattice Constant	88
	4.4.2 Atom Configuration in Alloys	90
	4.4.3 Dielectric Constants	90
	4.4.4 $Ga_{0.47}In_{0.53}As$	91
	4.4.5 Effects of Strain	93
	4.4.6 Epitaxial Growth	93
4.5	AlInAs and AlGaInAs	93
4.6	GaAsSb	96
4.7	Heteroepitaxy	96

InP-Based Materials and Devices: Physics and Technology, Edited by Osamu Wada and Hideki Hasegawa.
ISBN 0-471-18191-9 © 1999 John Wiley & Sons, Inc.

    4.7.1 InP/GaInAsP   98
    4.7.2 AlInAs/GaInAs   99
    4.7.3 AlInAs/InP   100
4.8 Optical Properties   100
    4.8.1 GaInAsP   100
    4.8.2 InP/GaInAs/InP Quantum Wells   102
4.9 Summary   103
References   104

4.1 INTRODUCTION

In this chapter, InP itself and those III-V alloy semiconductors that lattice match to InP are taken as the InP-based semiconductors, and their material physics, such as thermodynamic properties, crystal structures, atom configurations, band structures, effective masses, thermal properties as basic physics, and electrical and optical properties of bulk materials, are described. Materials covered in this chapter are InP, GaInAs, GaInAsP, AlInAs, AlGaInAs, and GaAsSb. Due to limited pages, some of the properties are listed as parameters in tables without description. Physics and properties of the heterostructures are described briefly. Details of growth and applications of a variety of heterostructures are presented in appropriate chapters in this book. A list of general reference books for further study is provided at the end of the chapter.

4.2 III-V COMPOUND SEMICONDUCTORS

In this section, general properties of III-V compound semiconductors with the zincblende structure are described since InP and related III-V binary compounds such as AlP, AlAs, GaP, GaAs, GaSb, and InAs have their common properties. Specific properties of each compound, alloy, and heterostructure between them are described separately.

4.2.1 Crystal Structure and Bond

Most III-V compounds have the zincblende structure, as shown in Figure 4.1. Coordination of the nearest-neighbor atoms around a central atom can be represented as shown in Figure 4.2, where the thin solid lines form a cube. At the center of the cube there is one atom, and four nearest-neighbor atoms are located at four corners out of eight. The atom at the left-back corner in Figure 4.1 (O atom) is taken as the central atom in Figure 4.2 (O atom) at the origin to have the common coordinates. The angle θ between each pair of bonds from the central atom is all equal and calculated from the equation $\theta = 2\tan^{-1}\sqrt{2}$, which is $109.47°$. This bond is known as the sp^3-hybridized bond. The s-symmetry electrons and p-symmetry electrons of the outer orbits; for example, in GaAs, Ga: $(4s)^2(4p)^1$ and As: $(4s)^2(4p)^3$ are hybridized to

4.2 III-IV COMPOUND SEMICONDUCTORS

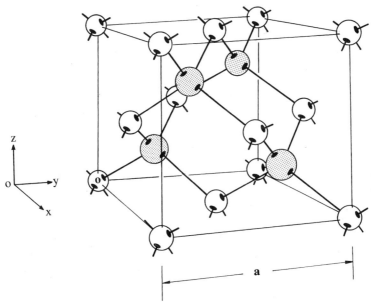

FIGURE 4.1 Zincblende structure. This structure contains four group III atoms and four group V atoms in the cube shown by thin solid lines; a is the lattice constant. The white atom (O) at the left back corner is taken as the origin of the Cartesian coordinate system.

form four bonds, whose directions are [111], [1$\bar{1}\bar{1}$], [$\bar{1}$1$\bar{1}$], and [$\bar{1}\bar{1}$1], as shown in Figure 4.2. Using the wavefunctions $|s\rangle$, $|p_x\rangle$, $|p_y\rangle$, and $|p_z\rangle$), the four sp^3-hybridized bonds can be expressed as

$$\psi(1,1,1) = \tfrac{1}{2}(|s\rangle + |p_x\rangle + |p_y\rangle + |p_z\rangle)$$

$$\psi(1,-1,-1) = \tfrac{1}{2}(|s\rangle + |p_x\rangle - |p_y\rangle - |p_z\rangle) \quad (1)$$

$$\psi(-1,1,-1) = \tfrac{1}{2}(|s\rangle - |p_x\rangle + |p_y\rangle - |p_z\rangle)$$

$$\psi(-1,-1,1) = \tfrac{1}{2}(|s\rangle - |p_x\rangle - |p_y\rangle + |p_z\rangle).$$

In the zincblende structure, each atom tends to keep its bond radius, and the bond radii attributed to various atoms are listed in Table 4.1. Binary compounds with the zincblende structure have the bond lengths calculated from each set of the two radii in Table 4.1. For example, InP is calculated to have a bond length d of 1.405 Å (In) + 1.128 Å (P) = 2.533 Å, which gives a lattice constant a of 5.850 Å from the equation $a = d \cdot 4/\sqrt{3}$. 5.850 Å is comparable to the real value of 5.8694 Å, as listed in Table 4.3. Nitrides such as GaN and InN usually have the hexagonal structure [2].

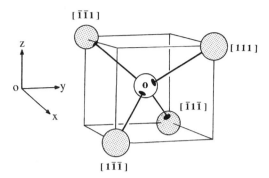

FIGURE 4.2 One central atom and four nearest-neighbor atoms of an sp^3-hybridized bond. The thin lines form a cube. The origin of the coordinates is set at the central atom (O), which is the same as the white atom (O) in Fig. 4.1. Then the other four atoms are in {111} directions, as shown. The top and bottom surfaces of the cube are parallel to the (001) surface.

4.2.2 {001} and {110} Surfaces

Most III-V compounds and alloys are grown on the (001) surface because cleavage surfaces of {110} are perpendicular to {001} surfaces and also to {110} surfaces themselves. A schematic drawing of the (001) surface with (110) and ($1\bar{1}0$) cleavage surfaces as illustrated in Figure 4.3 may be useful for understanding anisotropy in the growth habits and etching habits on the (001) surface. InP is taken as an example. As shown in Figure 4.3, the cleavage surfaces {110} maintain charge neutrality by each pair of nearest-neighbor atoms. The surfaces are drawn as cut. Some of the dangling bonds are bonded with each other and form surface-reconstructed structures [3].

Distribution of valence electron density on the ($1\bar{1}0$) plane in Ge, GaAs, and

TABLE 4.1 Rationalized Tetrahedral Radii (Å)

Period	Group						
	I	II	III	IV	V	VI	VII
2		Be	B	C	N	O	F
		0.975	0.853	0.774	0.719	0.678	0.672
3		Mg	Al	Si	P	S	Cl
		1.301	1.230	1.173	1.128	1.127	1.127
4	Cu	Zn	Ga	Ge	As	Se	Br
	1.225	1.225	1.225	1.225	1.225	1.225	1.225
5	Ag	Cd	In	Sn	Sb	Te	I
	1.405	1.405	1.405	1.405	1.405	1.405	1.405

Source: After Ref. 1.

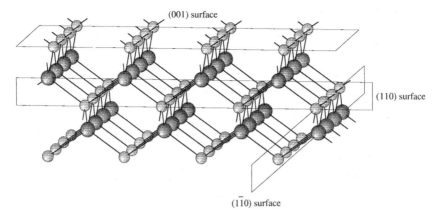

FIGURE 4.3 Schematic drawing in atomic scale of a (001) surface with {110} cleavage surfaces of a zincblende structure.

ZnSe is shown in Figure 4.4. The three semiconductors are composed of elements in the fourth row of the periodic table. Ge forms perfect covalent bonds where the electron density is highest at the center of the bond. The deviation of the electron distribution increases and hence the ionicity increases in going from III-V to II-VI compounds. The degree of bond ionicity increases the energy gap, easiness of cleavage, and chemical activity of the bond.

Figure 4.5a is a cross section of Fig. 4.3 along the ($1\bar{1}0$) surface with the [110] monolayer step on the (001) surface, and Fig. 4.5b is that of the (110) surface with the [$\bar{1}10$] monolayer step. The surfaces are assumed to be covered by P atoms, as in the case of normal growth from a vapor-phase or molecular beam process. The growth is usually known to start at the step edge. In Fig. 4.5a, the In atom at the edge is bonded to three bonds from P atoms, and in Fig. 4.5b the In atom at the edge is bonded to two bonds from P atoms. Therefore, it is natural to assume that there should be a difference in the probability of In atom incorporation at the [110] and [$\bar{1}10$] steps. In the organometallic vapor-phase epitaxy (OMVPE) it was shown that the growth rate toward [110] is higher than that toward [$\bar{1}10$] [5].

In the etching process of the opening along [$1\bar{1}0$] and [110], etching solution can be chosen to attack P atoms preferentially. As shown in Fig. 4.4, in a binary AB compound, the valence electrons highly populated at B atoms (As atom in GaAs and Se atom in ZnSe) are ready to take part in a reaction. Thus the {$\bar{1}\bar{1}\bar{1}$} B surfaces are very reactive compared with the {111} A surfaces. For InP the $Br_2 + CH_3OH$ solution attacks the {$\bar{1}\bar{1}\bar{1}$} B phosphorus surface and the remaining surface is the {111} A indium surfaces [6]. Then, unetched part forms a mesa shape, as shown in Fig. 4.6a, and in Fig. 4.6b it is inverse. Similar anisotropy is observed in other III-V compounds and alloys with the zincblende structure. Etching anisotropy is often used to find the crystal orientation and to structure substrate surfaces for device fabrication with a variety of etchants.

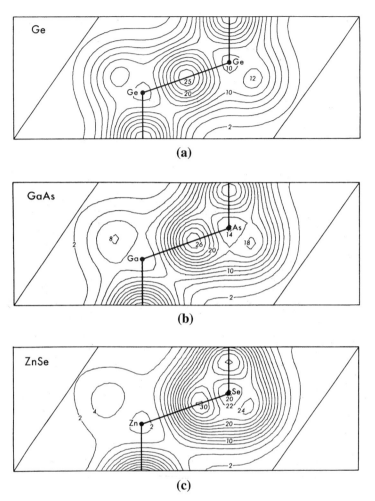

FIGURE 4.4 Distribution of valence electron density on the ($1\bar{1}0$) plane in (*a*) Ge, (*b*) GaAs, and (*c*) ZnSe. Contours are in units of electrons per unit cell volume. The increasing ionic nature from Ge to ZnSe is clearly shown. (After Ref. 4.)

4.2.3 Valence-Force-Field Model

In the calculation of mechanical properties in the atomic level, the valence-force-field (VFF) model, which is a semiclassical model, may be useful. With an externally applied stress to the bulk or a stress due to a dislocation, the location of atoms is shifted. This could be expressed by the change of bond lengths and the change of the bond angles if the distortion is within the elastic limit. The elastic energy U due to the change of the atom location in the zincblende structure as shown in Fig. 4.7 can be calculated as [7]

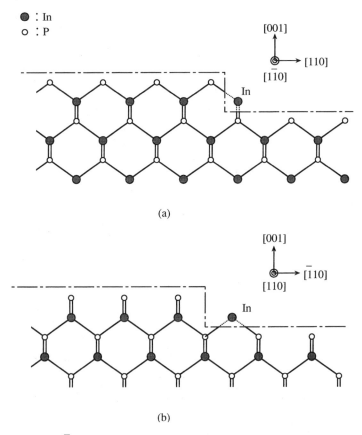

FIGURE 4.5 (a) $(1\bar{1}0)$ cross section with [110] step on a (001) surface of InP. At the corner three danglingn bonds are available for the In atom. (b) (110) cross section with $[\bar{1}10]$ step on (001) surface. At the corner two dangling bonds are available for the In atom.

$$U = \frac{1}{2}\alpha\left(\frac{3}{4b^2}\right)\sum_{i=1}^{4}[\Delta(\mathbf{b}_i^A\cdot\mathbf{b}_i^A)]^2 + \frac{1}{2}\sum_{s=1}^{2}\beta^s\left(\frac{3}{4b^2}\right)\sum_{i,j>i}[\Delta(\mathbf{b}_i^s\cdot\mathbf{b}_j^s)]^2 \quad (2)$$

where α is the elastic constant of bond stretching and β^s is the elastic constant of bond bending around the atom A or B (s = A or B). $\mathbf{b}_i^s$ and $\mathbf{b}_j^s$ (i, j = 1 to 4) are the bond vectors about the atom s and at equilibrium $|\mathbf{b}_i^s| = |\mathbf{b}_j^s| = b$, which should be equal to $a\sqrt{3}/4$. $\Delta(\mathbf{b}_i^s\cdot\mathbf{b}_j^s)/b^2$ represents the angle variations from equilibrium values. $(\mathbf{b}_1^A\cdot\mathbf{b}_4^A)/b^2 = \cos\theta_{14}^A$, as indicated in Fig. 4.7. Those values for α and β are listed in Table 4.2.

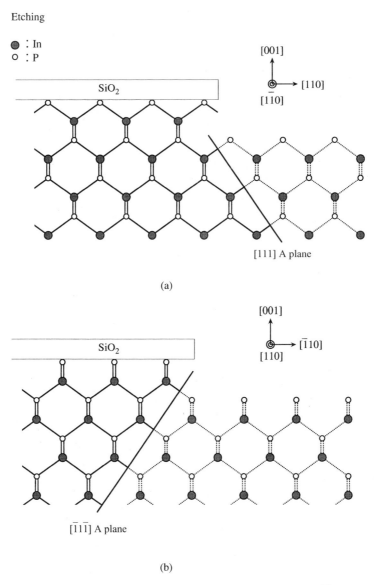

FIGURE 4.6 (*a*) When the (001) surface of InP with an opening along [1$\bar{1}$0] is etched with $Br_2 + CH_3OH$, the remaining surface is an (111)A In surface. (*b*) When the (001) surface with an opening along [110] is etched, the remaining surface is also an In surface ($\bar{1}\bar{1}\bar{1}$)A. Hence, the etched shape is inverted.

TABLE 4.2 Theoretical Values of α and β in Eq. (2)

Semiconductor	α	β
AlSb	35.35	6.77
GaP	47.32	10.44
GaAs	41.19	8.95
GaSb	33.16	7.22
InP	43.04	6.24
InAs	35.18	5.50

Source: After Ref. 7.

4.2.4 Energy Band Structure

The energy band structure of a semiconductor gives us a lot of information on the semiconductor properties. For example, the energy band structures (E–k dispersion curves) of GaAs, InP, and GaP are shown in Fig. 4.8a, b, and c, respectively [4]. Symbols such as Γ, X, and L represent high symmetry points in group theory. Specifically, Γ is $(0,0,0)2\pi/a$, X is $\{1,0,0\}2\pi/a$, and L is $\{\frac{1}{2},\frac{1}{2},\frac{1}{2}\}2\pi/a$ in k-space. In GaAs and InP, the lowest conduction band minimum is Γ_6, the next-lowest is L_6, and the third-lowest is X_6. The energy differences $L_6 - \Gamma_6$ and $X_6 - \Gamma_6$ are denoted as $\Delta E_{\Gamma L}$ and $\Delta E_{\Gamma X}$, respectively, in Table 4.4. Since the shape of the conduction band minima looks like a valley, they are sometimes called as Γ valley, L valley, and so on. The valence band maximum is Γ_8, and two bands (the heavy-hole band and the light-hole band) are degenerated at this point. To specify the conduction and valence bands, the symmetry points are sometimes denoted as Γ_{6C}, L_{6C}, and Γ_{8V} or as Γ_6^C, L_6^C, and Γ_8^V. The split-off energy $\Gamma_{8V} - \Gamma_{7V}$ is customarily represented as Δ_0. Needless to say, E_g is equal to $\Gamma_{6C} - \Gamma_{8V}$ in direct-gap InP and GaAs. In Fig. 4.8c for GaP, the split-off energy is neglected and $E_g = X_{1C} - \Gamma_{15V}$, which means "indirect gap."

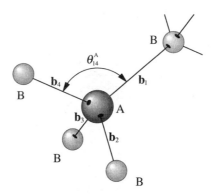

FIGURE 4.7 Valence-force field model for the zincblende structure with A and B atoms. $\mathbf{b}_i^s$ is a bond vector from atom s (s = A or B), and θ_{ij}^s is the angle between two bonds $\mathbf{b}_i^s$ and $\mathbf{b}_j^s$ ($i \neq j$).

TABLE 4.3 Materials Parameters of InP and Related Binary III-V Compounds[a]

Semiconductor	Crystal Structure	Lattice Constant (Å)	Density (g cm^{-3})	Linear Thermal Expansion Coefficient (K^{-1})	Thermal Conductivity (W cm^{-1} K^{-1})
AlP	Zincblende	5.4635	2.40	—	0.9
AlAs	Zincblende	5.6611	3.74	5.20×10^{-6}	0.91
GaP	Zincblende	5.4505	4.138	$(5.3–5.81) \times 10^{-6}$	1.1
GaAs	Zincblende	5.6533	5.3176	6.0×10^{-6}	0.54
GaSb	Zincblende	6.096	5.614	5.7×10^{-6}	0.35
InP	Zincblende	5.8694	4.787	4.5×10^{-6}	0.7
InAs	Zincblende	6.0583	5.667	4.52×10^{-6}	0.26

Semiconductor	Specific Heat (cal g^{-1} K^{-1})	Melting Point, T_m (°C)	Dissociation Pressure at T_m (atm)	Debye Temperature (K)	Piezoelectric Coefficient (C cm^{-2})
AlP	0.114 (400 K)	2,550	8	588 (0 K)	—
AlAs	0.108	1,740	1.4	417 (0 K)	—
GaP	5.243	1,457	35	446	−0.17
GaAs	5.46	1,240	1	376	−0.28
GaSb	5.776	712	$<10^{-3}$	240	−0.22
InP	5.32	1,062	25	420	—
InAs	5.67	942	0.3	280	−0.08

Semiconductor	Phonon Energies (meV)				Elastic Constants (10^{11} dyn cm^{-2})			Sound Velocity (km·s)
	TO	LO	LA	TA	C_{11}	C_{12}	C_{44}	
AlP	54.4	62			$\begin{cases} 18.83 \\ 14.59 \end{cases}$	6.71 / 8.44	3.69 / 4.24	(estimates)
AlAs	45	50	27	13	12.02	5.70	5.89	(interpolation)
GaP	45.4(Γ) / 43.8(L) / 44.1(X)	50.6(Γ) / 50.3(L) / 45.5(X)	31.0(X)	10.3(L) / 12.9(X)	14.05	6.20	7.03	$\begin{cases} 6.28 \, (110) \, l \\ 4.13 \, (100) \, t \end{cases}$
GaAs	33.1(Γ) / 32.4(L) / 31.2(X)	35.3(Γ) / 29.5(L) / 29.8(X)	25.9(L) / 28.1(X)	7.7(L) / 9.8(X)	11.9	5.38	5.99	$\begin{cases} 4.73 \, (100) \, l \\ 3.34 \, (100) \, t \end{cases}$
GaSb	27.7(Γ) / 27.0(L) / 27.0(X)	28.8(Γ) / 25.3(L) / 26.0(X)	19.2(L)	5.7(L) / 6.9(X)	8.83	4.02	4.32	$\begin{cases} 3.97 \, (001) \, l \\ 2.77 \, (001) \, t \end{cases}$
InP	37.6(Γ) / 39.3(L) / 40.1(X)	42.7(Γ) / 42.2(L) / 41.2(X)	20.7(L) / 24.0(X)	6.8(L) / 8.5(X)	10.11	5.61	4.56	$\begin{cases} 5.13 \, l \\ 3.10 \, t \end{cases}$
InAs	26.9(Γ) / 26.8(L) / 26.8(X)	29.6(Γ) / 25.2(L) / 25.2(X)	17.3(L) / 19.8(X)	5.5(L) / 6.6(X)	8.33	4.53	3.96	$\begin{cases} 4.35 \, (110) \, l \\ 2.64 \, (001) \, t \end{cases}$

Source: Values collected from Refs. 2, 34, and 57.
[a] At 300 K if not specified.

The sharper the valley shape, the lighter the effective mass m^*, because $m^* = \hbar^2/(\partial^2 E/\partial k^2)$.

4.2.5 Material Parameters

In Table 4.3, the material parameters related to crystal properties of InP and the binary compounds AlP, AlAs, GaP, GaAs, GaSb, and InAs, which compose the alloys GaInAs, GaInAsP, AlInAs, AlGaInAs, and GaAsSb, are listed. Parameters of the energy band structures and parameters of the electrical properties of those binary compounds are listed in Tables 4.4 and 4.5, respectively. Composition dependences of the energy gaps of ternary and quaternary alloys are listed in Table 4.6. Composition dependences of electron effective mass in the Γ valley for some of the alloys are listed in Table 4.7.

4.3 InP

InP itself had not received much interest as a very useful semiconductor until it was used as the substrate for GaInAsP as lasers [8] and for GaInAs as transport devices [9,10] because except for the peak drift velocity, the electrical and optical properties of InP were thought to be similar or even inferior to those of GaAs, of which crystal growth technique was much advanced.

4.3.1 InP and Alloys

The lattice constant of InP at 300 K is 5.8694 Å, which is in the middle between GaAs (5.6533 Å) and InAs (6.0583 Å). The alloy between GaAs and InAs lattice matches to InP at a composition of $Ga_{0.47}In_{0.53}As$. The quaternary alloy $Ga_xIn_{1-x}As_yP_{1-y}$ lattice matches to InP at the condition $x(1.032 - 0.032y) = 0.47y$, which could be approximated by $y \approx 2.2x$. Under this lattice-matching condition, this quaternary alloy reduces to the ternary alloy $Ga_{0.47}In_{0.53}As$ at $y = 1$, and at $y = 0$ is the binary InP. As shown in Fig. 4.13 and described in detail later, this quaternary alloy can cover the energy gap range from 0.75 eV ($Ga_{0.47}In_{0.53}As$) [9] to 1.35 eV (InP), and the corresponding wavelength range 1.65 to 0.92 μm at 300 K. This wavelength range is important in silica-fiber-based optical communication, where the lowest-loss window is between 1.24 and 1.56 μm. This is one of the biggest reasons why InP is important [11].

Similarly, other alloys, such as AlGaInAs, GaAsSb, and GaInTlP, can be grown lattice matched to InP. When materials are extended to II-VI alloys such as CdSSe, ZnSeTe, and ZnCdSe and their quaternary alloys, and chalcopyrite alloys such as that between $AgInS_2$ and $AgInSe_2$ and that between $CuInSe_2$ and $AgInSe_2$, can be grown lattice matched to InP. In this chapter, only III-V semiconductors are discussed.

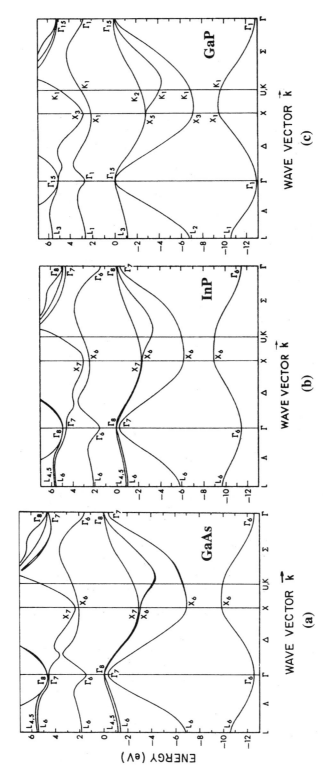

FIGURE 4.8 Energy band structures of (a) GaAs, (b) InP, and (c) GaP calculated by the pseudopotential method. (After Ref. 4.)

TABLE 4.4 Parameters for Energy Band Structure[a]

	Energy Band Structure									
	Energy Gap (eV)		Temperature Coefficient (10^{-4} eV/K)	Conduction Band Energy Minima Difference (eV)		Split-off Energy, Δ_0 (eV)	Effective Mass, m_0			Refractive Index, $h\nu \sim E_g$	Electron Affinity (eV)
Semiconductor	0 K	300 K		$\Delta E_{\Gamma L}$	$\Delta E_{\Gamma X}$		Electron	Heavy Hole	Light Hole		
AlP	2.52	2.45	−2.58	−0.3	−1.2	0.06	$\|3.67(X)$ $\perp 0.212(X)$	$0.513\|[100]$ $1.372\|[111]$	$0.211\|[100]$ $0.145\|[111]$	3.03	
AlAs	2.25	2.15	−4.0	−0.66	−0.865	0.29	$\|1.1(X)$ $\perp 0.19(X)$	$0.409\|[100]$ $1.022\|[111]$	$0.153\|[100]$ $0.109\|[111]$	3.178	2.62
GaP	2.350	2.272	−6.2	−0.12	−0.54	0.08	$\|0.254$ $\perp 4.8$	$0.67\|[111]$	$0.17\|[100]$	3.452	4.0
GaAs	1.51922	1.4248	−3.95	0.30	0.46	0.34	0.067	0.50	0.082	3.655	4.07
GaSb	0.822	0.75	−3.7	0.06	0.43	0.80	0.041	0.283	0.050	3.82	4.06
InP	1.4205	1.351	−2.9	0.5	0.85	0.11	0.08	0.56	0.12	3.45	4.40
InAs	0.4105	0.354	−3.5	0.78	1.47	0.38	0.023	0.41	0.026	3.52	4.90

Source: Values collected from Refs. 2, 34, and 57.
[a] At 300 K if not specified.

TABLE 4.5 Parameters for Transport Properties

Semiconductor	Mobility (cm^2 V^{-1} s^{-1})			Specific Dielectric Constant		Deformation Potential (eV)	Effective Charge (Longitudinal) (e)	Ionicity
	Electron (300 K)	Electron (77 K)	Hole (300 K)	ε_0	ε_∞			
AlP	60	30	—	9.8	1.3	—	—	0.307
AlAs	294	18	—	10.06	8.16	—	—	0.274
GaP	160	2,500	135	11.11	9.11	13.0	0.24	0.327
GaAs	9,200	244,000	402	13.18	10.9	8.6	0.20	0.310
GaSb	7,700	10,000	1,400	15.69	14.44	8.3	0.13	0.261
InP	5,370	131,600	150	12.56	9.61	6.8	0.28	0.421
InAs	33,000	120,000	450	15.15	12.25	5.8	0.22	0.357

Source: Values collected from Refs. 1, 2, 34, and 57.
[a]At 300 K if not specified.

TABLE 4.6 Energy Gaps as Functions of Composition in InP-Related Ternary and Quaternary Alloys[a]

Alloy	Energy Gap, E_g^b (eV)	Composition at Crossover
Ternaries		
$Al_xIn_{1-x}As$	$E_g(\Gamma) = 0.37 + 1.91x + 0.74x^2$	$x_C(\Gamma - X) = 0.68$
	$E_g(X) = 1.82 + 0.4x$	
$Ga_xIn_{1-x}As$	$E_g(\Gamma) = 0.324 + 0.7x + 0.4x^2$	
$GaAs_{1-x}Sb_x$	$E_g(\Gamma) = 1.43 - 1.9x + 1.2x^2$	
$InAs_{1-x}P_x$	$E_g(\Gamma) = 0.356 + 0.675x + 0.32x^2$	
Quaternaries		
$Ga_xIn_{1-x}As_yP_{1-y}$	$E_g(\Gamma) = 1.35 + 0.668x - 1.068y + 0.758x^2 + 0.078y^2$	
	$\quad - 0.069xy - 0.322x^2y + 0.03xy^2$	
	$E_g(\Gamma) = 1.35 - 0.775y + 0.149y^2$ ($x \doteq 0.47y$:lattice matched with InP)	
$Al_xGa_yIn_{1-x-y}As$	$E_g(\Gamma) = 0.36 + 2.093x + 0.629y + 0.577x^2 + 0.436y^2$	
	$\quad + 1.013xy - 2.0xy(1-x-y)$	
	$E_g(\Gamma) = 0.764 + 0.495z + 0.203z^2$ ($0.98x + y = 0.47$, $x = 0.48z$:lattice matched with InP)	

Source: Ref. 2.
[a] At 300 K.
[b] $E_g(\Gamma) = \Gamma_{6C} - \Gamma_{8V}$, $E_g(X) = X_{6C} - \Gamma_{8V}$, $E_g(L) = L_{6C} - \Gamma_{8V}$.

4.3.2 Velocity-Field Characteristics and Surface Properties

Other noticeable properties of InP are the large energy separation of the conduction band minima and the lower surface activity.

Velocity-Field Characteristics In the conduction band of InP, the energy separation $\Delta E_{\Gamma L}$ is 0.5 eV and $\Delta E_{\Gamma X}$ is 0.85 eV, as listed in Table 4.4. Those values in GaAs are $\Delta E_{\Gamma L} = 0.30$ eV and $\Delta E_{\Gamma X} = 0.46$ eV. Because of this smaller separation 0.30 eV in GaAs, the electrons in the Γ-valley, which have a smaller effective mass, are transferred, under a high electric field, to the L-valley, which has a heavier effective mass. This electron transfer causes negative differential resistivity (hence nega-

TABLE 4.7 Electron Effective Masses as Functions of Composition[a]

Alloy	Effective Mass[b] (in units of m_0)
$Ga_xIn_{1-x}As_yP_{1-y}$	$m_e^*(\Gamma) = 0.077 - 0.050y + 0.014y^2$
	($x \doteq 0.47y$:lattice matched with InP)
$GaAs_{1-x}Sb_x$	$m_e^*(\Gamma) = 0.0634 - 0.0483x + 0.0252x^2$

Source: Ref. 2.
[a] At 300 K.
[b] $m_e^*(\Gamma)$ is the effective mass of electron in Γ valley.

tive differential velocity), which induces the Gunn oscillations [12]. The peak velocity in GaAs is 2.2×10^7 cm s^{-1} at the electric field of 3.5 kV cm^{-1}. The larger energy separation in InP makes the peak velocity and the electric field higher (i.e., 2.9×10^7 cm s^{-1} at 11 kV cm^{-1}). Because of these characteristics, together with the higher thermal conductivity, InP may be suitable for power devices. Those velocity-field characteristics of InP and GaAs are compared in Fig. 4.9.

Surface Properties The surface of InP is known, from experience, to be more stable in the air than that of GaAs. In lasers that use cleaved {110} surfaces as mirrors, the surface degradation due to laser operation is less in InP-based lasers than in GaAs-based lasers [14,15]. The Schottky barrier height between Au and n-type InP is 0.52 eV, and that between Au and n-type GaAs is 0.9 eV [16]. On the other hand, the Schottky barrier height between Au and p-type InP is 0.76 eV, and that between Au and p-type GaAs is 0.42 eV [16]. These barrier heights result in a good Schottky diode for n-type GaAs and a poor diode for n-type InP. The same reason works in reverse for ohmic contacts; it is easier to form good ohmic contacts to p-type GaAs and n-type InP, and less easy for n-type GaAs and p-type InP. Lower-resistivity ohmic contacts are still looked for with n-type GaAs. Interface Fermi-level positions for Au in contact with various III-V compounds and alloys are shown in Fig. 4.10. These barrier height characteristics may be explained from the distribution curves of the surface-state density at the insulator–GaAs and insulator–InP in-

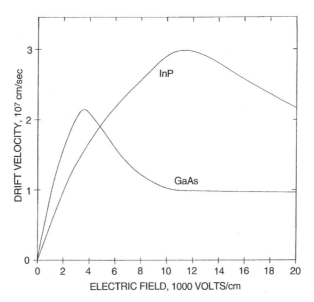

FIGURE 4.9 Monte Carlo calculation of velocity-field curves for InP and GaAs. InP has a higher peak velocity and a higher electric field at the velocity peak than does GaAs. (After Ref. 13.)

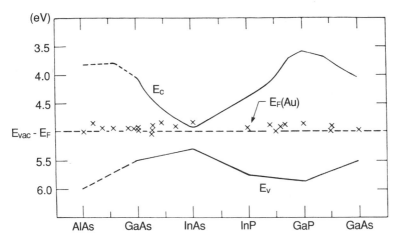

FIGURE 4.10 Interface Fermi-level positions (crosses) for Au in contact with various III-V compounds and alloys. Work functions ($E_{vac} - E_F$) are close to the dashed horizontal line at 5.0 eV. (After Ref. 3.)

terfaces, as shown in Figs. 4.11a and b, although there are discussions on the distribution shapes [17; see also Chapter 8 of this volume]. The location of the minima is thought to be related to the barrier heights. The density at minimum is lower in InP.

4.3.3 Bulk and Epitaxial Growth

Bulk Growth The liquid-encapsulated Czochralski (LEC) technique is commonly used for bulk growth of InP. Wafers 3 and 4 in. in diameter are commercially available. At the melting points, the phosphorous-based III-V compounds have a higher vapor pressure, as shown in Table 4.3. This requires a high-pressure-resistant furnace for LEC. On the other hand, GaAs, GaSb, and InAs can be grown in a quartz tube because of their lower vapor pressure; n-type InP wafers are Sn-doped or S-doped, and p-type wafers are Zn-doped. Fe-doped SI (semi-insulating) wafers are also available. Undoped InP is usually n-type. The dislocations in the active region of GaInAsP lasers grown on InP are known to have no such effects on degradation as dark-line defects, which are a serious problem in AlGaAs lasers on GaAs [15,18]. The dislocation density is usually on the order of 10^4 to 10^5 cm^{-2} in InP wafers. InP bulk crystal growth is described in detail in Chapter 5.

Epitaxial Growth High-quality epitaxial layers of InP have been grown by VPE (vapor-phase epitaxy), LPE (liquid-phase epitaxy), MOCVD [metal-organic vapor phase deposition (MOVPE or OMVPE)], and MOMBE (metal-organic molecular beam epitaxy). Some of these are described in Chapters 6 and 7.

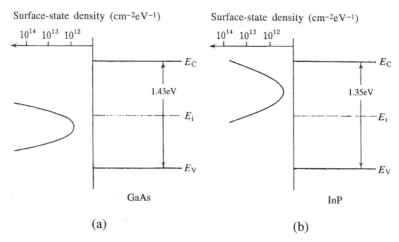

FIGURE 4.11 Energy distribution of surface-state density at a GaAs surface (*a*) and an InP surface (*b*). InP has a lower density than GaAs at a shallower level from the conduction band edge.

4.4 GaInAs AND GaInAsP

4.4.1 Energy Band Structure Versus Lattice Constant

The chart of the energy gap versus lattice constant of GaInAsP at 300 K is shown in Fig. 4.12. The boundary curves are the energy gaps of ternary alloys, which were calculated using the equations presented in Table 4.6. It should be noted that the equations are quoted from various sources and that there are some discrepancies among the endpoint values (of binary compounds) of energy gap. For example, at $x = 0$ in $Al_xIn_{1-x}As$ and $Ga_xIn_{1-x}As$, the energy gap equations give slightly different values for the same InAs.

As described in Section 4.3, $Ga_xIn_{1-x}As_yP_{1-y}$ can cover the energy gap from 0.75 to 1.35 eV at 300 K by keeping the lattice-matching condition with InP as shown by the thin dashed line in Fig. 4.12. Similarly, as can be seen in the figure, this quaternary alloy can cover the energy gap from 1.43 to 1.9 eV by keeping the lattice-matching condition with GaAs. The lowest conduction band minima are at Γ symmetry over the entire composition range in both alloys under lattice-matching conditions. The direct gap energy $E_g(\Gamma)$ and the split-off energy Δ_0 in $Ga_xIn_{1-x}As_yP_{1-y}$ on the dashed line in Fig. 4.12 are shown in Fig. 4.13.

When the double heterostructure (DH) is fabricated with InP as the clad layer and GaInAsP as the active layer, keeping a good carrier confinement in the active layer, the upper limit of the photon energy is about 1.2 eV. Since most of the lasers fabricated with this alloy are applied to the optical communication using silica fibers, the energy gap is adjusted precisely at a wavelength of 1.30 or 1.55 μm. Therefore, the active layers are composed of quaternary alloys for the DH lasers.

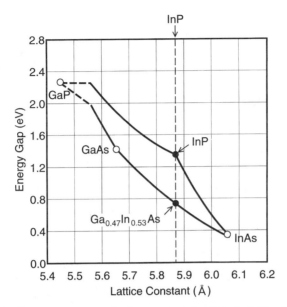

FIGURE 4.12 Energy gap versus lattice constant for GaInAsP. The boundaries are for ternary alloys, and open circles are for binaries. Solid circles are InP and ternary at the lattice matching to InP. Solid and dashed lines of the boundaries indicate direct and indirect gaps, respectively.

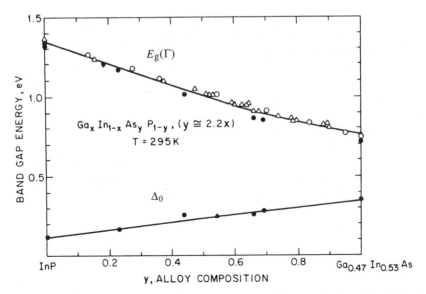

FIGURE 4.13 Direct-gap energy $E_g(\Gamma)$ versus As composition y and spin-orbit splitting Δ_0 versus y in $Ga_xIn_{1-x}As_yP_{1-y}$ at the lattice-matching condition. Experimental data and best-fit curves. (After Ref. 19.)

4.4.2 Atom Configuration in Alloys

The alloy compositions are quite often determined from x-ray diffraction using Vegard's law, where the lattice constant of a binary alloy or a pseudobinary alloy is a linear function of the alloy composition [i.e., in $Ga_xIn_{1-x}As$, the lattice constant a in units of Å is expressed as $a(x) = 6.058 - 0.405x$]. This relation is applied only to alloys whose lattice is not distorted (e.g., in the case of $Ga_xIn_{1-x}As$, it is cubic). Alloys are in most cases grown epitaxially on appropriate substrates of which lattice constant does not always match to that of the alloys. Even with a lattice mismatch, if the thickness is smaller than a critical thickness h_c, the epitaxial layer is grown coherently on the substrate although the lattice is elastically distorted. The distortion can be estimated from x-ray diffraction using symmetric and asymmetric reflections [20].

It has been believed from these experiences that the bond lengths in the alloys are also linear functions of the composition and that the atoms are randomly distributed on the lattice points (e.g., a Ga atom or an In atom occupies a group III sublattice at the probability of x or $1-x$), respectively. Thus a virtual crystal approximation had quite often been used as the first-order approximation for understanding the properties of alloys. In the virtual crystal approximation for $Ga_xIn_{1-x}As$, for example, an atom Z that has average properties as Ga_xIn_{1-x} forms a binary compound ZAs of the cubic lattice with $a(x) = 6.058 - 0.405x$. Effects of the randomness were then taken into consideration as the next-order perturbation. These simple images were disputed by two findings: the determination of the bond lengths Ga–As and In–As in $Ga_xIn_{1-x}As$ alloys by EXAFS measurement [21], and the observation of high ordering of Ga and In atoms in $Ga_{0.5}In_{0.5}P$ [22–25].

Figure 4.14 shows the composition dependence of bond lengths in $Ga_xIn_{1-x}As$. The Ga–As and In–As bonds tend to keep their original lengths. They vary linearly with the composition, but not as much as expected from x-ray diffraction measurements. Since the peak signals of x-ray diffraction are produced by multiple interference of x-rays reflected from many atoms, the only periodic portion of the reflections remains and other unperiodic reflections are canceled out. Thus the average lattice spacings are detected from the x-ray diffraction peak. Similar observations have been made in other ternary alloys and quaternary alloys.

Figure 4.15 is an atomic resolution cross-sectional TEM photograph of ordered $Ga_{0.5}In_{0.5}P$ grown on GaAs. This photograph clearly shows the CuPt ordering of Ga and In atoms [i.e., the ordering is along the $\langle 111 \rangle$ directions] [25]. This ordering does affect the energy gap of $Ga_{0.5}In_{0.5}P$. The $Ga_{0.5}In_{0.5}P$ random alloy has an energy gap of 1.90 eV at room temperature, and the $Ga_{0.5}In_{0.5}P$ ordered alloy has an energy gap as small as 1.85 eV [24]. Mechanisms for ordering and effects of ordering on the energy band structures have been considered in theory [26]. In Ref. 26, ordering in other semiconductors is also described.

4.4.3 Dielectric Constants

Dielectric constants are important for both electrical and optical applications. The static dielectric constant ε_0 may be used to estimate the capacitance of a device. The

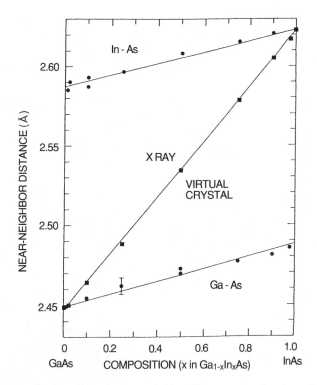

FIGURE 4.14 Ga–As and In–As near-neighbor distances as a function of alloy composition. Middle curve is VCA cation–anion bond length calculated from the measured x-ray lattice constants. (After Ref. 21.)

capacitance and resistance quite often determine the upper limit of the switching speed or the maximum frequency of operation of transport devices. The dielectric constant ε_∞ is related to the refractive index n as $n = \sqrt{\varepsilon_\infty}$ in the frequency region of light. The difference in n values between two layers is an important factor for optical confinement in lasers and optical guides.

The composition dependences of the dielectric constants are calculated and given as [27]

$$\varepsilon_0 = 12.40 + 1.5y \quad (3a)$$

$$\varepsilon_\infty = 9.55 + 2.2y \quad (3b)$$

at the lattice-matching composition $y = 2.2x$.

4.4.4 $Ga_{0.47}In_{0.53}As$

The lowest-energy endpoint of GaInAsP alloys lattice matched to InP is $Ga_{0.47}In_{0.53}As$. This ternary alloy has two important applications; one is the pho-

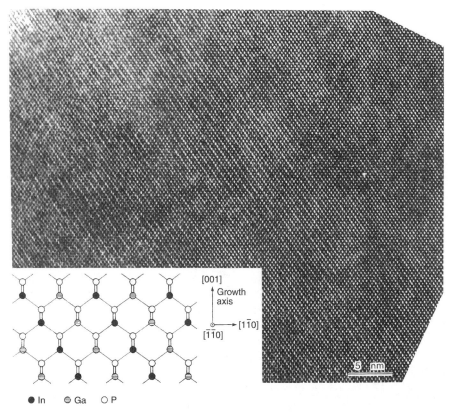

FIGURE 4.15 Atomic resolution cross-sectional TEM photograph of ordered $Ga_{0.5}In_{0.5}P$ on (001) GaAs. Inset is a schematic drawing of the atom configuration. (Courtesy of Dr. Ueda, Fujitsu Laboratories.)

todetector for the optical fiber communication mentioned above and another is the active layer of electron transport devices. A photodetector composed of $Ga_{0.47}In_{0.53}As$ with InP as a window layer can cover the wavelength region between approximately 0.92 and 1.65 μm, which corresponds to the energy range in Fig. 4.13 from $y = 0$ to 1.0. The detector can be a pin photodiode and an APD (avalanche photodiode). In both cases $Ga_{0.47}In_{0.53}As$ is used as an absorption layer to convert the photons to electrons and holes, and InP as the high-field region to extract generated carriers or to accelerate them for avalanche multiplication. Thus photodetectors made of $Ga_{0.47}In_{0.53}As$ with InP windows are indispensable devices in optical fiber communication systems and OEICs (optoelectronic integrated circuits) [11,28–32].

Because of the higher electron mobility (ca. 12,000 $cm^2 V^{-1} s^{-1}$) and a higher electron drift velocity ($\geq 2.5 \times 10^7$ $cm\ s^{-1}$) than those of GaAs, the intrinsic characteristics of $Ga_{0.47}In_{0.53}As$ transport devices should be superior to those of GaAs. Basically, smaller effective mass electrons have a higher electron mobility. They are

$0.08m_0$ in InP, $0.067m_0$ in GaAs, and $0.041m_0$ in $Ga_{0.47}In_{0.53}As$, and low-field electron mobilities at 300 K are about 5700 cm^2 V^{-1} s^{-1} in InP, about 8500 cm^2 V^{-1} s^{-1} in GaAs, and about 12,000 cm^2 V^{-1} s^{-1} in $Ga_{0.47}In_{0.53}As$, respectively, which are limited by the LO phonon scattering in very pure materials at room temperature [33–35]. The As composition dependence of effective masses in GaInAsP is shown in Fig. 4.16.

The peak drift velocity of electrons in a semiconductor that exhibits negative differential resistivity due to the electron transfer effect is governed by the low-field electron effective mass and the energy separation between the lowest Γ valley and the next-higher valley. Those energy separations are 0.30 eV in GaAs, 0.5 eV in InP, and 0.55 eV in $Ga_{0.47}In_{0.53}As$. At a much higher electric field, the velocity may saturate due to LO phonon emission, or the velocity-field curve will break due to the impact ionization of electrons and holes from valence band. With the use of InP or AlInAs as a wide gap layer for the gate or emitter and a $Ga_{0.47}In_{0.53}As$ layer as the transport channel or the base, this type of heterostructure is applied to high-performance FETs, HEMTs, and HBTs [36], as described in detail in Chapters 3, 10, and 11.

4.4.5 Effects of Strain

For most of the device applications, GaInAsP and GaInAs are designed to closely lattice match to InP. However, within a certain limit of the mismatch, GaInAsP and GaInAs can be grown coherently or pseudomorphically on InP. This induces a two-dimensional distortion of the lattices (compressive or tensile depending on the sign of the mismatch) and modifies the band structure as illustrated in Fig. 4.17. It was calculated that strain removes the degeneracy at the top of valence band, changing the energy gap, and that the heavy-hole or light-hole valence band then rises to the top [37–39]. This effect is important in the application to HFETs (see Chapter 10) and lasers (see Chapter 12).

4.4.6 Epitaxial Growth

High-quality GaInAsP and GaInAs layers were grown on InP by LPE, followed by VPE. The high-reliability lasers in the long-wavelength (1.3 and 1.55 μm) region were all grown by LPE until growth techniques of MOCVD and MOMBE were improved to produce high-quality epitaxial layers. To fabricate photodetectors, high-quality epitaxial layers are required to sustain a high electric field and to provide a composition-graded interface for carriers to move smoothly. For precise control of the thickness and interface abruptness in MQWs (multiple quantum wells), MOCVD and MOMBE are used presently.

4.5 AlInAs AND AlGaInAs

These ternary and quaternary alloys also lattice match to InP as shown in Fig. 4.18. By changing the composition in $Al_xGa_yIn_{1-x-y}As$ while keeping the lattice-matching

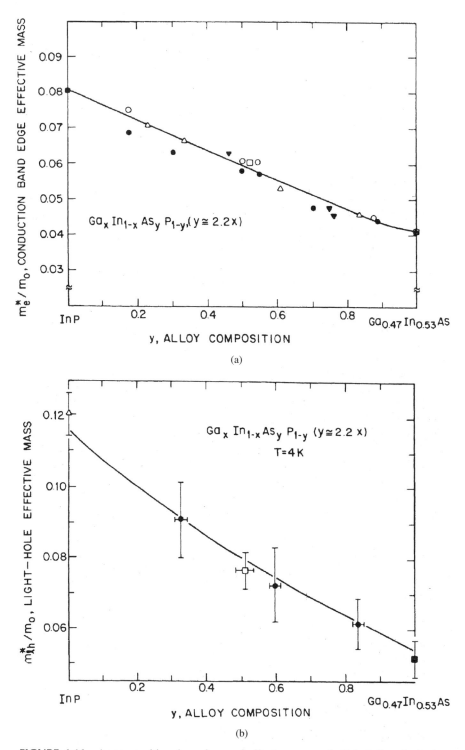

FIGURE 4.16 As composition dependence of effective masses in GaInAsP at the lattice-matching condition; experimental data and the best-fit curve (*a*) for electrons in Γ valley and (*b*) for light holes. (After Ref. 19.)

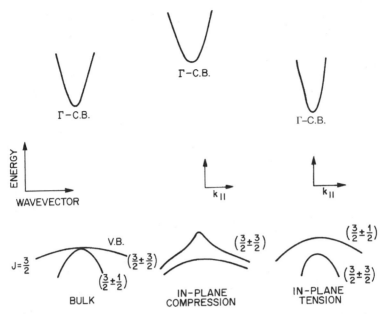

FIGURE 4.17 Distortion of valence band structure and change of energy gap by applied two-dimensional stress. The wave vector $k_\parallel$ lies in a direction normal to the growth direction, which is presumed to be [001]. (After Ref. 37.)

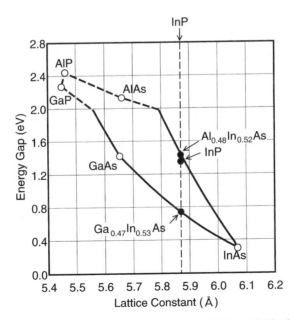

FIGURE 4.18 Energy gap versus lattice constant for AlGaInAsP. The boundaries are for ternary alloys, and open circles are for binaries. Solid circles are InP and ternaries at the lattice matching to InP. Solid and dashed lines of the boundaries indicate direct and indirect gaps, respectively.

condition, the energy gap can be varied from 0.75 eV in $Ga_{0.47}In_{0.53}As$ to 1.46 eV in $Al_{0.48}In_{0.52}As$. The energy gap of $Al_{0.48}In_{0.52}As$ is slightly larger than that of InP. However, the conduction band discontinuity between $Al_{0.48}In_{0.52}As$ and $Ga_{0.47}In_{0.53}As$ is twice larger than that between InP and $Ga_{0.47}In_{0.53}As$ (i.e., 0.52 eV for the first pair and 0.25 eV for the second pair) [30,36]. This makes $Al_{0.48}In_{0.52}As$ a better semiconductor as the barrier layer for electrons in $Ga_{0.47}In_{0.53}As$. The Schottky barrier height of Au on n-type $Al_{0.48}In_{0.52}As$ is about 0.8 eV [40]. This value is comparable to that of n-type GaAs and can be used as the gate of heterojunction FETs (HFETs) and high-electron-mobility transistor (HEMT) devices, as detailed in Chapter 10. The Schottky barrier height of Au to n-type $Ga_{0.47}In_{0.53}As$ is as low as 0.2 to 0.3 eV [30] and cannot be used for the gate at room temperature.

Although the energy gap of $Al_xGa_yIn_{1-x-y}As$ can be varied with the composition, the widest gap, $Al_{0.48}In_{0.52}As$, is used most often. It is known to contain a high density of deep levels, probably due to the high Al composition of the alloy, which degrades the photoluminescence (PL) efficiency and the reverse current of Schottky diodes. There is a report that the deep levels are related to oxygen in $Al_{0.48}In_{0.52}As$, although it has been decreasing, owing to recent efforts to purify growth source materials [41]. $Al_xGa_yIn_{1-x-y}As$ is grown by either MBE or MOCVD.

4.6 GaAsSb

The quaternary alloy $Al_xGa_{1-x}As_ySb_{1-y}$ can be designed to lattice match with InP, as shown in Fig. 4.19. However, $GaAs_ySb_{1-y}$ is a well-known immiscible alloy with a large immiscible region in the phase diagram from $y = 0.35$ to 0.7, as shown in Fig. 4.20. Thermodynamic calculation predicts that most of the region of this quaternary alloy is thermodynamically unstable, as shown in Fig. 4.21. In this figure, unstable regions of other alloys covered in this chapter are also shown. The dotted area is unstable at temperatures at 800°C or at 600°C. However, there are experiments that demonstrate growth of single-phase $GaAs_{1-x}Sb_{1-x}$ over the entire composition range by MBE [44] or MOCVD [45]. The alloy $GaAs_{0.5}Sb_{0.5}$, which lattice matches to InP, has an especially high quality. A PL spectrum at 4.2 K peaking at 0.782 eV with a FWHM of 7.2 meV, room-temperature mobility of 3800 cm^2 V^{-1} s^{-1}, and 77 K mobility of 6200 cm^2 V^{-1} s^{-1} at $N_d = 8.0 \times 10^{15}$ cm^{-3} is reported [46].

The refractive index is expected to be large because that of GaSb is one of the largest among the InP and related semiconductors listed in Table 4.4. The layered structure between this high-refractive-index alloy and a lower-index InP may be useful for optical applications.

4.7 HETEROEPITAXY

As described in Section 4.1, InP was used as a substrate on which to grow alloys. Growth of GaInAs or GaInAsP on InP especially requires a change in the group V atoms from P to As or from As to P. This makes it necessary to use LPE or VPE for

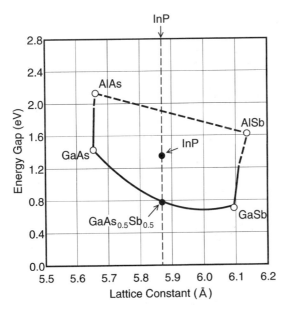

FIGURE 4.19 Energy gap versus lattice constant for AlGaAsSb. The boundaries are for ternary alloys, and open circles are for binaries. Solid circles are InP and ternaries at the lattice matching to InP. Solid and dashed lines of the boundaries indicate direct and indirect gaps, respectively.

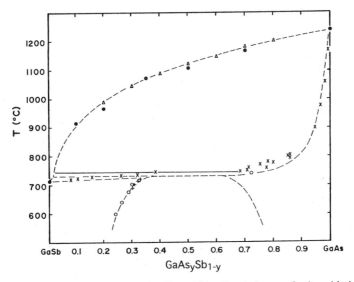

FIGURE 4.20 Phase diagram of $GaAs_ySb_{1-y}$. This alloy is known for its wide immiscible region at around 700°C between $y = 0.35$ and 0.7 at thermal equilibrium. (After Ref. 42.)

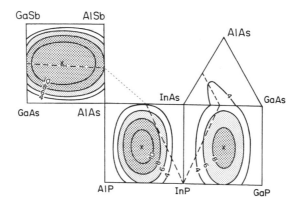

FIGURE 4.21 Unstable regions in alloys at 800°C (heavy-dotted area) and at 600°C (light-dotted area) are shown. Heavy dashed lines are the lattice-matching composition to InP. (After Ref. 43.)

growth. MBE was not good at continuous growth from InP to GaInAs(P), and using MOCVD, it took a long time to obtain good control of the composition. Abrupt change from P to As or from As to P is still a problem. In this section, band lineup in heterostructures and MQWs is described. Specific properties at each interface are also indicated.

4.7.1 InP/GaInAsP

The DH structure of this combination (i.e., InP/GaInAsP/InP) is very well known for long-wavelength-region optoelectronic devices. The band lineup of InP/Ga$_{0.47}$In$_{0.53}$As is shown in Fig. 4.22. The conduction band discontinuity ΔE_C and the valence band discontinuity ΔE_V are extremely important parameters in heterostructures. $\Delta E_C + \Delta E_V = \Delta E_g$, where ΔE_g is the difference in bandgaps of InP and GaInAsP. At the InP/GaInAsP interface, $\Delta E_C \approx 0.4 \times \Delta E_g$ and $\Delta E_V \approx 0.6 \times \Delta E_g$. At

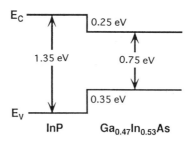

FIGURE 4.22 Band lineup of InP/Ga$_{0.47}$In$_{0.53}$As. There are some scatters in the values of band discontinuity in the literature.

InP/GaInAsP with the energy gap of 1.0 eV (which corresponds to a wavelength of 1.24 μm, due to band-to-band emission), ΔE_C is about 0.14 eV. This small value had been discussed to cause the electron overflow at a high current injection in laser operations and hence to be the reason for the low characteristic temperature T_0 of the laser threshold current.

In the MQW structure, the well layer and barrier layer may be as thin as 20 to 100 Å, which corresponds to 8 to 40 MLs (monolayers). This means that only a difference of 1 ML in the well layer causes more than 10% fluctuation in the 20-Å well thickness. Therefore, a very abrupt change in composition is required for well-controlled MQWs. Especially at InP/GaInAs and GaInAs/InP interfaces, the changes from As to P and from P to As are a problem since As(P) tends to replace the already bonded P(As) in the As(P) atmosphere [47,48]. A quick change in the source gas supply and quick extraction of the remaining gas out of the reactor are necessary techniques for the formation of abrupt interfaces. MOCVD or MOMBE are used almost exclusively for MQW growth at present since this abruptness is required, although LPE has long been used for DH lasers. InP/GaInAs can also be used for HBTs (heterojunction bipolar transistors) with InP as a wide-gap emitter, GaInAs as a base, and GaInAs or InP as a collector [36], as discussed in detail in Chapter 11.

4.7.2 AlInAs/GaInAs

This DH structure does not contain different group V atoms and can be fabricated by changing only the group III atoms while keeping the same As overpressure in MOCVD or MBE. The band lineup is shown in Fig. 4.23. The larger ΔE_C value is favorable for electron confinement in the conduction band. With the larger ΔE_C and the higher Schottky barrier height of AlInAs, this combination has been used for HFETs and HEMTs. HBTs have also been fabricated with AlInAs as the wide-gap emitter and GaInAs as the base and collector [36]. Due to a wider range of ΔE_C values in AlGaInAs and its thickness controllability by MBE, many structures in resonant-tunneling hot-electron transistors have been fabricated [36].

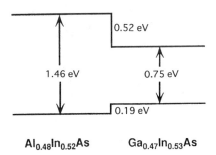

FIGURE 4.23 Band lineup of $Al_{0.48}In_{0.52}As/Ga_{0.47}In_{0.53}As$. There are some scatters in the values of band discontinuity in the literature.

4.7.3 AlInAs/InP

This heterostructure has a unique property from the two discussed in Sections 4.7.1 and 4.7.2. The band lineup is "staggered," as shown in Fig. 4.24. When an MQW of AlInAs/InP/AlInAs is fabricated, the resulting energy structure is shown in Fig. 4.25a, which is called a type II structure. Since the MQW of InP/GaInAs/InP is of type I, as illustrated in Fig. 4.25b, the MQW of $Al_xGa_yIn_{1-x-y}As/InP/Al_xGa_yIn_{1-x-y}As$ can be designed continuously from type I to type II.

4.8 OPTICAL PROPERTIES

All the alloys described in this chapter—GaInAs, GaInAsP, AlInAs, AlGaInAs, and GaAsSb—are direct-gap semiconductors. The direct-gap band structure is essentially important for both optical and high-mobility transport devices. Those alloys, especially GaInAs and GaInAsP, have attracted great interest at the beginning of research because of their direct-gap band structure [8,9,50] and their energy range, extending into the 1-μm region [8,50]. Only the photoluminescence properties of GaInAsP and InP/GaInAs/InP quantum wells are described in this section. Detailed photoluminescence and optoelectronic properties can be found in Refs. 11, 38, 39, and 51, and relevant chapters in this book.

4.8.1 GaInAsP

Theoretically, spontaneous emission from a nondegenerate direct-gap bulk semiconductor (where carriers have three-dimensional freedom of motion) has the spectral shape [52]

$$I(h\nu) \propto \nu^2(h\nu - E_g)^{1/2} \exp\left(-\frac{h\nu - E_g}{kT}\right) \qquad (4)$$

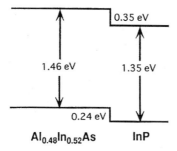

FIGURE 4.24 Staggered band lineup of $Al_{0.48}In_{0.52}As/InP$. There are some scatters in the values of band discontinuity in the literature.

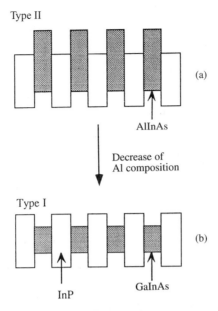

FIGURE 4.25 (a) Type II energy structure of an MQW of AlInAs/InP/AlInAs; (b) type I energy structure of an MQW of InP/GaInAs/InP. (After Ref. 49.)

where $I(h\nu)$ is the intensity per unit range of photon energy, h the Planck constant, ν the frequency of the light, $h\nu$ the photon energy, k the Boltzmann constant, and T the absolute temperature. This energy dependence of the intensity comes mostly from the density of states for three-dimensional carriers and the energy distribution of carriers. This formula indicates that the spectrum has a sharp cut in the lower-energy side at $h\nu = E_g$, has an exponential tail to the higher-energy side with the peak at about $h\nu = E_g + kT/2$. It has the same shape for any nondegenerate semiconductors at the same temperature T when the spectra are normalized at the peak height.

Figure 4.26a and b show photoluminescence spectra from GaInAsP layers lattice matched with InP at various compositions measured at 300 and 77 K, respectively. In the figure the compositions x and y are taken as $Ga_xIn_{1-x}As_{1-y}P_y$. One of the most important features of the figures is that the energy at the peak of spectrum is controlled by the alloy composition while lattice matching to InP is kept. Another important observation in the figures is that although the peak position is controlled as expected, the half-width of the spectra varies largely in contrast to the expectation from Eq. (4) where E_g is assumed to be perfectly constant. In real semiconductors E_g is known to fluctuate due to impurities and defects, resulting in tails in the lower-energy side [54]. In addition to this effect, there is a composition fluctuation in the alloys along the growth direction (especially when grown by LPE [55]) and due to spinodal fluctuation [56].

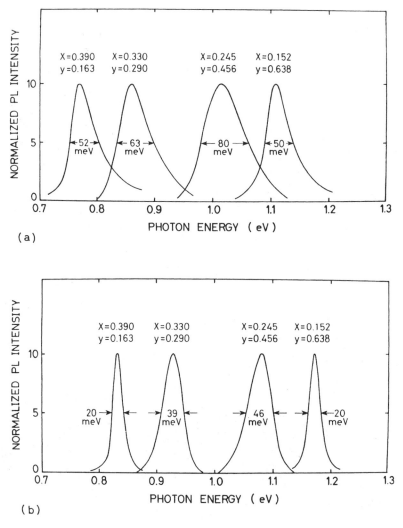

FIGURE 4.26 Photoluminescence spectra of lattice-matched $Ga_xIn_{1-x}As_{1-y}P_y$ layers for various alloy compositions x and y measured at (a) 300 K and (b) 77 K. Control of the energy at peak by alloy compositions is demonstrated, although there are variations in the half-width of the spectra. (After Ref. 53.)

4.8.2 InP/GaInAs/InP Quantum Wells

For optoelectronic applications, GaInAsP alloys have been used for the active region of DH lasers and GaInAs alloys for absorption region of photodetectors in the 1-μm wavelength region. For these devices, relatively thick (≥ 0.1 μm) layers are used. For high-efficiency and high-power lasers, MQW structures are fabricated with InP as the barrier layer and GaInAs(P) as the well layer. Because of the two-dimensional quantum confinement effect of carriers, the spectrum should be sharper and the density of states is higher, resulting in higher efficiency of luminescence. There is so much interesting physics in MQW structures and superlattices [39], and they play an important role in device applications, particularly in MQW lasers (see Chapter 12).

In the growth of InP/GaInAs/InP quantum wells by MOCVD, the difficulty lies in abrupt change in the P- and As-containing layers, as described in Section 4.7. By a careful design of the growth system and the gas switching sequence, extremely sharp photoluminescence spectra from InP/GaInAs/InP single quantum wells with a variety of well thicknesses were demonstrated, as shown in Fig. 4.27.

4.9 SUMMARY

Materials physics and bulk properties of InP-based compound semiconductors and alloys were described in this chapter. Since AlGaAs/GaAs heterostructures, which are a lattice-matching combination over the entire Al composition range, had been used successfully in devices before the advent of InP-based compounds, lattice-matched heteroepitaxy between InP and other alloys was of interest from the beginning of InP-based materials research. Several combinations of heterostructures were discussed in this chapter and band lineups at the interface were described. Optical properties of GaInAsP alloys and InP/GaInAs/InP quantum wells were described briefly. Many of the materials parameters were listed in tables referred to in this and other chapters in this book.

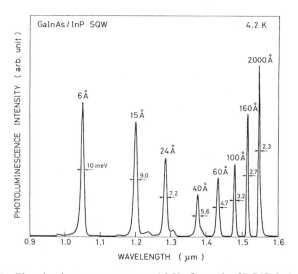

FIGURE 4.27 Photoluminescence spectra at 4.2 K of a stack of InP/GaInAs/InP SQWs and a GaInAs thick layer (2000 Å). Sharp peaks are observed even at the thinnest well of 6 Å. (After Ref. 47.)

REFERENCES

1. J. C. Phillips, *Bonds and Bands in Semiconductors,* Academic Press, San Diego, Calif., 1973, p. 22.
2. O. Madelung, ed., *Semiconductors: Group IV Elements and III-V Compounds,* Springer-Verlag, Berlin, 1991.
3. L. J. Brillson, "Surfaces and interfaces: atomic-scale structure, band bending and band offsets," in *Handbook on Semiconductors,* ed. T. S. Moss, Vol. 1, *Basic Properties of Semiconductors,* vol. ed. P. T. Landsberg, Elsevier Science Publishers, Amsterdam, 1992, pp. 281–417.
4. J. R. Chelikowsky and M. L. Cohen, "Nonlocal pseudopotential calculations for the electronic structure of eleven diamond and zincblende semiconductors," *Phys. Rev. B.,* **14,** 556–582 (1976).
5. H. Asai, "Anisotropic lateral growth in GaAs MOCVD layers on (001) substrates," *J. Cryst. Growth,* **80,** 425–433 (1987).
6. S. Adachi and H. Kawaguchi, "Chemical etching characteristics of (001) InP," *J. Electrochem. Soc.,* **128,** 1342–1349 (1981).
7. R. M. Martin, "Elastic properties of ZnS structure semiconductors," *Phys. Rev. B.,* **1,** 4005–4011 (1970).
8. J. J. Hsieh, J. A. Rossi, and J. P. Donelly, "Room temperature CW operation of GaInAsP/InP double-heterostructure diode lasers emitting at 1.1 μm," *Appl. Phys. Lett.,* **28,** 709–711 (1976).
9. Y. Takeda, A. Sasaki, Y. Imamura, and T. Takagi, "Electron mobility and energy gap of $In_{0.53}Ga_{0.47}As$ on InP," *J. Appl. Phys.,* **47,** 5405–5408 (1976).
10. A. Sasaki, Y. Takeda, N. Shikagawa, and T. Takagi, "Liquid phase epitaxial growth, electron mobility and maximum drift velocity of $In_{1-x}Ga_xAs$ ($x \simeq 0.5$) for microwave devices," *Jpn. J. Appl. Phys.,* Suppl. **16-1,** 239–243 (1977).
11. T. P. Pearsall, ed., *GaInAsP Alloy Semiconductors,* Wiley, Chichester, West Sussex, England, 1982.
12. J. B. Gunn, "Microwave oscillations of current in III-V semiconductors," *Solid-State Commun.,* **1,** 88–91 (1963).
13. L. W. James, J. P. Van Dyke, F. Herman, and D. M. Chang, "Band structure and high-field transport properties of InP," *Phys. Rev. B,* **1,** 3998–4004 (1970).
14. Y. Suematsu, K. Iga, and K. Kishino, "Double-heterostructure lasers," in *GaInAsP Alloy Semiconductors,* ed. T. P. Pearsall, Wiley, Chichester, West Sussex, England, 1982, pp. 341–378.
15. H. C. Casey, Jr. and M. B. Panish, *Heterostructure Lasers,* Part B, *Materials and Operating Characteristics,* Academic Press, San Diego, Calif., 1978, pp. 277–313.
16. S. M. Sze, *Physics of Semiconductor Devices,* Wiley, New York, 1981, p. 291.
17. H. Hasegawa and H. Ohno, "Unified disorder induced gap state model for insulator–semiconductor and metal–semiconductor interfaces," *J. Vac. Sci. Technol. B,* **4,** 1130–1138 (1986).
18. W. D. Johnston, Jr., "Defect motion and growth of extended non-radiative defect structures in GaInAsP," in *GaInAsP Alloy Semiconductors,* ed. T. P. Pearsall, Wiley, Chichester, West Sussex, England, 1982, pp. 169–188.

19. T. P. Pearsall, "Electronic structure of $Ga_xIn_{1-x}As_yP_{1-y}$ alloys lattice-matched to InP," in *GaInAsP Alloy Semiconductors,* ed. T. P. Pearsall, Wiley, Chichester, West Sussex, England, 1982, pp. 295–312.
20. T. Hattanda and A. Takeda, "Direct measurement of internal strains in liquid phase epitaxial garnet film on gadolinium gallium garnet (111) plate," *Jpn. J. Appl. Phys.,* **12,** 1104–1105 (1973).
21. J. C. Mikkelsen, Jr. and J. B. Boyce, "Extended x-ray absorption fine-structure study of $Ga_{1-x}In_xAs$ random solid solutions," *Phys. Rev. B,* **28,** 7130–7140 (1983).
22. A. Gomyo, T. Suzuki, K. Kobayashi, S. Kawata, I. Hino, and T. Yuasa, "Evidence for the existence of an ordered state in $Ga_{0.5}In_{0.5}P$ grown by metalorganic vapor phase epitaxy and its relation to band-gap energy," *Appl. Phys. Lett.,* **50,** 673–675 (1987).
23. O. Ueda, M. Takikawa, J. Komeno, and I. Umebu, "Atomic structure of ordered InGaP crystals grown on (001) GaAs substrates by metalorganic chemical vapor deposition," *Jpn. J. Appl. Phys.,* **26,** L1824–L1827 (1987).
24. T. Suzuki, A. Gomyo, S. Iijima, K. Kobayashi, S. Kawata, I. Hino, and T. Yuasa, "Band-gap energy anomaly and sublattice ordering in GaInP and AlGaInP grown by metalorganic vapor phase epitaxy," *Jpn. J. Appl. Phys.,* **27,** 2098–2106 (1988).
25. O. Ueda, M. Takikawa, M. Takechi, J. Komeno, and I. Umebu, "Transmission electron microscopic observation of atomic structure of InGaP crystals grown on (001) GaAs substrates by metalorganic chemical vapor deposition," *J. Cryst. Growth,* **93,** 418–425 (1988).
26. A. Zunger, and S. Mahajan, "Atomic ordering and phase separation in epitaxial III-V alloys," in *Handbook on Semiconductors,* ed. T. S. Moss, Vol. 3b, *Materials, Properties, and Preparation,* vol. ed. S. Mahajan, Elsevier Science Publishers, Amsterdam, 1994, pp. 1399–1514.
27. S. Adachi, "Refractive indices of III-V compounds: key properties of InGaAsP relevant to device design," *J. Appl. Phys.,* **53,** 5863–5869 (1982).
28. R. K. Willardson and A. C. Beer, eds., *Semiconductors and Semimetals,* Vol. 22, *Lightwave Communication Technology,* Part D, *Photodetectors,* vol. ed. W. T. Tsang, Academic Press, San Siego, Calif., 1985.
29. R. K. Willardson and A. C. Beer, eds., *Semiconductors and Semimetals,* Vol. 22, *Lightwave Communication Technology,* Part E, *Integrated Optoelectronics,* vol. ed. W. T. Tsang, Academic Press, San Diego, Calif., 1985.
30. K. Heime, *InGaAs Field-Effect Transistors,* Wiley, New York, 1989.
31. A. Katz, ed., *Indium Phosphide and Related Materials: Processing, Technology, and Devices,* Artech House, Norwood, Mass., 1992.
32. O. Wada, ed., *Optoelectronic Integration: Physics, Technology, and Applications,* Kluwer Academic Publishers, Boston, 1994.
33. C. M. Wolfe, G. E. Stillman, and W. T. Lindley, "Electron mobility in high-purity GaAs," *J. Appl. Phys.,* **41,** 3088–3091 (1970).
34. D. L. Rode, "Low-field electron transport," in *Semiconductors and Semimetals,* ed. R. K. Willardson and A. C. Beer, Vol. 10, *Transport Phenomena,* Academic Press, San Diego, Calif., 1975, pp. 1–89.
35. Y. Takeda, "Low-field transport calculations," in *GaInAsP Alloy Semiconductors,* ed. T. P. Pearsall, Wiley, Chichester, West Sussex, England, 1982, pp. 213–241.
36. R. K. Willardson, A. C. Beer, and E. R. Weber, eds., *Semiconductors and Semimetals,*

Vol. 41, *High Speed Heterostructure Devices,* vol. eds. R. A. Kiel and T. C. L. Gerhard Sollner, Academic Press, San Diego, Calif., 1994.

37. R. People and S. A. Jackson, "Structurally induced states from strain and confinement," in *Semiconductors and Semimetals,* ed. R. K. Willardson and A. C. Beer, Vol. 32, *Strained-Layer Superlattices: Physics,* vol. ed. T. P. Pearsall, Academic Press, San Diego, Calif., 1990, pp. 119–174.

38. J. Singh, "Strain-induced bandstructure modifications in semiconductor heterostructures: consequences for optical properties and optoelectronic devices," in *Handbook on Semiconductors,* ed. T. S. Moss, Vol. 2, *Optical Properties of Semiconductors,* vol. ed. M. Balkanski, Elsevier Science Publishers, Amsterdam, 1994, pp. 235–284.

39. B. D. McCombe and A. Petrou, "Optical properties of semiconductor quantum wells and superlattices," in *Handbook on Semiconductors,* ed. T. S. Moss, Vol. 2, *Optical Properties of Semiconductors,* vol. ed. M. Balkanski, Elsevier Science Publishers, Amsterdam, 1994, pp. 285–384.

40. H. Ohno and J. Barnard, "Field-effect transistors," in *GaInAsP Alloy Semiconductors,* ed. T. P. Pearsall, Wiley, Chichester, West Sussex, England, 1982, pp. 437–455.

41. M. Kamada, "MOVPE-grown GaInAs, AlInAs and selectively doped AlInAs/GaInAs heterostructures: a review," *Curr. Topics Cryst. Growth Res.,* **1,** 99–119 (1994).

42. M. F. Gratton and J. C. Woolley, "Investigation of two- and three-phase fields in the Ga–As–Sb system," *J. Electrochem. Soc.,* **127,** 55–62 (1980).

43. K. Onabe, "Unstable regions in III-V quaternary solid solutions composition plane calculated with strictly regular solution approximation," *Jpn. J. Appl. Phys.,* **21,** L323–L325 (1982).

44. H. Sakaki, L. L. Chang, R. Ludeke, Chin-An Chang, G. A. Sai-Halasz, and L. Esaki, "$In_{1-x}Ga_xAs$–$GaSb_{1-y}As_y$ heterojunctions by molecular beam epitaxy," *Appl. Phys. Lett.,* **31,** 211–213 (1977).

45. M. J. Cherng, G. B. Stringfellow, and R. M. Cohen, "Organometallic vapor phase epitaxial growth of $GaAs_{0.5}Sb_{0.5}$," *Apl. Phys. Lett.,* **44,** 677–679 (1984).

46. Y. Nakata, T. Fujii, A. Sandhu, Y. Sugiyama, and E. Miyauchi, "Growth and characterization of $GaAs_{0.5}Sb_{0.5}$ lattice-matched to InP by molecular beam epitaxy," *J. Cryst. Growth,* **91,** 655–658 (1988).

47. H. Kamei and H. Hayashi, "OMVPE growth of GaInAs/InP and GaInAs/GaInAsP quantum wells," *J. Cryst. Growth,* **107,** 567–572 (1991).

48. M. Tabuchi, N. Yamada, K. Fujibayashi, Y. Takeda, and H. Kamei, "Group-V atoms exchange due to exposure of InP surface to AsH_3 (+ PH_3) revealed by x-ray CTR scattering," *J. Electron. Mater.,* **25,** 671–675 (1996).

49. Y. Kawamura, H. Kobayashi, and H. Iwamura, "InGaAlAs/InP type II multiple quantum well structures grown by gas source molecular beam epitaxy," in *Gallium Arsenide and Related Compounds 1993,* ed. H. S. Rupprecht and G. Weimann, IOP Publishing, Bristol, Gloucestershire, England, 1994, pp. 391–396.

50. G. A. Antypas, R. L. Moon, L. W. James, J. Edgecumbe, and R. L. Bell, "III-V quaternary alloys," in *Gallium Arsenide and Related Compounds 1972,* Institute of Physics, London, 1973, pp. 48–54.

51. F. Pollak, "Modulation spectroscopy of semiconductors and semiconductor microstructures," in *Handbook on Semiconductors,* ed. T. S. Moss, Vol. 2, *Optical Properties of Semiconductors,* vol. ed. M. Balkanski, Elsevier Science Publishers, Amsterdam, 1994, pp. 527–635.

52. A. Mooradian and H. Y. Fan, "Recombination emission in InSb," *Phys. Rev.,* **148,** 873–885 (1966).
53. K. Nakajima, A. Yamaguchi, K. Akita, and T. Kotani, "Composition dependence of the band gaps of $In_{1-x}Ga_xAs_{1-y}P_y$ quaternary solids lattice matched on InP substrates," *J. Appl. Phys.,* **49,** 5944–5950 (1978).
54. J. I. Pankove, *Optical Processes in Semiconductors,* Prentice-Hall, Upper Saddle River, N.J., 1971, pp. 130–131.
55. J. Matsui, K. Onabe, T. Kamejima, and I. Hayashi, "Lattice mismatch study of LPE-grown InGaPAs on (001)-InP using x-ray double-crystal diffraction," *J. Electrochem. Soc.,* **126,** 664–667 (1979).
56. K. Takahei, H. Nagai, and S. Kondo, "Immiscible domain in low-temperature LPE of $In_{1-x}Ga_xAs_{1-y}P_y$ on InP," in *Gallium Arsenide and Related Compounds 1981,* ed. T. Sugano, Institute of Physics, London, 1982, pp. 53–58.
57. M. Neuberger, "III-V semiconducting compounds," in *Handbook of Electronic Materials,* Vol. 2, IFI/Plenum, New York, 1971.

REFERENCE BOOKS FOR FURTHER STUDY ON InP-BASED SEMICONDUCTORS

1. T. P. Pearsall, ed., *GaInAsP Alloy Semiconductors,* Wiley, Chichester, West Sussex, England, 1982.
2. R. K. Willardson and A. C. Beer, eds., *Semiconductors and Semimetals,* Vol. 22, *Lightwave Communication Technology,* vol. ed. W. T. Tsang, Academic Press, San Diego, Calif., 1985 (Part A, *Material Growth Technologies*; Part B, *Semiconductor Injection Lasers I*; Part C, *Semiconductor Injection Lasers II: Light-Emitting Diodes*; Part D, *Photodetectors*; Part E, *Integrated Optoelectronics*).
3. K. Heime, *InGaAs Field-Effect Transistors,* Wiley, New York, 1989.
4. A. Katz, ed., *Indium Phosphide and Related Materials: Processing, Technology, and Devices,* Artech House, Norwood, Mass., 1992.
5. O. Wada, ed., *Optoelectronic Integration: Physics, Technology, and Applications,* Kluwer Academic Publishers, Boston, 1994.
6. R. K. Willardson, A. C. Beer, and E. R. Weber, eds., *Semiconductors and Semimetals,* Vol. 40, *Epitaxial Microstructures,* vol. ed. A. C. Gossard, Academic Press, San Diego, Calif., 1994.
7. R. K. Willardson, A. C. Beer, and E. R. Weber, eds., *Semiconductors and Semimetals,* Vol. 41, *High Speed Heterostructure Devices,* vol. ed. R. A. Kiel and T. C. L. Gerhard Sollner, Academic Press, San Diego, Calif., 1994.

CHAPTER FIVE

InP Bulk Crystal Growth and Characterization

DAVID FRANCIS BLISS
U.S. Air Force Research Laboratory

5.1	Introduction	110
5.2	Crystal Growth Techniques	111
	5.2.1 Historical Development	111
	5.2.2 Liquid-Encapsulated Czochralski and Vertical Gradient Freeze Crystal Growth Methods	112
	5.2.3 Issues for Crystal Growth	113
5.3	InP Synthesis	116
	5.3.1 Polycrystal Synthesis	116
	5.3.2 Direct Synthesis: Injection Method	118
5.4.	Crystal Growth Methods	120
	5.4.1 Crystal Pulling from the Melt	121
	5.4.2 Container Growth	130
	5.4.3 New Directions in Crystal Growth	134
5.5.	Characterization of Bulk InP	137
	5.5.1 Electrical Properties of Bulk InP	137
	5.5.2 Wafer Mapping Techniques	146
5.6.	Summary and Future Challenges	156
	Acknowledgments	156
	References	157

InP-Based Materials and Devices: Physics and Technology, Edited by Osamu Wada and Hideki Hasegawa.
ISBN 0-471-18191-9   © 1999 John Wiley & Sons, Inc.

5.1 INTRODUCTION

Indium phosphide is a III-V compound semiconductor crystallizing in the sphalerite structure, which has a fortuitous lattice match to alloys with bandgaps coinciding with the 1.3- and 1.55-μm wavelength windows in optical fiber. The revolution in optical fiber communications has swept InP into a dominant position in optoelectronics. For lattice-matched growth of ternary alloys, GaInAs and AlInAs, and quaternary alloys, GaInAsP and AlGaInAs, InP is the substrate of choice. Heterostructure devices based on these alloys, by virtue of their bandgaps, provide a strong driving force for bulk InP crystal growth development.

During the past 25 years, as the growth of InP single crystals has gone from a laboratory curiosity to a commercial process, many new applications for InP substrates have emerged. The mainstay of demand continues to be in the field of telecommunications, but other uses for InP material involving high-speed electronic and photonic circuits have arrived. In addition to high-frequency wireless communications [1], broadband gigahertz radar has been achieved using InP photoconducting antennas [2]; recent extension to terahertz waves has been implemented [3] using femtosecond laser pulses. In addition, some workers have exploited the unique optical behavior of InP for photorefractive waveguides [4] and infrared beam confinement [5].

The promise of such a material, which operates at high frequencies and emits light at infrared wavelengths, is of course based on the development of high-quality bulk crystals, with uniform properties and a homogeneous distribution of defects. Since the control of defects is what gives semiconductors their useful electrical and optical properties, this chapter covers the important technological aspects of bulk InP crystal growth, defect control, and characterization methods, emphasizing the evaluation of materials properties by determining defect concentrations and mapping their uniformity.

In this chapter we survey the current state of the art in bulk InP crystal growth and characterization, with reference to the history of its development. Section 5.2 provides an overview of the various processes and how they were developed. Sections 5.3 and 5.4 deal with the state-of-the-art synthesis and crystal growth methods. Section 5.4 covers the characterization of defects by various means, for the ultimate purpose of controlling the electrical and optical properties of InP wafer substrates. Two general classes of InP material are discussed, semiconducting and semi-insulating InP. The most recent advances in development are found in the discussion of semi-insulating material, where control of the iron compensation mechanism is an emerging issue for the manufacturing environment. The final summary raises the perspective of new developments in wafer quality, dimensional expansion, and analytical methods for defect control, which are the challenges for the future.

5.2 CRYSTAL GROWTH TECHNIQUES

5.2.1 Historical Development

Historically, the development of InP material has been driven by the commercial telecommunications market for laser structures used in fiber-optic communications. Because of the small size of these discrete devices, the demand for larger-diameter wafers has been slow to gain momentum. At present, 2- and 3-in.-diameter InP wafers are available from commercial sources, and the major barrier to the production of larger sizes is the lack of a market for 4-in. wafers. The reader may note, however, that the extension of InP material to new applications has accompanied the advancement of crystal growth to larger diameters, and one may conclude that scale-up to still-larger-diameter wafers is imminent.

Liquid Encapsulation Before a successful technique was developed to grow InP single crystals, two principal problems had to be overcome: containment of a highly volatile melt and synthesis of a stoichiometric InP compound. Melt volatility is a common problem not only in III-V compounds but also in the II-VI group, as well as a range of others. Early work on crystal growth from volatile melts narrowed the possibilities to two alternatives. Either the crystal must be grown in a completely sealed hot-wall ampoule (Bridgman or gradient freeze method), or the crystal can be pulled using some kind of pressure balancing technique. For the latter case, the first breakthrough came in 1962, when Metz and a group at Westinghouse [6] developed the liquid-encapsulated Czochralski (LEC) method. Although they were growing ZnSe at that time, the LEC method was later developed for growth of InP using B_2O_3 as an encapsulant [7–9].

Synthesis The problem of how to synthesize a stoichiometric compound of InP can also be solved in two ways: either by synthesis in a hot-wall sealed ampoule with a horizontal boat, or by direct injection of phosphorus vapor into an encapsulated indium melt. The boat method was the first to be developed and is still in commercial production. A group working at Battelle Institute in the 1950s [10] developed a horizontal boat apparatus operating at atmospheric pressure. At 1 atm of phosphorus pressure, the indium melt contains only about 30% atomic phosphorus, but with sufficiently steep temperature gradients and slow freezing rates, very nearly stoichiometric polycrystalline ingots can be frozen from such liquids. Improvements in design have occurred, so that fairly rapid high-pressure synthesis in the horizontal Bridgman configuration now results in large-volume high-purity InP polycrystal [11]. The primary disadvantage of this method is the long preparation time and associated high capital cost. A potentially less expensive method that can be carried out in situ in the crystal growth apparatus was developed by Fischer and Pruss [12] at RCA in the early 1970s. The injection process is shown schematically in Fig. 5.1, a sketch from a 1974 U.K. patent specification.

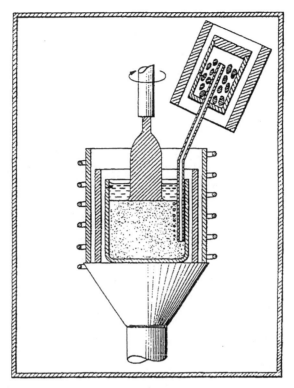

FIGURE 5.1 Schematic of apparatus for in situ injection technique for synthesis of InP. (From the 1974 U.K. patent specification 1,353,917 from Fischer at RCA.)

5.2.2 Liquid-Encapsulated Czochralski and Vertical Gradient Freeze Crystal Growth Methods

The state of the art today for InP crystal growth is divided between two competing technologies: top-seeded crystal pulling with liquid encapsulation [e.g., LEC, pressure-controlled PCLEC, vapor-pressure-controlled Czochralski (VCZ) or magnetically stabilized MLEC, etc.] and vertical growth in a container [VGF, or vertical Bridgman (VB)] with bottom seeding. These two general categories have many variants, stemming from unique laboratory research efforts. The pulling method has generally been the most cost-effective, but its disadvantage is the high dislocation density, caused by high levels of strain during growth. On the other hand, vertical container growth offers a very low dislocation density because of its low-stress environment, but it is plagued by yield problems due to twinning and interface breakdown in heavily doped crystals. The various adaptations of each technique outlined in Section 5.3 are engineering solutions designed to overcome these deficiencies.

5.2.3 Issues for Crystal Growth

Twinning Twins and twin lamella are yield-limiting defects in InP crystal growth. A twin represents two regions in a crystal which are mirror reflections of each other across a lattice plane that is common to both regions. A volume of the crystal bounded by two parallel twin planes separated by only a few lattice spacings is called a *twin lamella* [13], which for sphalerite structures occurs on the {111} plane. The {111} planes in the sphalerite structure have a stacking sequence of ABCABC.... The stacking sequence for a twin where the C plane is the twin plane is ABCBA. The formation of a twin is inherently tied to the stacking fault [14]. A stacking fault is a flaw in the stacking sequence of atomic planes in the lattice. For example, if we examine an intrinsic stacking fault where one plane A is missing, we get

which is the same as two twinning operations with C and B as twin planes. Since there is such a close connection between stacking faults and twins, the probability of twinning is large when the stacking fault energy is low.

The energy of stacking faults, γ, as shown in Fig. 5.2 for InP, is low compared with GaAs and silicon. This may explain why the formation of twins during crystal growth is much more probable in InP than in other III-V compound semiconductors. The problem is worsened by the conical growth angle [16] of the typical LEC

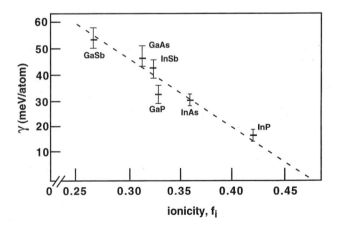

FIGURE 5.2 Energy of stacking faults as a function of ionicity. (From Ref. 15.)

ingot, growth from indium-rich melts [11], or low axial temperature gradients [18]. Instabilities in the melt–crystal interface shape may also contribute to twinning. These may result from thermal fluctuations, mechanical vibrations, or changes in surface tension. The viscosity of the encapsulant is also a factor. B_2O_3 with low moisture content is used as a means of raising the encapsulant's viscosity. More recent analysis by Hurle [17] suggests that undercooling near a facet can create a situation where a twinned nucleus is thermodynamically favored. The prediction is that for a range of conical growth angles the liklihood of twin formation is high; for (100) InP crystal growth, this twinning range is from 0.5° (very narrow) to 47°, while for wider angles (flatter crowns) the probability of twinning is lower. Various approaches are used to reduce the incidence of twins and twin lamella, as discussed in Section 5.3.

Dislocation Density A high density of dislocations is indicative of thermoelastic stress during crystal growth. It is known that dislocations degrade the performance of certain photodetectors [19] and may have a detrimental effect on the properties of other InP-based devices [20]. Dislocations may alter the mechanical properties of InP and contribute to point defect and impurity migration [21]. The preferred substrates for photodetectors are low-dislocation S-doped wafers. On the other hand, studies of epitaxial growth on InP have shown that dislocations propagating from InP substrates do not cause "dark line" defects in laser diodes as they do in GaAs. Most of the InP-based laser diodes have been built on substrates with dislocation densities in excess of 10^4 cm^{-2} unlike GaAs-based lasers, which typically require nearly dislocation-free Bridgman or gradient-freeze substrates. Commercial production of InP-based devices continues to expand despite the high dislocation densities of these substrates. However, the demand for reduced dislocation density is increasing, because structural defects may limit the performance of advanced monolithic microwave integrated circuits (MMICs) and optoelectronic integrated circuits (OEICs).

High dislocation density is a problem associated with pulled crystals but not with VGF or VB crystals. Densities of less than 500 cm^{-2} are generally reported for VGF crystals, less than can be counted using an optical microscope. Thermal gradients at the solid–melt interface are on the order of 10 to 30°C cm^{-1} in the VGF process, very low compared to typical LEC growth at 130°C cm^{-1} or higher. For LEC growth, steep temperature gradients provide stability to the crystal growth environment. Without such steep gradients, diameter control is more difficult, twinning probability is increased, and surface decomposition occurs as the crystal grows out of the encapsulant.

Despite conditions of very steep thermal gradients, silicon crystals are nevertheless grown dislocation-free because the use of Dash [22] seeding eliminates dislocations emanating from the seed. A similar technique [23] attempted for InP growth has demonstrated that dislocation-free InP can be grown, but only up to 15 mm in diameter. For larger crystals, dislocations are generated at the periphery of the boule, due to tensile hoop stresses, and at the center, due to compressive stresses.

Reduction of thermal stresses to reduce the density of dislocations in InP has been the goal of considerable effort during the past decade. The primary effort involves a major redesign of the growth apparatus to reduce the axial temperature gradient and associated thermal fluctuations. This is discussed in Section 5.4.

Doping Uniformity All of the commercial uses of InP material require impurity doping in the crystal. Unlike GaAs semi-insulating bulk material, where electronic properties are controlled by the intrinsic deep level EL2, semi-insulating undoped InP material has not been produced. Iron, the midgap acceptor, is the single impurity species that can compensate the residual shallow donors of InP to produce semi-insulating material with resistivity at least 10^7 Ω·cm. The use of impurity doping requires control of uniformity on both the macroscale and microscale. Dopant incorporation into the growing crystal is a function of the distribution coefficient, k, defined as $k = C_s/C_l$, where C_s and C_l are the concentrations of impurities in the solid and melt, respectively. The dopant concentration along the growing crystal is distributed or "segregated" for the case of complete mixing according to the normal freezing equation [24]

$$C = C_0 k (1-g)^{k-1} \qquad (1)$$

where C_0 is the initial composition of the melt and g is the fraction of the melt solidified. For dopant impurities, most of which have distribution coefficients $k < 1$, the concentration will increase exponentially from seed to tail. Macrosegregation of impurities can lead to yield problems; if maximum solubility is reached, precipitate formation, or interface breakdown due to constitutional supercooling will occur. Fortunately, the dopant distribution can be modified somewhat by controlling the growth conditions, such as altering the degree of mixing at the crystal–melt interface. Process controls can be used to increase the yield of usable material per ingot. Some of these are discussed in Section 5.4.

On the microscale, doping nonuniformities show up as striations. The striae follow the shape of the crystal–melt interface, which is generally not perfectly planar. The explanation for these striations is that dopant fluctuations due to random variations in the microscopic growth rate cause local strain from lattice distortion. That is, rapid changes of the dopant concentration in the melt boundary layer adjacent to the growing crystal cause rapid changes in crystal doping concentration, which in turn appear as striations when viewed by infrared (IR) absorption or preferential-etch microscopy. Cross-sectional images by IR transmission or x-ray topography and Nomarski images of etched surfaces show clearly the shape and frequency of striations. Because of the curvature of the crystal–melt interface, striations manifest themselves as nearly circular nonuniformities on the wafer surface. Striations are more pronounced in crystals with higher doping concentrations, and near the upper limit of solubility precipitates may nucleate on striae. Some laboratories have instituted engineering controls to reduce the magnitude of striations, such as the use of thermal baffles or the application of an axial magnetic field during crystal growth.

5.3 InP SYNTHESIS

Efficient and rapid synthesis of InP is a requirement for large-scale low-cost commercialization of high-quality InP wafers. However, the vapor pressure of phosphorus is too high to allow in situ mixing of red phosphorus and indium in the crucible. Red phosphorus has a vapor pressure (shown in Fig. 5.3) of well over 100 atm at 1000°C [25], whereas the dissociation pressure over InP at the melting point of 1062°C is much lower (27 atm). For this reason, most synthesis techniques involve transporting phosphorus vapor into an indium melt utilizing a multizone furnace. A high-pressure synthesis furnace is designed so that solid phosphorus can be held at about 550°C in equilibrium with an InP melt at about 1100°C. The stoichiometry of the InP charge can then be controlled by changing the phosphorus temperature. Synthesis is performed either in situ inside the crystal puller or in another high-pressure chamber, in which case the polycrystal charge is removed and shaped to fit the growth crucible.

5.3.1 Polycrystal Synthesis

For industrial production, the horizontal Bridgman (HB) technique is presently preferred for polycrystal synthesis [11]. A typical HB process is depicted in Fig. 5.4. Generally, a sealed silica ampoule is employed inside a high-pressure chamber with three heater zones: a narrow indium melt zone, a synthesis zone, and a phosphorus vapor control zone. The HB method provides a high degree of flexibility for control of polycrystal purity and stoichiometry, but because of the time required for synthe-

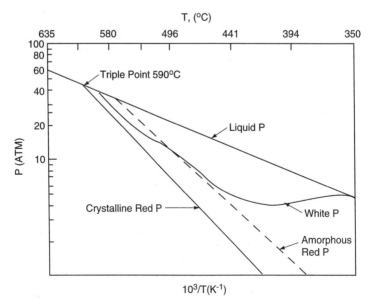

FIGURE 5.3 Vapor pressure of phosphorus versus reciprocal temperature. (From Ref. 128.)

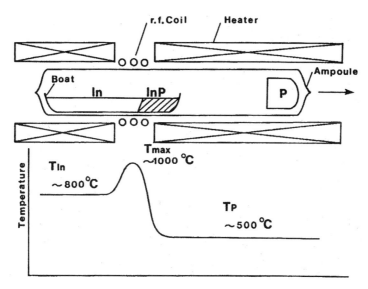

FIGURE 5.4 Horizontal Bridgman furnace for the synthesis of InP polycrystals. Here T_{in} is the temperature of the indium melt zone, T_{max} that of the synthesis zone, and T_P that of the phosphorus vapor control zone. (From Ref. 11.)

sis and capital expenditure, it has a large production cost. The flexibility of the HB system stems from the many variables that can be controlled. Three temperature zones and the ampoule transfer rate must be carefully optimized to achieve the following desired polycrystal properties: (1) no indium inclusions or intergranular precipitates, (2) low residual impurity concentrations as determined by chemical analysis, and (3) electron transport measurements indicating low free carrier concentration (less than 1×10^{15} cm^{-3}) and high 77 K mobility (greater than 40,000 cm^2/V·s).

To reduce contamination from silicon by the reaction between indium vapor and the silica fixtures, the synthesis zone temperature must be maintained as low as possible given the indium-rich solution, well below the congruent melting temperature of InP. A lower temperature is better as far as impurity contamination is concerned, since the reaction between SiO_2 and indium vapor is thermally driven. However, a low synthesis temperature reduces the phosphorus content of the melt, and consequently the polycrystal growth rate must be slowed down to avoid indium inclusions. During polycrystal synthesis, the indium melt is saturated with phosphorus, and as it cools down, InP precipitates out of the supersaturated solution at the liquidus line. Unless the ampoule translation rate is slower than the precipitation rate, the growth front is likely to break down and indium inclusions or intergranular precipitates are formed. These defects, easily seen by optical microscopy, obviously result in off-stoichiometric indium-rich melts during the crystal growth step. Indium inclusions can be reduced by raising the synthesis temperature and phosphorus pressure, by steepening the temperature gradient at the solidification front, or by reducing the ampoule translation rate. For example, with a synthesis temperature of 1000°C as shown in Fig. 5.5, the maximum growth rate would be about 4 mm/h without indium inclusions. At this rate, synthesis time for a 4-kg InP polycrystal charge would be about 100 h.

The phosphorus zone temperature can be raised to increase the phosphorus content of the melt, but raising the phosphorus pressure requires balancing the chamber pressure as the phosphorus temperature approaches the triple point. The vapor pressures of the various condensed phases of phosphorus vary considerably, and the sublimation rates are quite different, so it is not possible to rely upon phosphorus vapor pressure data as a function of temperature. For this reason, polycrystal synthesis is performed with a phosphorus temperature of about 500°C, corresponding to a pressure of about 20 atm using only the red phase.

5.3.2 Direct Synthesis: Injection Method

Direct injection of phosphorus can be accomplished in a high-pressure crystal growth furnace prior to LEC growth. The main advantage of direct synthesis over HB polycrystal synthesis is the lower production cost. The time required to synthesize a 2-kg melt is only about 1 h. Crystal purity is also a benefit of this technique, since pure materials are not contaminated by removal from the furnace after synthesis. After the experimental work of Fischer and Pruss [12] at RCA in the early 1970s the concept of direct synthesis was reduced to industrial practice by Antypas and Hyder [26,27], who reported a high-purity in situ process using a silica ampoule injector, where crystal growth was initiated immediately after injection. They also reported slightly higher impurity levels for the two-step process, where a presynthesized polycrystal charge was used to grow a single crystal [28]. Other variations of the injection process have also been proposed [29], but the question of reproducibility has lingered because of the difficulty of controlling the volatility of such a rapid reaction.

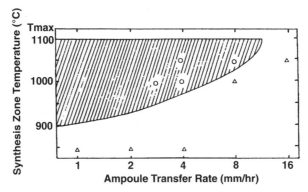

FIGURE 5.5 Occurrence of indium inclusions as a function of polycrystal synthesis conditions. For polycrystals synthesized in the hatched region, indium inclusions are rarely observed. (From Ref. 11.)

A schematic of a state-of-the-art phosphorus injection system [30] is shown in Fig. 5.6. Here the design of the phosphorus-containing ampoule incorporates an annular shape to allow independent motion of the seed and injector. Phosphorus is injected into the encapsulated indium melt through a transfer tube. The injector is placed near the crucible to heat the phosphorus by radiation and convection. Red phosphorus is transformed to a vapor, and the increased pressure forces it down the transfer tube along with the inert gas N_2. A physical model [31] of the injection process shows that the role of the inert gas is critical in successful synthesis. The mass flow rates of P_4 and N_2 gases during synthesis were compared with the phosphorus content of the melt as determined from Fig. 5.7. It was found that the P_4 pressure builds up slowly toward its thermal equilibrium value in the injector, and the melt absorbs phosphorus very rapidly, so the actual pressure inside the injector is significantly lower than the equilibrium pressure of phosphorus. This creates a built-in safety margin to maintain a steady flow of P_4 gas despite thermal perturbations.

The in situ synthesis and crystal growth process allows growth of InP from a P-rich melt. By comparison, growth from presynthesized polycrystal charges, even stoichiometric ones, often results in In-rich melt formation, because of P vapor loss during heating. Analysis of GaAs crystals grown from As-rich melts [32] has shown

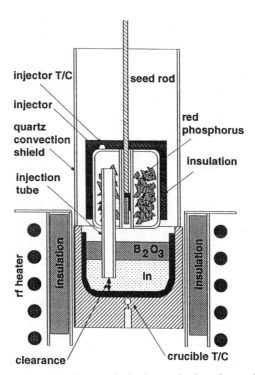

FIGURE 5.6 Schematic of system for in situ synthesis and crystal growth.

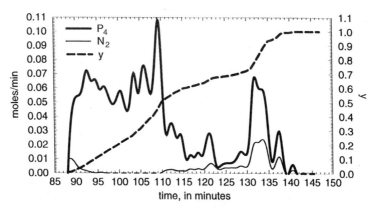

FIGURE 5.7 Mole fraction of phosphorus (dashed line, y) during injection and the injection rates of P_4 and N_2 in moles per minute, as a function of time, calculated using real-time injector temperature data. (From Ref. 31.)

that an existence region is found where the solid can be As rich because of gallium vacancies. The high concentrations of arsenic antisite defects is thought to account for the semi-insulating property of undoped GaAs [33]. By analogy with GaAs, an excess group V component in the crystal may account for deep-level traps and semi-insulating electrical properties. Analysis of InP crystals grown from P-rich melts [34] showed the presence of a P-related defect as an electron trap with an activation energy of 0.4 eV, but the electrical transport was still dominated by shallow donors.

5.4 CRYSTAL GROWTH METHODS

The two basic methods of InP crystal growth, LEC and VGF, are distinguished by opposite melt convection behavior. The two methods are illustrated schematically in Figs. 5.8 and 5.9. Top-seeded LEC melts are thermally unstable, characterized by turbulent or oscillatory flows, whereas bottom-seeded VGF is a thermally stable configuration, the melt Grashof number being several orders of magnitude lower. Optimization of each technique has led to a number of variants, but almost all have one thing in common: The reduction and control of thermal gradients and transient temperature fluctuations is the key to improved crystal quality. Crystal growth under such low-gradient conditions requires utmost control of the growth environment: reduction of turbulent fluid flows and elimination of mechanical perturbations. A variety of approaches have been engineered to accomplish the goal of process control, each of which adds a level of sophistication to the crystal growth hardware. All of these approaches are aimed at twin-free growth of crystals; different approaches may be combined to optimize the growth conditions. Thermal baffles, vapor pressure control, optimized rotation conditions, stoichiometry control, magnetic field stabilization, and crystal shape control are all active areas of research and development in crystal growth laboratories.

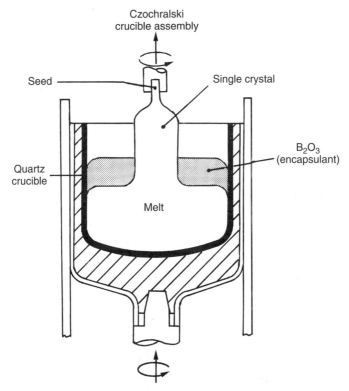

FIGURE 5.8 Schematic of LEC crystal growth of InP.

5.4.1 Crystal Pulling from the Melt

Liquid-encapsulated crystal pulling methods have historically been the most cost effective means of producing high-quality InP single crystals. Commercial producers have been supplying 2-in. InP wafers since the early 1980s, and 3-in. wafers since the early 1990s. Improvements in process control have made available higher-purity lower-defect density wafers for advanced device processing. However, the trade-off to achieve these goals is in the increasingly complex hardware setup and higher cost factor for manufacturing. In the quest to reduce thermal gradients and control temperature fluctuations, hardware and software must be added to suppress convection in a normally chaotic fluid system. The low gradient raises an additional demand for vapor pressure control, since thermal decomposition of the growing crystal above the encapsulant becomes a yield-limiting factor.

Vapor-Controlled LEC Vapor-pressure-controlled Czochralski (VCZ) [35] and pressure-controlled LEC (PC-LEC) [36] are two processes designed to overcome the problem of crystal growth in a low gradient. Both of these methods use an encapsulant with a hot-wall containment vessel which is sealed to maintain a partial

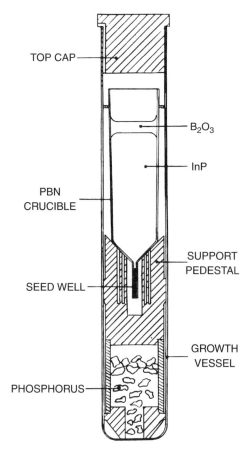

FIGURE 5.9 Schematic of VGF crystal growth of InP. (From Ref. 68.)

pressure of phosphorus over the encapsulated melt sufficient to prevent crystal decomposition. The axial temperature gradient in the B_2O_3 is reduced from about 150°C cm^{-1} to 30°C cm^{-1}, so that the temperature of the InP crystal just above the encapsulant is about 980°C. At such a high temperature, the equilibrium vapor pressure of phosphorus P_P over InP is about 1 atm. To prevent dissociation of InP to P_4 gas plus liquid indium, a vapor pressure of P_4 inside the hot wall container is created with a quantity of heated solid phosphorus. In the VCZ method [37] a liquid B_2O_3 seal is used to prevent leakage of phosphorus vapor from the hot-wall container, while for PC-LEC [36] a solid seal is used. Both methods have been used successfully to reduce the dislocation density of 3-in. Fe-doped wafers to the range of 2000 to 8000 cm^{-2} and S-doped wafers below 10^3 cm^{-2}.

Magnetic Stabilization Techniques It was suggested in the 1960s [38,39] that the application of an external magnetic field may be used to damp time-dependent turbulent convective flows in melts during the growth of semiconductors. At some research labs [40,41], this concept has been under development for the growth of InP for many years. In principle, two advantages result from magnetic field growth: (1) thermal fluctuations are reduced, stabilizing the microscopic growth rate; consequently (2) the size of the diffusion boundary layer increases, enhancing the uniformity of dopant distribution.

Convection is always present during crystal growth, and solute distribution is directly affected by the magnitude and direction of fluid flow. Fluid flows in electrically conducting liquids are restricted by the Lorentz force, **L**, when crossing the magnetic lines of flux:

$$\mathbf{L} = \mathbf{j} \times \mathbf{B} \tag{2}$$

where **j** is the ionic current density vector and **B** is the applied magnetic field. The net effect is an increase in the effective viscosity of the melt. The magnetic viscosity is related to the rate of fluid transport, thus affecting the momentum boundary layer thickness. The following analysis comes from Burton et al. [42]. A diffusion boundary layer, δ, is assumed, beyond which convection maintains a uniform melt composition and inside of which transport is by diffusion only. After steady state is reached, for an infinite liquid one can define an effective distribution coefficient, k_{eff}:

$$\frac{C_s}{C_l} = k_{\text{eff}} = \frac{k_0}{k_0 + (1 - k_0)e^{-(R\delta/D)}} \tag{3}$$

where R is the growth rate and D is the diffusion coefficient for the solute in the melt. This is a useful expression because it relates the composition of the growing crystal to convection conditions and the growth rate. It describes the dopant distribution provided that the thickness δ of the boundary layer is small compared to the extent of the crucible. For example, the effective distribution coefficient tends toward unity for any increase in the growth rate or boundary layer thickness. In the case of a strong enough applied magnetic field, the diffusion boundary layer thickness increases significantly, and hence k_{eff} is closer to 1. This is a benefit for achieving a more uniform incorporation of a dopant element.

Magnetic Czochralski growth has been modeled by several authors [43, 44], who have pointed to the significant enhancement of the effective segregation coefficient due to an axial magnetic field. On the other hand, for a cusped magnetic field, where there is only a radial component of magnetic field at the melt surface, only axial flows are damped. But the boundary layer thickness should still increase when an electric current is applied between the melt and crystal, improving the dopant uniformity in the crystal [43]. For crystal growth from nonstoichiometric melts, the situation depends on the compound being grown [45]. Because of the strong solutal convection in metallic solutions of III-V compounds, certain solutions cannot be magnetically stabilized. Striations indicating turbulent flows are present in the material system In–InP, whereas they are suppressed in the systems Ga–GaSb and

Ga–GaAs. The different behavior can be explained by the concentration gradients and the resulting solutal convection effects.

Crystal growth of InP in an applied magnetic field has been explored by a number of laboratories. Miyairi et al. [40] showed that with an axial magnetic field of 1000 G, thermal fluctuations in the melt are reduced from 9°C to less than 1°C, and random, time-dependent striations disappeared. The suppression of compositional variations in the melt was considered to be a reason for the disappearance of dislocation clusters in the material. Ozawa et al. [46] used a programmed reduction of applied magnetic field to achieve a homogeneity improvement in Fe-doped InP. The possibility of the use of axial magnetic fields as a practical method of growth control has great appeal.

Magnetically stabilized liquid-encapsulated Kyropoulos (MLEK) growth is a variation of LEC using magnetic stabilization of the melt to control the growth process. In MLEK, the crystal is grown initially without pulling the seed, creating a characteristic flat-top shape. After the crystal has reached full diameter, the seed is slowly raised to pull a cylindrical (100) crystal. A study comparing MLEK to standard LEC growth of InP showed that there were some important distinctions [47]. To investigate the effect of magnetic fields on interface shape and dopant uniformity, two tin-doped crystals were grown, one by LEC and the other by MLEK. Both were grown from melts weighing about 1 kg, having tin concentrations of 5×10^{19} cm^{-3}. The MLEK crystal was grown in an axial magnetic field of 2 kG, while the LEC crystal was grown without a magnetic field in the standard cone shape. After growth, these crystals were cross-sectioned as shown in Fig. 5.10 and sliced into samples representing each fraction of the grown crystal for glow discharge mass spectrometry (GDMS). The concentration ratio C_s/C_l taken from the first slice of each crystal was used to determine the effective distribution coefficient k_{eff} for each process. A comparison of the axial doping profiles for each crystal, as shown in Fig. 5.11, reveals a distribution coefficient for the MLEK process that is a factor of 2.5 larger than for LEC. For MLEK the measured tin concentration starts higher and re-

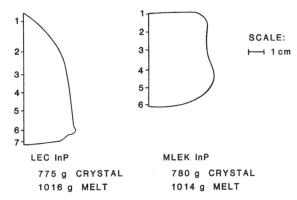

FIGURE 5.10 Cross section of tin-doped LEC and MLEK crystals of similar mass. (From Ref. 47.)

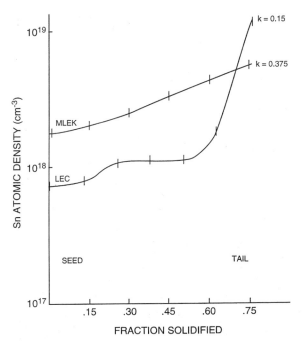

FIGURE 5.11 Axial doping profile of LEC and MLEK crystals as determined by GDMS. (From Ref. 47.)

mains higher until the tail, where the nonmagnetic LEC exceeds the MLEK. (A factor-of-2 error bar is expected for GDMS measurements, which may explain the bump in the LEC data.) The difference in k_{eff} is an indication that axial magnetic fields increase the boundary layer thickness δ at the melt–solid interface. Considering the BPS analysis in Eq. (3), one can see that k_{eff} is an exponential function of δ since the growth rate R is the same in both cases, and the value of diffusivity D_L is assumed to be independent of the applied magnetic field. Using this analysis, the boundary layer thickness is found to be 3.5 times larger for MLEK growth than for standard LEC.

Crystal Shaping Techniques Research at several laboratories has taken advantage of crystal shaping as a means to minimize the incidence of twinning. MLEK growth is one such technique [48], where the crystal is grown with a flat top before initiation of pulling, to reduce the incidence of twins. When pulling begins, the crystal has already reached full diameter, and the growth angle changes from perpendicular to parallel to the growth axis. At that moment the growth angle must pass through the critical range, but facet formation is reduced because thermal gradients near the periphery are steeper than at the center. As the flat top shape cannot readily be achieved in the presence of convective melt instabilities, this method relies on magnetic stabilization of the melt to suppress turbulent flows at the solid–liquid interface. Named after Spyro Kyropoulos [49], this method commemorates the pioneering work of this German scientist, who devised the technique to avoid the cracking problems arising from container growth of large alkali halide crystals.

Another flat-top shaping technique is described as controlled supercooling [50], a variation of the LEC method, used to restrain the generation of twin boundaries from the crystal shoulder. The authors refer to supercooling the melt between 7 and 10°C to obtain a square or "isotropic" facet formation around the seed. Lateral growth proceeds until the full 2-in. diameter is reached, and then the crystal is pulled as a cylinder at a rate of 10 to 15 mm/h. This method does not employ a magnetic field for melt stabilization; still, reproducible single crystals have been grown with low etch-pit densities (< 2000 cm^{-2}) for Zn-doped InP (3×10^{18} cm^{-3}).

A similar principle has been applied to vertical Bridgman growth using a flat-bottomed crucible [51]. Twin-free (100) crystals were grown using a seed that has the same diameter as the crystal. Dislocation densities in the undoped crystals were less than 10^4 cm^{-2}, an order of magnitude less than in the seed, and S-doped crystals were dislocation-free in the center 20-mm-diameter core.

Facet formation on the periphery of bulk InP crystals is indicative of a low radial temperature gradient. A close correlation has been found [17,52] between the formation of edge facets and the generation of twins. At the other extreme, Hirano et al. [53,54] have reported single crystals that appear almost square in shape because of the large edge facets. These S-doped crystals exhibit low dislocation densities (less than 500 cm^{-2} over 90% of the wafer, and in fact, large facet formation may be characteristic of low-dislocation-density material. Where facets form in doped crystals, however, a problem arises due to dopant segregation behavior on and off the facet. Since impurity incorporation within the facet is much higher than off-facet, a nonuniform dopant distribution limits the effective diameter of the crystal. The depth of the facet can be controlled by increasing the radial temperature gradient, but at the cost of increased thermal stress and consequent higher dislocation density. A better result is obtained by reducing the crystal pulling velocity. Hirano reported that a 20% reduction in growth rate was effective in reducing the facet depth.

Modeling High-Pressure LEC Numerical simulation of LEC growth first centered on the modeling of GaAs crystal growth. Because LEC growth of GaAs takes place at a moderate pressure, most of these models have either neglected the gas convection or have assumed a fixed gas temperature and used an arbitrary value of heat transfer coefficient to calculate the energy loss from the encapsulant surface. Modeling by Jordan et al. [55] has contributed an estimate of physical and thermal parameters for InP, as well as a thermoelastic strain model based on the concept of a critical resolved shear stress (CRSS). Völkl and Müller [56] developed a more sophisticated model for numerical simulation of dislocation formation in InP crystals. Their model takes into account the thermally activated nature of the plastic deformation process (i.e., the generation and multiplication of dislocations).

In the case of InP growth, the chamber pressure is very high, much above 2.83 MPa, and the flow and heat transfer interaction among the recirculating gas, the encapsulant surface, and the crystal is extremely complex. Indeed, the gas flow in a InP growth furnace is turbulent and oscillatory [57]. Recently, Zhang et al. [58,59] modeled the high-pressure (HP) LEC growth of InP. They developed a high-resolution computer model based on a body-fitted coordinate system, multizone adaptive grid generation, and curvilinear finite volume discretization. Using this model they showed that the HPLEC growth conditions are greatly influenced by gas convection in the furnace. Their computations indicate that a multicellular flow pattern may exist in the gas which may either heat or cool the crystal, depending on the direction of gas flow (Fig. 5.12). This, in turn, can change the melt–crystal interface shape and modify the temperature distribution in the encapsulant and the crystal. The gas flow also enhances heat transfer from both the encapsulant surface and the crystal (when it emerges out of the B_2O_3 layer).

The heat transfer and thermal stress calculations of Zou et al. [60] demonstrate a strong correlation between the heat transfer and thermal stress generation in the crystal. The highest stress is introduced primarily by the distorted temperature field in the crystal, where a sudden change of the cooling condition takes place, especially as the crystal first emerges from the encapsulant. As the crystal grows, the effect of this sudden variation in the cooling effect from encapsulant to the gas is reduced

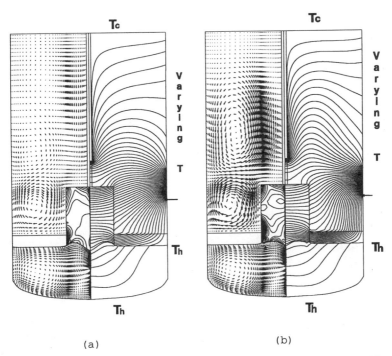

FIGURE 5.12 Distribution of temperature, stress, and stream function for HPLEC growth at various gas Grashof numbers: (a) $Gr_g = 10^7$; (b) $Gr_g = 10^8$. Melt Grashof number, $Gr_m = 10^6$; crystal and crucible rotation Reynolds numbers, $Re_s = -10^2$ and $Re_c = 10^2$. (From Ref. 63.)

and the isotherms in the crystal become flatter, leading to lower thermal stress. There are two stress maxima at the periphery of the crystal, as shown in Fig. 5.13, one just above the encapsulant layer and the other near the top of the crystal. The maximum stress spot near the gas–encapsulant interface is the most likely to cause dislocations because the CRSS in this region is low due to a high crystal temperature. In other areas, where the isotherms appear flat, very small stress is generated. The large thermal stress may be attributed to the radial temperature gradient in the crystal. Figure 13 indicates two other interesting phenomena. First, the locations where the maximum stresses appear do not change with the growth, and second, the magnitude of these maxima keeps decreasing as the crystal grows longer. This kind of variation in the maximum stress may explain the observations by x-ray topography [61] of the crystal defects. For example, generation of slip bands on the edge of the crystal may be due to these large stresses. The notable decrease in the dislocation density along the growth direction also agrees qualitatively with the reduction predicted in the peak values of the stress. Figure 14 shows the predicted von Mises stress at the melt–crystal interface as the crystal height is increased during growth. The maximum stress in the center region reaches a von Mises stress of about 9.5 MPa, while its value near the edge of the interface is about 5.0 MPa. The maximum stress at the edge of the crystal just above the encapsulant is, however, as high as

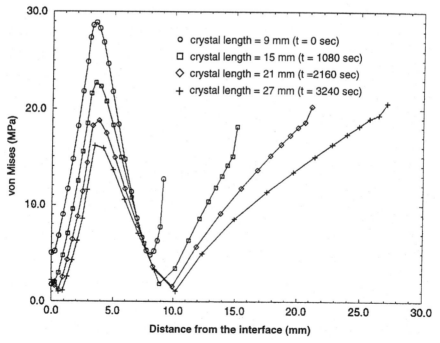

FIGURE 5.13 Von Mises stress along the periphery of the crystal during growth for various crystal lengths. (From Ref. 63.)

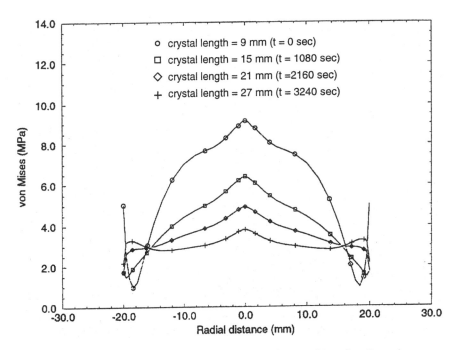

FIGURE 5.14 Predicted von Mises stress along the melt–crystal interface for various crystal lengths. (From Ref. 63.)

29.0 MPa. These values must be compared with the CRSS to obtain excess stress. As is well known, the CRSS reduces rapidly with an increase in temperature. Since the temperature at the melt–solid interface is the highest anywhere in the crystal, the CRSS value will be lowest there, and hence the excess stress near the interface may be very high, leading to the generation of dislocations. Either one of the two high stress points (at the melt interface or at the three-way interface of crystal–gas–encapsulant) may be the main source for generation and multiplication of dislocations. The answer will depend on the difference between the stress values at those points and the CRSS at the corresponding temperature.

The stress variation as the crystal grows predicted by Zou et al. [60] also agrees with the observation of Völkl [62], who suggested that all of the growing crystal passes through an almost-unchanged high stress zone. Volkl also found a maximum stress spot near the three-way interface of the crystal–gas–encapsulant using a simplified heat transfer model. The HPLEC model has also been used to study the effect of segregation [63]. As the crystal grows, more and more solute is rejected from the crystal–melt interface. This solute accumulates in the melt near the interface and is transferred away from the interface region by both diffusion and convection. From Fig. 5.15 it is evident that the solute transport is dominated by melt convection and is strongly accumulated in the region where the two larger cells meet. This

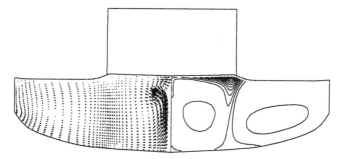

FIGURE 5.15 Flow pattern (left) and solute concentration distribution (right) in the melt. (From Ref. 63.)

can be explained by the flow structure in the melt. The rotation of the crystal has a pumping effect on the melt, which brings the melt up from the bottom along the centerline of the system. This flow of melt washes the solute away from the interface area as it turns outward radially. On the other hand, the melt flow induced by the buoyancy force brings the melt in the outer region inward from the crucible sidewall. This opposing flow of the melt pushes the solute toward the central portion of the interface. As a result, depending on the strength of these two convective rolls, different solute patterns will be formed. In the case of Fig. 5.15, these two cells meet close to the edge of the crystal, and most of the solutes therefore accumulate in that area, showing the highest concentration along the melt–crystal interface. A similar interaction between two convective rolls is also seen in the central region of the interface where the stagnant flow small cell meets the medium-sized cell, but with a much smaller effect on solute distribution. The radial distributions of the solute along the melt–crystal interface corresponding to four growth heights are shown in Fig. 5.16. As the crystal grows, the solute concentration at the interface increases because of the solute rejection from the interface. Meanwhile, the radial segregation also develops because of complex flow patterns in the melt. Severe radial solute segregation is therefore predicted by the model. For all locations, the solute concentration keeps increasing as crystal grows, with the largest rate of increase at the location where the two bigger convection cells meet, especially during the early stages of the growth.

Although HPLEC simulations are still in the elementary stages, they have shown interesting features of high-pressure growth and have provided important insights into InP growth phenomena. They have also demonstrated that the thermal conditions and mechanisms for defects and dislocation generation and propagation in a high-pressure system are quite different from those in the lower-pressure GaAs LEC furnace.

5.4.2 Container Growth

In 1986 the first successful growth of 2-in. InP single crystals by VGF was reported by Gault et al. [64]. This process used a two-zone heater system (see Fig. 5.9), one

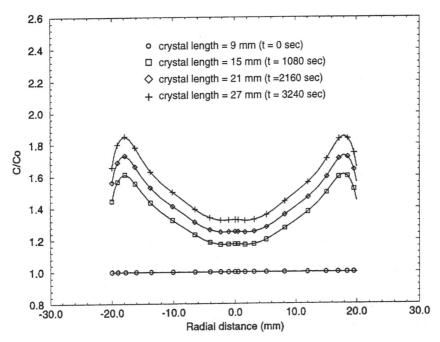

FIGURE 5.16 Dopant concentration in the melt near the crystal–melt interface for various crystal lengths. (From Ref. 63.)

zone to control the phosphorus vapor pressure and the other to provide a thermal gradient for driving crystal growth. With no encapsulation over the melt, the InP charge and the phosphorus reservoir were contained in a semitight ampoule. Crystals grown by this technique had measured dislocation densities of less than 500 cm^{-2}, demonstrating the effect of a low thermal gradient on crystal growth. The axial gradients were 8°C and 40°C cm^{-1} in the crystal and seed, respectively. Stoichiometry was maintained by holding the phosphorus reservoir at 550°C to control the vapor pressure at about 27 atm. This temperature happens to be near the phase transformation point [65] between two alternative forms of phosphorus with different vapor pressures; hence a semitight seal was needed to avoid explosions.

Later work by Monberg and others [66] showed defect-free regions in the center of S-doped crystals and dislocation densities of less than 100 cm^{-2} at the wafer edges (Fig. 5.17). For iron-doped crystals [67,68], dislocation density was reported between 800 and 4000 cm^{-2}, and for undoped crystals below 1000 cm^{-2}. It soon became clear that the VGF technique offered an unprecedented capability for production of low-density material. However, crystallographic orientation (111 growth direction), imperfections (twins and twin lamellas), and seed-to-tail Fe-dopant distribution still limited the yield.

For optimized VGF growth control, Müller et al. [69] built an advanced multizone furnace to new design specifications. This furnace has a high degree of ther-

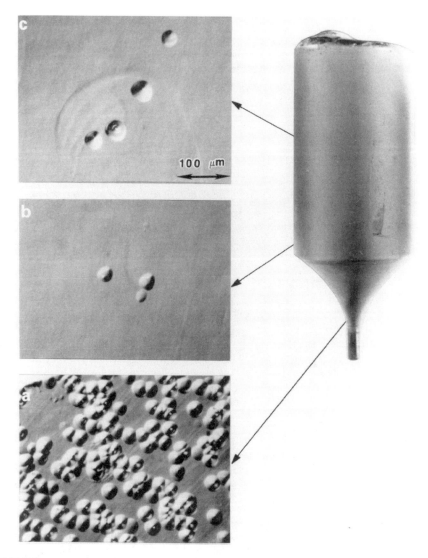

FIGURE 5.17 Micrographs showing etch-pit densities for a 2-in. diameter VGF-grown InP crystal. (From Ref. 68.)

mal flexibility for control of temperature gradients; each segment is cooled and heated separately, and furnace construction is of materials that resist phosphorus contamination. Two suppliers of this type of furnace have emerged. The Mellen Company builds the electrodynamic gradient (EDG) furnace, in which each element is water cooled, and Linn High Therm GmbH has designed a gas-cooled system called the multizone cold wall (MZCW) furnace. Each furnace can be made with any number of elements, with typical spacing between zones of 2 to 3 cm. The greater flexibility of these furnaces over standard VGF means that there are vastly more process variables to control the heat and mass transport during crystal growth. Simulation models have been written [70] to predict the position and shape of the solid–liquid interface and to calculate the content of the boundary layer adjacent to the interface for each time segment of crystal growth. In fact, the investment in software, sensors, and controls is a significant part of the capital cost of these state-of-the-art furnaces.

Despite the engineering improvements afforded by the multizone furnace, successful single-crystal growth was accomplished only for <111>-oriented crystals. Yield problems due to twinning have limited growth in the <100> direction. Work on a related technique [51] has shown that twin-free <100> InP single crystals can be grown by a liquid encapsulated vertical Bridgman (LE-VB) method. The Bridgman method utilizes translation of the charge through a fixed temperature gradient, in this case about 23°C cm^{-1}. To avoid the problem of twinning in the cone section, a full-diameter seed is used, so that the entire crystal grows in cylindrical shape. The use of B_2O_3 as an encapsulant also differentiates this technique from the original VGF growth practice., and it eliminates the need for a separate phosphorus heater zone.

A recent report describes another commercial source of VGF InP wafers [71]. American Crystal Technology, Inc. has licensed the VGF process from AT&T and with modifications is manufacturing 2-in. wafers grown in <100> direction. The dislocation density is comparable to that reported earlier [66] for <111> growth, but few details are available yet concerning this growth process.

Another dilemma for VGF growth is the dopant segregation problem illustrated theoretically in Fig. 5.18, from the work of Favier [72]. Here the normalized con-

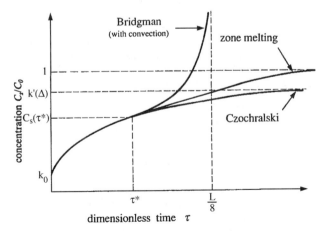

FIGURE 5.18 Normalized concentration versus dimensionless time for directional solidification of finite melt volumes (Bridgman and zone melting) and infinite volume (Czochralski). $k'(\Delta)$ is equal to k_{eff} according to BPS, after the initial transient time τ^*; L is the dimension of the melt in the direction of solidification. (From Ref. 72.)

centration, C_s/C_0, is plotted against dimensionless time. For the case of Czochralski growth, a gradual transition to the steady state occurs where segregation proceeds according to the model of Burton et al. [42]. For zone melting, the ratio goes to unity (i.e., the effective distribution coefficient $k_{\text{eff}} = 1$). However, for Bridgman growth under realistic conditions, with $k_{\text{eff}} \ll 1$ for most impurities in InP and where a finite volume of melt is contained with some degree of convection, an increase in dopant concentration begins immediately after the starting transient. Hence a steady state of constant composition may not be achieved. A delay in this transient behavior could in principle be obtained if one had an infinite melt or if one could prevent convection in the melt. Although the VGF configuration limits buoyancy-driven convection, other forces, such as surface tension and mechanical vibrations, can induce flows in the melt. Any source of convection is a problem for the yield of usable material within the doping range, and a buildup of impurities in the boundary layer brings about the onset of constitutional supercooling.

Short of crystal growth in space, some engineering solutions have been suggested to reduce the segregation effect in VGF growth. Müller [73] has patented a process that allows only axial heat flow through the melt. With only vertical temperature gradients, horizontal temperature and density gradients do not disturb the hydrostatic stability of the melt, and a planar interface is maintained. A flat longitudinal doping profile is generated in the crystal, similar to that calculated for pure diffusive transport.

Another approach has been suggested by Ostrogorsky and Müller [74]. The submerged heater method shown in Fig. 5.19 is a vertical growth method where the crystal is grown from a small zone into which melt and impurities are replenished continuously. One advantage of this system is that the submerged heater acts as a baffle to reduce destabilizing buoyancy forces in the melt. Also, in the narrow-zone-melt region, diffusion-controlled growth is established; that is, the concentration of impurities does not change with time. In diffusion-controlled normal freezing with an infinite melt, the concentration of impurities does not change rapidly because the rate at which convection sweeps the solute from the growth interface is negligibly small compared to the rate at which solute is incorporated into the crystal. Although InP has not been grown by this method, Ge and InSb crystals grown by this method [75] show axial doping profiles very uniform even for dopants with very small k_{eff} values. Similar to the zone leveling scheme of Pfann [24], this technique holds promise for higher yields for VGF growth [76].

5.4.3 New Directions in Crystal Growth

The challenge for the future in crystal growth is to advance new methods for control of quality and uniformity of bulk crystals. In the past, the demand was for larger and better crystals of elemental or binary semiconductors. These materials have congruent melting points, which makes them relatively easy to grow. Today there is a growing interest in ternary or multicomponent crystal systems. For example, the bandgap energy can be engineered to allow increased freedom for tuning the wavelength of optical devices. Similarly, the lattice parameter of bulk substrates can be

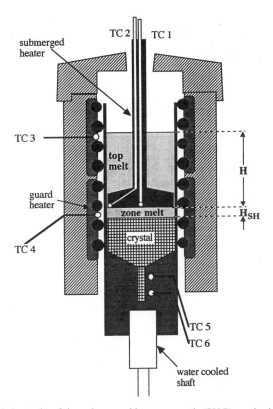

FIGURE 5.19 Schematic of the submerged heater growth (SHG) method. (From Ref. 74.)

varied by changing the alloy composition. Since the more forgiving materials systems have already been exploited, new approaches will be needed to develop novel methods for the noncongruent melt growth of these new materials.

Bonner et al. [77] have pioneered the growth of III-V ternary alloys by the LEC process, growing $In_xGa_{1-x}As$ over the composition range $0.02 \leq x \leq 0.12$. The drawback of the standard LEC process is the marked compositional variation from seed to tail and the rapid onset of constitutional supercooling and polycrystallinity. This is clearly illustrated in Fig. 5.20, showing the axial variation of In content for several of Bonner's crystals.

InGaAs is one of the most important ternary III-V alloys because its lattice parameter varies with In content and can be lattice matched to alloy substrates for optical device applications [78]. Kusunoki and others [79] have developed a melt replenishment system by melting a supply rod of GaAs to maintain composition while the crystal is grown. Although they were able to grow a small homogeneous crystal of $In_{0.14}Ga_{0.86}As$, large thermal fluctuations due to melt convection resulted in crystals with large striations and many cracks. Perhaps for this reason, the more recent work reported by this group [80] has focused on the zone melting technique to grow

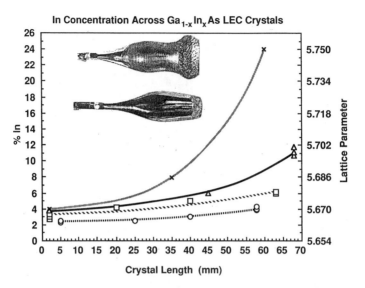

FIGURE 5.20 Indium concentration versus length for LEC growth of GaInAs alloy crystals. (From Ref. 77.)

ternary InGaAs. In their configuration, the container is held stationary while the molten zone migrates up the temperature gradient. This method overcomes the difficulty of thermal fluctuations by reducing the aspect ratio and volume of the melt. They applied the zone melting technique to grow $In_xGa_{1-x}As$ crystals at compositions $x = 0.25$ to 0.08 with good radial uniformity. However, since the alloy composition depends on the growth temperature, the axial In content varies by a factor of 2 from seed to tail.

Experiments by Ware et al. [81] showed there are two conflicting requirements for double crucible growth of InGaAs. The capillary diameter should be as large as possible to allow melt flow and as small as possible to prevent back diffusion of indium. For flow through a hole, the hydrostatic pressure across the hole must exceed the surface tension force:

$$r^2 h(d_{GaAs} - d_{oxide}) > 2\pi r \sigma \quad (4)$$

where r is the radius of the hole, h the depth of the hole below the melt surface, σ the surface tension of GaAs (442 dyn/cm), and d_{GaAs} and d_{oxide} are the densities of GaAs and B_2O_3, respectively. This reduces to the inequality $rh > 0.21$. If this condition is not met, the liquid from the outer crucible will stop flowing into the inner crucible, and the melt composition will change.

On the other hand, as Matare [82] pointed out, the requirement to prevent back diffusion is expressed as

$$vl \gg D \quad (5)$$

where v is the linear velocity through the hole of length l and D is the diffusion coefficient of indium in the InGaAs melt. Since the linear velocity v is inversely related to the square of the hole size r, the design of the double crucible must allow for a long path length l, preferably on the order of several centimeters. Ware et al. [81] designed a new crucible according to these criteria, and demonstrated successful growth of an 800-g 40-mm-diameter crystal of nearly constant composition, 3 mol% InAs.

Ashley et al. [83] have applied the double crucible method to the growth of $In_{1-x}Ga_xSb$ crystals with $0 < x \leq 0.11$. Narrow-gap semiconductors such as InSb have many interesting optical properties because of their band structure. For research into uncooled infrared detectors and emitters, it is necessary to reduce the cutoff wavelength below the 7-μm room-temperature cutoff wavelength of InSb, into the 3- to 5-μm atmospheric window. For this purpose, InAlSb layers can be grown lattice matched to InGaSb substrates; the latter are chosen to grow as bulk crystals because, of the two ternaries, InGaSb has the narrower liquidus–solidus separation.

The design of the double crucible must prevent freezing of the outer melt as it joins the inner melt through the channel. The wall of the inner crucible must minimize heat transfer from the outer crucible, increasing the temperature of the outer melt at equilibrium. This was accomplished in Ashley et al.'s [83] double crucible design, incorporating an insulating collar above the melt surface and a thick-walled silica inner crucible. These workers demonstrated successful growth of InGaSb with composition variation within 10% of the target [83]. To grow single crystals, the most successful approach was to initiate growth from a pure InSb melt with an InSb seed, and then grade the solution up to the desired alloy composition.

5.5 CHARACTERIZATION OF BULK InP

5.5.1 Electrical Properties of Bulk InP

Semiconducting InP Indium phosphide has an intrinsic carrier concentration at room temperature on the order of $n_i \simeq 10^7$ cm^{-3}. However, because all crystals contain unintentional impurities such as Si and S that are ionized at room temperature, the measured free carrier concentration is always far above the intrinsic level. Since both Si and S are common impurity donors, the free electron concentration n for nominally undoped InP crystals is usually in the range 10^{14} to 10^{15} cm^{-3}. The required doping range for most conducting substrates is 10^{18} to 10^{19} cm^{-3}, substantially higher than the residual impurity concentration. Hence the issue of compensation is not as important in the selection of conducting substrates as it is for semi-insulating material. However, control over electrical uniformity and control over dopant incorporation remain as yield-limiting issues for crystal growth.

InP crystals with n-type conductivity can be grown by adding shallow donors such as Sn, which substitutes for In, and S, which substitutes for P [84]. The free carrier concentration can be varied continuously up to the limit determined by the solid solubility of that particular donor species in InP, as long as no compensating acceptor impurities are introduced. In either case, where n increases or where the compensation ratio increases, μ_n decreases due to ionized impurity scattering, in accordance with calculated [85] data. There seems to be some disagreement between the earlier calculations of Walukiewicz et al. [85] and the later mobility data of Benzaquen et al. [86], with the latter overestimating the value of mobility for uncompensated samples. One possible explanation may be that the earlier data fit bulk material more closely, whereas the later data are derived from epitaxial material, which has a lower intrinsic defect concentration.

Bulk InP with p-type conductivity is grown by adding group II elements such as Zn, Cd, Be, or Hg, which substitute for In. The range of resistivity for p-type doping is not as large as for n-type doping, because the mobility μ_p falls off rapidly as a function of free carrier concentration. This effect is even more pronounced at 77 K, where a partial freeze-out occurs because of the relatively deep energy level of acceptor impurities. For p-doped samples, with $p < 1 \times 10^{18}$ cm^{-3}, the free carrier concentration decreases by an order of magnitude between 300 and 77 K, while the mobility increases, leaving the resistivity constant within a range of 25%. Even more heavily doped samples, $p \geq 2 \times 10^{18}$ cm^{-3}, do not exhibit reduced resistivity because of the rapid fall-off in mobility.

As-grown InP bulk crystals, whether n-type or p-type, are usually nonuniform in resistivity over their length; that is, the resistivity decreases continuously along the length from seed to tail. The dopant concentration C incorporated into the growing crystal varies according to Eq. (1), depending on the distribution coefficient for each dopant species. The values of the effective distribution coefficients, $k \equiv C_s/C_l$ for a variety of dopants have been compiled in a monograph by Iseler [84]. Short-range nonuniform distribution of dopant results in striations, particularly in LEC growth, where steep thermal gradients induce natural thermal convection in the melt.

Semi-insulating InP One of the key prerequisites for the development of high-speed devices and OEICs is semi-insulating substrate material. Because of their bandgaps, semi-insulating wafers of both GaAs and InP are available with resistivity greater than 10^7 Ω·cm. Semi-insulating LEC GaAs can be grown by taking advantage of the deep donor level EL2, which is related to the arsenic antisite defect. This defect compensates the residual carbon acceptor impurity, the dominant shallow-level impurity in crystals grown in pyrolytic BN crucibles with controlled stoichiometry. In contrast, semi-insulating behavior in InP is obtained through the use of intentional dopants because there are no suitable native defects. Undoped (or unintentionally doped) InP is usually n-type conductive due to the Si and S donors. Carbon does not act as an acceptor but rather as a weak donor in InP. Moreover, the intrinsic defect concentrations in crystals grown from P-rich melts are relatively low [34] compared to EL2 in GaAs, and the energy level associated with this defect is ≃ 0.4 eV, too low to create semi-insulating properties. As a result, iron is the single deep-level acceptor that can be added to give resistivity greater than 10^7 Ω·cm.

However, the distribution of iron must be controlled both during crystal growth and afterward to prevent diffusion during high-temperature processing. The distribution of dopants in the growing crystal has been discussed in Section 5.4. Redistribution of Fe during epitaxial growth is a problem that has been studied by several authors [87,88]. Figure 5.21 illustrates the problem of iron redistribution for liquid-phase epitaxy. A comparison of Fe profiles of unheated and preheated substrates shows that Fe accumulates near the surface of the substrate. The depth of the accumulation region for samples treated at 650°C ranged from 0.1 to 0.2 μm, as shown. Note that the amount of Fe accumulated near the surface seemingly exceeds the amount depleted from the bulk. Although the Fe profile seems to level off at a depth of 0.6 μm, Fe must have been transported over a distance of many micrometers from the surface. Redistribution of Fe has two possible effects: first, an epitaxial layer deposited on the surface may become contaminated with iron, and second, the Fe-depleted substrate may develop a conducting layer. Both of these effects affect device performance and reliability.

The instability of Fe in InP has led many authors to seek alternatives to Fe dop-

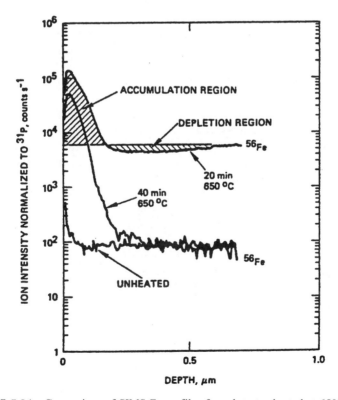

FIGURE 5.21 Comparison of SIMS Fe profiles for substrates heated at 650°C for 0, 20, and 40 min. Note the different Fe concentrations in the bulk substrates that were used. (From Ref. 87.)

ing [89–91]. Chromium behaves as a very deep acceptor (0.35 eV from the conduction band) in n-type InP, and like Fe, it becomes Cr^{2+} when ionized, resulting in resistivities of 10^3 to 10^5 $\Omega \cdot$cm. In p-type material, however, Cr behaves as a deep donor and becomes Cr^{4+} when ionized. Cobalt introduces an acceptor level located closer to the valence band than the Fe level, resulting in p-type material with low conductivity but not semi-insulating [91]. Rhodium is a deep acceptor that has a much lower diffusion coefficient than Fe, but no bulk semi-insulating material has been produced using Rh [90]. Co-doping with two elements has also been attempted. The most promising combination seems to be a shallow acceptor (Hg or Cd) and a deep donor (Ti or Cr), which results in resistivities of 10^4 to 10^5 $\Omega \cdot$cm and low mobility due to compensation [91]. Although diffusion of Hg and Ti is very low compared to Fe, because the Ti level is less than 0.5 eV from the conduction band, the resistivity is not sufficiently high to support isolation and capacitance requirements for OEICs.

Compensation Mechanism of InP:Fe The compensation mechanism controlling the semi-insulating property of InP material is an active area for research. Several questions concerning the compensation mechanism have been resolved only recently. It has long been known that the Fe content necessary to fix resistivity in as-grown crystals at 10^7 $\Omega \cdot$cm is much greater than the residual shallow donor concentration [92]. Typically, the required Fe concentration as measured by mass spectrometric analysis is about 1×10^{16} cm^{-3}, whereas the total residual donor concentration (Si and S) is on the order of 10^{15} cm^{-3}. Crystals grown with less than the required amount of iron exhibit low resistivity. A further complication occurs when low-resistivity InP is converted to semi-insulating material by annealing. It has been argued that either Fe is not always on an active site [93], or that there are some "intrinsic" shallow donors that cannot be measured by chemical analysis [94]. Although there is not total agreement on the charge compensation mechanism of iron, some salient facts have been firmly established.

It is known that Fe substitutes for In in InP and adopts primarily the neutral Fe^{3+} state (so-called to denote its oxidation state). However, some of the iron is compensated by shallow donors to form Fe^{2+}, and the ratio between the ionized and neutral concentrations, Fe^{2+}/Fe^{3+} varies linearly (see Fig. 5.22) with the free carrier concentration, n [92]. A linear dependence of n on Fe^{2+}/Fe^{3+} is expected since both quantities are proportional to the term in the Fermi distribution function, $\exp(-E_F/kT)$, where E_F is the Fermi energy and k is the Boltzmann constant. The free carrier concentration can be expressed as

$$n \simeq N_C \exp\left(-\frac{E_C - E_A}{kT}\right) \frac{[Fe^{2+}]}{[Fe^{3+}]} \frac{g(Fe^{3+})}{g(Fe^{2+})} \qquad (6)$$

where N_C is the density of states in the conduction band and g represents the degeneracies of the two iron defect states. The energy difference $E_C - E_A$ is simply the change in Gibbs free energy associated with the transfer of an electron from the occupied iron acceptor level to the conduction band. The Gibbs free energy can be ex-

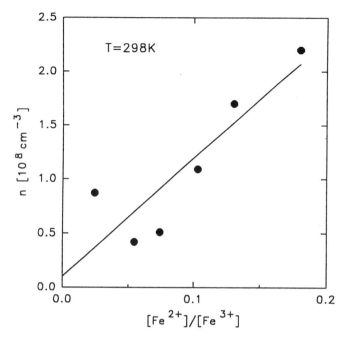

FIGURE 5.22 Experimental data for semi-insulating InP crystals showing a linear dependence of free carrier concentration, n, on the ratio between ionized and neutral iron (Fe^{2+}/Fe^{3+}) concentrations. The solid line is a best fit to the data. (From Ref. 92.)

pressed in terms of the enthalpy H and entropy S as $G = H - TS$. Equation (6) can then be rewritten as

$$n \simeq N_C \exp\left(-\frac{H}{kT}\right)\exp\left(\frac{S}{k}\right)\frac{[Fe^{2+}]}{[Fe^{3+}]}\frac{g(Fe^{3+})}{g(Fe^{2+})} \qquad (7)$$

The slope of an Arrhenius plot of n versus $1/T$ as in Fig. 5.23 gives the change in enthalpy H, or the activation energy extrapolated to temperature $T = 0$, the same value that is measured by the temperature-dependent Hall effect or by deep-level transient spectroscopy (DLTS). But the entropy S can also be extracted to give the temperature shift of the iron acceptor level. Zach determined this value and found the temperature shift to be different from the temperature shift of the bandgap [92]. This is illustrated in Fig. 5.24, showing the relative shift of the bandgap versus the deep defect level. The results indicate that the thermodynamic position of the Fe^{2+}/Fe^{3+} level at room temperature is 0.49 eV below the conduction band, and it is this energy that determines the carrier concentration. For example, referring to Eq. (7), the room-temperature carrier concentration for a half-occupied iron acceptor level, $Fe^{2+}/Fe^{3+} = \frac{1}{2}$, would be $n \simeq 10^9$ cm^{-3}. In other words, with Fe doping levels even twice as high as residual shallow donors, at room temperature InP material is still

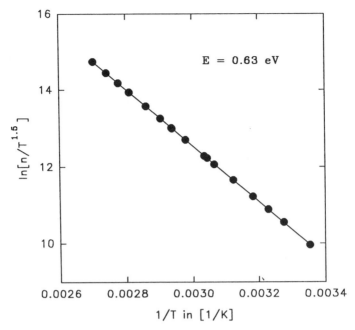

FIGURE 5.23 Activation energy for the iron acceptor in InP as determined from an Arrhenius plot of n versus $1/T$. (From Ref. 92.)

conducting. The Fermi energy can only be reduced to the midgap level as the occupancy of the iron acceptor level decreases, so that for $Fe^{2+}/Fe^{3+} = \frac{1}{10}$, the free carrier concentration will be $n \simeq 10^7$. The experimental data in Fig. 5.22 illustrate this point. In fact, for typical semi-insulating InP with residual donor concentrations of 1 to 3×10^{15} cm^{-3}, a minimum Fe doping level of 1 to 3×10^{16} cm^{-3} is required.

Wafer Annealing for Semi-insulating InP Several authors [95–99] have exploited the possibility of annealing high-purity InP with only a small amount of Fe, to obtain semi-insulating material. The best results have been obtained using very low (e.g., ca. 10^{15} cm^{-3}) Fe content material, which has a free carrier concentration of 1 to 2×10^{15} cm^{-3} as grown. Slices of this material are annealed at 950°C for 20 to 80 h under a partial pressure of phosphorus of a few atmospheres and slowly cooled. The annealed slices are then fabricated into wafers by grinding and polishing the surfaces. The results in Fig. 5.25 show that high-quality semi-insulating wafers can be produced in this fashion. After annealing, the purity of annealed InP wafers produced by this method is very high: Mobility is approximately 4000 cm^2/V·s and photoluminescence spectra prove that the material purity has not deteriorated. These wafers are thermally stable and may be suitable for ion implantation processing [100].

The predominant effect of annealing was proven to be a reduction in the concentration of shallow donors [94]. In addition, Fe is incorporated in concentrations of 1

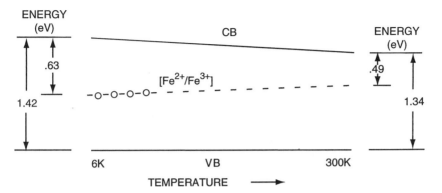

FIGURE 5.24 Energy band diagram versus temperature for InP:Fe showing the shift in bandgap energy (slope = -3.8×10^{-4} eV/K), and the shift in the Fe acceptor ionization energy (slope = 4.7×10^{-4} eV/K).

to 2×10^{15} into the samples during annealing [101]. Both of these occurrences are related to diffusion effects. An experiment [102] performed to determine the diffusion profile of the effects showed that a layer of 170, 225, and 280 μm is affected by annealing for 2, 6, and 24 h, respectively. Such a time dependence of depth can give an estimated diffusion coefficient of $D \simeq 1 \times 10^{-8}$ cm²/s for both the in-diffusion of Fe and the out-diffusion of the "intrinsic" defect. The published data for Fe diffusion [87] in InP give $D_{Fe}(900°C) \simeq 1$ to 5×10^{-9}, and the data for self-diffusion [103] of In and P in InP are several orders of magnitude lower than the diffusion effects ob-

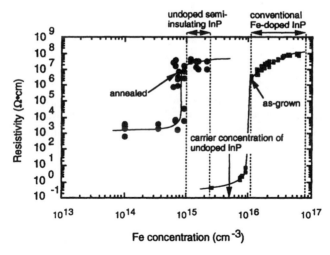

FIGURE 5.25 Relationship between resistivity and Fe concentration of predoped annealed InP and as-grown Fe-doped LEC InP. (From Ref. 100.)

served. Hirt et al. [94] proposed a fast interstitial diffusion mechanism and a subsequent turnover of the diffused atoms onto lattice sites. This hypothesis allows for diffusion coefficients consistent with the "kick-out" model [104] which are in agreement with the diffusion profiles observed. With this model one can speculate that due to the phosphorus overpressure, outdiffusion of In atoms shifts the local equilibrium of intrinsic defects over to the phosphorus side. The generation of V_{In} defects enhances the turnover of Fe atoms from interstitial sites to electrically active substitutional sites.

Annealing experiments performed in vacuum have also resulted in conversion of high-purity undoped InP to semi-insulating [97]. Annealing for 200 h at 910°C either under vacuum or in P ambient gives rise to a drop in the free carrier concentration of bulk InP. The reduction in carrier concentration is small in samples that are cooled quickly, whereas it is much larger for samples cooled at 30°C h^{-1}. For the best of these samples, the resistivity was $4 \times 10^7 \Omega\cdot$cm after annealing, with a mobility of $\simeq 4000$ cm^2/V·s. In the case of vacuum annealing there is no excess phosphorus; in fact, P vacancies are more likely to be created than eliminated. To explain the compensation mechanism, these authors argued that new shallow acceptors were introduced or generated during the anneal. Although SIMS analysis did not reveal an increase in the common shallow acceptors, it did not rule out intrinsic or complex defects. Fornari et al. concluded that when the sample was quenched, the diffused Fe atoms were frozen on interstitial sites, and when the cooling rate was sufficiently slow, Fe has the possibility of occupying In sites, with consequent generation of deep acceptor levels.

The kick-out hypothesis assumes that the effect of annealing is to allow more iron atoms to occupy In sites, so that more Fe becomes available to compensate shallow donor impurities, reducing the free carrier concentration. This implies that after annealing, the ionized acceptor concentration Fe^{2+} increases. In fact, as shown below, the opposite is true; after annealing, the Fe^{2+} concentration is reduced.

Role of Hydrogen   High-purity bulk InP crystals have measured free carrier concentrations which are higher than residual net donor concentrations, determined by glow discharge mass spectroscopy (GDMS). This implies that additional residual donors, not detected by impurity analysis, are present in these samples. Similarly, in semi-insulating material the concentration of ionized acceptors, Fe^{2+}, appears to be higher than the measured concentrations of Si and S. (In most cases, the shallow acceptor concentration is negligible compared to the shallow donor concentration.) The residual donor concentrations are not large enough to account for the total Fe^{2+} concentration. Therefore, it has been proposed that another donor must be present which contributes to the ionized Fe^{2+} concentration [105].

Several studies [106,107] of the absorption spectra of hydrogen in InP have established the structure of the hydrogen-related complexes. The local vibrational mode (LVM) absorption at 2315.6 cm^{-1} in LEC InP is due to the P–H stretching modes in an environment that has tetrahedral symmetry [108]. Deuterium and hydrogen co-doping was used [105] to investigate the LVM in InP. The presence of two different H isotopes causes splittings in the hydrogen-only spectra, helping to determine the atomic arrangement in the lattice and thus its electronic behavior. The D–H spectra were consistent with a defect having tetrahedral symmetry, and the authors proposed that the absorption is due to a fully passivated indium vacancy [$V_{In}(PH_4)$]. Investigation of the electrical properties of the hydrogen defect in InP [105,109] showed that this defect has a characteristic donor behavior. A model of the $V_{In}H_4$ defect, shown schematically in Fig. 5.26, illustrates the donor behavior that arises from the tetrahedral formation of H around an indium vacancy. Four electrons are contributed by hydrogen atoms to a site whose valence bonds are satisfied by three electrons.

The $V_{In}H_4$ defect disappears during annealing at ≃900°C under a phosphorus overpressure, as shown in Fig. 5.27. For undoped bulk InP, disappearance of the LVM absorption peak is accompanied by a reduction in free carrier concentration. This effect is consistent with the other published results [94,110], where "intrinsic" donors are reduced by annealing. However, if the drop in free carrier concentration were due to the presence of additional compensating impurities rather than hydrogen, a corresponding decrease in sample mobility should be observed. To differentiate donor reduction from compensation, the mobilities of InP samples were measured before and after annealing [129]. An increase in mobility after annealing was observed, consistent with a decrease in the donor related $V_{In}H_4$ defect.

For the case of Fe-doped material, similar annealing experiments [129,130] showed the role of hydrogen-related donors in controlling resistivity. In these experiments the Fe^{2+} absorption peak was measured quantitatively to determine the residual shallow donor concentration. For semi-insulating Fe-doped material, annealing reduces the Fe^{2+} absorption peak, as shown in Fig. 5.28. The change occurs without any measurable increase in Fe^{3+} by diffusion or other mechanisms. These experimental results confirm the donor nature of the H defect and the important role of hydrogen in the compensation mechanism of Fe-doped InP. Understanding the

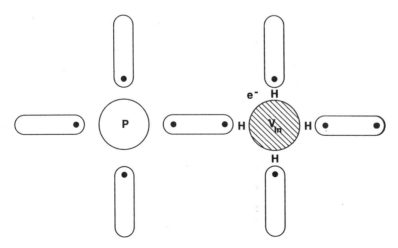

FIGURE 5.26 Schematic of the $V_{In}H_4$ defect. (From Ref. 105.)

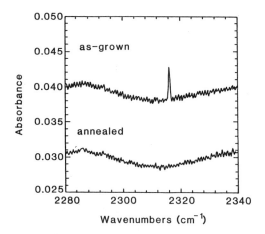

FIGURE 5.27 Infrared absorption spectra in the range of the InP:H LVM line before and after annealing. (From Ref. 105.)

$V_{In}H_4$ defect is crucial for annealing undoped InP to generate semi-insulating material.

5.5.2 Wafer Mapping Techniques

As the demand for high-quality InP wafers has increased, wafer characterization has increased in its level of sophistication and resolution. Various complementary mapping techniques provide information on the following properties: (1) uniformity of dopant distribution across the wafer and from wafer to wafer, (2) dislocation density and distribution, (3) residual strain concentration and distribution, (4) precipi-

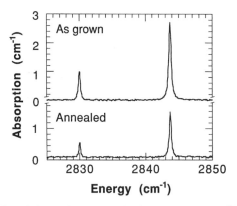

FIGURE 5.28 Infrared absorption spectra in the range of the Fe^{2+} intracenter absorption before and after annealing. (From Ref. 129.)

tates and other volume defects, and (5) distribution and concentration of surface states. The sophistication of mapping techniques has improved dramatically over the past decade; improved spatial resolution allows the fine structure of many defects to be revealed. At the same time, computerized instrumentation has provided capabilities for image analysis. The tools for wafer mapping can now be used in mass-production wafer fabrication. The key to successful commercial application has been rapid nondestructive automated evaluation throughout the production process. Tools that can meet those requirements are essential to wafer processing, especially for device structures grown on InP substrates. Performance characteristics of these structures depend critically on the bulk and surface properties, the conductivity of the substrate, and the presence of impurities or surface states at the interface prior to epitaxial growth. Variations in polishing or conductivity from lot to lot or within lots can cause drastic reduction in the device yield. To meet the demands of commercial production lines, wafer mapping for measurement of bulk properties and for control of surface conditions is necessary.

Several mapping techniques are available today that were in prototype stages as recently as 1990: (1) scanning photoluminescence (sPL), (2) spatially resolved photoreflectance (PR), (3) double crystal x-ray diffraction mapping (DCDM), (4) resistivity mapping by time-dependent charge measurement (TDCM), and (5) residual strain distribution by scanning infrared polariscope. Each technique offers information about the uniformity and concentration of defects in the substrate. In general, the apparatus required for these mapping techniques is quite expensive compared to infrared absorption methods, and the information obtained is somewhat difficult to quantify, but the whole-wafer view is invaluable for selection and qualification of substrates. Time is also a factor. The acquisition of a complete image, scanning at a typical rate of 1 s per line, could take from 1 to 10 min, depending on the resolution of the image. The question is raised [111] as to how much sampling time is necessary to assess uniformity accurately. In other words, what is the minimum sampling density needed to discern and locate the defects? Naturally, the answer is that the number of points needed to resolve the pattern depends on the details of the distribution, which is determined by the crystal growth conditions.

Mapping is a valuable diagnostic tool when crystal growth processes are being optimized, where detailed knowledge of the pattern is required. But in a manufacturing environment, in which wafers proceed through multiple fabrication steps, a simple estimate of the number of defects on each wafer may be a sufficient process control. In that case it should be possible to reduce the number of sampling points and decrease the inspection time per wafer.

Scanning Photoluminescence When electron–hole pairs are created in a semiconductor by optical excitation, the resulting radiative recombination is called photoluminescence (PL). A setup for scanning photoluminescence is shown schematically in Fig. 5.29. This laser scanner system consists of a high-powered argon laser at the excitation wavelength 514.5 nm, polygon rotating mirror, scanning lens, and precision translation stage. Other laser sources operating at lower power may also be used as excitation sources. The y-axis is defined by the scanning laser beam; the

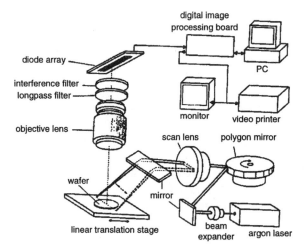

FIGURE 5.29 Schematic of a scanning photoluminescence system. (From Ref. 112.)

x-axis is provided by the translation stage. The imaging system collects the luminescence through an objective lens to a linear Si-photodiode array. Light passing through the lens is filtered to prevent scattered laser light from reaching the detector.

The PL spectrum for undoped InP recorded at low temperature shown in Fig. 5.30 exhibits two peaks, located at 1.418 and 1.38 eV. At or near liquid He temperature and at low excitation intensities, the injected electrons and holes may combine to form free excitons, or the electron–hole pair can be bound to an impurity to form bound excitons. The higher-energy line is due to bound excitons (BEs), together with the free exciton line. The lower-energy line is usually associated with the conduction band-to-acceptor transition (eA). At low temperature ($\simeq$ 30 K) these two lines can be imaged separately by sPL, so that two images are generated. The BE

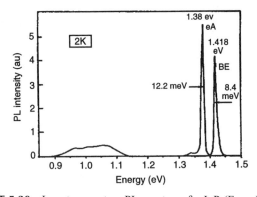

FIGURE 5.30 Low-temperature PL spectrum for InP. (From Ref. 113.)

image is due to radiative recombination of electron–hole pairs generated by the argon laser, which has a penetration depth of $\simeq 1000$ Å. Because the excitation of electron–hole pairs is restricted to a region very close to the surface, BE images clearly show the detailed optical structure of the material. In contrast, the eA image appears slightly smeared because it originates from a larger volume of material. (The eA energy level is about 40 meV below the BE emission. Since the absorption coefficient of InP at the energy of BE emission is fairly low, the PL emission is reabsorbed and eA excitation occurs over a depth of several micrometers.)

For Fe-doped wafers, a significant feature of low-temperature sPL images obtained at the energy of the BE line is the appearance of bright-spot defects [i.e., regions with higher (ca. 20%) PL intensity than that of the surrounding areas]. Typical dimensions of the bright spots are 10 to 20 μm. The bright spots are associated with points of intersection of the dislocations and the sample surface [114]. The increase in PL intensity in the vicinity of the dislocation is explained by the gettering effect of the dislocation, which locally reduces the Fe concentration. (Iron causes nonradiative recombination, thus reducing the PL intensity.) The bright spots are not usually observable in eA images because of the greater probed depth, a further confirmation that the the BE emission image is highly surface sensitive.

At room temperature the PL spectrum exhibits one broad line centered around the bandgap energy, corresponding to band-to-band transitions. A typical sPL image of an Fe-doped wafer is shown in Fig. 5.31. The most obvious feature of this image is the spiral striation pattern. The brightness in the pattern has been shown to be an-

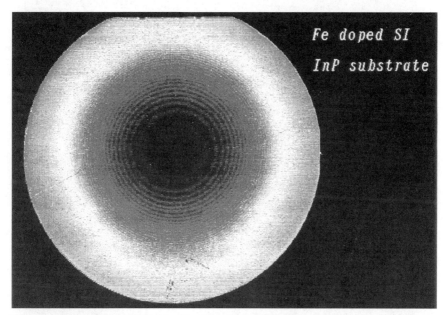

FIGURE 5.31 Wafer map of 2-in. diameter InP:Fe wafer from room-temperature sPL image. (From Ref. 115.)

150 InP BULK CRYSTAL GROWTH AND CHARACTERIZATION

ticorrelated with the Fe concentration by SIMS. The PL intensity I is inversely proportional to the resistivity ρ. A qualitative expression has been derived [113] from the equation of charge neutrality:

$$\rho \propto \frac{1}{\mu}\left(\frac{N_{Fe}}{N_D - N_A} - 1\right) \propto I^{-1} \tag{8}$$

where N_{Fe} is the concentration of the deep iron level ($\sim Fe^{3+}$), μ the mobility, and $N_D - N_A$ the net residual donor concentration ($= Fe^{2+}$). Since neither the mobility nor the residual donor concentration was found to vary significantly across the wafer, the sPL image shows the Fe distribution and provides a qualitative map of resistivity. More precisely, the PL intensity is dependent on the ratio of Fe^{2+}/Fe^{3+} because

$$\frac{1}{\rho} \propto n \propto \frac{Fe^{2+}}{Fe^{3+}} \propto I \tag{9}$$

since $Fe^{3+} \gg Fe^{2+}$.

Another feature of room-temperature sPL images of Fe-doped InP is the bright-line defect seen in Fig. 5.32. Like the bright-spot defects observed at low temperature, these defects are associated with the Fe distribution around the dislocations.

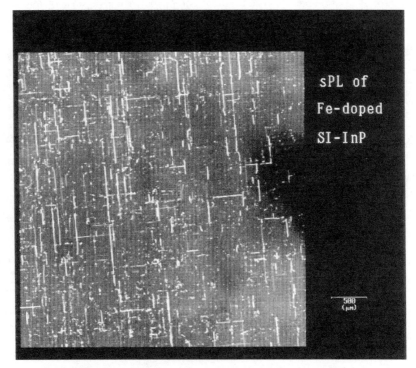

FIGURE 5.32 Bright-line defects as seen in InP:Fe. (From Ref. 115.)

One difference is that all the bright lines appear to be oriented in one of the [110] directions. As a screening technique, sPL has proven to be invaluable in assessing substrate material. Lattice-matched InGaAs epitaxial layers grown on wafers with a high density of bright-line defects are susceptible to the formation of sliplike defects. The presence of iron is necessary for the formation of bright-line defects. However, the density of bright-line defects does not correlate with the iron concentration as it increases along the length of the boule. Clearly, there is more than one variable involved in the formation of bright-line defects; perhaps the thermally induced growth stress plays a role, and if so, a map of residual strain would show significant crystallographic orientation along easy glide planes. Other mapping techniques are necessary to determine the other variables responsible for these defects.

X-ray Topography Large area images of bulk InP material have been obtained using x-ray transmission or reflection topography. For constructive interference, the geometrical alignment must comply with the Bragg condition $\lambda = 2d \sin \theta$, and for monochromatic x-ray sources there is only one Bragg angle. However, for a white x-ray source such as the synchrotron, a separate image appears at all the allowed **g**-vectors. Information about the various types of defects can be obtained by selecting a **g**-vector image appropriate to the defect. For example, among images of slip bands where the crystallographic direction of strain is characterized by a vector **s**, the appropriate image is found where $\mathbf{s} \cdot \mathbf{g} \neq 0$. Similarly, the lattice dilation strain caused by striations can be observed only in certain images. Dopant striations are visible only in images where $\mathbf{s} \cdot \mathbf{g}$ has a finite value, that is, where the diffraction vector lies perpendicular to the striations. Figure 5.33 is an example of synchrotron white beam x-ray topography (SWXRT) showing a cross section of a S-doped InP LEK crystal. This transmission topograph shows both the striation pattern and the slip bands.

Resistivity Mapping: TDCM Spatial fluctuations of the dopant concentration across the substrate surface or between wafers in the boule can cause undesirable reductions in device yield. Dopant fluctuations degrade the uniformity of resistivity, a parameter important to device processing. Ideally, a technique is needed for process control that is quantitative, reproducible, rapid, and nondestructive. For this, the best method is based on capacitance relaxation. A group at Fraunhofer Institute [116] has developed an instrument exploiting time-dependent charge measurement (TDCM). The semi-insulating substrate is inserted between two charged plates so that the two air gaps form two capacitors. The relaxation time, τ, is evaluated by measuring the charge flowing into the capacitors, using a charge-sensitive amplifier. The space-charge relaxation time τ in an *RC* circuit gives the resistivity ρ as

$$\tau = \varepsilon \varepsilon_0 \rho \tag{9}$$

where ε is the dielectric constant and ε_0 is the permittivity of vacuum. By ensuring that only the area defined by the capacitor electrode contributes to the charge measurement, and with precise positioning of the *xy* stage, the spatial resolution of

FIGURE 5.33 Transmission x-ray topograph of an axial slice of an S-doped 2-in. diameter InP MLEK crystal, showing striations and slip bands. (From Ref. 61.)

TDCM is on the order of 1 mm. The instrument correlates well with other techniques: between 10^6 and 10^9 $\Omega \cdot$cm the results agree with the measured Hall resistivity.

As a measurement tool to assess and optimize the quality of semi-insulating InP substrates, TDCM provides adequate geometric resolution, free of complications arising from contact preparation, which may introduce erroneous information. A pair of maps in Fig. 5.34 showing a comparison between scanning PL and TDCM demonstrates the correlation between the two techniques. In the core region where the PL intensity is low, the resistivity is high, and conversely, the wafer periphery shows a bright PL and low resistivity. This wafer is typical of Fe-doped wafers where striations are present. It shows the effect of iron variations on nonradiative recombination and resistivity. Interestingly, in the bright peripheral ring there is a high density of iron precipitates, and because of gettering effects [118], the overall resistivity is reduced in that region.

Residual Strain Mapping An infrared polariscope has been used [62] to estimate the strain-induced birefringence in thick InP slices grown by LEC. The sensitivity was improved using a computer-controlled infrared polariscope developed by Yamada et al. [119], and high-resolution imaging techniques were added to visualize the fine structure of residual strain in LEC- and VCZ-grown InP [121]. This photoelastic method makes it possible to quantify the absolute difference of tensile residual strain components along the radial and tangential directions. The light source is an LED emitting at $\lambda = 1.3$ μm, which was modulated at a frequency of about 1 kHz to

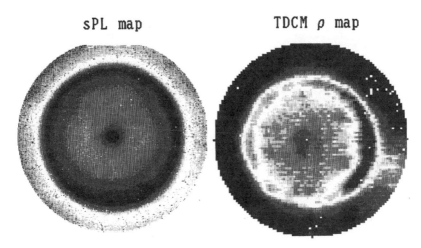

FIGURE 5.34 Comparison between sPL (left) and TDCM (right) maps of 2-in. diameter InP:Fe, showing radial segregation and a high density of iron precipitates near the periphery. (From Ref. 117.)

utilize a phase-sensitive detection. The images of LEC InP typically show sliplike patterns that are aligned along the <011> crystallographic directions. The instrument can separately resolve the shear component $2|S_{yz}|$ from the tensile component $|S_{yy} - S_{zz}|$. The sliplike defect pattern appears clearly in the $2|S_{yz}|$ maps, indicating the dominant origin of sliplike defects to be the shear strain between the crystallographic y and z directions. The patterns are also consistent with the glide directions in the $\{111\}/\langle 110\rangle$ slip system for zincblende structures. These sliplike defects are oriented in the same planes as the bright line defects observed in sPL images by Miner et al. [115], corroborating the link between shear strain and Fe precipitate formation. This leads to the conclusion that growth-induced residual strain contributes to defects that observably degrade the quality of epitaxial layers.

A comparison of residual strain between LEC- and VCZ-grown InP crystals has been made by Yabuhara et al. [120]. The same comparative result, using high-resolution imaging, is shown in Fig. 5.35. Both crystals are Fe doped at a concentration of about 10^{16} cm^{-3}, which is too low for impurity hardening effects. The VCZ crystal is grown with the lower axial thermal gradient at 30 to 50°C cm^{-1}, compared to the LEC crystal at 100 to 130°C cm^{-1}. Because of the lower thermal gradient, the VCZ crystal should theoretically exhibit lower residual strain and lower dislocation density. In fact, work by Fukuzawa and Yamada [121] confirms this theory: the average residual strain value averaged over the entire wafer is reduced by a factor of 4, and the EPD is reduced by nearly an order of magnitude.

IR Striation Mapping A very simple method for imaging the shape of the growth interface using infrared-sensitive film was developed by Barns [122] for studying InP bulk growth. Images like that of Fig. 5.36 reveal the solid-liquid interface shape

154   InP BULK CRYSTAL GROWTH AND CHARACTERIZATION

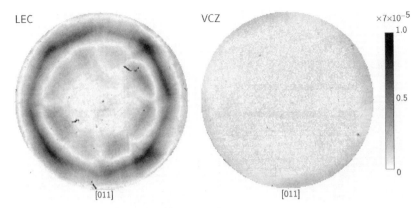

FIGURE 5.35 Comparison of residual strain between LEC and VCZ grown 2-in. InP crystals, using an infrared polariscope. (From Ref. 121.)

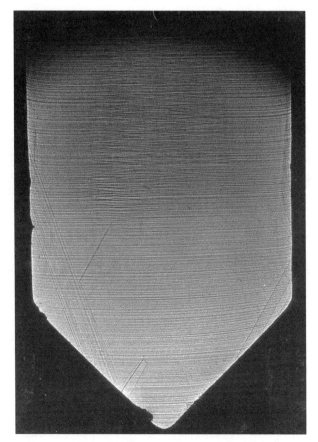

FIGURE 5.36 IR striagraph showing the striations in a 2-in. VGF InP crystal. (From Ref. 68.)

for a VGF crystal through doping striations. The contrast is produced either by absorption [123,124] or by local changes in refractive index, n [125]. In either case, the IR striagraph is useful for modeling heat and mass transport in bulk crystal growth, because it gives qualitative information about the position and shape of the interface. To make this technique quantitative, some questions about the origin of the IR contrast must be clarified. Two arguments are at issue. On the one hand, the transmitted light is especially sensitive to refractive index gradients, since light is scattered away from the beam by the gradient. Since lattice dilation results from dopant incorporation [126], the refractive index varies with doping striations. Contrast in the image, therefore, is purely a result of scattering. In this view, the absorption is irrelevant. On the other hand, a linear dependence of the absorption coefficient at 1 μm on free carrier concentration was found for homogeneous InP material without striations [124,127]. For bulk material with striations on the order of 10 to 50 μm (large with respect to a 1-μm wavelength), the gray-level contrast can be calibrated to the carrier concentration. With this experimentally determined relationship, quantitative absorption measurements yield dopant variations across the striae.

An application of this technique to the analysis of segregation behavior in LEK crystal growth is shown in Fig. 5.37 using a video IR transmission microscope. Here the ($\lambda \approx 1$ μm) light intensity distribution in the video image is quantified with a digital image processing system, digitizing it into a 512 × 512 pixel image with a 256-gray-level dynamic range. As with any detector system, the computer-

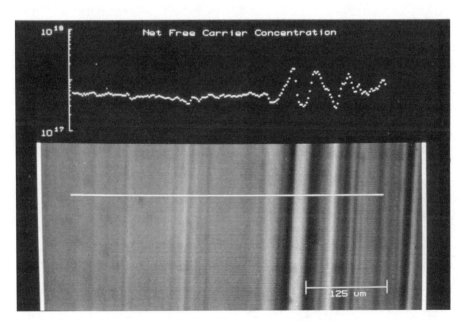

FIGURE 5.37 Segregation analysis along the length of an InP:Sn crystal using IR transmission microscopy showing the transition region between the applied magnetic field and conventional nonmagnetic growth. On the left side a 2-kG magnetic field was applied, and on the right it was turned off.

sensed gray-level values must be calibrated to actual transmittances. By utilizing neutral density filters, the digitized light intensity can be equated to a given transmittance. The figure contains an IR micrograph (bottom of figure) and an analysis of microsegregation measured along the white line (superimposed on the micrograph). The figure shows the transition region between conventional growth and growth with an applied magnetic field of $2kG$. The observed striations are due to turbulent natural thermal convection in the melt. This analysis shows the dramatic increase in dopant uniformity upon application of a magnetic field.

5.6 SUMMARY AND FUTURE CHALLENGES

Over the past decade, crystal growth of bulk InP has progressed in step with demand for wafers of improved quality and uniformity. Issues such as twinning, doping uniformity, and dislocation density have been addressed through engineering controls and modification of the crystal growth environment. These measures have been highly effective in improving the quality and availability of InP substrate material. Enabling this development, are reproducible synthesis methods and control of the crystal growth environment through magnetic stabilization, phosphorus vapor pressure control, and crystal shape control.

In developing InP material, laboratories and commercial suppliers have contributed to a better understanding of the semiconductor's electronic and optical properties. Such issues as the compensation mechanism for Fe-doped semi-insulating InP and the nature of the hydrogen-related defect have been studied. Investigating the role of the defects that determine electronic properties has, in turn, led to better tools for measuring and mapping the wafers. Wafer mapping techniques such as scanning PL, spatially resolved resistivity and strain, and segregation analysis by x-ray topography and IR transmission striagraphs, have aided producers in advancing the reproducibility and reliability of InP substrates.

Future developments in bulk crystal growth will continue in response to the needs of commercial users. With the growth of new applications for InP-based microwave and millimeter-wave devices, emphasis will be on larger-diameter low-cost wafers. Wafer surface preparation and control of surface states are paramount to all epitaxial users of bulk substrates. There is also a need arising for substrates that are lattice matched to alloys based on InP. New techniques are being investigated to grow alloys of constant composition from noncongruent solutions. The key to these technological advances will be, of course, close communication between suppliers and users. Based on progress to date, it is likely that the future challenges for bulk crystal growth will be met by improved technology.

ACKNOWLEDGMENTS

The author thanks Gerald Iseler for his support and for the help rendered by his broad understanding of the subject; and Georg Müller for his encouragement to continue in an area of research where he has already contributed so much. The author's work has been supported by the Air Force Office of Scientific Research.

REFERENCES

1. M. Abe, "Ultrahigh-speed InGaAs-based HEMT technology," *Proc. 4th Int Conf. InP and Related Materials*, Newport, R.I., 1992, pp. 3–5.
2. D. Liu, J. Thaxter, and D. Bliss, "Gigahertz planar photoconducting antenna activated by picosecond optical pulses," *Opt. Lett.*, **20,** 1544–1546 (1995).
3. A. Rice, Y. Jin, X. Ma, X.-C. Zhang, D. Bliss, J. Larkin, and M. Alexander, "Terahertz optical rectification from ⟨110⟩ zinc-blende crystals," *Appl. Phys. Lett.*, **64,** 1324 (1994).
4. D. Herve, J. Viallet, S. Salaun, M. Chauvet, A. Le Corre, and B. Mainguet, "Photorefractive singlemode waveguide on InP:Fe substrate," *Proc. 6th Int. Conf. InP and Related Materials*, Santa Barbara, Calif., 1994, pp. 480–483
5. M. Chauvet, S. Hawkins, G. Salamo, M. Segev, D. Bliss, and G. Bryant, "Self trapping of planar optical beams by use of the photorefractive effect in InP:Fe," *Opt. Lett.*, **21,** 1333–1335 (1996).
6. E. Metz, R. Miller, and R. Mazelsky, "A technique for pulling single crystals of volatile materials," *J. Appl. Phys.*, **33,** 2016–2017 (1962).
7. J. B. Mullin, R. Heritage, C. Holiday, and B. Straughan, "Liquid encapsulation crystal pulling at high pressures," *J. Cryst. Growth*, **3/4,** 284 (1968).
8. K. Bachmann and E. Buehler, "The growth of InP crystals from the melt," *J. Electron. Mater.*, **3,** 279 (1974).
9. R. L. Henry and E. M. Swiggard, "InP growth and properties," *J. Electron. Mater.*, **7,** 647–657 (1978).
10. T. Harman, J. Jenco, W. Allred, and H. Goering, "Preparation and some characteristics of single crystal indium phosphide," *J. Electrochem. Soc.*, **105,** 731 (1958).
11. O. Oda, K. Katagiri, K. Shinohara, S. Katsura, Y. Takahashi, K. Kainosho, K. Kohiro, and R. Hirano, "InP crystal growth, substrate preparation and evaluation" in *Semiconductors and Semimetals*, ed. R. K. Willardson and A. C. Beer, Vol. 31, *Indium Phosphide: Crystal Growth and Characterization*, Academic Press, San Diego, Calif., 1990, Chap. 4, pp. 93–174.
12. A. Fischer and T. Pruss, "Experiences with crystal growth using liquid encapsulation," presented at American Association for Crystal Growth Conf., Gaithersburg, Md., 1969.
13. V. Swaminathan and A. Macrander, *Materials Aspects of GaAs and InP Based Structures*, Prentice Hall, Upper Saddle River, NJ, 1991, pp. 43–77.
14. J. Hirth and J. Lothe, *Theory of Dislocations*, 2nd ed., Wiley, New York, 1982, p. 309.
15. H. Gottschalk, G. Patzer, and H. Alexander, "Stacking fault energy and ionicity of cubic III–V compounds," *Phys. Status Solidi A.*, **45,** 207 (1978).
16. W. Bonner, "Reproducible preparation of twin-free InP crystals using the LEC technique," *Mater. Res. Bull.*, **15,** 63–72 (1980).
17. D. Hurle, "A mechanism for twin formation during Czochralski and encapsulated vertical Bridgman growth of III-V compound semiconductors," *J. Cryst. Growth*, **147,** 239–250 (1995).

18. G. Iseler, "Advances in LEC growth on InP crystals," *J. Electron. Mater.* **13**, 989–1011 (1984).
19. E. Beam, H. Temkin, and S. Mahajan, "Influence of dislocation density on I–V characteristics of InP photodiodes," *Semicond. Sci. Technol.*, **7**, A229-A232 (1992).
20. R. K. Jain and D. Flood, "Influence of the dislocation density on the performance of heteroepitaxial InP solar cells," *IEEE Trans. Electron Devices,* **40**, 1928–1933 (1993).
21. T. Lee and C. Burrus, "Dark current and breakdown characteristics of dislocation-free InP photodiodes," *Appl. Phys. Lett.*, **36**, 587–589 (1980).
22. W. C. Dash, "Single crystals free of dislocations," *J. Appl. Phys.*, **29**, 736–737 (1958).
23. S. Shinoyama, C. Uemura, A. Yamamoto, and S. Tohno, "Growth of dislocation-free undoped InP crystals," *Jpn. J. Appl. Phys.*, **19**, L331–L334 (1980).
24. W. Pfann, *Zone Melting*, 2nd ed., Wiley, New York, 1966, p. 11.
25. J. R. van Wazer, *Phosphorus and Its Compounds,* Vol. I, *Chemistry*, Interscience, New York, 1958, pp. 93–125.
26. G. Antypas, "Synthesis of polycrystalline InP and single crystal growth," *Proc. 2nd NATO Workshop on Materials Aspects of InP*, Lancaster, Lancashire, England, 1983, No. 8.
27. S. Hyder and G. Antypas, "Synthesis and 'in-situ' LEC growth of (100) oriented InP single crystals," *Proc. 3rd NATO Workshop on Materials Aspects of InP*, Harwichport, Mass., 1986, No. 2.
28. G. Antypas, "Preparation of high purity bulk InP," *Inst. Phys. Conf. Ser.*, **33b**, 55–59 (1977).
29. J.- P. Farges, "A method for the 'in-situ' synthesis and growth of indium phosphide in a Czochralski puller," *J. Cryst. Growth*, **59**, 665–668 (1982).
30. D. Bliss, R. Hilton, and J. Adamski, "In-situ synthesis and crystal growth of high purity InP," *Proc. 4th Int. Conf. InP and Related Materials*, Newport, R.I., 1992, pp. 262–265.
31. G. Iseler, A. Anselmo, D. Bliss, G. Bryant, and R. Lancto, "A model for rapid synthesis of large volume InP melts," *Proc. 8th Int. Conf. InP and Related Materials*, Schwäbisch-Gmünd, Germany, 1996 pp. 50–52.
32. H. Wenzl, K. Mika, and D. Henkel, "Phase relations and point defect equilibria in GaAs crystal growth," *J. Cryst. Growth*, **100**, 377–394 (1990).
33. D. Holmes, R. Chen, K. Elliot, and C. Kirkpatrick, "Stoichiometry-controlled compensation in liquid encapsulated Czochralski GaAs," *Appl. Phys. Lett.*, **40**, 46–48 (1982).
34. D. Bliss, J. Adamski, W. Higgins, V. Prasad, and F. Zach, "Phosphorus-rich InP grown by a one step in-situ MLEK crystal growth process," *Proc. 54th Int. Conf. InP and Related Materials*, Paris, 1993, pp. 66–68.
35. M. Tatsumi, T. Kawase, T. Araki, N. Yamabayashi, T. Iwasaki, Y. Miura, S. Murai, K. Tada, and S. Akai, " Growth of low-dislocation-density InP single crystals by the VCZ method," *SPIE Proc., Vol.* 1144, *Indium Phosphide and Related Materials*, 1989, p. 18.
36. K. Kohiro, M. Ohta, and O. Oda, "Growth of long-length 3 inch diameter Fe-doped InP single crystals," *J. Cryst. Growth*, **158**, 197–204 (1996).
37. M. Tatsumi, Y. Hosokawa, T. Iwasaki, N. Toyoda, and K. Fujita, "Growth and characterization of III–V materials grown by vapor-pressure-controlled Czochralski method: comparison with standard liquid-encapsulated Czochralski materials," *Mater. Sci. Eng. B*, **28**, 65–71 (1994).

38. H. Utech and M. Flemings, "Elimination of solute banding in indium antimonide crystals by growth in a magnetic field," *J. Appl. Phys.*, **37**, 2021–2024 (1966).
39. H. Chedzey and D. Hurle, "Avoidance of growth-striae in semiconductor and metal crystals grown by zone-melting techniques," *Nature* **210**, 933–934 (1966).
40. H. Miyairi, T. Inada, M. Eguchi, and T. Fukuda, "Growth and properties of InP single crystals grown by the magnetic field applied LEC method," *J. Cryst. Growth*, **79**, 291–295 (1986).
41. S. Bachowski, D. F. Bliss, B. Ahern, and R. M. Hilton, "Magnetically stabilized Kyropoulos and Czochralski growth of InP," *Proc. 2nd Int. Conf. InP and Related Materials*, Denver, Colo., 1990, pp. 30–34.
42. J. Burton, R. Prim, and W. Slichter, "The distribution of solute in crystals grown from the melt: I. Theoretical," *J. Chem. Phys.*, **21**, 1987–1991 (1953).
43. T. Hicks, and N. Riley, "Boundary layers in magnetic Czochralski crystal growth," *J. Cryst. Growth*, **96**, 957–968 (1989).
44. D. Hurle and R. Series, "Effective distribution coefficient in magnetic Czochralski growth," *J. Cryst. Growth*, **73**, 1–9 (1985).
45. A. Danilewsky, P. Dold, and K. Benz, "Growth of III-V semiconductors from metallic solutions by applying magnetic fields," *Cryst. Prop. Prep.*, **36/38**, 298–305 (1991).
46. S. Ozawa, T. Kimura, J. Kobayashi, and T. Fukuda, "Programmed magnetic field applied liquid encapsulated Czochralski crystal growth," *Appl. Phys. Lett.*, **50**, 329–331 (1987).
47. D. Bliss, R. Hilton, and J. Adamski, "MLEK crystal growth of large diameter (100) indium phosphide," *J. Cryst. Growth*, **128**, 451–456 (1993).
48. D. Bliss, R. Hilton, S. Bachowski, and J. Adamski, "MLEK crystal growth of (100) indium phosphide," *J. Electron. Mater.*, **20**, 967–971 (1991).
49. S. Kyropoulos, "Ein Verfahren zur Herstellung grosser Kristalle," *Z. Anorg. Allg. Chem.*, **154**, 308–313 (1926).
50. S. Yoshida, S. Ozawa, T. Kijima, J. Suzuki, and T. Kikuta, "InP single crystal growth with controlled supercooling during the early stage by a modified LEC method," *J. Cryst. Growth*, **113**, 221–226 (1991).
51. F. Matsumoto, Y. Okano, I. Yonenaga, K. Hoshikawa, and T. Fukuda, "Growth of twin-free <100> InP single crystals by the liquid encapsulated vertical Bridgman technique," *J. Cryst. Growth*, **132**, 348–350 (1993).
52. J. Tower, R. Tobin, P. Pearah, and R. Ware, "Interface shape and crystallinity in LEC GaAs," *J. Cryst. Growth*, **114**, 665–675 (1991).
53. R. Hirano, T. Kanazawa, Y. Itoh, T. Itokawa, H. Onodera and M. Nakamura, "Crystal growth of low EPD S-doped <100> InP by facet formation," *Proc. 5th Int. Conf. InP and Related Materials*, Paris, 1993, pp. 648–651.
54. R. Hirano, and M. Uchida, "Reduction of dislocation densities in InP single crystals by the LEC method using thermal baffles," *J. Electron. Mater.*, **25**, 347–351 (1996).
55. A. S. Jordan, A. R. Von Nieda, and R. Caruso, "The theoretical and experimental fundamentals of decreasing dislocations in melt grown GaAs and InP," *J. Cryst. Growth*, **76**, 243–262 (1986); and A. S. Jordan, "Some thermal and mechanical properties of InP essential to crystal growth modeling," *J. Cryst. Growth*, **71**, 559–565 (1985).
56. J. Völkl and G. Müller, "A new model for the calculation of dislocation formation in

semiconductor melt growth by taking into account the dynamics of plastic deformation," *J. Cryst. Growth*, **97**, 136–145 (1989).

57. V. Prasad, D. F. Bliss, and J. A. Adamski, "Thermal characterization of the high pressure crystal growth system for in-situ synthesis and growth of InP crystals," *J. Cryst. Growth*, **142**, 21 (1994),

58. H. Zhang and V. Prasad, "A multizone adaptive process model for low and high pressure crystal growth," *J. Cryst. Growth*, **155**, 47–65 (1995).

59. H. Zhang, V. Prasad, and D. F. Bliss, "Modeling of high pressure, liquid-encapsulated Czochralski growth of InP crystals," *J. Cryst. Growth*, **169**, 250–260 (1996).

60. Y. F. Zou, H. Zhang, and V. Prasad, "Dynamics of melt-crystal interface and coupled convection-stress predictions for Czochralski crystal growth processes," *J. Cryst. Growth*, **166**, 476–482 (1996).

61. H. Chung, W. Si, M. Dudley, D. F. Bliss, A. Maniatty, H. Zhang, and V. Prasad, "Characterization of defect structures in liquid encapsulated Kyropoulos grown InP single crystals," *J. Cryst. Growth*, **181**, 17–25 (1997).

62. J. Volkl, in *Handbook of Crystal Growth*, ed. D. T. J. Hurle, North-Holland, Amsterdam, 1994, Vol. 2b, Chap. 14, p. 821.

63. Y. F. Zou, G.-X. Wang, H. Zhang, V. Prasad, and D. F. Bliss, "Macro-segregation, dynamics of interface, and stresses in high pressure LEC grown crystals," *J. Cryst. Growth*, **180**, 524–533 (1997).

64. W. A. Gault, E. M. Monberg, and J. E. Clemans, "A novel application of the vertical gradient freeze method to the growth of high quality III-V crystals," *J. Cryst. Growth*, **74**, 491–506 (1986).

65. W. L. Roth, T. DeWitt, and A. Smith, " Polymorphism of red phosphorus," *J. Am. Chem. Soc.*, **69**, 2881–2885 (1947).

66. E. Monberg, H. Brown, S. Chu, and J. Parsey, "The growth of low defect density semi-insulating InP," *Proc. 5th Conf. Semi-insulating III-V Materials*, Malmö, Sweden, 1988, pp. 459–464.

67. E. Monberg, W. Gault, F. Dominguez, F. Simchock, S. N. G. Chu, and C. M. Stiles, "The growth and characterization of large size, high quality, InP single crystals," *J. Electrochem. Soc.*, **135**, 500–503 (1988).

68. E. Monberg, "Bridgman and related growth techniques," in *Handbook of Crystal Growth*, Vol. 2, ed. D. T. J. Hurle, North-Holland, Amsterdam, 1994, Chap. 2, pp. 53–97.

69. G. Müller, D. Hofmann, and N. Schafer, "Perspectives of the VGF growth process for the preparation of low-defect InP substrate crystals," *Proc. 5th Int. Conf. InP and Related Materials*, Paris, 1993, pp. 60–65.

70. D. Hofmann, Ph.D. thesis, Universität Erlangen-Nürnberg, 1992.

71. R. Tobin, M. Young, H. Helava, and X. Hu, "Vertical gradient freeze growth of III-V materials," *Final Report,* Contract F33615-92-C-5921, USAF (WL/MLPO), Dayton, OH, 1997.

72. J. J. Favier, "Macrosegregation: I. Unified analysis during non-steady state solidification," *Acta Metall.*, **29**, 197–204 (1981).

73. G. Müller, *Crystal Growth from the Melt*, Springer-Verlag, Berlin, 1988, pp. 57–60.

74. A. Ostrogorsky and G. Müller, "Normal and zone solidification using the submerged heater method," *J. Cryst. Growth*, **137**, 64–71 (1994).

75. A. Ostrogorsky, H. Sell, S. Scharl, and G. Müller, "Convection and segregation during growth of Ge and InSb crystals by the submerged heater method," *J. Cryst. Growth*, **128**, 201–206 (1993).
76. P. Gille, S. Scharl, and G. Müller, "A generalized description of solute distribution in melt growth by the submerged heater method," *J. Cryst. Growth*, **148**, 183–188 (1995).
77. W. Bonner, R. Nahory, H. Gilchrist, and E. Berry, "Semi-insulating single crystal $Ga_{1-x}In_xAs$: LEC growth and characterization," in *Semi-insulating III-V Materials*, Toronto, 1990, ed. A. Milnes and C. Miner, Adam Hilger, Bristol, Gloucestershire, England, pp. 199–204.
78. H. Ishikawa, "Theoretical gain of strained quantum well grown on an InGaAs ternary substrate," *Appl. Phys. Lett.*, **63**, 712–714 (1993).
79. T. Kusunoki, C. Takenaka, and K. Nakajima, "Growth of ternary $In_{0.14}Ga_{0.86}As$ bulk crystal with uniform composition at constant temperature through GaAs supply," *J. Cryst. Growth*, **115**, 723–727 (1991).
80. T. Suzuki, K., Nakajima, T. Kusunoki, and T. Katoh, "Multicomponent zone melting growth of ternary InGaAs bulk crystal," *J. Electron. Mater.*, **25**, 357 (1996).
81. R. Ware, R. Puechner, M. Tiernan, and M. Morris, "Bulk growth of $In_xGa_{1-x}As$," *Final Report*, Contract DAAH04-93-C-0045, DARPA, Arlington Va., 1995.
82. H. Mataré, "General considerations concerning the double-crucible method to grow uniformly doped germanium crystals of high perfection," *Solid State Electron.*, **6**, 163–167 (1963).
83. T. Ashley, J. Beswick, B. Cockayne, and C. Elliott, "The growth of ternary substrates of indium gallium antimonide by the double crucible Czochralski technique," in *Narrow Gap Semiconductors*, ed. J. L. Reno, Inst. Phys. Conf. Ser., Vol. 144, 1995, pp. 209–213.
84. G. W. Iseler, "Resistivity of bulk InP" in *Properties of InP*, ed. J. Brice and S. Adachi, INSPEC, London, 1991, pp. 25–32.
85. W. Walukiewicz, J. Lagowski, L. Jastrzebski, P. Rava, M. Lichtensteiger, C. H. Gatos, and H. C. Gatos, "Electron mobility and free-carrier absorption in InP: determination of the compensation ratio," *J. Appl. Phys.*, **51**, 2659–2668 (1980).
86. M. Benzaquen, K. Mazuruk, D. Walsh, A. Springthorpe, and C. Miner, "Determination of donor and acceptor impurity concentrations in n-InP and n-GaAs," *J. Electron. Mater.*, **16**, 111–117 (1987).
87. D. Holmes, R. Wilson, and P. Yu, "Redistribution of Fe in InP during liquid phase epitaxy," *J. Appl. Phys.*, **52**, 3396–3399 (1981).
88. R. Fornari, "Influence of iron content on electrical characteristics and thermal stability of LEC indium phosphide," *J. Electron. Mater.*, **20**, 1043–1048 (1991).
89. Y. Toudic, R. Coquille, M. Gauneau, G. Grandpierre, L. Marechal, and B. Lambert, "Growth of double doped semi-insulating InP single crystals," *J. Cryst. Growth*, **83**, 184–189 (1987).
90. A. Näser, A. Dadgar, M. Kuttler, R. Heitz, D. Bimberg, J. Hyeon, and H. Schumann, "Thermal stability of the midgap acceptor rhodium in indium phosphide," *Appl. Phys. Lett.*, **67**, 479–481 (1995).
91. G. Iseler, "Properties of InP doped with Fe, Cr, or Co," *Inst. Phys. Conf. Ser.*, Vol. 45, 1979, Chap. 2, pp. 144–153.

92. F. X. Zach, "New insights into the compensation mechanism of Fe-doped InP," *J. Appl. Phys.*, **75**, 7894 (1994).

93. R. Fornari, A. Brinciotti, E. Gombia, R. Mosca, A. Huber, and C. Grattepain, "Annealing-related compensation in bulk undoped InP," *Proc. 8th Conf. Semi-insulating III–V Materials*, Warsaw, 1994, ed. M Godlewski, World Scientific, Singapore, pp. 283–286.

94. G. Hirt, D. Wolf, and G. Müller, "Quantitative study of the contribution of deep and shallow levels to the compensation mechanisms in annealed InP," *J. Appl. Phys.*, **74**, 5538–5545 (1993).

95. K. Kainosho, O. Oda, G. Hirt, and G. Müller, "Semi-insulating behavior of undoped InP," *Mater. Res. Soc. Symp. Proc.*, Vol. 325, 1994, pp. 101–112.

96. P. B. Klein, R. L. Henry, T. A. Kennedy, and N. D. Wilsey, "Semi-insulating behavior in undoped LEC InP after annealing in phosphorus," in *Defects in Semiconductors*, ed. H. J. von Bardeleben, Vol. 10–12, *Materials Science Forum*, Trans Tech Publications, Aedermannsdorf, Switzerland, 1986, pp. 1259–1264.

97. R. Fornari, A. Brinciotti, E. Gombia, R. Mosca, and A. Sentiri, "Preparation and characterization of semi-insulating undoped indium phosphide," *Mater. Sci. Eng. B*, **28**, 95–100 (1994).

98. K. Kainosho, H. Shimakura, H. Yamamoto, and O. Oda, "Undoped semi-insulating InP by high pressure annealing," *Appl. Phys. Lett.*, **59**, 8 (1991).

99. D. Wolf, G. Hirt, and G. Müller, "Control of low Fe content in the preparation of semi-insulating InP by wafer annealing," *J. Electron. Mater.*, **24**, 93 (1995).

100. K. Kainosho, M. Ohta, M. Uchida, M. Nakamura, and O. Oda, "Effect of annealing conditions on the uniformity of undoped semi-insulating InP," *J. Electron. Mater.*, **25**, 353–356 (1996).

101. G. Hirt, D. Wolf, B. Hoffmann, U. Kretzer, G. Kuhnel, A. Woitech, D. Zemke, and G. Müller, "Mesoscopic nonuniformity of wafer-annealed semi-insulating InP," *J. Electron. Mater.*, **25**, 363 (1996).

102. G. Hirt, D. Wolf, and G. Müller, "Diffusion mechanisms controlling the preparation of annealed semi-insulating InP," presented at 8th Conf. Semi-insulating III–V Materials, Warsaw, 1994.

103. B. Goldstein, "Diffusion in compound semiconductors," *Phys. Rev.*, **121**, 1305–1311 (1961).

104. H. Zimmermann, U. Gösele, and T. Y. Tan, "Diffusion of Fe in InP via the kick-out mechanism," *Appl. Phys. Lett.*, **62**, 75–77 (1993).

105. F. X. Zach, E. E. Haller, D. Gabbe, G. Iseler, G. G. Bryant, and D. F. Bliss, "Electrical properties of the hydrogen defect in InP and the microscopic structure of the 2316 cm^{-1} hydrogen related line," *J. Electron. Mater.*, **25**, 331–335 (1996).

106. See, for example, J. Pankove and N. Johnson, eds., *Hydrogen in Semiconductors*, Academic Press, San Diego, Calif., 1991.

107. B. Pajot, J. Chevallier, A. Jalil, and B. Rose, "Spectroscopic evidence for hydrogen–phosphorus pairing in zinc-doped InP containing hydrogen," *Semicond. Sci. Tecnnol.*, **4**, 91–93 (1989).

108. R. Darwich, B. Pajot, B. Rose, D. Robein, B. Theys, R. Rahbi, C. Porte, and F. Gendron, "Experimental study of the hydrogen complexes in indium phosphide," *Phys. Rev. B*, **48**, 48 (1993).

109. C. Ewels, S. Oberg, R. Jones, B. Pajot, and P.Briddon, "Vacancy- and acceptor-H complexes in InP," *Semicond. Sci. Technol.*, **11,** 502–507 (1996).
110. R. Fornari, B. DeDavid, A. Sentiri, and M. Curti, "Reproducible thermal annealing processes for the preparation of semi-insulating undoped InP," *Proc. 8th Conf. Semi-insulating III–V Materials*, Warsaw, 1994, ed. M Godlewski, World Scientific, Singapore, pp. 35–38.
111. C. J. Miner, "Pre-processing epitaxial layer evaluation: what do you really have to measure?" *Proc. 5th Conf. Defect Recognition in Processing*, Santander, Spain, 1993 Inst. Phys. Conf. Ser., Vol. 135, pp. 147–156.
112. T. Vetter and A. Winnacker, "Characterization of InP wafers by use of a system for high resolution photoluminescence imaging," *J. Mater. Res.*, **6,** 1055–1060 (1991).
113. M. Erman, G. Gillardin, J. Le Bris, M. Renaud, and E. Tomzig, "Characterization of Fe-doped semi-insulating InP by low temperature and room temperature spatially resolved photoluminescence," *J. Cryst. Growth*, **96,** 469–482 (1989).
114. T. Vetter, R. Treichler, and A. Winnacker, "Investigation of striations in doped InP wafers by scanned photoluminescence and spatially resolved SIMS," *Semicond. Sci. Technol.*, **7,** 150–153 (1992).
115. C. J. Miner, D. G. Knight, J. M. Zorzi, and M. Ikisawa, "Characterization of slip-like defects in InGaAs epitaxial layers grown on Fe doped semi-insulating InP," *Proc. 5th Conf. Defect Recognition in Processing*, Santander, Spain, 1993, Inst. Phys. Conf. Ser., Vol. 135, pp. 181–186.
116. R. Stibal, J. Windscheif, and W. Jantz, "Contactless evaluation of semi-insulating GaAs wafer resistivity using the time-dependent charge measurement," *Semicond. Sci. Technol.*, **6,** 995–1001 (1991).
117. C. J. Miner, D. G. Knight, J. M. Zorzi, D. A. Macquistan, R. Mallard, and M. Ikisawa, "Photoluminescence and double crystal diffraction mapping of semi-insulating InP," *Proc. Semi-insulating III-V Materials Conference*, Warsaw, 1995, World Scientific, Singapore, pp. 151–155.
118. M. Martin, J. Jimenez, M. Gonzales, L. Sanz, M. Chafai, and M. Avella, "Characterization of the homogeneity of semi-insulating InP by the spatially resolved photocurrent," *Mater. Sci. Eng. B*, **20,** 105–108 (1993).
119. M. Yamada, M. Fuzukawa, Y. Yabuhara, and M. Yokogawa, "Quantitative photoelastic characterization of residual strains in LEC-grown indium phosphide (100) wafers," *Proc. 5th Int. Conf. InP and Related Materials*, Paris, 1993, pp. 69–72.
120. Y. Yabuhara, S. Kawarabayashi, N. Toyoda, M. Yokogawa, K. Fujita, and M. Yamada, "Development of low-dislocation-density semi-insulating long 2-inch InP single crystals by the VCZ method," *Proc. 5th Int. Conf. InP and Related Materials*, Paris 1993, pp. 309–312.
121. M. Fukuzawa and M. Yamada, "Fine structures of residual strain distribution in Fe-doped InP-(100) wafers grown by the LEC and VCZ methods," *J. Electron. Mater.*, **25,** 337–342 (1996).
122. R. L. Barns, "Infra-red striagraph topography for imaging defects in semiconductor crystals," *J. Electron. Mater.*, **18,** 703–309 (1989).
123. D. Carlson and D. Bliss, "Near infrared microscopy for the determination of dopant distributions and segregation in n-type InP," *Proc. 4th Int. Conf. InP and Related Materials*, Newport, R.I., 1992, pp. 515–517.

124. W. P. Dumke, M. R. Lorenz, and G. D. Pettit, "Intra- and interband free-carrier absorption and the fundamental absorption edge in n-type InP," *Phys. Rev., B,* **1,** 4668–4673 (1970).
125. J. Donecker, "Origin of infrared contrasts in InP:S and problems of optical transmission mapping," *Phys. Status Solidi A*, **127,** 275–280 (1991).
126. K. Sugii, H. Koizumi, and E. Kubota, "Precision lattice parameter measurements on doped indium phosphide single crystals," *J. Electron. Mater.* **12,** 701–712 (1983).
127. O. K. Kim and W. A. Bonner, "Infrared reflectance and absorption of n-type InP," *J. Electron. Mater.*, **12,** 827–836 (1983).
128. K. Bachmann and E. Buehler, "Phase equilibria and vapor pressures of pure phosphorus and of indium/phosphorus system and their implications regarding crystal growth of InP," *J. Electrochem. Soc.*, **121,** 835–845 (1974).
129. J. Wolk, G. Iseler, G. Bryant, E. Bourret-Courchesne, and D. Bliss, "Annealing behavior of the hydrogen-related defect in LEC indium phosphide," *Proc. 9th Int. Conf. InP and Related Materials*, Hyannis, Mass., 1997, pp. 408–411.
130. A. Zappettini, R. Fornari, and R. Capelletti, "Electrical and optical properties of semi-insulating InP obtained by wafer and ingot annealing," *Mater. Sci. Eng. B*, **45,** 147–151 (1997).

CHAPTER SIX

Metal-Organic Chemical Vapor Deposition of InP-Based Materials

TAKASHI FUKUI
Hokkaido University

6.1	Introduction	166
6.2	InP and InGaAs Growth	166
	6.2.1 New Precursors	166
	6.2.2 Surface Reconstruction During MOCVD	167
	6.2.3 Surface Morphology	168
6.3	Heteroepitaxial Growth and Interface	170
6.4	Selective-Area Growth	175
6.5	Self-Organized Growth	177
	6.5.1 InAs Quantum Dots by the Stranski–Krastanow Growth Mode	178
	6.5.2 InGaAs Quantum Disks on GaAs (311)B Substrate	178
6.6	Summary	183
	References	183

InP-Based Materials and Devices: Physics and Technology, Edited by Osamu Wada and Hideki Hasegawa.
ISBN 0-471-18191-9 © 1999 John Wiley & Sons, Inc.

6.1 INTRODUCTION

Metal-organic chemical vapor deposition (MOCVD) is one of the most important epitaxial growth techniques for all III-V compound semiconductors and their alloy compositions. Epitaxial wafers for practical InP-based devices, such as high-electron-mobility transistors (HEMTs) and optoelectronics devices, are grown primarily by MOCVD. All AlGaInP and GaInN short-wavelength laser diodes, especially, are fabricated using the MOCVD technique. MOCVD, developed in the late 1960s, has also been called metal-organic vapor-phase epitaxy (MOVPE). Now MOCVD systems for high-throughput production are commercially available from a number of speciality companies. To avoid gas turbulence, low-pressure horizontal or vertical reactor systems are used. A typical MOCVD reactor used in research is composed of a radio-frequency heated susceptor in a cold-wall quartz reactor. Trimethylgallium (TMGa), trimethylindium (TMIn), and trimethylaluminum (TMAl) are generally used as group III precursors. They are supplied to the reactor from vessels immersed in a temperature-controlled bath, through a hydrogen carrier gas. AsH_3 and PH_3 are typically used as the group V source gases. For InP epitaxial growth, a combination of TMIn and PH_3 is commonly used. The purity of these sources has been improved enormously during the past 20 years. High-purity InP has been reported by a number of authors [1,2]. Mobilities at 77 K are in the range 70,000 to 140,000 $cm^2/V \cdot s$ with a carrier concentration of 10^{14} to 10^{13} cm^{-3}. Boud et al. reported peak mobilities of 400,000 $cm^2/V \cdot s$ at 40 K [2]. For n-type conduction, SiH_4 and H_2S are used as doping gases, while for p-type conduction, CCl_4 is used as a C doping gas.

In Section 6.2, new precursors, surface reconstruction during MOCVD, and surface morphology for InP and InGaAs growth are discussed. In Section 6.3, the interfaces of InGaAs/InP and InGaP/GaAs heteroepitaxial growth are investigated. In Section 6.4, the growth mechanism and applications of selective-area MOCVD are discussed. In Section 6.5, self-organized growth of InAs and InGaAs quantum dots is described. Section 6.6 is a summary of the MOCVD of InP-based materials.

6.2 InP AND InGaAs GROWTH

6.2.1 New Precursors

For InP and InGaAs high-quality epitaxial layer growth, TMGa, TMIn, AsH_3, and PH_3 are commonly used. However, alternative precursors have also been investigated, for a variety of reasons. For example, for group III precursors, cycropentadienylindium (C_5H_5In) was studied as a In precursor to prevent serious polymerization problems between conventional trialkylindium (R_3In) and group V hydrides. Cyclopentadienyl In decomposes rapidly to metallic indium and cyclopentadiene (C_5H_6). Therefore, it has the advantage of less carbon contamination into epitaxial layers. Lattice-matched InGaAs/InP heterostructures grown on InP exhibited a quality similar to that achieved using standard precursors. The electron mobility at 77K is more than 100,000 $cm^2/V \cdot s$ at a carrier concentration of 3.9×10^{11} cm^{-2} [3].

For group V precursors, tertiarybutylarsine (TBAs) and tertiarybutylphosphine (TBP) are becoming popular as alternative sources because they pyrolyze more rapidly at lower temperature, and they are less toxic. The handling of these liquid-source bubblers is much easier, and low V/III ratios and low-temperature growth can be achieved. High-quality InP and InGaAs were obtained with TBP and TBAs. 77 K electron mobilities are more than 50,000 cm^2/V·s for InP and 84,000 cm^2/V·s for InGaAs [4,5]. These values will be improved by further purification of precursors.

6.2.2 Surface Reconstruction During MOCVD

Surface phase diagrams of P-stabilized (001) InP during metal-organic vapor-phase epitaxy were investigated as a function of substrate temperature and PH$_3$ partial pressure using a surface photoabsorption (SPA) technique [6]. In situ characterization procedures are as follows: A vertical MOCVD reactor with AsH$_3$ and PH$_3$ was used. The substrates were (001)-oriented GaAs and InP. Linearly p-polarized light from a Xe lamp irradiated the substrate surface at an incidence angle of 70°, which is close to the Brewster angle, and SPA spectra were measured using an optical multichannel analyzer. SPA spectra were also measured for well-established c(4 × 4) and (2 × 4) reconstruction surfaces on (001) GaAs grown by molecular beam epitaxy (MBE). From a comparison of the SPA spectra to those for MOCVD grown surfaces, surface reconstructions were investigated.

Figure 6.1 shows phase diagrams of GaAs and InP surfaces during growth as a function of the substrate temperature and partial pressure of AsH$_3$ and PH$_3$. For GaAs, the substrate temperature was varied from 550 to 700°C, and AsH$_3$ partial

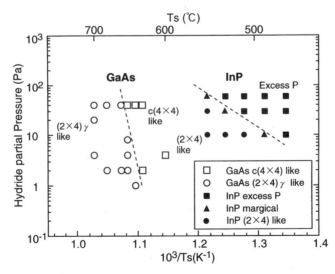

FIGURE 6.1 Phase diagram as a function of substrate temperature and hydride partial pressure. (After Ref. 6.)

pressure was varied from 1 to 40 Pa. For InP, the substrate temperature was varied from 470 to 550°C, and the PH$_3$ partial pressure was varied from 10 to 60 Pa. For GaAs, the (2 × 4)-like surface is stable above 660°C, and it evolves to a c(4 × 4)-like surface as the substrate temperature decreases and the AsH$_3$ partial pressure increases. In contrast, the InP surface is (2 × 4)-like reconstruction when the growth temperature is 550°C and PH$_3$ partial pressures are between 10 and 30 Pa. As the substrate temperature decreases and PH$_3$ partial pressure increases, it evolves to an excess-P surface, due to amorphous P excessively adsorbing onto the (2 × 4)-like surface. It should be noted that a c(4 × 4)-like surface having a P bilayer structure did not appear in InP MOCVD. The experimental results also indicate that the (2 × 4)-like surface is stable at substrate temperatures around 600°C, which is the temperature level generally used for InP MOCVD, while the c(4 × 4)-like surface is stable in GaAs MOCVD.

The difference in the P structure during InP MOCVD influences the epitaxial layer qualities. Although the conduction was n-type for all samples, higher Hall mobility wafers with lower carrier concentration were obtained under (2 × 4)-like reconstruction conditions. Suppression of excessive P adsorption is needed to obtain a high-quality InP epitaxial layer.

6.2.3 Surface Morphology

The surface morphologies of InP, InGaAs, and InAlAs grown by MOCVD were investigated using atomic force microscopy (AFM) [7]. Figures 6.2 and 6.3 show InP, InGaAs, and InAlAs surfaces grown on a (001) slightly misoriented substrate (0.08°) and a 1°-misoriented substrate at 550°C. Schematic illustrations and the degree of the interface roughness (height and size) estimated from the surface morphology are also shown.

As shown in Figs. 6.2 and 6.3, regular step arrays are observed on slightly misoriented (001) substrate, while rough surfaces due to step bunching were observed on the 1°-misoriented substrate. During growth of the InP layer, regular monolayer step arrays whose terrace width is determined by the substrate misorientation angle are formed as shown in Figs. 6.2a and 6.3a. As the growth temperature increases, the steps became smooth and two-dimensional islands disappear. Terraces about 1 μm wide without two-dimensional islands are formed during growth on the (001) slightly misoriented substrate. These results are similar to that of GaAs [8] and suggest that the surface diffusivity of In atoms is as large as that of Ga atoms under a MOCVD atmosphere. However, for the higher misorientation angle of the substrate, rough surfaces on MOCVD-grown InP caused by step bunching have often been observed by other groups [9]. The discrepancy probably arises from differences in the gas flow dynamics and the purity of the carrier gas.

Details of the surface morphologies of InGaAs layers were also observed [7]. Although both InP and GaAs grow in a step-flow mode (Fig. 6.2b), as mentioned above, InGaAs grows as a mixture of step flow and two-dimensional nucleation, forming microscopic islands and valleys less than 25 nm across. Kohl et al. [10] reported similar results observed using high-resolution transmission electron mi-

6.2 InP AND InGaAs GROWTH   169

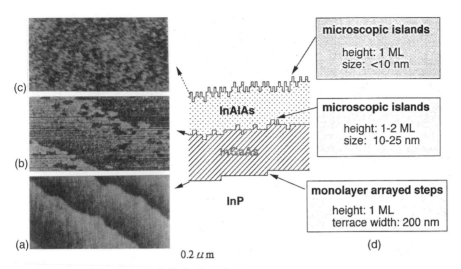

FIGURE 6.2 Interface morphologies of InAlAs/InGaAs grown at 550°C on the nominally (001) oriented (0.08° off) substrate. (*a*), (*b*), (*c*) AFM images of the surfaces of InP, InGaAs, and InAlAs, respectively; (*d*) schematic illustration of each interface (After Ref. 7.)

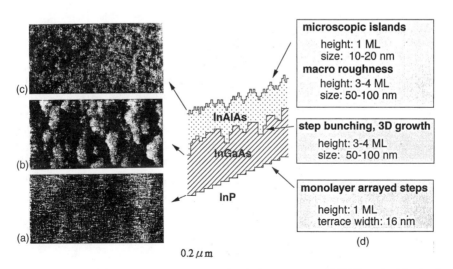

FIGURE 6.3 Interface morphologies of InAlAs/InGaAs grown at 550°C on the 1°-misoriented substrate. (*a*), (*b*), (*c*) AFM images of the surfaces of InP, InGaAs, and InAlAs, respectively; (*d*) schematic illustration of each interface. (After Ref. 7.)

croscopy (TEM), and inferred that the roughness at the interface (InP on InGaAs) was due to imperfect two-dimensional growth of InGaAs. From the results noted above, the interface (InP on InGaAs) is found to be rougher than for the inverted interface (InGaAs on InP). These morphologies are not strongly dependent on the growth temperature (550 to 650°C) and V/III ratio (70 to 220).

Since the surface diffusivity of individual In and Ga atoms is large, as described above, an attractive interaction is thought to cause some of the In atoms to combine with Ga atoms on the surface, thus decreasing surface migration length and making it easy for two-dimensional nuclei to form. The reason that the growth proceeds in a step-flow mode despite the generation of microscopic islands is thought to be the long migration length of individually migrating Ga and In atoms.

The roughness depends strongly on the substrate misorientation angles because of the change of growth mode from the mixture of step flow and two-dimensional growth to the mode of step bunching and three-dimensional growth. For InGaAs growth on highly misoriented substrates, three-dimensional growth occurs easily, with localized bunched steps resulting in a roughness 3 to 4 monolayers in height. Similar thicknesses of the transition layer (3 to 5 monolayers) between InGaAs and InP have been reported for a heterostructure grown on a 3°-misoriented InP substrate [11]. In this case, however, arsenic carryover is thought to be the most likely explanation. Interface abruptness is discussed in the next section.

These trends are different from those observed in AlGaAs growth, in which the step-flow growth mode is dominant under a wide range of growth conditions [12]. For AlGaAs, Al and Ga atoms are assumed to migrate individually even after colliding with each other.

6.3 HETEROEPITAXIAL GROWTH AND INTERFACE

For optical and transport device applications, such as semiconductor lasers and HEMTs, heteroepitaxial growth is essential, and device performance is influenced directly by the abruptness and quality of heterojunction interfaces. For InGaAs/InP and InGaP/GaAs heteroepitaxial growth, the control of gas-flow sequences is very important to obtain an abrupt heterojunction interface, particularly for group V atoms, because the group V elements As and P have deposition and desorption processes simultaneously. The atom replacements of As/P and P/As were investigated using various methods. For example, interface atomic species can be estimated from the photoluminescence peak shift for very thin quantum wells. However, these methods are rather indirect measurements. Here a novel characterization technique, the total reflection mode of the extended x-ray absorption fine structure method (EXAFS), is introduced [13]. EXAFS is used to detect atomic species on both sides of the Ga marker atomic plane in InAs/GaAs monolayer structure grown on InP by MOCVD.

Two types of samples, SL1 and SL2, were grown on (001) InP substrates at 500°C using TEGa, TEIn, AsH_3, and PH_3 as test samples to measure interface abruptness using EXAFS. Their idealized atomic structures and gas-flow sequences are shown in Fig. 6.4. Both structures include only one monolayer of Ga, designated

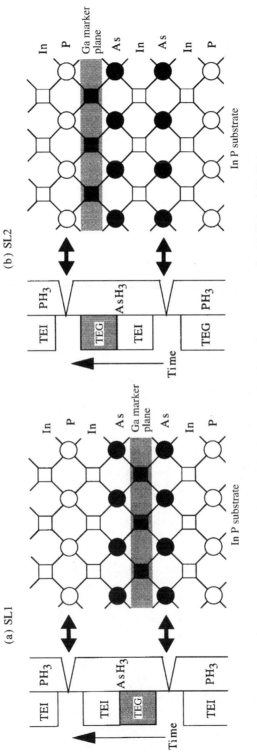

FIGURE 6.4 Idealized atomic arrangements and gas-flow sequences for (*a*) SL1 and (*b*) SL2.

as the Ga marker plane. SL1 consists of (1) an InP buffer layer (600 Å), (2) one GaAs monolayer (3 Å), (3) one InAs monolayer (3 Å), and (4) an InP cap layer (17 Å). All layers are undoped. To form an abrupt heterojunction between InAs/GaAs monolayers and InP, crystal growth is interrupted between InAs/GaAs monolayers and InP layer growth. During this interruption, the group III organometallic gas (TEIn and TEGa) flows to the bypass line, and the group V hydride gas in the reactor alternates between AsH_3 and PH_3.

The nearest-neighbor atomic species on both sides of the Ga marker layer are identified using the total reflection mode of EXAFS. This technique has one-monolayer resolution. Synchrotron radiation from the 2.5-GeV storage ring at the photon factory (the High Energy Institute at Tsukuba) is used. The sample surface is tilted 0.6° from the incident beam for total reflection measurement. The Ga atomic layer under the 17-Å InP cap layer is excited using x-rays near the Ga K-edge photon energy (10.37 keV). Ga K-fluorescence was detected using a Si(Li) solid-state detector and a multichannel analyzer. The Ga K-fluorescence signals are normalized to the incident beam intensity and are measured as a function of photon energy. In energy regions higher than the Ga K-edge, the intensity oscillations (EXAFS oscillations) are observed. The atomic species on both sides of the Ga layer are identified from Fourier transforms of the Ga K-edge EXAFS spectra.

These Fourier transforms for SL1 and SL2 are shown in Fig. 6.5. Based on a comparison with the Fourier amplitudes of Ga–As and Ga–P bonds measured for standard $In_{1-x}Ga_xAs_{1-y}P_y$ quaternary alloys, it is estimated that the peak at 2.1 Å in Fig. 6.5a and b corresponds to a Ga–As bond, and the shoulder near 1.8 Å in Fig. 6.5b to a Ga–P bond.

To form the lower interface (L1) of SL1, the gas composition inside the reactor is alternated between PH_3 and AsH_3 every 2 s during a 4-s interval. The atoms neighboring the Ga marker plane are all As atoms in Fig. 6.5a. This means that the Ga marker plane is sandwiched between two As atomic planes. Therefore, As atoms replace primarily surface-adsorbed P atoms within a 2-s interval.

The upper interface (U2) of the SL2 sample is formed by changing the gas composition inside the reactor from AsH_3 to PH_3 within a 2-s interval. However, the Ga K-EXAFS spectrum for SL2 (not shown in Fig. 6.5) is almost the same as that for SL1, indicating that the neighboring atoms are all As. This means that the surface-adsorbed As atoms on the Ga marker plane do not change P atoms within 2 s. Next, the gas composition is changed from AsH_3 to PH_3 with a 6-s interval for SL2 sample growth. In this case, both Ga–As and Ga–P bonds are observed, as shown in Fig. 6.5b. Based on comparison with the Fourier amplitudes of As and P atoms measured for standard $In_{1-x}Ga_xAs_{1-y}P_y$ quaternary alloys, it is estimated that As and P atoms exist in approximately equal amounts. Therefore, the surface-adsorbed As atoms completely desorb during the long growth interruption. P atoms then adsorb on top of the Ga atomic plane.

Using these results, the incorporation processes of group V atoms are discussed. The total number of surface sites for group V atoms is denoted as N_s. The numbers of these sites occupied by As and P are $N_{As}(t)$ and $N_P(t)$, respectively. The differential equations describing the covering of a group V atom monolayer are given in

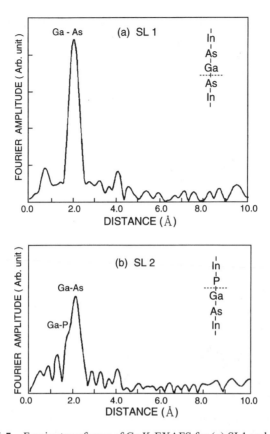

FIGURE 6.5 Fourier transforms of Ga K-EXAFS for (a) SL1 and (b) SL2.

Ref. 13, eqs. (1)–(4). The surface-covering ratio of group V atoms, $[N_{As}(t) + N_P(t)]/N_s$, is assumed to be almost unity. This is due to the high AsH_3 and PH_3 pressures. Boundary conditions for forming the lower interface in SL1 are given as $P(AsH_3) = 0$ for $t < 0$, and $P(PH_3) = 0$ for $t > 0$, where P is the partial pressure in the gas phase.

From these differential equations and boundary conditions, the solid composition of P for $t > 0$ is obtained in the following simplified form:

$$\frac{N_P(t)}{N_s} = \exp(-d_P t) \quad (1)$$

The desorption rate, d_P, of P determines the solid composition of P. The surface adsorption time, τ_P ($= 1/d_P$), of P atoms is less than 2 s for SL1. A similar approach was used for SL2. The surface adsorption time, τ_{As} ($= 1/d_{As}$) of As atoms is 2 to 6 s for SL2. This is longer than τ_P. In these discussions, the delay times between gas-line switching and gas composition change at the crystal surface have been assumed to be negligible. The incorporation process for the upper heterojunction formation for SL2 is shown schematically in Fig. 6.6.

To form an abrupt heterojunction, the growth interval must be longer than the surface adsorption time. The longer residence time of As on the surface than that of P is probably caused by the higher bonding energy of As to Ga. These phenomena are commonly observed for the growth of InGaAs/InP heterostructures and show significant effects, especially for forming thin InGaAs(P)/InP quantum-well structures.

Next, the effects of interface abruptness to band offset are investigated. As an example, InGaP/GaAs heterostructures are discussed here briefly. Band offset is also an important parameter in fabricating optical and high-speed devices. The effects of AsH_3 gas-flow sequences after growth of an underlying GaAs on the band offset

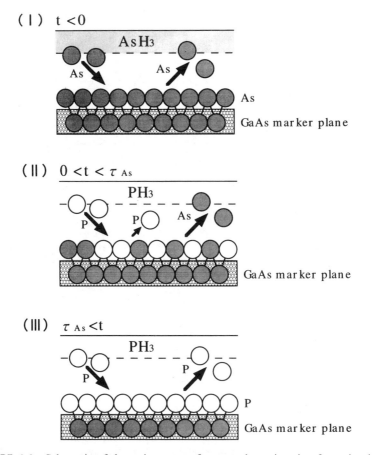

FIGURE 6.6 Schematic of dynamic process for upper heterojunction formation in SL2. Group V gas-flow changes from AsH_3 to PH_3.

and interface charge density were studied for InGaP/GaAs heterointerfaces by using capacitance–voltage (C–V) measurements [14]. When AsH_3 cover time increases from 3 s to 150 s, the conduction band offset at the InGaP/GaAs heterointerface is reduced from 0.2 eV to about 0 eV, while the interface charge density increases from 6×10^{10} to about 8×10^{11} cm^{-2}. In these experiments, the PH_3 purge time is 150 s before InGaP growth. A two-dimensional simulation reveals the interface charge to be shallow donor-type impurity. An atomically flat terrace structure is observed for long AsH_3 cover time samples by AFM.

Variation in the interface morphology of GaAs arises from excess As atoms left by the long AsH_3 cover time during the growth interval and have also induced variation in conduction band offset and interface charge density. Although the amount and location of As in InGaP are not clear, As might originate from the lattice strain, which makes donorlike defects. The reason for the conduction-band offset reduction might be the ordering of InGaP at the interface formed on atomically flat GaAs terraces or by interface strain. These phenomena do not depend on the AsH_3 flow rate. The results show that an appropriate cover time of AsH_3 after the GaAs growth is also important for obtaining high-quality InGaP/GaAs interfaces.

6.4 SELECTIVE-AREA GROWTH

Selective-area (SA) MOCVD is a promising technique for novel photoelectronic integrated device and quantum device fabrication using masked or patterned substrates. To obtain lateral dimensional control of the growing films, SA MOCVD has been investigated first for GaAs and AlGaAs, and then for InP, GaInAs, and GaInP. There are a lot of contributions in this research field [15–24]. Until now, SA MOCVD has been used widely in a variety of applications, such as for low-dimensional quantum confinement structures, integration of waveguides with lasers and detectors, formation of source and drain contacts in field-effect transistors, and fabrication of tapered waveguides.

The ability to control lateral dimensions has been used to achieve quantum wire and quantum dot structures in both GaAs- and InP-based material systems [15–18]. Another feature of SA MOCVD is lateral thickness and composition variation along the opening in a mask, which is applied to fabricate waveguides with tapered thickness for efficient coupling of fibers to detectors or lasers [19]. It has also been used for selective deposition of the layers for field-effect transistors (FETs) and of the layers needed for low-resistance source and drain contacts for FETs [20]. Selective-area epitaxy has also been used in regrowth steps in the fabrication of buried heterostructure (BH) lasers [21]. Here we discuss only the basics of SA MOCVD rather than details of a variety of examples.

In MOCVD, growth takes place far from equilibrium between the solid and vapor phases. Therefore, selectivity can be obtained only under limited growth conditions and mask sizes. In general, SA MOCVD has been carried out under conditions of lower reactor pressure, lower reagent flux, and higher growth temperature. Nevertheless, deposition on the masked region is frequent. The mask materials used are SiO_2 and SiN_x, and the quality of the mask also influences the selectivity.

Figure 6.7 shows a scanning electron microscope image of a typical example of selective-area growth of GaAs on (001) GaAs substrate pertially masked with SiN_x. Square openings are formed on a masked substrate by using a combined electron beam lithography and wet chemical etching technique. Uniform pyramid structures were observed. The pyramidal structures are used to form the quantum dots [22].

The detailed mechanisms of SA MOCVD have been investigated by many researchers. There are two possible mechanisms for selective-area growth. One proposes that the diffusion of group III atomic species on the mask surface is the main cause, and the other proposes that the main cause is desorption and gas-

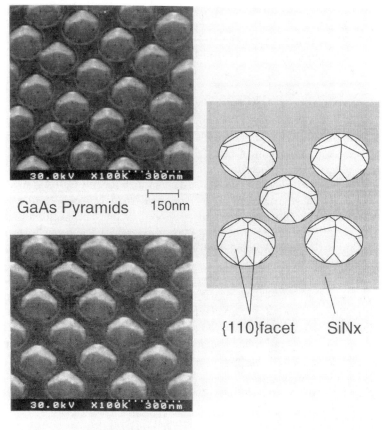

FIGURE 6.7 Scanning electron microscope image of selective-area growth of GaAs pyramid structures on (001) GaAs substrate partially masked with SiN_x.

phase diffusion. Lateral concentration gradients of group III materials enhance the lateral diffusion in the gas phase. The growth rate in the opening area is, therefore, higher than that on unmasked substrate. The enhancement is higher the smaller the area ratio of the opening. To clarify the gas-phase diffusion effects, the two-dimensional diffusion equation was solved numerically by Galeuchet, assuming that the gas velocity is zero in this region and neglecting chemical reactions occurring in the gas phase and on the substrate surface [23]. The analysis clearly shows that the gas-phase concentration is strongly affected by the presence of nongrowing dielectric regions. The lateral concentration gradient depends on the fill factor and the concentration gradient normal to the surface. The results show that growth-rate enhancement is caused primarily by gas-phase diffusion, especially for a mask size wider than several micrometers. However, if the mask size is small enough (less than 2 μm), the surface diffusion on the mask also contributes greatly to the growth-rate enhancement.

SA MOCVD of lattice-mismatched ternary materials such as GaInAs and GaInP encounters significant problems in that substantial compositional deviations occur in the lateral direction [23,24]. Differences in either the surface diffusion rates on the mask or the gas-phase diffusion rates among various species are thought to be the origin of compositional deviations.

6.5 SELF-ORGANIZED GROWTH

Nanostructure fabrication of laterally confined semiconductors, such as quantum wires and quantum dots, is one of the major challenges in atomic-scale engineering and in the field of microstructure materials science. If this can be done successfully, quantum wire and quantum dot devices or single-electron transistors can be realized, which will have an important impact on future microelectronics and optoelectronics. Many researchers have tried to fabricate quantum structures by lateral patterning of two-dimensional heterostructures with lithographic techniques [25,26] or by crystal growth on masked and prepatterned substrates, as described in Section 6.4 [15–18].

Reports of the natural formation of quantum structures during crystal growth have also attracted considerable interest, because of no processing damage being introduced in the resulting quantum structures. Quantum wires and quantum dots have been grown using step structures on vicinal surfaces [27,28] and the breakup of high-index surfaces into arrays of nanometer-scale facets [29,30]. Independently, the three-dimensional growth mode of thin layers in lattice-mismatched systems has been examined for the fabrication of quantum dots. Various structures have been grown in InGaAs- and InP-related material systems by molecular beam epitaxy (MBE) and MOVPE [31,32]. In the following section, two promising methods, InAs and InGaAs self-organized quantum dots using the Stranski–Krastanow growth mode and InGaAs quantum disks on a GaAs high-index surface, are introduced.

6.5.1 InAs Quantum Dots by the Stranski–Krastanow Growth Mode

InAs quantum dots were formed on (001) GaAs by using a lattice-mismatch-induced three-dimensional growth mode, which is called the Stranski–Krastanow (S-K) growth mode. InAs quantum dot structures are formed spontaneously during InAs thin-layer growth on GaAs. Above the critical average thickness of InAs deposition (1.6 to 1.7 monolayers), growth-mode transition occurs from two-dimensional layer growth to three-dimensional island growth. The lateral size of well-developed InAs dots is from 10 to 140 nm. Distribution of these quantum dots is not random but aligned notably in a certain crystal orientation. Figure 6.8 shows typical AFM images of InAs dots on (001) GaAs grown by MOCVD. Dot density depends strongly on InAs coverage and is as high as 10^{10} cm^{-2}. Figure 6.9 shows a photoluminescence (PL) spectrum of InAs quantum dots buried in GaAs measured at 4 K. A broad spectrum is observed for emission from quantum dots. This is caused by large size fluctuations. InAs/GaAs and InGaAs/GaAs quantum dots are studied primarily by molecular beam epitaxy (MBE) and partly by MOCVD [34,35]. Regardless of the size distribution of quantum dots, QD lasers with reasonably low threshold current densities (below 100 A/cm^2) and high characteristic temperature (425 K) have been reported [34].

6.5.2 InGaAs Quantum Disks on GaAs (311)B Substrate

A novel self-organizing phenomenon in the MOCVD growth on high-index GaAs (n11)B substrates, introduced by Nötzel, overcomes many difficulties associated with the direct synthesis of quantum dot structures [36,37]. After the growth of an epitaxial InGaAs layer over an AlGaAs buffer layer, the strained InGaAs film naturally arranges into nanometer-scale islands that are buried beneath AlGaAs sponta-

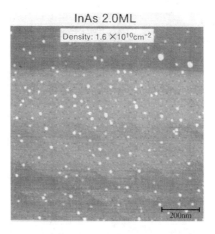

FIGURE 6.8 AFM image of InAs quantum dots grown on (001) GaAs using an S-K growth mode. (From Ref. 33.)

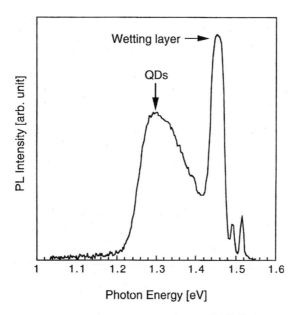

FIGURE 6.9 PL spectrum of InAs quantum dots on (001) GaAs measured at 4 K.

neously, due to lateral mass transport. During this process, well-ordered and high-density arrays of AlGaAs microcrystals containing disk-shaped InGaAs quantum dots are formed.

The microcrystals are formed after the growth of nominal-3.5-nm-thick $In_{0.4}Ga_{0.6}As$ layers over 50-nm-thick $Al_{0.5}Ga_{0.5}As$ buffer layers at a temperature of 720°C. The homogeneity in size and the ordering are optimum for the (311)B surface orientation. This exceptional role of (311)B planes could be connected to its nominal composition of equal units of the singular (100) and (111) planes that might provide the highest degree of anisotropy of surface diffusion and/or atomic arrangement, which is an important prerequisite for ordering phenomena.

The dependence of the size of the AlGaAs microcrystals and InGaAs disks on the In composition is shown in the AFM images in Fig. 6.10a and b for large and small disks with nominal In composition of 0.2 and 0.4, respectively. The size of microcrystals is characterized by the base width measured at half-height along the [−233] direction and the height itself. The base width at half-height is found to be equal to the average base width deduced from scanning electron microscopy (SEM) measurements and therefore provides a good measure of the size. Figure 6.10a shows the AlGaAs microcrystals with a 220-nm base width formed by nominal-10-nm-thick $In_{0.2}Ga_{0.8}As$ layers grown at 800°C. The appearance of the faceted surface of the microcrystals and their strong positional alignment are clearly observed. The insets show the schematic and cross-sectional SEM image after stain etching of the AlGaAs microcrystals coating the InGaAs disks with a diameter of 150 nm. From this image the formation mechanism of the disks is easily understood. Since no

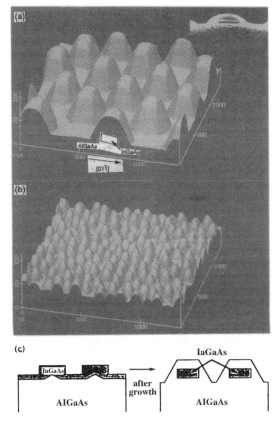

FIGURE 6.10 AFM image of (*a*) AlGaAs microcrystals with 220-nm base width formed by nominally 10-nm-thick $In_{0.2}Ga_{0.8}As$ (311)B, and (*b*) AlGaAs microcrystals with 70-nm base width formed by nominally 3.5-nm-thick $In_{0.4}Ga_{0.6}As$ (311)B. The insets show the schematic and cross-sectional SEM image of the InGaAs containing structure. (*c*) Schematic of the self-organizing InGaAs disk formation. (After Ref. 36.)

AlGaAs has been grown after the InGaAs layer, InGaAs, rearrange themselves into islands that are buried spontaneously beneath the AlGaAs after growth, due to lateral mass transport from the buffer layer. The growth mechanism is discussed in detail in Refs. 33 and 34. The direction of alignment is about 45° off the [01-1] azimuth and does not correspond to any specific low-index azimuts of the (311)B plane. Hence this seems to indicate that the alignment is due primarily to the appearance of crystal facets of AlGaAs microcrystals, providing further anisotropy for surface migration during the self-organizing process rather than arising from the specific microscopic structure of the initial surface. The small microcrystals in Fig. 6.10*b* are formed by 3.5-nm-thick $In_{0.4}Ga_{0.6}As$ layers grown at 720°C. No dependence of microcrystal size on the growth temperature is observed. The base width is 70 nm, and this corresponds to the InGaAs quantum disk with the average diameter of 30 nm. The AFM images in Figs. 6.10a and b demonstrate that under optimized growth conditions, the high density and positional alignment of the microcrystals are maintained down to the quantum-size regime.

The uniformity in size of the microcrystals is maintained upon size reduction. This is clearly seen in the histograms shown in Figs. 6.11a and b of the base width and of the height of the large and small microcrystals imaged in Figs. 6.10a and b. The full scale is twice the average base width in both cases, for comparison. For both samples, the size distribution of the base width and of the height is well within 10% of the average values. This means that the relative size distribution of the microcrystals does not depend on their actual size. Moreover, the shape of the microcrystals characterized here by the ratio of base width and height does not change with size. This behavior highlights the advantages of this natural formation phenomenon over conventional processing methods, where relative size fluctuations increase significantly and control of the shape of microstructures is more and more difficult as their size is reduced.

To investigate the area distribution of AlGaAs microcrystals, the average distance (square root of the area density) and base width versus the nominal thickness of the InGaAs layer are observed. The fraction of In composition is kept at 0.4. The average distance of the microcrystals decreases with increase in InGaAs layer thickness, while the average base width and height remain almost unchanged. On the other hand, for the same InGaAs layer thickness, the average distance of the microcrystals does not depend on the In composition (i.e., the size) [33]. Therefore, for InGaAs quantum disks, the size and distance can be controlled independently by the In composition and thickness of the InGaAs layer, respectively.

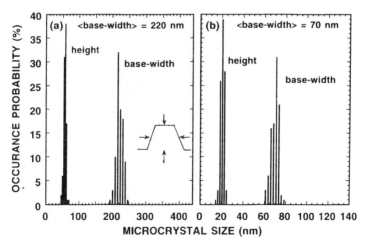

FIGURE 6.11 Histogram of the base width measured at half-height and of the height of AlGaAs (311)B microcrystals with (a) 220 nm and (b) 70 nm average base width. (After Ref. 36.)

The formation of zero-dimensional microcrystals is also observed on InP (311)B substrates for $Ga_{0.2}In_{0.8}As$ layers grown over $Al_{0.48}In_{0.52}As$ buffer layers lattice matched to InP. For GaInAs layers grown directly on InP buffer layers, the microcrystals exhibit less pronounced faceting and ordering. Altogether, the natural evolution of strained quantum disks seems to be a rather common feature in the heteroepitaxial growth of strained layers on high-index semiconductor surfaces.

Figure 6.12a shows the PL and photoluminescence exitation (PLE) spectra at room temperature of 200-nm pitch-coupled $In_{0.25}Ga_{0.75}As$ disks and a reference (100) quantum well grown side by side at a reduced temperature of 750°C on GaAs (311)B and (100) substrates. For the PL measurements, the InGaAs quantum disks are overgrown by 50-nm $Al_{0.5}Ga_{0.5}As$ after 3 min of growth interruption. The linewidth of the disks is only 13 meV, compared to the 27-meV linewidth of the quantum well. First, this narrow disk linewidth indicates smoothing and ordering of the interfaces during disk formation. This behavior is opposite to that of (100) surfaces, where coherent islanding is known to result in PL line broadening. This ordering on (311)B surfaces seems to be caused by the strong interactive behavior of the buffer layer and the strained InGaAs film during disk formation. A small linewidth at room temperature (13 meV) indicates reduced thermal broadening, due to efficient lateral localization and confinement of the photogenerated carriers in the disks. This is consistent with energy dispersive spectroscopy during transmis-

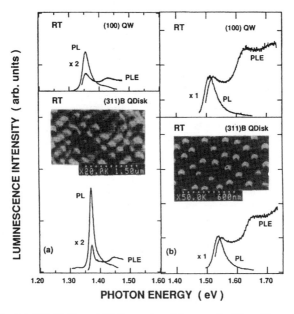

FIGURE 6.12 (a) 300-K PL and PLE spectra of the coupled $In_{0.25}Ga_{0.75}As$ (311)B quantum disks and reference (100) quantum well. The SEM image shows the surface of the coupled disks. (b) 300-K PL and PLE spectra of the 30-nm-diameter $In_{0.4}Ga_{0.6}As$ (311)B quantum disks and reference quantum well. The inset shows the SEM image of AlGaAs microcrystals coating the InGaAs disks. (After Ref. 36.)

sion electron microscopy, which shows the In composition in the AlGaAs microcrystals decreasing from the center to the side and top. This In profile, presumably established due to In segregation during island formation, suggests that the confining area is smaller than the disk diameter as determined from SEM.

Figure 6.12b presents the room-temperature PL and PLE spectra of the 30-nm-diameter isolated disks and reference quantum well. Even for disks with a diameter of 30 nm, the luminescence efficiency is not reduced compared to that of the reference (100) quantum well. The linewidth of the disks is still smaller than that of the quantum well, although broadened compared to that of the larger disks. This is attributed to the fact that alloy fluctuations are enhanced for larger In composition and correspondingly become more significant for reduced disk size. The well-resolved resonances in PLE, however, reveal that the structural perfection of the disks is maintained for diameters as small as 30 nm. The PL line of the disks is always blue shifted compared to that of the reference quantum well, but a pronounced red shift is expected from the enhanced layer thickness in the disks and the elastic strain relaxation during islanding. This red shift, could, however, be compensated due to enhanced In segregation of the In-containing structure during formation of the disks. Therefore, we cannot separate the lateral quantum confinement effect that is assumed to contribute to the observed blue shift of the PL lines.

Studies of strained disk lasers have also been reported. The active region has been made from 6-nm-thick $In_{0.25}Ga_{0.75}As$ layers. As expected from the excellent optical properties, the disk lasers operate in continuous-wave mode at room temperature with a threshold current of 23 mA, which is considerably lower than that (27 mA) of the quantum well laser on (100) [38].

6.6 SUMMARY

In this chapter we have described the growth of InP and related materials by MOCVD. The advantages of this growth method are that high quality and highly uniform III-V epitaxial layers can be obtained over a large area. MOCVD plays an important role in both optoelectronic and transport device technologies, and practical application of MOCVD is expected to widen with time to contribute to the industry by producing as much as 25% of epitaxially grown devices by the year 2000 [39]. Details of optoelectronic and transport devices such as lasers, HEMT, and HBT fabricated using MOCVD techniques are discussed in succeeding chapters. MOCVD is also important for the fabrication of future quantum nanodevices, due to the promising features of the selective-area growth method and/or self-organized growth mode.

REFERENCES

1. M. Razeghi, Ph. Maurel, M. Defour, F. Omnes, G. Neu, and A. Kozachi, "Very high purity InP epilayer grown by metalorganic chemical vapor deposition," *Appl. Phys. Lett.,* **52,** 117–119 (1988).

2. J. M. Boud, M. A. Fisher, D. Lancefield, A. R. Adams, E. J. Thrush, and C. G. Cureton, "Low temperature electron transport properties of exceptionally high purity InP," *Inst. Phys. Conf. Ser.*, **91**, 801–804 (1988).

3. M. Usuda, K. Sato, R. Takeuchi, K. Onuma, and T. Udagawa, "High-mobility $Ga_{0.47}In_{0.53}As/InP$ heterostructure by atmospheric-pressure MOVPE using cyclopentadienyl indium," *Proc. 7th Int. Conf. InP and Related Materials*, Sapporo, Japan, 1995, p. 835.

4. R. Beccard, D. Schmitz, J. Knauf, G. Lengrling, F. Schulte, and H. Jurgensen, "MOVPE growth of GaAs and InP based compounds in production reactors using TBAs and TBP," *Proc. 8th Int. Conf. InP and Related Materials*, Schwäbisch-Gmünd, Germany, 1996, p. 507.

5. H. Protzmann, F. Hohnsdorf, Z. Spika, W. Stolz, E. O. Gobel, M. Muller, and J. Lorberth, "Properties of $(Ga_{0.47}In_{0.53})As$ epitaxial layers grown by metalorganic vapor phase epitaxy (MOVPE) using alternative arsenic precursors," *J. Cryst. Growth*, **170**, 155 (1997).

6. Y. Kobayashi and N. Kobayashi, "Chemical structure of InP surface in MOVPE studied by surface photo-absorption," *Proc. 7th Int. Conf. InP and Related Materials*, Sapporo, Japan, 1995, p. 225.

7. M. Shinohara, H. Yokoyama, and K. Wada, "Roughness on resonant tunneling characteristics," *Proc. 8th Int. Conf. InP and Related Materials*, Schwäbisch-Gmünd, Germany, 1996, p. 400.

8. M. Shinohara, M. Tanimoto, H. Yokoyama, and N. Inoue, "Behavior and mechanism of wide terrace formation during metalorganic vapor phase epitaxy of GaAs and related materials," *J. Cryst. Growth*, **145**, 113 (1994).

9. V. Thévenot, V. Souliere, H. Dumont, Y. Monteil, J. Bouix, P. Regreny, and T.-M. Duc, "Behaviour of vicinal InP surfaces grown by MOVPE: exploitation of AFM images," *J. Cryst. Growth*, **170**, 251 (1997).

10. A. Kohl, S. Juillaguet, B. Fraisse, R. Schmedler, F. Royo, H. Peyre, F. Bruggeman, K. Wolter, K. Leo, H. Kurz, and J. Camassel, "Growth and characterization of $In_{0.53}Ga_{0.47}As/In_xGa_{1-x}As$ strained-layer superlattices," *Mater. Sci. Eng. B Solid-State Mater. Adv. Technol. B*, **21(2/3)**, 244 (1993).

11. K. W. Carey, R. Hull, J. E. Fouquet, F. G. Kellert, and G. R. Trott, "Structural and photoluminescent properties of GaInAs quantum wells with InP barriers grown by organometallic vapor phase epitaxy," *Appl. Phys. Lett.* **51**, 910 (1987).

12. T. Fukui, "Growth and characterization of InAs/GaAs monolayer structures," *J. Cryst. Growth*, **93**, 301 (1988).

13. L. Samuelson, P. Omling, H. Titze, and H. G. Grimmeiss, "Organometallic epitaxial growth of $GaAs_{1-x}P_x$," *J. Phys.*, **43**, C5–323 (1982).

14. Y. K. Fukai, F. Hyuga, T. Nittono, K. Watanabe, and H. Sugahara, "Improvement of InGaP/GaAs heterointerface quality by controlling AsH_3 flow conditions," presented at 1996 MRS Fall Meeting.

15. S. Ando and T. Fukui, "Facet growth of AlGaAs on GaAs with SiO_2 gratings by MOCVD and applications to quantum well wires," *J. Cryst. Growth*, **98**, 646 (1989).

16. T. Fukui, S. Ando and Y. K. Fukai, "Lateral quantum well wires fabricated by selective metalorganic chemical vapor deposition," *Appl. Phys. Lett.*, **57**, 1209 (1990).

17. T. Fukui, S. Ando, and Y. Tokura, "GaAs tetrahedral quantum dot structures fabricated using selective area metalorganic chemical vapor deposition," *Appl. Phys. Lett.*, **58**, 2018 (1991).

18. Y. D. Galeuchet, H. Rothuizen, and P. Roentgen, "In situ buried GaInAs/InP quantum dot arrays by selective area metalorganic vapor phase epitaxy," *Appl. Phys. Lett.,* **58,** 2423 (1991).
19. E. Colas, A. Shahar, J. B. D. Soole, W. J. Tomlinson, J. R. Hayes, C. Caneau, and R. Bhat, "Lateral and longitudinal patterning of semiconductor structures by crystal growth on nonplanar and dielectric-masked GaAs substrates," *J. Cryst. Growth,* **107,** 226 (1991).
20. I. Imamura, N. Yokoyama, T. Ohnishi, S. Suzuki, K. Nakai, H. Nishi, and A. Shibatomi, "A WSi/TiN/Au gate self-aligned GaAs MESFET with selectively grown n^+-layer using MOCVD," *Jpn. J. Appl. Phys.,* **23,** L342 (1984).
21. A. W. Nelson, W. J. Devlin, R. E. Hobbs, C. G. D. Lenton, and S. Wong, "High-power, low-threshold BH lasers operating at 1.25 μm grown entirely by MOVPE," *Electron. Lett.,* **21,** 888 (1985).
22. K. Kumakura, K. Nakakoshi, J. Motohisa, T. Fukui, and H. Hasegawa, "Novel formation method of quantum dot structures by self-limited selective area metalorganic vapor phase epitaxy," *Jpn. J. Appl. Phys.,* **34,** 4387 (1995).
23. Y. D. Galeuchet and P. Roentgen, "Selective area MOVPE of GaInAs/InP heterostructures on masked and nonplanar (100) and {111} substrates," *J. Cryst. Growth,* **107,** 147 (1991).
24. O. Kayser, B. Opitz, R. Westphalen, U. Niggebrügge, K. Schneider, and P. Balk, "Selective embedded growth of GaInAs by low pressure MOVPE," *J. Cryst. Growth,* **107,** 141 (1991).
25. M. A. Reed, J. N. Randall, J. R. Aggarwal, R. J. Matyi, T. M. Moore, and A. E. Wetsel, "Observation of discrete electronic states in a zero-dimensional semiconductor nanostructure," *Phys. Rev. Lett.,* **60,** 535–537 (1988).
26. K. Kash, B. P. Van der Gaag, D. D. Mahoney, A. S. Gozdz, L. T. Florez, J. P. Harbison, and M. D. Sturge, "Observation of quantum confinement by strain gradients," *Phys. Rev. Lett.,* **67,** 1326–1329 (1991).
27. P. M. Petroff, A. C. Gossard, and W. Wiegmann, "Structure of AlAs–GaAs interfaces grown on (100) vicinal surfaces by molecular beam epitaxy," *Appl. Phys. Lett.,* **45,** 620–622 (1984).
28. T. Fukui and H. Saito, "Natural superstep formed on GaAs vicinal surface by metalorganic chemical vapor deposition," *Jpn. J. Appl. Phys.,* **29,** L483–L485 (1990).
29. R. Nötzel, N. Ledentsov, L. Däweritz, M. Hohenstein, and K. Ploog, "Direct synthesis of corrugated superlattices on non-(100) oriented surfaces," *Phys. Rev. Lett.,* **67,** 3812–3815 (1991).
30. R. Nötzel, L. Däweritz, and K. Ploog, "Topography of high- and low-index GaAs surfaces," *Phys. Rev. B,* **46,** 4736–4743 (1992).
31. J. M. Moison, F. Houzay, F. Barthe, L. Leprince, E. Andr, and O. Vatel, "Self-organized growth of regular nanometer-scale InAs dots on GaAs," *Appl. Phys. Lett.,* **64,** 196–198 (1994).
32. Y. Nabetani, T. Ishikawa, S. Noda, and A. Sasaki, "Initial growth stage and optical properties of a three-dimensional InAs structure on GaAs," *J. Appl. Phys.,* **76,** 347–351 (1994).
33. T. Umeda, K. Kumakura, J. Motohisa, and T. Fukui, "InAs quantum dot formation on GaAs pyramids by selective area MOVPE," submitted.
34. D. Bimberg, N. Ledentsov, M. Grundmann, N. Kirstaedter, O. G. Schmidt, M. H. Mao, V. M. Ustinov, A. Yu. Egorov, A. E. Zhukov, P. S. Kopev, Zh. I. Alferov, S. S. Ruvimov, U.

Gösele, and J. Heydenreich, "InAs–GaAs quantum dots: from growth to lasers," *Phys. Status Solidi.,* **194,** 159 (1996).

35. J. Oshinowo, M. Nishioka, S. Ishida, and Y. Arakawa, "Highly uniform In–GaAs/GaAs quantum dots (~15nm) by metalorganic chemical vapor deposition," *Appl. Phys. Lett.,* **65,** 1421–1423 (1994).

36. R. Nötzel, J. Temmyo, and T. Tamamura, "Self-organized growth of strained InGaAs quantum disks," *Nature,* **369,** 131–133 (1994).

37. R. Nötzel, J. Temmyo, and T. Tamamura, "Self-organization of boxlike microstructures on GaAs(311)B surfaces by metalorganic vapor-phase epitaxy," *Jpn. J. Appl. Phys.,* **33,** L275–L278 (1994).

38. J. Temmyo, E. Kuramochi, M. Sugo, T. Nishiya, H. Kamada, R. Nötzel, and T. Tamamura, "A novel disk laser with active region from strained InGaAs nanostructure self-organizing during MOVPE," *Proc. 14th IEEE Int. Semiconductor Laser Conf.,* Hawaii, 1994, postdeadline paper PD4.

39. R. L. Moon, "MOVPE: is there any other technology for optoelectronics," *J. Cryst. Growth,* **170,** 1 (1997).

CHAPTER SEVEN

InP and Related Compound Growth Based on MBE Technologies with Gaseous Sources

HARALD HEINECKE
BMW AG

7.1	Material Systems and Growth Technologies	188
7.2	MOMBE Growth System	189
	7.2.1. Overview of the Growth System	189
	7.2.2. Specific MOMBE System Design Criteria	192
7.3.	Surface-Selective Growth in MOMBE	198
	7.3.1 Orientation Dependence of Growth	205
	7.3.2 Planar Selective-Area Growth	206
	7.3.3. Embedded Growth	213
	7.3.4. Doping Using Gas Sources	215
7.4.	Technology Push	222
	7.4.1 New Precursors	222
	7.4.2 Multiwafer Growth	227
	7.4.3 Novel Device Development and Integration Capability	229
7.5.	Summary and Future Challenges	237
	Acknowledgments	238
	References	238

InP-Based Materials and Devices: Physics and Technology, Edited by Osamu Wada and Hideki Hasegawa.
ISBN 0-471-18191-9 © 1999 John Wiley & Sons, Inc.

7.1 MATERIAL SYSTEMS AND GROWTH TECHNOLOGIES

For applications in device structures on InP substrates, basically any mixtures of the binary systems InAs, GaP, InSb, and AlAs can be utilized for tuning the semiconductor characteristics of the individual films in multilayer structures. However, the accumulation of strain in thick (0.5 to 4 μm) structures can lead to a relaxation process and hence a degradation of the structure due to the difference in lattice constants. For $Ga_{0.47}In_{0.53}As$ (E_{gap} = 0.755 eV) and $Al_{0.48}In_{0.52}As$ (E_{gap} = 1.453 eV) [1] the lattice constant is identical to the binary substrate material InP. By creating mixtures of both ternary materials, the entire energy range between 0.755 and 1.453 eV is accessible using the quaternary materials GaInAlAs. On the other hand, any mixture of $Ga_{0.47}In_{0.53}As$ and InP enables a similar energy range using GaInAsP layers [2,3]. The material systems GaInAlAs/InP and GaInAsP/InP both require ultimate control over the incorporation rate of the chemical elements to maintain lattice-matched conditions in thicker layered structures. Assuming the linear incorporation behavior of growth elements, this implies a material flux control of better than 0.2% of the flux selected for growth. This reveals the demanding effort for the design of the crystal growth setup to be used. For thin layers below a critical thickness, an additional degree of freedom is frequently utilized to tune the band structure of the semiconductor material. Intentionally strained layers and strain compensation layers are utilized to increase the conduction band offset [4–7].

Apart from the highly accurate material flux adjustment, this means that the flux for all elements in the ternary/quaternary layered structures have to be modulated instantaneously with low transients. Fast element switching along with a suppressed memory in the growth setup is mandatory to achieve abrupt material transitions with a high modulation depth. Such fast control is achievable through control by gas fluxes. In metal-organic vapor-phase epitaxy (MOVPE = MOCVD = metal-organic chemical vapor deposition), carrier gases such as H_2 or N_2 are used for the transport of reactive molecules such as trimethylindium (TMI), trimethylgallium (TMG), arsine, and phosphine to the reaction zone in laminar flow reactors (see Chapter 6). This intrinsically allows gas-phase and surface reactions at the substrate. In the early 1980s the practice was begun of injecting these gases in an ultrahigh-vacuum system leading to molecular beams, and hence suppressing any prereactions prior to arrival at the growth front [8–11]. Based on such a concept, several techniques have been developed. The first is the use of hydrides as group V sources and element evaporation for group III sources [12,13]. This method is of advantage for the growth of Al-containing layers, since the use of alkyls increases carbon and oxygen uptake in the layers. Transfer of the full MOVPE process in a molecular beam setup is of the highest importance for the GaInAsP/InP material system, where in multilayer structures fast switching in the group V elements is required and safe handling of the phosphorus must be ensured. From the environmental viewpoint, the low consumption of toxic gases such as the hydrides by more than a factor of 10 [13,14] with respect to the MOVPE process acts as a significant driving force for the development of gas-source MBE technologies.

Important results from the development of state-of-the-art devices such as multi-quantum-well (MQW) lasers, photodetectors, waveguides, and heterojunction bipolar transistors (HBTs) in the InP-based material system were achieved as well in metal-organic MBE (MOMBE or CBE = chemical beam epitaxy) using group III alkyls and group V hydrides or organic molecules, as in gas-source MBE with group III effusion cells [15–30]. This development is documented impressively in the proceedings of conferences on chemical beam epitaxy and related growth techniques and the books of Panish and Foord et al. [13,31–36]. For the development and fabrication of sophisticated device structures where overgrowth or localized growth for material confinement is often required, GaInAsP/InP has a clear advantage, due to the lower stability of the natural oxides. Any new growth technology here has to compete with the well-developed MOVPE process.

This chapter focuses on the most recent developments in MOMBE, with the aim to demonstrate that in addition to state-of-the-art fabrication of devices, this technology encourages the development of novel pathways for photonic and electronic circuits.

7.2 MOMBE GROWTH SYSTEM

7.2.1 Overview of the Growth System

Figure 7.1 presents the growth system developed particularly for the MOMBE process in an industrial environment [14,37–39]. The sample holders are molybdenum platens for up to three 2-in. wafers. The platens are introduced in six-platens

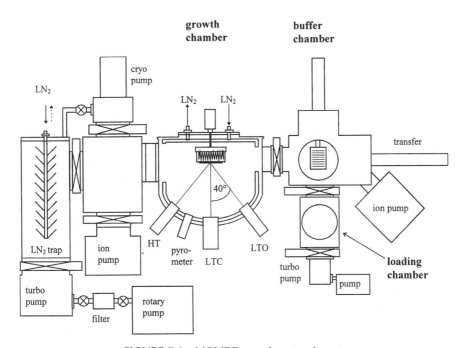

FIGURE 7.1 MOMBE growth system layout.

cassettes into the loading chamber, where a first outgasing takes place. The buffer chamber includes an area for the storage of platens so that growth series can be executed without interference from action taking place in the loading chamber. The entire transfer of platens is automated and the system is fully integrated in a gas cabinet, as in MOVPE systems. The platens are transferred into the growth chamber with the wafers face down by simple linear movements. These movements and the associated valve actions are interlocked electronically to simplify the operation. The growth chamber is pumped by a turbomolecular pump and a special cryo pump [40]. Between the turbomolecular pump and the growth chamber is a cryo trap enclosed between two gate valves (see Fig. 7.1). The cryo panel (LN_2) in the growth chamber has to be warmed up regularly to evaporate unreacted gas and to remove small amounts of white phosphorus. This evaporation converts a major part of the phosphorus modification into red P. Condensation then takes place in the cryo trap. In a second step the gate valve between the growth chamber and the trap is closed and this trap is warmed up. This leads to further P modification conversion. This procedure is essential for safe system operation. The pressure in the growth chamber during growth is in the range 10^{-5} to 3×10^{-4} torr, depending on the growth process. This pressure comes primarily from hydrogen produced during the pre-cracking of group V molecules in the high-temperature injector (HT).

The group III alkyls are injected via two different types of injectors. One is normal to the wafer, the other shows a tilt angle of about 40° with respect to this position. The overall reaction is given by the relation

$$q(\text{TMI}) + q(\text{TEG}) + q(\text{AsH}_3(\text{precracked})) + q(\text{PH}_3(\text{precracked})) \xrightarrow{T} $$
$$\text{GaInAsP} + \cdots \qquad (1)$$

or

$$q(\text{TMI}) + q(\text{TEG}) + q(\text{As-alkyl}) + q(\text{P-alkyl}) \xrightarrow{T} \text{GaInAsP} + \cdots \qquad (2)$$

where the relative value of the material fluxes q determines the material composition. For abrupt material transitions for each precursor, two independent gas lines are provided so that the material flux used in a subsequent layer is always stabilized in a special vent chamber. Such a twin-gas-line layout is mandatory for the fabrication of strained and unstrained MQW structures. The layout of one of these gas lines is shown in Fig. 7.2 as an example of group III lines. The material is evaporated from a stainless steel container. The vapor pressure of the alkyl is monitored by pressure transducers P_{MO} (Baratron). This allows immediate detection of source depletion as important information in any fabrication process. This vapor pressure is the prepressure of a control valve driven by a high-precision electronic (Sigmann) 16-bit controller. The input pressure p_1 at the flow element is set to a preselected value. Downstream of the flow element is a vent/run valve switching manifold, which is integrated directly in the gas injector mounted at the growth chamber. The low dead volume in the manifold, along with the design of high-conductance injec-

7.2 MOMBE GROWTH SYSTEM

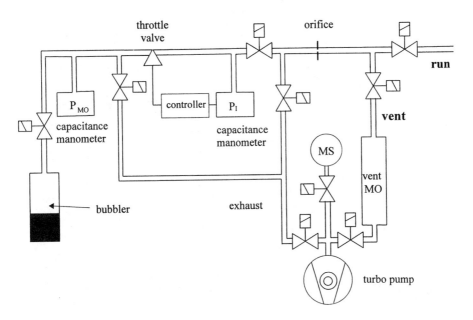

FIGURE 7.2 Gas injection system.

tors, enables an almost transient free molecule modulation at the growth front of the substrate surface. Up to four gas lines can be connected to each injector of the growth system. The high-conductance design yields crosstalk values between adjacent lines of less than 0.5% of the reading. This is important since the material flux into the system is determined by the adjusted pressure drop along the flow element,

$$q(\text{alkyl}) = f(p_1 - p_{\text{cell}}) \tag{3}$$

where p_{cell} is the pressure built up in the cell due to the actual conductance and the adjusted material flux.

To achieve ultimate flexibility for the growth of InP-based multilayer structures, the system should be equipped with two low-temperature injectors (e.g., each with TMI/TEG gas lines), one doping low-temperature injector (e.g., for diethylzinc injection), two high-temperature cells (e.g., each with AsH_3/PH_3 gas lines), and one extra high-temperature injector for n-type doping using disilane, for example. This allows fast switching fronts [e.g., between GaInAs(P) and InP], where any As memory in InP can lead to significant strain. The most crucial point of group V element switching can be improved further by implementing mechanical cell shutters. To date, system layout requires a sophisticated standard for safe operation in an industrial environment. Figure 7.3 presents a photograph of the MOMBE setup as presently developed, including a safety interlock system and an air-extracting cabinet housing the entire growth system.

FIGURE 7.3 MOMBE system integration in the laboratory.

7.2.2 SPECIFIC MOMBE SYSTEM DESIGN CRITERIA

As in MBE and MOVPE, the accuracy and stability of the material fluxes to the growing surface have to be ensured in MOMBE. Formation of the molecular beam and the flux density at the surface of the wafer can be calculated in a manner identical to that for MBE [41]. The gas source layout for growth in multiwafer systems has to be designed very carefully to guarantee a sufficient growth rate and uniformity in such a multiwafer process. Since most metal organics have vapor pressures in the range 0.5 to 5 torr at 300 K [42], the pressure drop along the feed pipework has to be minimized. The use of at least $\frac{1}{2}$-in.-inner-diameter tubing is mandatory since the alternative of heating the precursors to obtain higher vapor pressures is limited by their low thermal stability. As can be seen from Refs. 30, 39, 40, and 43–48, there are two additional very critical parameters in the gas source and MOMBE growth of InP-based materials, which we discuss next.

Cracking Group V Molecules in the High-Temperature Injector The high-temperature injector must have the following characteristics: (1) maximum cracking efficiency for hydrides and, in a different design, for organic group V precursors; (2) flux density better than 1% across the 5 in. holder (3- by 2-in. process); (3) yielding beam pressure up to 2×10^{-4} torr; (4) high conductance value (see the discussion above); (5) suppressed back diffusion of elements from the cracked gases. The element back diffusion has to be avoided since these atoms condense in the manifold area and thus are drained from the growth flux. The design has to be a high-conductance layout with the integration of back-diffusion baffles.

In MOMBE, the hydrides AsH_3 and PH_3 are generally utilized as group V precursors, but replacement precursors such as TBAs ($C_4H_9AsH_2$) and TBP ($C_4H_9PH_2$) are under investigation. In decomposition studies, the molecule mass peak intensities of the precursors being investigated can be evaluated as a function of the cracker temperature in comparison with the starting signal of the uncracked material, as plotted in Figs. 7.4 and 7.5. At the maximum cracker temperature of 1100°C, the hydrides show a decomposition of about 90% and the monotertiary butyl compounds' (TBAs and TBP) signals are reduced by more than 97%. Pyrolysis of TBAs and TBP takes place at temperatures 100 to 150°C lower than those required for their corresponding hydrides. The decomposition of the precursors examined has reached 50% at 820°C (AsH_3), 700°C (TBAs), 675°C (PH_3), and 540°C (TBP).

For the evaluation of effective As/P production for material growth, the cracking products of the precursors have to be characterized carefully. The identification of the cracking products of decomposed hydrides is rather obvious [49,50]; only hydrogen H_2, As_2/P_2, and As_4/P_4 appear (see Fig. 7.4). The dimers As_2/P_2 are the major cracking species of the hydrides, except for the temperature range 250 to 450°C, where significant As_4 production is observed. The tetramer P_4/As_4 signals measured

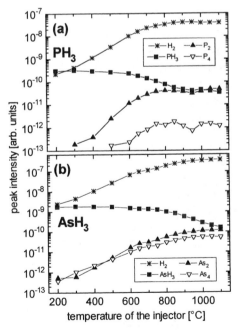

FIGURE 7.4 Peak intensities of the source material and the major cracking products of the hydrides (*a*) phosphine and (*b*) arsine versus the cracker temperature.

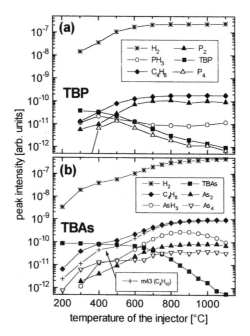

FIGURE 7.5 Peak intensities of the source material and the major cracking products of (*a*) tertiary butylphosphine and (*b*) tertiary butylarsine versus the cracker temperature.

at temperatures above 800°C (which are typically used for epitaxy in MOMBE) are probably due to recombination processes of the dimer molecules in the setup.

The cracking patterns turn out to be more complex for the alkyl compounds (see Fig. 7.5). Due to the additional tertiary butyl group, the main cracking products of TBP/TBAs are hydrogen (H_2), isobutene (C_4H_8), P_2/As_2, P_4/As_4, and the hydride PH_3/AsH_3. During the pyrolysis of TBAs, isobutane C_4H_{10} is also formed, detected by its characteristic fragment at mass 43 amu (see Fig. 7.5). However, we cannot exclude the possibility that this is an artifact of a self-dissociation process of TBAs [44,45].

In principle, the decomposition behavior of TBAs seems to correspond to the TBP data reported [51–53]. In the temperature region below 550°C, the tetramer As_4 is the dominant arsenic species. In the temperature range above, the As_4 production is suppressed or the As_4 is decomposed catalytically into As_2. With respect to this group V dimer production, the situation is comparable to that of hydride dissociation. However, the tertiary butyl As/P cracking is accompanied by production of PH_3/AsH_3. In the case of TBP, the PH_3 peak intensities detected are rather low, with a flat maximum at about 400°C (see Fig. 7.5*a*). Similar behavior was reported by Zahzouh et al. [51], but they observed a higher modulation of the phosphine signal of about 1 decade. For the decomposition of TBAs, our investigations more obviously reveal the existence of a reaction path producing arsine (see Fig. 7.5*b*). Starting at 300°C, the AsH_3 signal increases by more than 2 decades and reaches its maximum at about 900°C. Above 900°C, decomposition of the arsine takes over and the peak intensity diminishes. The total amount detected at this temperature is almost comparable to that measured for the injection of arsine itself. In consequence, decomposition of the arsine produced is the real limiting factor for the growth-relevant cracking efficiency of TBAs. Another key to evaluation of the extent of hydride production is the hydride/isobutene ratio (from data in Fig. 7.5). At 900°C this value is 1\10 for TBP but reaches $\frac{1}{3}$ for TBAs despite the additional production of C_4H_{10}.

To achieve more reliable data about cracking behavior under growth conditions, the beam equivalent pressures of precracked PH_3 and TBP have to be increased to standard growth values of about 10^{-4} torr and then the gas-phase composition detected by the multiple-ion monitoring mode of the mass spectrometer. In Fig. 7.6, the signal intensities of H_2, PH_3, P_2, and P_2H_4 are plotted, increasing the prepressure of TBP from 0.1 torr to 2.0 torr by 0.1-torr steps per minute. This corresponds to a beam equivalent pressure of about 1 to 20×10^{-6} torr and 1 to 83×10^{-4} torr in the cracker area of the injector, respectively. The cracker temperature was 1040°C. From the pyrolysis regime A at beam equivalent pressures equal or smaller than 10^{-6} torr, which are often used for mass spectrometric studies, a transition takes place to a pyrolysis regime B at growth-relevant pressures in the range 10^{-5} torr. The cracking behavior of TBP changes significantly by strong formation of PH_3, P_2H_4, and P_4 accompanied by reduced production of H_2 and P_2. The total decomposition of TBP drops from 99% to 80–85%. The first injection was performed after degassing the injector cell overnight at 1100°C, the second after waiting 70 min with a constant cracker temperature of 1040°C, and the third after waiting an additional 5

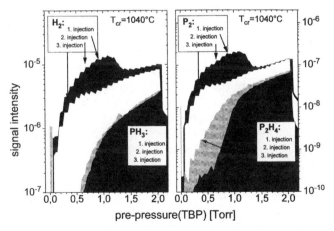

FIGURE 7.6 Investigation of the transition from pyrolysis regime A at beam equivalent pressures $\leq 10^{-6}$ torr to pyrolysis regime B at beam equivalent pressures $> 10^{-5}$ torr using the multiple-ion monitoring mode of the mass spectrometer. The signal intensities of H_2, PH_3, P_2, and P_2H_4 are presented for increasing the prepressure of TBP corresponding to a beam equivalent pressure of 1 to 20×10^{-6} torr (see the text for details).

min. Obviously, the transition does not take place at constant prepressures. The corresponding internal pressure in the cracking area was monitored during the second and third injections and found to be identical. Therefore, we can exclude that this transition of the pyrolysis from regime A to B is caused by gas-phase interactions in the cracker. It depends on the duration of degassing and in consequence on preparation of the molybdenum surfaces.

Ritter et al. found similar reversible behavior and concluded that there was a carbide formation on the surface of their tantalum cracker [54]. We observed this transitional effect only during injection of TBP, not for TBAs or PH_3. Therefore, we assume coverage of the molybdenum surface by adsorbed TBP and/or its cracking species, which reduces the catalytic effect of the cracker material and modifies the heterogeneous (surface) reactions. For example, Li et al. proposed a heterogeneous pyrolysis of TBP on a GaP surface with adsorbed PH and PH_2 species playing a dominant role [55]. As mentioned above, the geometrical conditions of our experimental setup probably prevent getting evidence of free radical species in the injected beam. Detection of the dihydride P_2H_4 indicates a radical decomposition process. The recombination processes forming PH_3 and P_2H_4 take place on the walls of the growth chamber or in the cracker. Due to the fact that P_2H_4 starts to decompose at temperatures higher than 300°C [56], the presence of active hydrogen on a growing-layer surface may be assumed when TBP is used in MOMBE.

Surface Temperature Control During Growth The growth chemistry in MOMBE is a sensitive function of the surface temperature. Hence the MOMBE set-up has to integrate sophisticated routines for surface temperature control. In addition, control has to be extended across a large surface area for applications in multi-wafer MOMBE machines [38,39]. The principle of such a strategy is presented in Fig. 7.7. The molybdenum susceptor is heated by a filament driven by a dc power supply. The temperature of the susceptor is evaluated by a thermocouple, and the temperature at the thermocouple is kept constant in a closed loop with a PID controller. Since coupling between the heater and substrate surface is a complex function of growth sequence as well as susceptor design and history, the surface temperature has to be measured in situ by a pyrometer unit. To suppress window coating, this measurement is recorded electronically at certain intervals. The PC growth control evaluates the surface temperature measured against the value requested and calculates a bias value for the thermocouple (TC) setpoint of the closed-loop control described above. The measurement intervals are a function of surface temperature deviation in a variable sampling-interval approach. These routines from statistical process control enable local surface temperature control of better than ±1°C. This is shown in Fig. 7.8 for the growth of an InP/GaInAsP heterostructure. The surface temperature (T_{pyro}) and thermocouple setpoint are plotted versus time starting at the growth onset. The pyrometer reading at the start is about 512°C in order to remove the surface oxides and then is set to the growth temperature of 505°C. This temperature is nicely stabilized during InP growth. At the transition to GaInAsP there is a short increase in surface temperature, due to the higher absorption of heater radiation in the lower-bandgap material.

7.2 MOMBE GROWTH SYSTEM   197

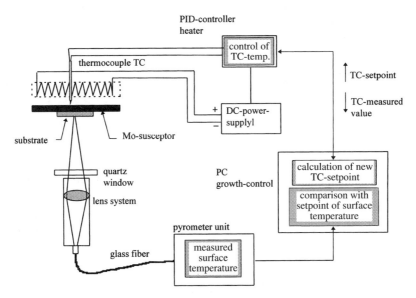

FIGURE 7.7 Schematic of surface temperature control system in a MOMBE setup.

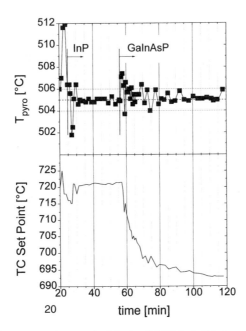

FIGURE 7.8 Surface temperature control during InP/GaInAsP heterostructure growth; impact on heater drive and thermocouple setpoint.

198 InP AND RELATED COMPOUND GROWTH

As described in Fig. 7.7 the setup decreases the TC setpoints instantaneously and the GaInAsP surface temperature is kept constant at 505°C. The TC setpoint shows a nearly exponential decay with time. Such behavior is indeed expected from Beer's absorption law, due to the increasing thickness of the low-bandgap material. Figure 7.8 demonstrates that accurate, well-designed surface temperature control is mandatory in MOMBE systems when InP-based multilayered structures are being fabricated.

Figure 7.9 gives details of a wafer holder design that enables uniform lateral temperature distribution. The task of the heat transfer plates (HTPs) is to maintain lateral constant wafer surface temperature. This means that the material and shape of the HTPs have to be such that heat transfer to any point on the wafer is kept constant, independent of wafer holder proximity. Some results of this are described in Section 7.4.2.

7.3 SURFACE-SELECTIVE GROWTH IN MOMBE

In the metal-organic growth technologies (MOMBE and MOVPE), the chemical reactions for forming the crystal elements are localized at the growing surface, so the nature of this surface plays an important role in the overall reactions. It was recog-

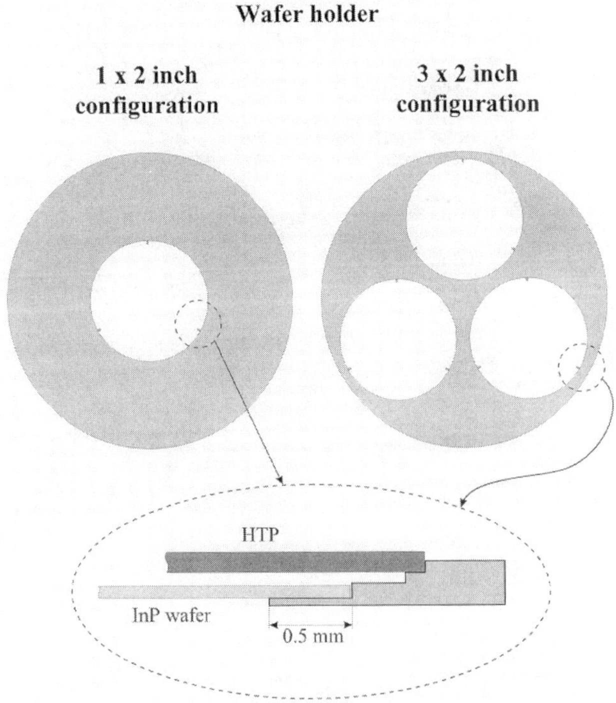

FIGURE 7.9 Wafer holder design for large-area growth.

nized early that the presence of indium at the surface leads to Ga-molecule loss during GaInAs growth [46]. For this material it was proven by means of modulated beam mass spectroscopy (MBMS) that the desorbing species is diethylgallium [48]. For GaInAsP growth, the incorporation mechanism of the elements is even more complex. Figure 7.10 shows the relative phosphorus (Fig. 7.10a) and gallium (Fig. 7.10b) incorporation in GaInAsP lattice matched to InP as a function of the relative mass transfer into the growth system of PH_3 and TEG. The material close to InP was grown using a constant PH_3 flux, and for compositions close to GaInAs, a constant AsH_3 flux was injected. The results for material compositions farther away from these endpoints are not plotted in Fig. 7.10 since they are not easily accessible with the constant fluxes mentioned before. The experimental data are systematically below the 1:1 relation between gas-phase and solid composition (see the dashed lines in Fig. 7.10). The relative incorporation efficiency of phosphorus varies between about 1 and 0.5, depending on the material composition.

Figure 7.10b reveals that starting with GaInAs growth (the highest Ga concentration in Fig. 7.10b), Ga incorporation efficiency drops drastically for the lower Ga concentrations, which means materials closer to InP and consequently, higher phosphorus content. This can be recognized by the relatively higher vertical distance of the data point with respect to the dashed line in Fig. 7.10b. The absolute Ga incorporation efficiency in GaInAs is about 68%, as a consequence of the above-mentioned DEG desorption. For $Ga_xIn_{1-x}As_yP_{1-y}$ with a gap wavelength of 0.95 μm ($x = 0.025$, $y = 0.06$), this efficiency, calculated from the data in Fig. 7.10b is only 11%. The absolute efficiency values cannot be taken for granted since measurement of the beam pressures of the reactants can be affected by the different cracking and ionization characteristics in the ion gauge. However, the data show clearly that the gallium incorporation efficiency is up to a factor of 6 lower in GaInAsP than in GaInAs. Since we know from analysis of the layer thickness and materials composition data that the InAs(P) growth rate (for the constant TMI flux used here) was kept constant over the full band of compositions between InP and GaInAs, it can be concluded that the Ga molecule desorption is enhanced when more phosphorus is transferred to the growing surface. The experimental data in Fig. 7.10 were all obtained on (100) InP wafers having a 2° misorientation toward (110). We reported earlier that the sticking coefficient of phosphorus on exactly oriented (100) wafers is substantially lower [57] and that the gallium incorporation is reduced for the GaInAsP ($\lambda_G = 1.05$ μm) growth with respect to misoriented wafers [27,58]. This means that the Ga incorporation efficiency in GaInAsP normalized to GaInAs is decreased further when using a (100) wafer with nominally exact orientation.

To determine the effect of growth temperature on Ga and P incorporation efficiencies, single GaInAsP, GaInAs, and InAsP layers were grown on InP substrates in the temperature range 450 to 520°C. For quaternary material and GaInAs, typical u-curves for the measured lattice mismatch $\Delta d/d$ and the PL wavelength λ using a fixed V/III ratio were recorded as a basis for the calibration curve in our growth system, as reported earlier. Figure 7.11 presents such a curve for GaInAsP with $\lambda_G = 1.55$ μm. The minimum of these $\Delta d/d$ u-curves is shifted in dependence on the material composition of the GaInAsP layers, which according to Refs. 2, and 4 is cal-

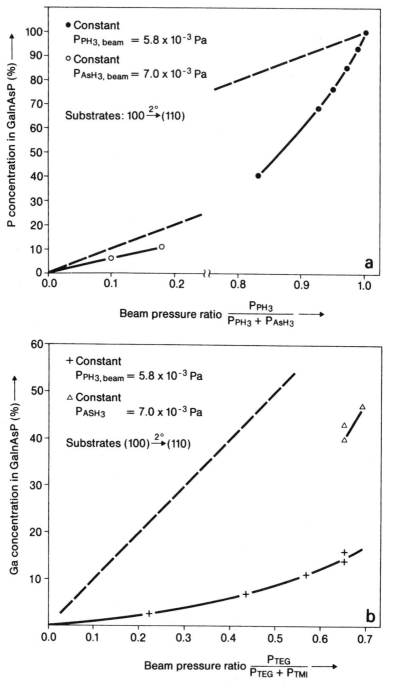

FIGURE 7.10 Element incorporation in GaInAsP lattice matched to InP: (*a*) phosphorus; (*b*) gallium.

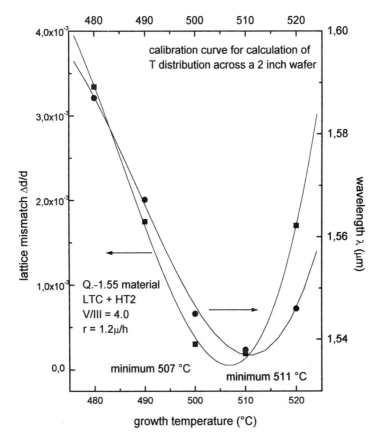

FIGURE 7.11 Effect of growth temperature on lattice mismatch and PL wavelength in GaInAsP with $\lambda_G = 1.55$ μm.

culated between 500°C ($\lambda_G = 1.05$ μm) and 510°C ($\lambda_G = 1.25$ μm) [30]. The shift is caused by the variation in the relative importance of the P incorporation efficiency and the loss of Ga molecules toward higher temperatures above 490°C for GaInAsP compositions. The strongest shift is observed, for the material yielding $\lambda_G = 1.05$ μm, where the lowest Ga incorporation efficiency is observed, so that for quaternary material, the optimized temperature was set at 500°C.

In Fig. 7.12 the relative Ga and P incorporation efficiencies calculated from the data in Fig. 7.11 are shown. The increasing P incorporation efficiency at $T_{gr} > 500$°C explains the different minima in $\Delta d/d$ and λ presented in Fig. 7.11.

Figure 7.13 focuses on the P incorporation efficiencies for quaternaries with 1.05 μm $\leq \lambda_G \leq$ 1.55 μm. The corresponding data for Ga are shown in Fig. 7.14. A comparison of these data sets demonstrates that selection of the growth temperature determines the difference in incorporation efficiencies for the various quaternary

202 InP AND RELATED COMPOUND GROWTH

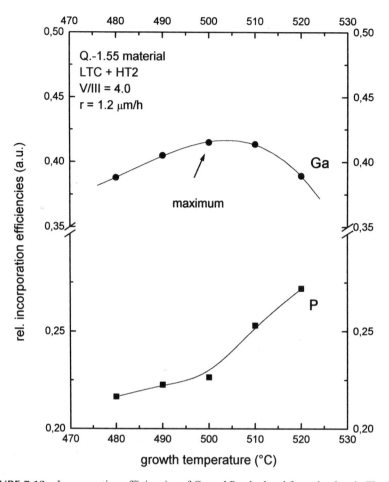

FIGURE 7.12 Incorporation efficiencies of Ga and P calculated from the data in Fig. 7.11.

material compositions. For optimum process control a growth temperature of 505 to 507°C is favorable, due to the lowest slope in the graphs of Figs. 7.13 and 7.14. The trade-off is, of course, a somewhat lower P incorporation efficiency. By increasing the V/III ratio at this temperature around 500°C we observe an increased Ga along with a reduced P incorporation efficiency for all quaternary compositions. This leads to a strong decrease in the lattice mismatch and a shift to higher wavelengths. Figure 7.15 focuses on this effect of the V/III ratio on the P/Ga incorporation efficiency in lattice-matched GaInAsP (λ_G = 1.05 μm) and compressively strained InAsP material grown at the selected temperature of 500°C with a fixed and precisely calibrated As/P ratio. With increasing V/III ratios a distinct decrease in the P-incorporation efficiency is observed for GaInAsP and InAsP materials. The P content in the layers is reduced from 0.815 to 0.722 for InAsP, and from 0.789 to 0.722

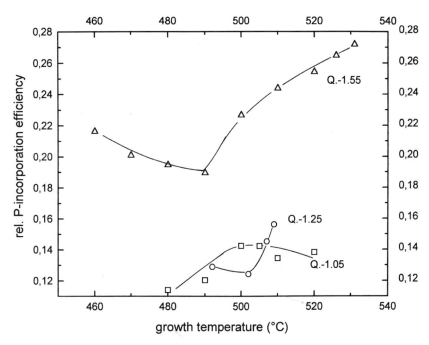

FIGURE 7.13 P incorporation efficiency versus growth temperature for GaInAsP.

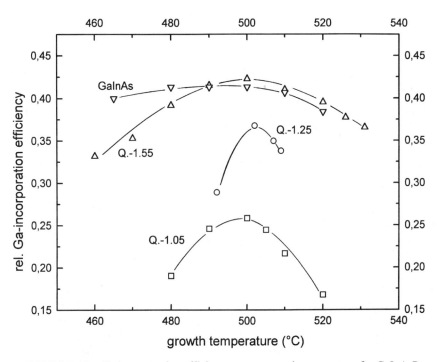

FIGURE 7.14 Ga incorporation efficiency versus growth temperature for GaInAsP.

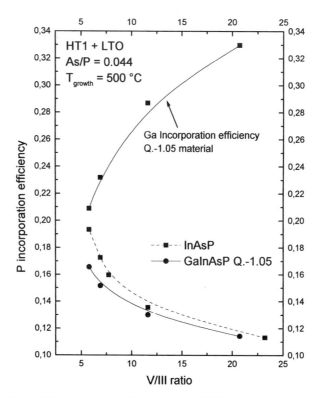

FIGURE 7.15 P incorporation efficiency versus V/III ratio in InAsP and GaInAsP.

for quaternary material. The P incorporation efficiencies for the InAsP material are higher than for quaternary material using the same V/III ratio, but this gap is reduced with higher V/III ratios. This decrease in the P incorporation efficiency could be caused by the clearly higher As sticking coefficients. In addition, for quaternary material we observe a clear increase in Ga incorporation efficiency with rising V/III ratios. From these data it is obvious that in ternary strained InAsP material, growth control should be more simplified, due to lack of Ga–P interaction observed in the GaInAsP system [30,58,59].

The phenomena described so far reveal the surface-selective process in MOMBE. Figure 7.16 presents an overview of the surface-selective growth conditions on patterned surfaces. The reactants injected in the growth chamber arrive at the growth front, where they undergo adsorption, reaction, surface diffusion, incorporation, and/or desorption processes. The balance between these rate-limiting steps can depend sensitively on the nature of the reaction surface, so that different desorption processes can rule the growth. A basic subdivision into four categories has to be established (see Fig. 7.16): surface orientation dependence, planar selective-area epitaxy (SAE), embedded SAE, and nonplanar growth.

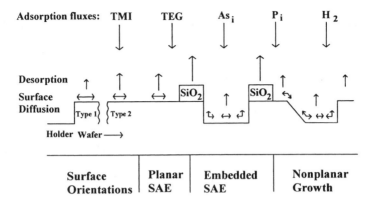

FIGURE 7.16 Surface-selective growth conditions in MOMBE.

7.3.1 Orientation Dependence of Growth

Figure 7.17 presents the Ga and P concentrations in GaInAsP for growth on various misoriented substrates toward (110) and (0$\bar{1}\bar{1}$). The GaInAs composition remains somewhat unchanged for the various angles in both directions (Fig. 7.17a). The quaternary materials close to GaInAs (Fig. 7.17b) show a slightly increased P incorporation value for the larger misorientation toward (110), whereas there is no effect for a tilt toward (0$\bar{1}\bar{1}$), having only one step type. Along with the higher P uptake, there is a small tendency to lower Ga incorporation. This is in line with the data from Fig. 7.10, where an increased P supply yields a lower Ga incorporation efficiency. The difference is that the P incorporation is not changed by the supply but by the degree of substrate misorentation and therefore the adsorption side step density. It can be argued that the phosphorus adsorption in the steps toward the (0$\bar{1}\bar{1}$) plane affects Ga incorporation during growth of the near-GaInAs material. However, these effects are rather small and even less pronounced during the growth of GaInAsP very close to InP. The situation is more complex for $Ga_{0.11}In_{0.89}As_{0.23}P_{0.77}$ ($\lambda_G = 1.05$ μm), which can be used for waveguide material [27]. We found empirically that the uniformity and repeatability of material growth is enhanced when using (100) wafers 2° off toward (110).

These observations can be explained on the basis of the data from Fig. 7.18, where the Ga and P concentrations in nominally $\lambda_G = 1.05$ μm quaternary material are plotted versus the misorientation toward (110). The reactant fluxes (beam pressures: $p_{PH3} = 5.8 \times 10^{-3}$ Pa, $p_{AsH3} = 3.3 \times 10^{-4}$ Pa, $p_{TMI} = 5.5 \times 10^{-4}$ Pa, $p_{TEG} = 7.3 \times 10^{-4}$ Pa) were identical for all data points. The P and Ga concentrations both increase with rising angle of misorientation, but the values for P incorporation saturate for angles larger than 1.5°, whereas Ga uptake seems to show this behavior only for angles above about 4°. With respect to tolerances of wafer surface orientations (accuracy of misorientation, wafer bending), the 2°-off material enables independent P incorporation, and the Ga incorporation curve presents an acceptable slope (see Fig. 7.18). This is an important factor for reproducible and large-area uni-

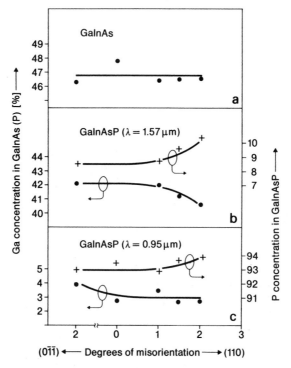

FIGURE 7.17 Effect of wafer misorientation on element incorporation in (a) GaInAs, (b) GaInAsP (λ_G = 1.57 μm), and (c) GaInAsP (λ_G = 0.95 μm).

form growth. A comparison of Figs. 7.17 and 7.18 shows immediately that surface orientation effects play a major role for GaInAsP farther away from InP and GaInAs material composition. For the material labeled in Fig. 7.17, this effect can be largely ignored, whereas the $Ga_{0.11}In_{0.89}As_{0.23}P_{0.77}$ is very sensitive to the nature of the wafer surface. These orientation effects are of major importance during the overgrowth of structured surfaces, where various surface orientations do occur. It is interesting to recall here that Isu et al. found during GaAs MOMBE μ-RHEED studies that Ga droplets formed intentionally move toward the $[0\bar{1}1]$ direction, having steps upward [60]. During this surface diffusion process, the droplets are decreasing in size, which means that Ga atoms are reactively trapped at steps into the $[01\bar{1}]$ or $[0\bar{1}1]$ directions. This result supports the findings during GaInAsP growth discussed above and both insights are now known to represent key mechanisms in the selective-area epitaxy process.

7.3.2 Planar Selective-Area Growth

The first report on the selective-area epitaxy (SAE) of III-V materials (GaAs) by MOMBE was published by Vodjdani et al. [9]. SAE of GaAs was also investigated

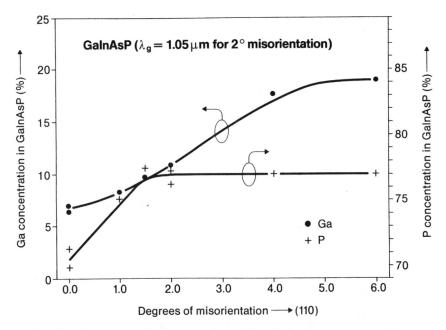

FIGURE 7.18 Effect of wafer misorientation toward (110) on GaInAsP material composition ($\lambda_G = 1.05$ μm for 2° misorientation; for details, see the text).

by other groups using MOMBE [61–63]. An important basic rule is that the selectivity is improved when the substrate temperature is increased. This observation of selective temperature-driven desorption from the mask holds true for all Ga-In-As-P-containing materials. It is worth mentioning that in MOVPE, higher growth temperatures are typically required for SAE growth than in MOMBE. The most challenging application of SAE is for InP-based materials where the MOVPE has significant problems to control the materials composition [64]. Some significant results on SAE of InP [65–69], GaInAs [64,66,70–72], and GaInAsP [68,73–75] have been reported in the literature. The real driving force for the development of SAE is the potential for future device development and integration concepts. For InP and GaInAsP, true selective-area growth, where no deposition takes place on the mask, was observed at standard MOMBE growth temperatures between 480 and 560°C. Figure 7.19 focusses on a grown InP bar. This micrograph proved for the first time that narrow InP bars can be grown that have vertical sidewalls and flat surfaces. Structures like these have been obtained on (100) wafers with a misorientation of, for example, 2° off toward (110). When using only (100)-oriented wafers, the structures grown under identical conditions exhibit (111) crystal planes [68].

As discussed above, it has been found that the phosphorus sticking coefficient and phosphorous incorporation are enhanced when using (100) wafers misoriented toward (110) [58,76]. These phenomena increase the effective V/III ratio at the growing front. This is an important factor for obtaining vertical sidewalls such as

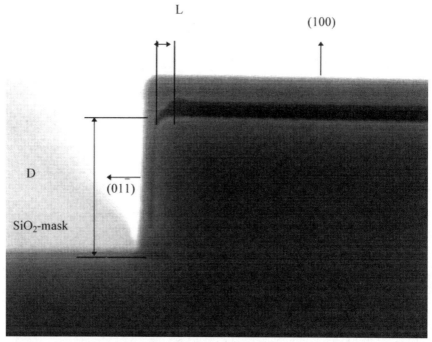

FIGURE 7.19 SEM micrograph of a InP(1 μm)/GaInAs(0.12 μm)/InP(0.2 μm) heterostructure. The lateral growth L into the [011] direction (showing to the left) and the vertical growth D is marked for the 1-μm-thick InP ridge.

those shown in Fig. 7.19. To suppress kinetic limitations on (111) planes, higher V/III ratios are mandatory, as observed earlier during low-pressure (LP) MOCVD growth of GaAs on (111) surfaces [77]. Since the (111)B plane is terminated by group V atoms, high effective V/III ratios are required to reduce the group III molecule surface diffusion and possible desorption. Consequently, the higher effective V/III ratio enables higher growth rates on the (111) planes leading, to a suppression of the formation of these planes. On the other hand, the growth rate in $\langle 01\bar{1}\rangle$ is still limited significantly by the surface kinetics, thus stabilizing these planes. This is of important technological interest since a perfect ridge structure can be achieved even under imperfect mask edges; the growth planes are stabilized by the growth kinetics [78]. Another important aspect is that under these conditions (high V/III ratio), the interfacet diffusion of the group III molecules is suppressed, on the one hand, due to a lower surface diffusion; on the other hand, the increase in kinetically limited growth rates leads to a reduction in the total element concentration on the various facets and thus helps to minimize interfacet group III element diffusion, which is driven by concentration gradients. Hence a decoupling of the growth mechanisms at the facets is observed, leading to planar surfaces. It was found that the growth of structures more than 2 μm thick with vertical sidewalls is possible when the InP growth rate is reduced to about 0.5 μm/h with the input V/III ratio set to 9 to 12 during growth on substrates off-oriented toward (110) [78], whereas under identical conditions on just-oriented wafers, (111) planes are always formed. From these results it can be speculated that the mechanism driving this facet growth is the group V element–molecule interfacet diffusion. The higher effective V/III ratio on the stepped (100) surface builds up an increased lateral group V concentration gradient between the facets, thus (according to Fick's diffusion laws), leading to a group V surface diffusion flux toward the sidewalls. As a consequence, the group III surface (also interfacet) diffusion and desorption are lowered. This enhances the (111) growth rate, so that the kinetically limited rates are then given on the (011) or ($0\bar{1}1$) planes.

In line with this discussion, the transfer of these MOMBE-developed parameters also enables the SAE of InP with vertical sidewalls and planar surfaces in LP MOVPE, so that the facet growth and diffusion mechanisms discussed above are of general validity [79]. There are also some reports in the literature on vertical sidewall growth or overgrowth using atomic layer epitaxy (ALE) in MOMBE or chloride VPE or conventional MOVPE on (111) substrates [80–82]. These studies concentrated on GaAs-based materials systems, and vertical planes are achieved by a surface-controlled reaction mechanism, as in MOMBE. The advantage of the MOMBE method used here is that the growth speed is significantly higher than in ALE systems and that it is applicable to ternary and quaternary InP-based materials, for which ALE seems to be difficult to achieve.

Figure 7.19 is a SEM micrograph of a planar SAE-grown ridge used for test heterostructures to investigate vertical sidewall growth. InP ridges 1 μm thick with vertical sidewalls were grown as starting layers in the windows of the SiO_2-masked InP substrate. On top of these ridges (1.7 to 250 μm in width) we grew GaInAs (0.12 μm)/InP (0.20 to 0.25 μm) heterostructures. After the growth, the samples were cleaved and stained for SEM investigations. The InP layers of the sample in Fig. 7.19 were grown at a substrate temperature of 515°C and a V/III ratio of 23.2. For a standard SAE-InP growth rate of 0.5 μm/h, the beam equivalent pressure of TMI measured by a Bayard Alpert flux gauge directly beneath the substrate was 8.3 10^{-6} mbar. The point of interest was the lateral growth rate of the vertical sidewalls of the InP/GaInAs ridges compared to the vertical sidewall on the (100) surface. The value of the mask overgrowth L, × 100, divided by the ridge height D (see Fig. 7.19) yields a percentage of the lateral growth rate (PLG) with respect to the vertical growth rate. The PLG value was evaluated for all structures in both the the [$01\bar{1}$] and [011] directions. For the standard substrate misorientation this means that a facet consisting of a ($0\bar{1}1$) plane (Fig. 7.19) exhibits an upward direction into the surface step. The starting layer ridge in Fig. 7.19 demonstrates a PLG of 11% in the downward [$01\bar{1}$] direction.

In Fig. 7.20 the PLG value of 250-μm-wide InP ridges grown with a central or conventional (tilted) injector (see Fig. 7.1) is plotted versus the width of the SiO_2 stripes. The stripes are oriented in the [$01\bar{1}$] direction. The structures show vertical sidewalls for growth with both types of injectors for sufficiently high V/III ratios, as reported earlier [68]. Lateral growth at the edges of SiO_2 stripes of width 4 to 100 μm is constant for each crystal plane and group III injector configuration used. The

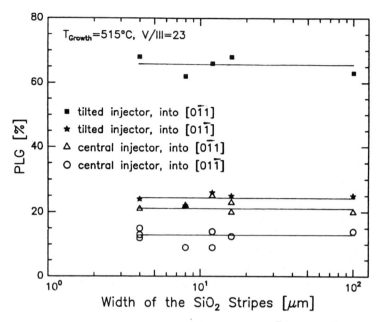

FIGURE 7.20 Percentage of lateral growth (PLG) into the [0$\bar{1}$1] and [01$\bar{1}$] directions versus the width of the SiO$_2$ stripes for 250 μm wide InP ridges grown with a central or conventional tilted injector.

data of Fig. 7.20 show clearly that lateral growth on the (0$\bar{1}$1) plane is systematically higher than on the opposite (01$\bar{1}$) plane at a constant growth temperature of 515°C. This observation holds true for both molecular beam geometries. As expected, the lateral growth rates on the vertical side facets in central cell geometry are about half that observed using a conventional tilted gas injector, since the beam flux impinging on these side facets approaches zero in central geometry. However, the PLG value, which is between 10 and 22% for the perpendicular geometry, is surprisingly high and can be explained only by interfacet diffusion. This argument is supported by the anisotropic side facet growth speed. As found earlier, there is a marked reactivity of upward surface steps, leading to an anisotropic molecule attachment in the [0$\bar{1}$1] direction [73] with increased trapping probability. Consequently, in the [0$\bar{1}$1] downward direction the diffusion flux from the tilted (100) plane to the (0$\bar{1}$1) ridge side facet is higher than the interfacet diffusion toward the (0$\bar{1}$1) crystal plane. Hence the minimum lateral growth is given by interfacet surface diffusion processes. In the case of conventional molecular beam geometry, this lateral growth is increased by roughly a factor of 2, due to the higher effective material flux arriving from the gas phase. In conventional geometry lateral growth is controlled by the interfacet diffusion and mass tranfer from the source.

On the basis of these considerations it is obvious that the ridge width is a critical parameter for extensions below about twice the average upward diffusion lengths for the molecules. Therefore, by changing the width of the area for material growth, the interfacet diffusion is affected as discussed. To study this effect in detail, the PLG value was analyzed as a function of the semiconductor stripe width in the central cell geometry. The data in Fig. 7.21 show that, as expected, there is nearly no effect for the $(0\bar{1}1)$ facets showing upward, but the growth rate on the $(01\bar{1})$ facet in the downward direction is clearly increased for ridge widths below 5 μm. This diagram reveals that the anisotropic surface diffusion into the $[0\bar{1}1]$ direction on the (100) surface loses importance on ridges below 5 μm in width. It is interesting to note that the fine surface ripples, which can be found near mask edges due to the anisotropic surface diffusion [69,78], described below, disappear on the small ridges. These anisotropic effects are reduced when lowering the substrate misorientation angle and therefore the surface step density [83].

Figure 7.22 presents the view from above of the structure shown in Fig. 7.19. The Nomarski micrograph reveals that there is a fine surface corrugation at the semiconductor–mask transition, only in the $\langle 01\bar{1}\rangle$ direction [78], when the substrate surface misorientation is directed toward the next plane. This corrugation was evaluated by alpha-step measurements and simulated by the solution of a nonlinear partial differential equation (see Fig. 7.23) using the proper boundary conditions for the semiconductor–mask transition [84]. As shown in Fig. 7.19, the structures are also grown with integrated layer sequences of GaInAsP layers. Because of the surface-selective growth, these materials show different material compositions on different surfaces. Hence it is expected that there is a material localization at the side facets

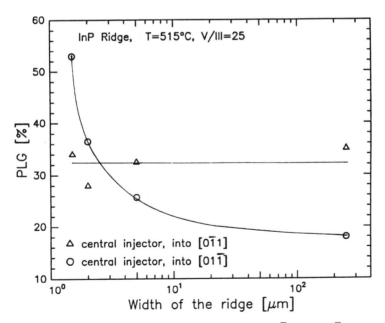

FIGURE 7.21 Percentage of lateral growth (PLG) into the $[0\bar{1}1]$ and $[01\bar{1}]$ directions versus the width of the InP ridges grown with the central injector.

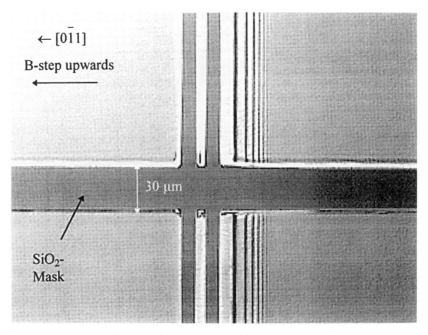

FIGURE 7.22 Micrograph of the surface corrugation in the semiconductor–mask transition region in the downward direction toward $(01\bar{1})$.

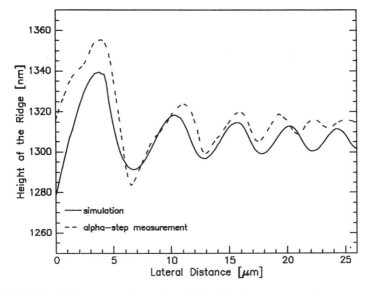

FIGURE 7.23 Surface corrugation in the height of the ridge versus the distance to the mask for the simulation and alpha-step measurement.

in ridge structures like that shown in Fig. 7.19. The corresponding energy gap can be recorded by spatially resolved photoluminescence (PL). In a standard measurement setup, excitation and luminescence detection are achieved by scanning across the wafer surface [74]. However, luminescence collection from the side facets can be significantly suppressed. Therefore, an improved technique of scanning across a cleaved facet was developed (see Fig. 7.24). Figure 7.25 presents in a summary plot the results from a ridge structure, including GaInAsP with $\lambda_G = 1.05$ μm. The wavelength-intensity plot reveals clearly that there is a [01$\bar{1}$] sidewall facet recombination at $\lambda_G = 1175$ nm. The original raw data (PL spectra) are shown in the inset. This localization of low-bandgap material at the side facets is an attractive path to novel low-dimensional structures. The energy confinement as a function of the [100] surface GaInAsP material composition is presently under investigation. The plot in Fig. 7.25 demonstrates that the material composition on the (100) surface remains fairly constant ($\Delta\lambda_G \leq 3$ nm) up to the facet transitions. This is an important result for the lateral coupling of device structures using embedded SAE.

7.3.3 Embedded Growth

For an optimum selective infill exhibiting planar growth fronts it is mandatory that the lateral growth rates on the sidewalls of tubs etched into the semiconductor material be substantially low. High lateral growth rates yield surface bending of the infills and significant "ear" formations at the transition point to the mask area. On the other hand, the lateral growth rates should not vanish, and growth conditions for which the (111) planes are the stabilized side facet should not be selected. Under such conditions gaps between the infill and the tub sidewalls would be created. Based on the insights from planar SAE, the lateral growth is minimized if the molecular beams arrive perpendicular to the basic wafer surface. This avoids shadowing effects and minimizes the beam flux density on the tub wall facets.

Figure 7.26 presents an example in which under these optimized conditions for planar SAE a tub has been selectively filled with an InP/GaInAsP/InP multilayer structure. Figure 7.26 focuses on the lateral junctions between a SiO$_2$-capped InP(d = 0.4 μm)/GaInAsP(d = 1.0 μm, $\lambda_G = 1058$ nm)/InP(d = 0.4 μm) grown in the first

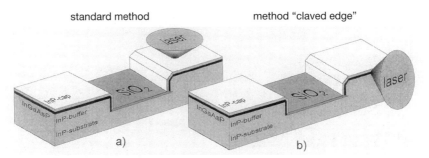

FIGURE 7.24 Schematic of μ-PL measurement applied to the edge of the structures: (*a*) standard method; (*b*) cleaved-edge method.

214 InP AND RELATED COMPOUND GROWTH

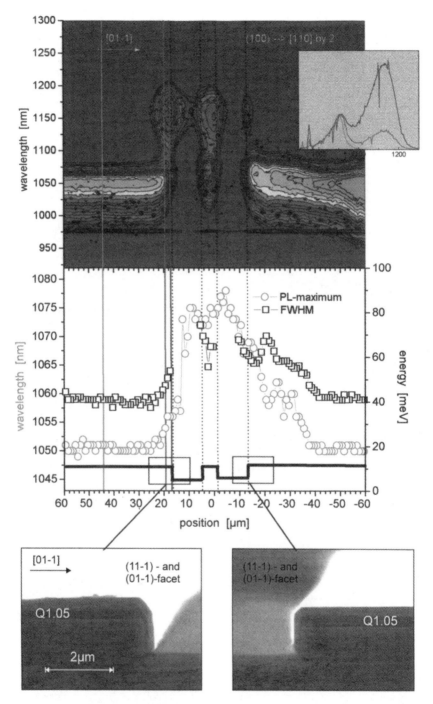

FIGURE 7.25 Spatially resolved PL measurements on InP/GaInAsP ($\lambda_G = 1.05$ μm) ridge structures focusing on the facet transitions. Inset presents a schematic and SEM micrograph of the ridge structure on a wafer 2° misoriented toward (110).

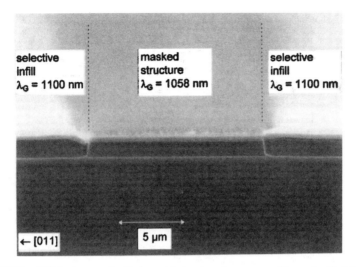

FIGURE 7.26 SEM micrograph of a lateral heterojunction prepared by embedded SAE (for details, see the text).

epitaxy with subsequent selective infill of a comparable double heterostructure (DH) with InP($d = 0.2$ μm)/GaInAsP($d = 1.1$ μm, $\lambda = 1100$ nm)/InP($d = 0.3$ μm). This SEM picture reveals that smooth lateral DH junctions are obtained by this method.

Figure 7.27 gives the results for spatially resolved photoluminescence measurements obtained for the structure shown in Fig. 7.26. The lateral position indicates the location of the excitation laser spot (ca. 5 μm in diameter). The PL intensity of the quaternary infill is on a high level up to the lateral junction. The PL intensity of the first-grown GaInAsP with $\lambda_G = 1.058$ μm is about half that of PL from the infill material at $\lambda_G = 1.1$ μm, due to the fact that the quaternary growth was further optimized between these two growth runs. Decay takes place over a distance of 5 μm, which coincides with lateral excitation of the photoluminescence setup. There is a sharp transition in wavelength at the junction. Both results reveal that the infill exhibits excellent material quality up to the junction zone. This is supported by spatially resolved x-ray measurements which prove that the embedded SAE in MOMBE for structures as presented in Fig. 7.26 are free of perturbations up to the lateral junctions [85].

7.3.4 Doping Using Gas Sources

For industrial acceptance of the MOMBE process it is important to overcome the problems associated with high-temperature effusion cells, which exhibit several drawbacks: (1) hot sources in the vacuum chamber may lead to additional contamination or to problems related to temperature control of source and substrate; (2) a contamination of the solid source, by oxygen or hydrocarbons for example, may re-

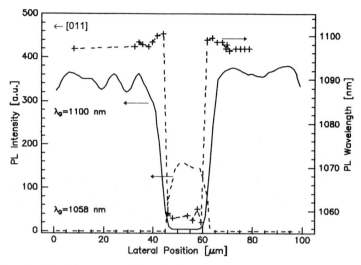

FIGURE 7.27 Spatially resolved photoluminescence from a structure as shown in Fig. 7.26. Note the sharp intensity and wavelength transitions in junction areas.

sult in problems concerning long-term stability and reproducibility [86–88]; and (3) more sophisticated dopant profiles, such as dopant ramps, are difficult to achieve. To overcome these drawbacks of solid sources, gaseous dopant sources have been proposed for MOMBE in analogy to metal-organic vapor-phase epitaxy (MOVPE) [89–94]. The handling of these sources outside the vacuum system makes possible a fast exchange and a flexible choice of dopant sources. Any required dopant profile in the grown layers, including dopant ramps and spikes, can easily be realized by fast pressure control of the dopant beam flux, making this type of source much more advantageous than conventional solid sources.

Especially for p-type doping of InP, the problems mentioned above concerning source contamination and low electrical activation have been reported in the literature for the solid source Be [87,88,95]. In a few studies, Zn doping from a gaseous source has been proposed for p-type doping [89–93,96]. Some n-type doping has been achieved by utilizing silane and disilane [15,30,97–99]. The Zn uptake from DEZn dissociation in InP is strongly dependent on the growth temperature as shown in Fig. 7.28. The Zn concentration of the samples was determined by SIMS, the electrical activation in these layers was determined by a comparison of SIMS data with Hall data. Figure 7.28 shows the marked effect of growth temperature T on electrical activation. The data are visualized by a dashed line. The solid line represents SIMS data for one set of experimental parameters. At high temperatures ($T > 460°C$) the incorporated Zn concentration is reduced (solid line) along with a low electrical activation (dashed line). At these temperatures the Zn incorporation is limited by desoption processes. Few Zn atoms probably compete with group III species for incorporation on substitutional lattice sites. The low electrical activation suggests Zn incorporation on interstitial sites. As the temperature is decreased, more

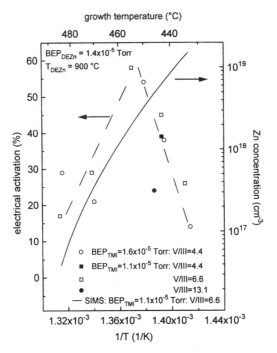

FIGURE 7.28 Electrical activation (left y-axis: dashed line) in Zn-doped InP as a function of the growth temperature. The DEZn beam flux pressure and DEZn cracker cell temperature were kept constant. Included are SIMS data (right y-axis: solid line) for one set of experimental parameters. (From Ref. 96.)

Zn atoms are incorporated substitutionally, leading to a maximum activation of almost 60% in the range 460 to 450°C. These values are about a factor of 2 higher than the activation values in Be-doped InP [87]. A further reduction in growth temperature results in a decrease in activation. This decrease for temperatures below 450°C (dashed line) might be caused by the still-increasing Zn concentration (solid line). Such behavior is reported for Zn and Mg doping of InP in MOVPE [100].

As in MOVPE, Si_2H_6 has been used for n-type doping of InP layers. For use in device structures, a sufficient uniformity in the doping process is required. To avoid clogging, a high-conductance, high-temperature cell was used. The injector is comparable to the hydride cracker, yielding a good 5-in. uniformity. The growth temperature was set to 510°C. The layer thickness was 1.3 μm (growth rate 1.3 μm/h, V/III = 4.5). Figure 7.29 gives the dependence of the precracking temperature on the net electron concentration in InP using 5% disilane diluted in H_2 and pure disilane. The dopant gas flux was held constant. The Si incorporation in InP starts at 600°C precracking temperature. A maximum of the net electron concentration is observed for 5% disilane at 900°C precracking temperature and for undiluted disilane at 1000°C precracking temperature. Above 1000°C in the cracker the electron concentration in the layer is reduced. We assume that at lower cracking temperatures (600 to

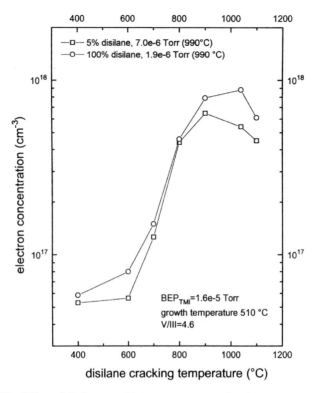

FIGURE 7.29 Effect of disilane cracking temperature on the electron concentration in Si-doped InP.

1000°C) the disilane is not completely precracked (SiH_2 formation supports Si incorporation in InP). Above 1000°C the disilane decomposes and Si deposits in the HTD injector, but the vapor pressure of Si at this temperature is too low to achieve significant dopant flux. The electrical activation calculated by a comparison of SIMS and Hall data is independent of the precracking temperature between 400 and 1100°C and reaches 85 to 90%. At a growth temperature of 450°C the electrical activation is reduced to 46%. Furthermore, as expected, a linear dependence of the electron concentration on the Si_2H_6 flux is measured in the range 10^{16} to 7×10^{17} cm^3 [30].

For the application to SAE and nonplanar overgrowth, it is mandatory to investigate dopant incorporation in such layered structures. The point of interest is in which way the Si or Zn incorporation is affected by the overgrowth on higher-index surfaces. Consequently, dopant incorporation on large-area (100) and (111) planes and on localized planes is described below.

Si Incorporation in (100) and (111) InP Si-doped samples showed n-type conduction for all surfaces investigated. No difference in incorporation behavior was observed using elemental silicon or precracked Si_2H_6 as a dopant source. Doping levels between 2.5×10^{17} cm^{-3} and 2×10^{18} cm^{-3} were measured on the tilted (100) structures. Figure 7.30 gives an overview of the free electron concentrations obtained from the (111)A and (111)B surfaces compared with (100). The electron concentrations obtained from these structures were typically about a factor 1.2 higher than for the (100) reference surfaces. PL measurements on the band-filling effect confirm this observation, as the PL wavelengths measured for (111)B structures fit the data measured for (100) structures [101]. The (111)A structures always reveal electron concentrations about a factor 1.5 to 2 lower than the corresponding tilted (100) structures, as determined by Hall and PL measurements. On the other hand, SIMS measurements confirm that the atomic Si concentrations are identical for (111)A and (100) structures. This means that the incorporation efficiency of Si into InP is identical for (100) and (111)A surfaces; the lower carrier concentration can be attributed either to Si atoms that are not electrically activated or to Si atoms that are incorporated on group V sites and therefore act as acceptors. Compensation ratios were determined to be $\Theta = N_A/N_D = 0.5$ to 0.7 from the electron mobilities [102]. This is a further indication of autocompensation by Si atoms on P-sites.

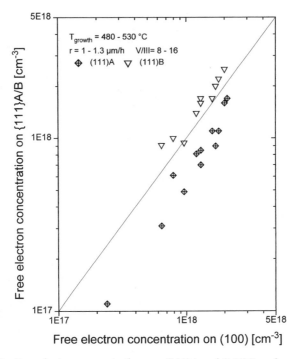

FIGURE 7.30 Free electron concentrations on (111)A and (111)B surfaces versus the corresponding electron concentrations from (100) samples.

Zn Incorporation in (100) and (111) InP   Zn doping of InP has been performed at growth temperatures between 450 and 480°C. The free hole concentration obtained by Zn doping has an Arrhenius dependency on the temperature [96]. The DEZn flux was adjusted for each temperature to obtain free hole concentrations $p \approx 5.0 \times 10^{17}$ cm^{-3} on the (100) structures. Table 7.1 summarizes the free hole concentrations measured for various orientations. For the (111)A orientation it is about an order of magnitude higher than for (100), whereas the structures on (111)B substrates reveal hole concentrations which are about a factor of 2 to 3 lower than that for (100). As the activation of Zn or InP (100) planes is about 60% at a growth temperature $T = 450$°C, the higher free hole concentration on (111)A can be attributed to higher Zn incorporation. The results concerning the orientation dependence of the free hole concentration are in agreement with the observations reported for MOVPE growth [103,104]. Kondo et al. [105] give a model for the incorporation of group II dopants in InP that explains the high incorporation rate by the stable adsorption sites on A surfaces, whereas B surfaces provide weak adsorption sites for the group II atoms.

Si and Zn Co-doping in (100) and (111) InP   The observations made for Si and Zn doping of InP lead to the conclusion that simultaneous doping should yield p- or n-type conduction, depending on the substrate orientation used. We therefore investigated the Si–Zn co-doping of InP at a growth temperature $T = 480$°C using precracked Si$_2$H$_6$ together with DEZn as dopant sources. Based on the results from the preceding sections, the doping source fluxes were adjusted such that a net n-type conduction was achieved on the tilted (100) and (111)B structures. The surface-selective doping mechanisms cause a net p-type conduction for the (111)A surface within the same growth process. The high electron concentrations along with the low hole concentrations on (100) and (111)B substrates result in an effective n-type conduction. On the other hand, lower electron concentrations on the (111)A substrates together with the high hole concentrations from Zn doping result in an effective p-type conduction. In Table 7.1, the free carrier concentrations measured for the various surfaces for Si, Zn, and Si–Zn co-doping are compared. For (100) and (111)B the free electron concentrations are about the difference of ionized donors and acceptors obtained from the single doped layers. On the other hand, co-doping on (111)A reveals a free hole concentration which is about an order of magnitude lower than expected from single Si and Zn doping. This suggests competing mecha-

TABLE 7.1   Carrier Concentrations Measured for Si–Zn Co-doped InP Layers Grown on Various Substrate Surfaces

Surface	InP:Si	InP:Zn	InP:Si:Zn
(100)	n = 1.0 × 10^{18} cm^{-3}	p = 7.8 × 10^{17} cm^{-3}	n = 2.0 × 10^{17} cm^{-3}
(111)A	n = 5.7 × 10^{17} cm^{-3}	p = 4.3 × 10^{18} cm^{-3}	n = 4.9 × 10^{17} cm^{-3}
(111)B	n = 1.3 × 10^{18} cm^{-3}	p = 3.1 × 10^{17} cm^{-3}	n = 8.2 × 10^{17} cm^{-3}

nisms for Si and Zn incorporation on (111)A, as both dopants preferably occupy group V sites.

Spatially Resolved PL Studies on Localized Structures

InP:Si As described in Ref. 101 the PL wavelength of n-type InP is a clear function of the free carrier concentration. Spatially resolved photoluminescence is therefore a powerful tool to determine the free electron concentration along the facets of patterned structures. Figure 7.31 shows the results taken from PL measurements along a (100)/(111)A/(100) transition, overgrown with InP:Si. On top of the (100) ridge and in the recess the luminescence is at $\lambda = 904$ to 905 nm, which corresponds to a free electron concentration $n = 1.3$ to 1.4×10^{18} cm^{-3}. On the facet the PL wavelength increases within 1 μm up to $\lambda = 910$ to 911 nm, which corresponds with a free electron concentration of $n = 1$ to 7×10^{17} cm^{-3}. This is an almost abrupt change in carrier concentration from ridge to facet by a factor of 2. On the facet the PL wavelength, and thus the free electron concentration, stays at a constant value. At a distance of about 1.5 μm from the edge to the groove, the PL wavelength decreases. It settles at a constant value $\lambda = 904$ to 905 nm 2 μm from the edge in the groove. This graded transition can be understood from the graded transition in film thickness at this edge. Here the dopant incorporation mechanisms change due to the gradual transition of the surface orientation. The results taken from the spatially re-

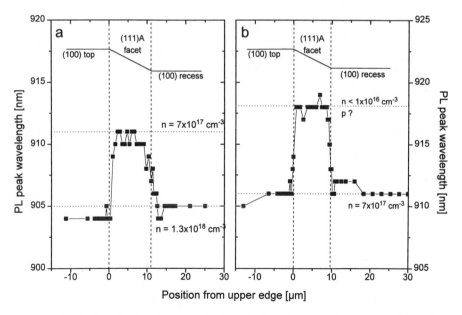

FIGURE 7.31 Results of spatially resolved PL studies: line scans along the [011] direction across (100)/(111)A/(100) transitions. Nonplanar structures overgrown with (*a*) InP:Si and (*b*) InP:Si:Zn.

solved PL measured on the nonplanar structure are in good agreement with the Hall measurements from the large-area references. The free electron concentrations were $n = 1.3 \times 10^{18}$ cm^{-3} for (100) and $n = 7 \times 10^{17}$ cm^{-3} for (111)A. These results demonstrate that the dopant incorporation mechanisms are the same for large-area growth as for the overgrowth of nonplanar structures.

InP:Si:Zn Figure 7.31*b* shows the results from spatially resolved PL measured across the (111)A facet of a co-doped InP:Si:Zn sample. The PL peak wavelength on the ridge is λ = 910 to 911 nm, which corresponds to a free electron concentration n = 7×10^{17} cm^{-3}. At the edge of the (111)A facet, the PL peak wavelength increases within 1 μm up to λ = 918 nm. This indicates an abrupt drop in free electron concentration of at least one order of magnitude. Large-area references give a free electron concentration n = 7×10^{17} cm^{-3} for the (100) and a free hole concentration p = 2×10^{17} cm^{-3} for the (111)A sample, and the results from PL spectroscopy are comparable. This implies that during the overgrowth of the nonplanar structure a lateral p-n-junction can be realized.

7.4 TECHNOLOGY PUSH

7.4.1 New Precursors

The industrial exploitation of the metal-organic growth process is accompanied by significant costs concerning safety requirements, storage of group V starting materials, and toxic waste. Consequently, the search for hydride replacement precursors is of enormous economic importance. Precursors such as TBP, TBAs, or ditertiary butyl As and P compounds (DTBP, DTBAs) are already in use by MOVPE. These compounds are very expensive, so their use in a growth method such as MOMBE, where the precursors are used up very efficiently, is a clear move toward a safe and cost-efficient III-V growth technology.

Growth of InP Using TBP and DTBP InP layers were grown with a thickness of 2 μm at a nominal growth rate of 1 μm/h under variation of substrate temperature, cracker temperature, and V/III ratio. The layers exhibited specular surfaces with good mophological quality. Figure 7.32 gives an overview of the electrical data of InP layers using three different TBP batches from supplier 1 (Epichem Ltd.) and one TBP batch from supplier 2 (sgs mochem). In all samples the free carrier concentration was n-type varying in the range 1×10^{14} cm^{-3} to 3×10^{15} cm^{-3}. Typical 300-K Hall mobilities were 2700 to 4500 cm^2/V·s with a best value of 5200 cm^2/V·s at n = 2.3×10^{14} cm^{-3}. The first TBP bubbler was used only for a few InP growth experiments, but with each of the other TBP bubblers, 77-K Hall mobilities better than 100,000 cm^2/V·s could be obtained, with a best value of 167,000 cm^2/V·s at n = 1.8×10^{14} cm^{-3}. These data are identical with the best value reported for MOVPE growth using TBP [106] and significantly better than those reported for MOMBE growth with TBP [51] so far (see Table 7.2). The best values for the Hall mobility of

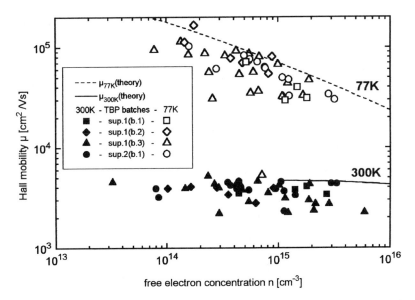

FIGURE 7.32 300- and 77-K electron Hall mobilities in InP layers versus the free electron concentration. The data are presented for three different TBP batches from the first supplier and one TBP batch from the second supplier. The theoretical curves for uncompensated material are from Walukiewicz et al. [102].

InP layers deposited with TMI + DTBP were 5100 cm^2/V·s with n = 6.3 × 10^{14} cm^{-3} at 300 K, and 77,300 cm^2/V·s with n = 5.1 × 10^{14} cm^{-3} at 77 K, respectively. The layers exhibit excellent optical quality. Figure 7.33 presents a PL spectrum originating from the above-mentioned InP sample with the best electrical data. The low-temperature spectrum recorded at 1.8 K is dominated by donor-bound excitonic

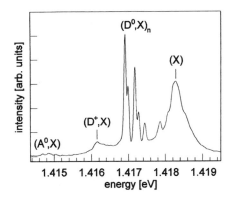

FIGURE 7.33 Low-temperature PL spectrum (1.8 K) of an InP layer grown with TMI + TBP, focusing on the exitonic transitions.

transitions $(D^0,X)_n$. The narrow linewidth of the donor-bound exciton of around 0.09 meV and the low recombination intensity of the neutral acceptor-bound exciton (A^0,X) emission confirm the very low impurity concentration, as indicated by the free electron concentration $n = N_D - N_A$ and compensation ratio N_A/N_D derived from Hall measurements. On the low-energy side of the PL spectrum, which is not plotted here, no additional significant transitions appear.

GaInAsP Growth Using Replacement Precursors Figure 7.34 presents the Ga and P incorporation efficiencies in GaInAsP when using TBAs/TBP in comparison with the hydride data in Figs. 7.13 and 7.14, represented by the gray curves and circles. It can be clearly seen that for increasing growth temperature, the P incorporation efficiency is increased. The hydride data for Q 1.05/Q 1.07 have a more pro-

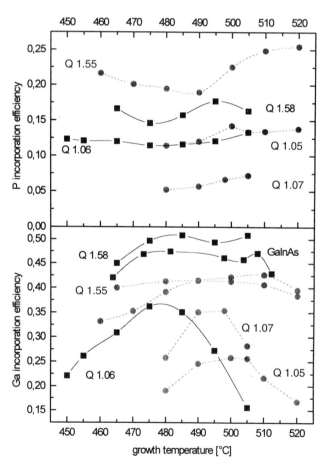

FIGURE 7.34 Effect of growth temperature on the Ga/P incorporation efficiency using hydrides and TBAs/TBP. (Circles, data from the hydride process; squares, data from the TBAs/TBP process.)

TABLE 7.2 Comparison of the Best Reported Values of the 77-K Electron Hall Mobilities in InP Layers Grown by MOVPE and MOMBE Using Different Phosphorus Precursors ($cm^2/V \cdot s$)

Method	PH3	TBP	DTBP
MOVPE [ref.]	300,000 [108]	167,000 [106]	59,600 [109]
MOMBE [ref.]	238,000 [107]	62,500 [51]	—
This work [45]	—	167,000	77,300

nounced slope for the P incorporation efficiency above 480°C than the TBAs/TBP data for Q 1.06. In addition, the maximum Ga incorporation efficiency is clearly enhanced for the TBAs/TBP process. In the hydride system it is observed that increasing the V/III ratio, on the one hand leads to a significant increase the Ga incorporation efficiency, and on the other hand to a reduction in the P incorporation efficiency (see above). From this observation it can be concluded that the effective V/III ratio on the growing surface is clearly higher in the TBAs/TBP process than in the hydride process. For clarifying to what extent the formation of elemental hydrogen on the growing surface affects this effective V/III ratio and therefore influences the temperature-dependent competition of group III alkyl dissociation, the process of group III and group V incorporation and desorption remains to be studied [110].

Laser Structures With all sets of precursors, DH laser structures have been grown and processed. In Table 7.3 the results are given for threshold current densities as well as the value for a MQW laser structure consisting of five GaInAs wells and Q 1.25 barriers with two separate confinement layers on both sides of the MQW. Independent of the precursor selected, the structures yield values between 2.2 and 2.7 kA/cm^2 for DH and 1.3 and 1.5 kA/cm^2 for the hydride-grown MQW structure. Parallel-grown localized (SAE) structures in general yield somewhat lower threshold current densities than the references for the DH structures. In Fig. 7.35, SEM micrographs of the cross section of localized grown DH laser ridges using the various precursor sets mentioned earlier are displayed. The structures exhibit a

TABLE 7.3 Threshold Current Densities of 1.55-μm Broad-Area Lasers[a]

Group V Precursors	Stucture/Quantum Wells	j_{th} (kA/cm^2) Ref.	j_{th} (kA/cm^2) SAE
AsH$_3$/PH$_3$	DH/DEZn	2.2–2.3	1.9–2.1
AsH$_3$/PH$_3$	MQW Q1.25 GaInAs/DEZn	1.3–1.5	1.5–1.8
TBAs/TBP	DH/Be	2.6–2.7	2.5–3.1
TBAs/TBP	DH/DEZn	2.4–2.5	2.3–3.0
DTBAs/DTBP	DH/DEZn	2.3–2.7	1.5–1.8

[a]50 or 100 μm × 400 μm.

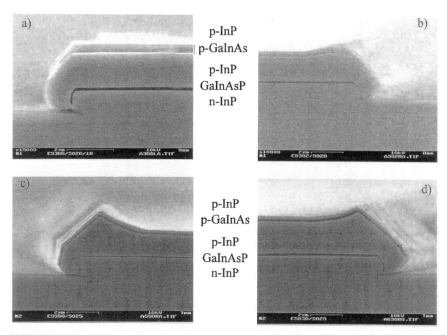

FIGURE 7.35 SEM micrographs of cross sections of laser structures selectively grown using different precursors: (a) TBAs/TBP Be; (b) TBAs/TBP Zn; (c) hydride Zn; (d) DTBAs/DTBP Zn. All structures grown with DEZn show a multifacet formation in the upper cladding layer at the position of the transition from growth to nongrowth zone, independent of the group V precursor used.

smooth vertical sidewall $(01\bar{1})$ for the lower cladding layer (InP:Si) with a small second facet $(11\bar{1})$ on top of the structure. This lower cladding layer represents the in situ–formed base ridge of the laser structure. The growth temperature in this part of the structure was set at 505°C, which led to minimum formation of a second facet with {111} orientation (see the discussion above). The quaternary active layer (lower dark line) covering the base ridge presents the typical growth phenomenon of reduced sidewall growth. This quaternary active layer is uniformly extended across the ridge structure. The structures (Fig. 7.35a and c) grown at conditions optimized for hydrides to achieve vertical sidewalls exhibit a very small {111} facet. In the samples using TBAs/TBP (Fig. 7.35b) and DTBAs/DTBP (Fig. 7.35d), a more pronounced facet formation has been found; here, further optimization of the growth temperature is required for improved vertical sidewall definition in the base ridge.

A different facet evolution has been found for the growth of the upper cladding layer of the DH laser structures. Here, the Be- and the Zn-doped layers exhibit different facet formation processes. For the Be-doped structure shown in Fig. 7.35a, the growth temperature was reduced to 480°C to reach the optimum activation condition for p-type doping. This temperature reduction leads to a uniform overgrowth and does not show a significant new facet formation. The structure formed in the lower cladding layer is reproduced with a somewhat increased {111} facet length due to the lowered growth temperature. For Zn doping a further reduction in growth temperature to 450°C is mandatory to optimize the doping condition. The structure of Fig. 7.35b with an InP:Zn layer grown at 450°C using TBAs/TBP shows formation of an additional facet leading to thickness enhancement at the ridge sidewalls. The structure in Fig. 7.35c was grown using hydrides, and the temperature for the Zn-doped InP was increased to 470°C. Again, this structure presents pronounced formation of {111} facets in the InP:Zn. All ridges grown at various temperatures below 480°C using different sets of precursors exhibit this inverted V shape in the Zn-doped upper cladding layer. Two factors can be responsible for the occurrence of this shape. First, the reduction in growth temperature is below the limit for the formation of vertical side facets found with the structure shown in Fig. 7.35a. In general, the temperature reduction leads to a higher effective V/III ratio and different diffusion phenomena in combination with a reduced decomposition of the group III species. Hence multifacets are observed at growth-to-nongrowth transitions. Second, there is an effect on the diffusion length of the growth species in the presence of DEZn on the surface.

7.4.2 Multiwafer Growth

The results in Section 7.3 have clearly revealed the requirement for lateral uniformity and stability of the growth temperature and V/III ratio for a large uniform growth process across a 2-in. wafer. To optimize the holder design, we grew lattice-matched GaInAsP material with different material compositions on a single 1- by 2-in. holder and on a 3- by 2-in. holder with no indium mounting of the substrates. The growth temperature was varied between 480 and 520°C, being controlled in the center of the substrate. From calculation of the beam flux distribution and earlier results, we can exclude an effect of nonuniform V/III, V/V, and III/III ratios across the wafer [38]. Consequently, the lateral nonuniformities in temperature can be calculated using the data presented in Section 7.3.1. Different heat transfer plates (HTPs; see Fig. 7.9 and Table 7.4) were used to adjust the temperature distribution and therefore the material composition across the 2-in. wafer (see Section 7.3.). The HTPs were mounted along with the wafer in a new holder design [39]. The lateral coupling between the wafer holder and the wafer was made as small as possible, as reported by Ando et al. [111]. The wafer was mounted via three pins with a gap between the wafer edge and the molybdenum holder. At the back side of the substrate the HTP is loaded. The upper side of the HTP is machined in various shapes, such as planar, concave, and convex. This design was used for the 1- by 2-in. and the 3- by 2-in. molybdenum holder configuration and allowed an intentional lateral variation in heat coupling to the manipulator heater.

Table 7.4 gives an overview of the variations in the susceptor geometries and growth parameters and the results for the 1- by 2-in. large area growth configuration. Without any HTP at the back side of the wafer, the temperature difference between the center and the edge of the wafer was reduced toward higher growth temperatures. The maximum value of the temperature is calculated on the edge of the

TABLE 7.4 Variation of Growth Conditions for Various HTP Shapes and Adjustment

Pos.	Material Composition GaInAsP (λ_G)	HTP Shape	T_{growth} in the Center (°C)	$\Delta\lambda$ (nm)	$\Delta(\Delta d/d)$	$\Delta T_{edge\text{-}center}$ (°C)	LT Cell	Configuration (in.)
1	1.55	No	480	7	6.6×10^{-4}	3.7	LTC	1×2
2	1.55	No	500	3	1.2×10^{-4}	2.7	LTC	1×2
3	1.55	No	520	2	3.0×10^{-4}	1.2	LTC	1×2
4	1.55	▭	500	1.5	6.0×10^{-5}	0.8	LTC	1×2
5	1.55	◖◗	500	2.0	1.2×10^{-4}	1.9	LTC	1×2
6	1.55	▭	500	4.0	-8.6×10^{-4}	−6.7	LTC	1×2
7	1.55	▭	500	1.5	1×10^{-4}	1	LTO	1×2
8	1.25	▭	500	4.0	2.2×10^{-4}	1	LTO	1×2
9	1.05	▭	500	1.5	3.0×10^{-4}	1	LTO	1×2
10	1.55	▭	500	2.0	1.0×10^{-4}	2	LTC	3×2

2-in. wafer. Toward higher growth temperatures the difference is reduced from 3.7°C (480°C center growth temperature) to 1.2°C (520°C center growth temperature) (see rows 1 to 3 in Table 7.4). Calculating the growth temperature gradient between the center (480°C growth temperature) and the edge of the wafer instead from the incorporation efficiencies (the data basis in Fig 7.11) also from the two u-curves in Fig. 7.11 yielded identical figures. This confirms that the measured variation in material composition is predominated by nonuniformities in the temperature, indicating that any beam pressure nonuniformity plays a minor role.

By using HPTs with planar, concave, or convex shapes, the temperature difference between the center and the edge can be tuned (see rows 4 to 6 in Table 7.4). We achieved optimized stability data in a wavelength of 1.5 nm and a lattice mismatch of $\Delta d/d$ of 6×10^{-5} using a HTP with a planar shape (see Fig. 7.36a). The lateral temperature difference across the 2-in. wafer was calculated to be around 0.8 °C from the growth result. This reveals a clear improvement with respect to the results obtained by the substrate holders of our first series [38]. There the growth temperature was 510°C for optimizing the wavelength stability only for the 1.55-μm material.

Rows 7 to 9 in Table 7.4 focus on the uniformity of GaInAsP material with different compositions using the low-temperature off-axis cell. For Q 1.55 materials (row 7) we achieved the same excellent uniformity across the 2-in. wafer as in the case of a low-temperature central cell. The wavelength stability yielded 1.5 nm for 1.55-μm material, 4 nm for 1.25-μm material, and again 1.5 nm for 1.05-μm material, since the center temperature of 500°C was set to the minimum in the u-curve of this material (see Fig. 7.11). The temperature difference between the center and the edge of the wafer was calculated to be 1°C for all material compositions. Row 10 in Table 7.4 gives an initial result for GaInAsP material grown in a 3- by 2-in. configuration which also revealed an excellent uniformity in PL wavelength of $\Delta\lambda = 2$ nm (see Fig. 7.36d) and a variation in lattice mismatch of less than $\Delta(\Delta d/d) = 1.0 \times 10^{-4}$. Consequently, the temperature gradient across the wafer was less than 2°C in this multiwafer process.

These insights and uniformity data for material compositions were transferred to the growth of MQW device structures [59]. The wavelength uniformity of an InAsP/GaInAsP MQW and GaInAsP/GaInAsP MQW is better than ±2.5 nm across more than 95% of the wafer area [59], which implies that the thickness uniformity across a 2-in. wafer is achieved in monolayer accuracy (see Fig. 7.36b and c). This was also confirmed by evaluation of x-ray mapping data. The data revealed the suitability of this novel wafer holder design for industrial applications.

7.4.3 Novel Device Development and Integration Capability

Generally, the design of the confinement region of an MQW laser is important since it affects optical confinement, filling factor, carrier injection, and carrier capture efficiency. In most long-wavelength lasers, this confinement region consists of several quaternary layers with different composition. [21,112–115]. Here the large number of heterointerfaces, however, may be detrimental to device performance due to

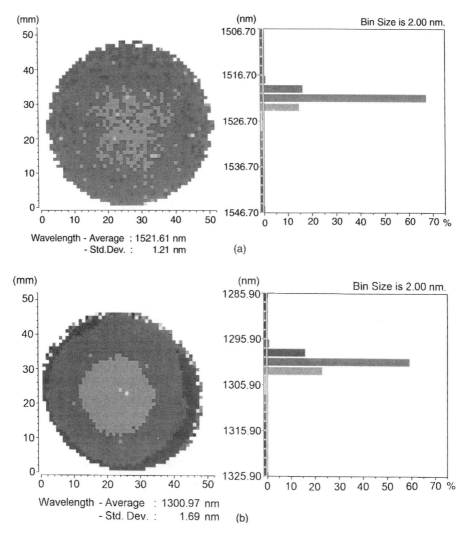

FIGURE 7.36 PL wavelength map across a 2-in. wafer: (*a*) 1.55-μm GaInAsP bulk layer grown in a 1- by 2-in. configuration; (*b*) 1.3-μm GaInAsP MQW structure; (*c*) 1.3-μm InAsP MQW structure; (*d*) 1.55-μm GaInAsP bulk layer grown in a 3- by 2-in. configuration.

increased internal losses, for example. A more elegant approach is the incorporation of $Ga_xIn_{1-x}As_yP_{1-y}$ confinement layers with continuously graded alloy compositions x and y [i.e., graded bandgap, or graded refractive index separate confinement heterostructure (GRINSCH)]. There are very few MOVPE reports on this topic. In GRINSCH MQW lasers prepared previously, a linear variation of the bandgap was

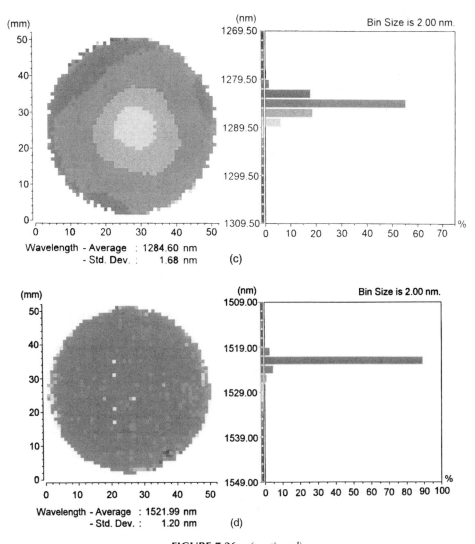

FIGURE 7.36 *(continued)*

realized [116,117]. MOMBE should be a promising technology to fabricate such structures because of its excellent flow control.

The growth procedure for the GRINSCH region is rather complex. To achieve $Ga_xIn_{1-x}As_yP_{1-y}$ confinement layers with continuously graded alloy compositions x and y, the incorporation behavior of the matrix elements Ga, In, As, and P had to be studied in detail. The GRINSCH region chosen covers the range Q 1.05 to Q 1.1 over a thickness of 20 nm. As a strategy we determined the Ga and P incorporation efficiencies in analogy to Ref. 30 for seven material compositions in the λ_G wavelength range 0.95 to 1.25 µm. A fitting function delivers the incorporation efficiencies for the wavelength region of the GRINSCH, which was subdivided into 20 sublayers having individual starting flow setpoints for materials between $\lambda_G = 1.05$ and 1.1 µm. These setpoints were calculated after the fitting functions for the group III and group V incorporation efficiencies. In each of the 20 sublayers with identical thickness of 1 nm, the flow values of all group III and group V precursors were linearly ramped to the setpoint of the subsequent sublayer. Consequently, in the entire GRINSCH region, all group III and group V flows were ramped synchronously for maintaining lattice-matching growth conditions. This growth process was carried out without any growth interruptions. It should be pointed out that the growth rate and V/III flow ratio were kept constant during the growth of this GRINSCH structure. The starting setpoints of the sublayers were set for a parabolic shape of the bandgap (see Fig. 7.37). Any other bandgap shape should be accessible by this procedure. For thicker GRINSCH structures, the number of sublayers should be increased.

The depth profiles of the matrix atoms Ga, As, and P in the GRINSCH region of the laser structure were analyzed by SIMS and are given in Fig. 7.38. These data clearly illustrate the continous compositional grading x, y of the $Ga_xIn_{1-x}As_yP_{1-y}$ structure. For a test of the quality of the GRINSCH structure and for an evaluation of the highly strained InAsP quantum wells, broad-area lasers (100 × 400 µm²) were fabricated and characterized. Dependent on device design and Zn profile, absolute values of j_{th} in the range 500 to 800 A/cm² were obtained in SCH MQW lasers results that are typical for these structures [30]. Comparable data were measured on InAsP MQW lasers and on GRINSCH MQW lasers. The results reveal the high material quality of MOMBE-grown $Ga_xIn_{1-x}As_yP_{1-y}$ structures with compositional grading. Such structures may be utilized in further advanced devices in the future.

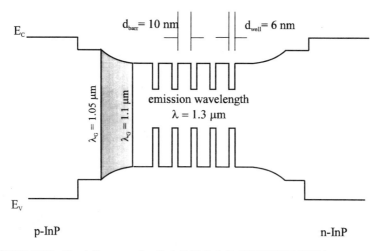

FIGURE 7.37 Band diagram of an InAsP/GaInAsP GRINSCH MQW laser structure.

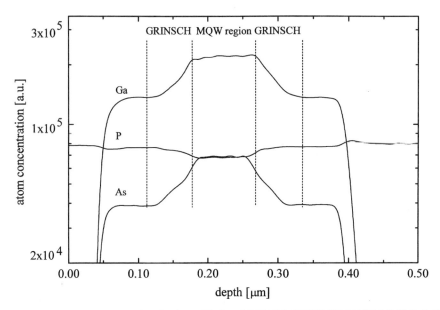

FIGURE 7.38 Ga, As, and P depth profile in a MOMBE GRINSCH MQW InP/GaInAsP laser structure measured by SIMS.

Across 2-in. wafers the relative threshold current density variation is $\Delta j_{th}/j_{th} \leq \pm 5\%$. Comparing structures with standard GaInAsP wells to those with InAsP wells, the latter shows an increase in characteristic temperature from approximately 68 K up to 80 K. In addition, there appears to be a slight increase in differential efficiency. Such behavior is expected to be due to the larger conduction band offset compared to that for quaternary wells. In MOMBE the growth temperature is typically lower than in MOVPE. Consequently, critical thickness in strained-layer MQW structures can be increased by growing the structure at lower temperaures [7,118,119]. The temperature for the growth of InP-based compounds is at least 100°C lower in MOMBE as compared to MOVPE [7], suggesting the high potential of MOMBE for preparing complex device structures. Figure 7.39 shows the effect of the well number on PL intensity of strain-compensated and uncompensated MOMBE-grown InAsP MQWs [7]. The strain and thickness of the well layers were kept constant at 1.5% and 5.5 nm, respectively. For compensation a tensile strain of 0.4% was selected in the 10-nm-thick GaInAsP barriers. It can be recognized that the PL intensity for uncompensated MQWs decreases when the well number is larger than 10, whereas high-quality material can be achieved up to 25 wells for compensated MQWs, which are thermally stable even at 620°C. From these studies it can be concluded that a combination of low growth temperatures with strain compensation can be utilized in MOMBE to grow highly strained MQW structures which have been proven to be difficult to grow by any other techniques.

For improved high-temperature operation, strained 1.3-μm InAsP MQW lasers

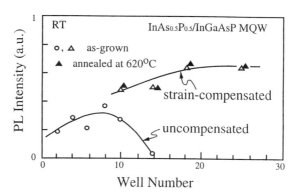

FIGURE 7.39 Effect of well number on PL intensity of strain-compensated and uncompensated MOMBE-grown InAsP MQWs. Annealing conditions were 620°C for 2.5 h. (From Ref. 7.)

have been developed by both MOVPE [120,121] and MOMBE [7,59]. By a comparison with MOVPE, the benefits of low-temperature growth were demonstrated for MOMBE [7]. A higher number of wells can be incorporated and higher operation temperatures can be reached in MOMBE. Figure 7.40 shows the maximum operating temperature T_{max} as a function of well number, in comparison to MOVPE data from Ref. 7. For broad-area 1.3-μm lasers with 15 wells, T_{max} = 155°C was achieved. BH lasers with one coated, facet for high reflectivity exhibited an output power of 37 mW at 90°C. With both facets coated, these lasers had a T_{max} value of 160°C. Long-term reliability was demonstrated [7]. To extend the emission wave-

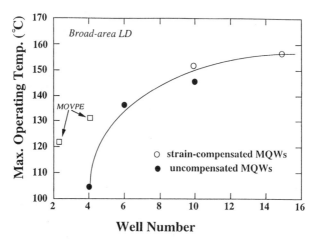

FIGURE 7.40 Maximum operating temperature of MOMBE-grown InAsP MQW lasers as a function of the well number. MOVPE data are included for comparison. (From Ref. 7.)

length to 1.55 μm, an even higher strain is required in the InAsP well layers. By reducing the growth temperature to the range 460 to 480°C, strain-compensated 1.55-μm lasers were obtained by MOMBE, showing the best device characteristic so far [122]. Up to 17 wells (1.7% compressive strain) were integrated in the compensated MQW structure without showing any degradation of the laser characteristics. Moreover, the easier composition control of the ternary InAsP layers [7,39] makes this material attractive for industrial applications.

Integrated Structures Figure 7.41a shows a Nomarski interference micrograph of a top view of reactive-ion-etched rib waveguides prepared perpendicularly across the lateral heterojunctions shown in Fig. 7.26. The horizontal 8- and 12-μm-wide lines represent the heterostructure with λ_G = 1058 nm in the first structure grown. In between there is an infill of DH structure with λ_G = 1100 nm. We have etched two different types of rib waveguides, as shown in the inset of Fig. 7.41b. In type 1 the wave is guided by a 3-μm-wide reactive-ion-etched ridge in the quaternary material. The second type of waveguide is obtained by a 4.5-μm-wide ridge etched in the InP cap layer. The precise determination of the loss of the waveguide–waveguide butt coupling was achieved with the cutback method using TE polarized laser light with a wavelength of 1540 nm. In Fig. 7.41b the coupling loss of these different types of waveguides is plotted versus the number of butt couplings. The small spread of the data points illustrates a high uniformity of couplings. Type 1 shows an optical loss of (0.22 ± 0.05) dB per coupling and type 2 the best value, of (0.12 ± 0.04) dB per coupling. The propagation loss of identical waveguides on a reference sample without couplings was measured to be (0.59 ± 0.06) dB/cm by the Fabry–Perot resonance method [123]. This low propagation loss is comparable to the best published data [124,125] and is still included in the cutback measurements.

The results of the hydride process for selective infill were applied to laser–waveguide butt couplings. Figure 7.42 presents a cross-section of a laser–waveguide butt coupling grown by the conventional hydride technique. There is a smooth contact between the strained multi-quantum well laser structure with two separate confinement layers [30] and the waveguide infill. The 2.5-μm-high infill shows neither any overgrowth over the mask nor ears at the junctions. During growth of the entire selective infill, the growth rate, growth temperature, and V/III ratio were kept constant. This convenient growth procedure is applicable only in perpendicular molecular beam geometry [37,69]. Masked islands of 100 μm × 600 μm were embedded by the waveguide DH structure with λ_G = 1050 nm, forming lateral coulings. Laser–waveguide chips were cleaved. The lasers (400 μm in length) are surrounded on three sides by the waveguide DH structure. The threshold current density of a laser–waveguide combination with a 300-μm-long waveguide DH structure is j_{th} = 1.28 kA/cm². The threshold current density decreases to j_{th} = 880 A/cm² if the waveguide DH structure in the direction of the laser cavity is cleaved off. This threshold current density is identical to the value of a reference sample without selective-area growth processing [30,110]. Consequently, the etching and SAE growth of the waveguide have not degraded laser performance. In a photonic integrated circuit (PIC), the electrical separation of device structures is also a central is-

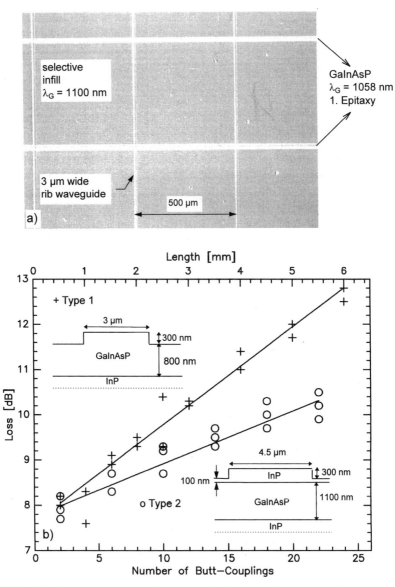

FIGURE 7.41 Cascades of waveguide–waveguide butt coupling: (*a*) interference micrograph of the top view of a coupling cascade; (*b*) optical loss versus the number of butt couplings for two different types of rib waveguides.

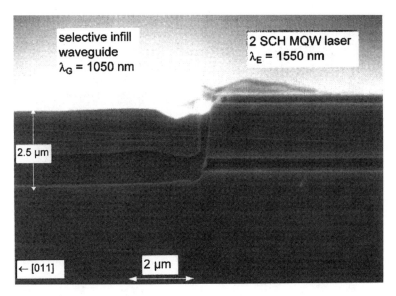

FIGURE 7.42 SEM cross section of lateral butt coupling of a two-SCH-MQW (λ_E = 1550 nm) laser with a waveguide (λ_G = 1050 nm) structure.

sue. The SAE of semi-insulating waveguides by Fe doping is a demanding process. Figure 7.43 presents a first realization of the SAE growth of GaInAsP (λ_G = 1.05 μm), Fe-doped around an etched laser mesa [126]. The SEM micrograph suggests that SAE semi-insulating waveguides can be implemented in advanced PICs.

7.5 SUMMARY AND FUTURE CHALLENGES

The MOMBE method started as a research tool about 16 years ago. It is the youngest III-V growth technology, and overlaps significantly with the MOVPE process. In past years this correlation has accelerated the development and innovation speed in both the MOVPE and MOMBE technologies. The MOMBE process is now on its way to being a production technology for sophisticated InP-based device structures. This new process offers unique pathways for the in situ generation of low-dimensional structures and the integration of complex device structures and concepts by utilizing surface-selective growth processes.

A growth technique such as MOMBE, enabling the efficient use of source materials, is particularly in demand for multiwafer equipment. This opens a pathway for precursors, which are currently rather expensive but are much less dangerous than the conventional ones for use in a production environment being targeted for a volume market. Thus in the future, MOMBE can play an important role in both the fabrication of sophisticated InP-based devices and the fulfillment of environmental requirements for semiconductor fabrication processes.

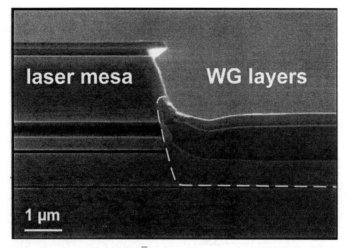

FIGURE 7.43 SEM showing the (01$\bar{1}$) cleavage plane of a selectively iron-doped waveguide layer in the vicinity of a laser mesa.

ACKNOWLEDGMENTS

The author is indebted to Eberhard Veuhoff for stimulating discussions and cooperation during the years of MOMBE development. I also thank Horst Baumeister, Bernd Baur, Amalie Milde, Reiner Höger, Robert Primig, Josef Rieger, Martin Wachter, Bastian Marheineke, Michael Popp, Rainer Butendeich, Dietmar Ritter, Markus Keidler, Martin Hurich, Jürgen Schneider, and Rudolf Lenz for their support in the developments described here. Parts of the work were funded by the German Ministry of Education and Research under contracts 01 BP 467/7 and 01 BM 412/2 and by direct support from Siemens AG and Riber during the time at the University of Ulm, Germany.

REFERENCES

1. J. M. Schneider, J.-T. Pietralla, and H. Heinecke," Control of chemical composition and bandgap energy in $Ga_xIn_{1-x-y}Al_yAs$ on InP during molecular beam epitaxy," *J. Cryst. Growth*, **175/176**, 184–190 (1997).
2. E. Kuphal, "Phase diagrams of InGaAsP, InGaAs and InP lattice matched to (100) InP," *J. Cryst. Growth*, **67**, 441–457 (1984).
3. Y. Yamazoe, T. Nishino, Y. Hamakawa, and T. Kariga, "Bandgap energy of InGaAsP quaternary alloys," *Jpn. J. Appl. Phys.*, **19**, 1473–1479 (1980).
4. H. Asai and K. Oe, "Energy band-gap shift with elastic strain in GaInP epitaxial layers on (001) GaAs substrates," *J. Appl. Phys.*, **54**, 2052–2056 (1983).
5. P. J. A. Thijs, L. F. Tiemeijer, P. I. Kaindersma, J. J. M. Binsma, and T. van Dongen,

"High-performance 1.5μm wavelength InGaAs-InGaAsP strained quantum well lasers and amplifiers," *IEEE Trans. Quantum Electron.*, **27,** 1426–1439 (1991).

6. H. Oohashi, T. Hirono, S. Seki, H. Sugiura, J. Nakamo, M. Yamamoto, Y. Tohmori, and K. Yokoyama, "1. 3 μm InAsP compressively strained multiple-quantum-well lasers for high-temperature operation," *J. Appl. Phys.*, **77,** 4119–4121 (1995).

7. H. Sugiura, "MOMBE of InAsP laser materials," *J. Cryst. Growth*, **164,** 434–441 (1996).

8. E. Veuhoff, W. Pletschen, P. Balk, and H. Lüth, "Metalorganic CVD of GaAs in a molecular beam system," *J. Cryst. Growth*, **55,** 30–34 (1981).

9. N. Vodjdani, A. Lemarchand, and H. Paradan, "Parametric studies of GaAs growth by metalorganic molecular beam epitaxy," *J. Phys. C*, **5,** 339–349 (1982).

10. W. T. Tsang, "Chemical beam epitaxy of InP and GaAs," *Appl. Phys. Lett.*, **45,** 1234–1236 (1984).

11. N. Pütz, E. Veuhoff, H. Heinecke, M. Heyen, H. Lüth, and P. Balk, "GaAs growth in metal-organic MBE," *J. Vac. Sci. Technol. B*, **3,** 671–673 (1985).

12. M. B. Panish, "Molecular beam epitaxy of GaAs and InP with gas sources for As and P," *J. Electrochem. Soc.*, **127,** 2729–2733 (1980).

13. M. B. Panish and H. Temkin, *Gas Source Molecular Beam Epitaxy*, Springer Ser. Mater. Sci. 26, ed. U. Gomser, A. Mooradin, R. M. Osgood, M. B. Panish, and H. Sakaki, Springer-Verlag, New York, 1993.

14. E. Veuhoff, "Potential of MOMBE/CBE for the production of photonic devices in comparison with MOVPE," *J. Cryst. Growth*, **188,** 237–246 (1998).

15. H. Ando, T. Fujii, A. Sandhu, T. Takahashi, H. Ishikawa, N. Okamoto, and N. Yokoyama, "Growth of carbon base GaAs/AlGaAs HBT by gas source MBE using TEG, TEA, TMG, AsH$_3$ and Si$_2$H$_6$," *J. Cryst. Growth*, **120,** 228–233 (1992).

16. C. R. Abernathy, F. Ren, S. J. Pearton, T. R. Fullowan, R. K. Montgomery, P. W. Wisk, J. R. Lothian, P. R. Smith, and R. N. Nottenburg, "Growth of GaAs/AlGaAs HBTs by MOMBE(CBE)," *J. Cryst. Growth*, **120,** 234–239 (1992).

17. P. A. Lane, C. R. Whitehouse, T. Martin, M. Houlton, G. M. Williams, A. G. Cullis, S. S. Gill, J. R. Dawsey, G. Ball, B. T. Hughes, M. A. Crouch, and M. B. Allenson, "CBE growth of GaAs/AlGaAs HBTs using the new DEAlH-NMe$_3$ precursor and all gaseous dopants," *J. Cryst. Growth*, **120,** 245–251 (1992).

18. A. Sandhu, T. Nakamura, H. Ando, K. Domen, N. Okamoto, and T. Fujii, "The electrical, optical and crystalline properties of GaAs:C grown by GSMBE using TMG and AsH$_3$ for application to HBTs," *J. Cryst. Growth*, **120,** 296–300 (1992).

19. W. T. Tsang, "From chemical vapor epitaxy to chemical beam epitaxy," *J. Cryst. Growth*, **95,** 121–131 (1989).

20. W. T. Tsang, "Progress in chemical beam epitaxy," *J. Cryst. Growth*, **105,** 1–29 (1990).

21. W. T. Tsang, F. S. Choa, M. C. Wu, Y. K. Chen, A. M. Sergent, and P. F. Sciortino, Jr., "Very low threshold single quantum well graded-index separate confinement heterostructure InGaAs/InGaAsP lasers grown by chemical beam epitaxy," *Appl. Phys. Lett.*, **58,** 2610–2612 (1991).

22. W. T. Tsang, F. S. Choa, M. C. Wu, Y. K. Chen, R. A. Logan, S. N. G. Chu, A. M. Sergent, P. Magill, K. C. Reichmann, and C. A. Burras, "Long wavelength InGaAsP/InP distributed feedback lasers grown by chemical beam epitaxy," *J. Cryst. Growth*, **124,** 716–722 (1992).

23. M. Nakao, R. Iga, T. Yamada, and H. Sugiura, "High performance 1.5-μm distributed feedback lasers with strained multi-quantum well structure grown by metalorganic molecular beam epitaxy (chemical beam epitaxy)," *Jpn. J. Appl. Phys.*, **31**, L1549-L1551 (1992).
24. T. Yamada, R. Iga, Y. Noguchi, and H. Sugiura, "CBE growth of low threshold 1.5 μm InGaAs/InGaAsP MQW lasers," *J. Cryst. Growth*, **120**, 177–179 (1992).
25. H. P. Meier, R. F. Broom, P. W. Epperlein, S. Hausser, A. Jakubowicz, and W. Walter, "Growth investigations of 1.3 μm GaInAsP/InP MQW laser diodes grown by chemical beam epitaxy," *J. Cryst. Growth*, **127**, 165–168 (1993).
26. L. Yang, A. S. Sudo, W. T. Tsang, P. A. Garbinski, and R. M. Camarda, "Monolythically integrated InGaAs/InP MSM-FET photo-receiver prepared by chemical beam epitaxy," *J. Cryst. Growth*, **105**, 162–167 (1990).
27. H. Heinecke, B. Baur, N. Emeis, and M. Schier, "Growth of highly uniform InP/GaInAs/GaInAsP heterostructures by MOMBE (CBE) for device integration," *J. Cryst. Growth*, **120**, 140–144 (1992).
28. N. Emeis, M. Schier, L. Hoffmann, H. Heinecke, and B. Baur, "High speed waveguide integrated photodiodes grown by MOMBE," *Electron. Lett.*, **28**, 344–345 (1992).
29. M. Keidler, M. Popp, D. Ritter, B. Marheineke, H. Heinecke, H. Baumeister, and E. Veuhoff, "Growth of 1.55 μm DH-laserstructures using TBAs and TBP in MOMBE," *J. Cryst. Growth*, **170**, 161–166 (1997).
30. M. Popp, H. Heinecke, H. Baumeister, and E. Veuhoff, "Full gaseous source growth of separate confinement MQW 1.55 μm laser structures in a production MOMBE," *J. Cryst. Growth*, **175/176**, 1247–1253 (1997).
31. Proc. 2nd Int. Conf. Chemical Beam Epitaxy and Related Growth Techniques, Houston, Texas, 1989, *J. Cryst. Growth*, **105**, (1990).
32. Proc. 3rd Int. Conf. Chemical Beam Epitaxy and Related Growth Techniques, Oxford, 1991, *J. Cryst. Growth*, **120**, (1992).
33. Proc. 4th Int. Conf. Chemical Beam Epitaxy and Related Growth Techniques, Nara, Japan, 1993, *J. Cryst. Growth*, **136**, (1994).
34. Proc. 5th Int. Conf. Chemical Beam Epitaxy and Related Growth Techniques, La Jolla, Calif., 1995, *J. Cryst. Growth*, **164**, (1996).
35. Proc. 6th Int. Conf. Chemical Beam Epitaxy and Related Growth Techniques, Montreux, Switzerland, 1997, *J. Cryst. Growth*, **188**, (1998).
36. J. S. Foord, G. J. Davies, and W. T. Tsang, eds., *Chemical Beam Epitaxy and Related Techniques*, Wiley, New York 1997.
37. M. Wachter and H. Heinecke, "Beam geometrical effects on planar SAE of InP/GaInAs heterostructures," *J. Cryst. Growth*, **164**, 302–307 (1996).
38. B. Marheineke, M. Popp, and H. Heinecke, "Growth of GaInAs(P) using a multiwafer MOMBE," *J. Cryst. Growth*, **164**, 16–21 (1996).
39. M. Popp, H. Baumeister, E. Veuhoff, and H. Heinecke, "Element incorporation in GaInAsP for uniform large area MOMBE," *J. Cryst. Growth*, **188**, 247–254 (1998).
40. H. Heinecke, "III-V crystal growth of novel layered structures using metalorganic molecular beam epitaxy," *Phys. Scr.*, **T49**, 742–747 (1993).
41. M. A. Hermann and H. Sitter, *Molecular Beam Epitaxy: Fundamentals and Current Status,* Springer-Verlag, Berlin, Heidelberg, 1989.

42. O. Kayser, H. Heinecke, A. Brauers, H. Lüth, and P. Balk, "Vapor pressures of MOCVD precursors," *Chemtronics*, **3**, 90–93 (1988).
43. M. B. Panish, "Gas source molecular beam epitaxy of GaInAs(P): gas sources, single quantum wells, superlattice p-i-n(s and bipolar transistors," *J. Cryst. Growth*, **81**, 249–260 (1987).
44. D. Ritter and H. Heinecke, "Evaluation of cracking efficiency of As- and P-precursors," *J. Cryst. Growth*, **170**, 149–154 (1997).
45. D. Ritter, M. Keidler, and H. Heinecke, "Growth of InP using TBP and DTPB in metalorganic molecular beam epitaxy," *J. Cryst. Growth*, **188**, 152–158 (1998).
46. T. H. Chiu, W. T. Tsang, J. E. Cunningham, and A. H. Robertson, Jr., "Gallium- and arsenic-induced oscillations of intensity reflection high-energy electron diffraction in the growth of (001) GaAs by chemical beam epitaxy," *J. Appl. Phys.*, **62**, 2302–2307 (1987).
47. A. Robertson, Jr., T. H. Chiu, W. T. Tsang, and J. E. Cunningham, "A model for the surface chemical kinetics of GaAs deposition by chemical-beam epitaxy," *J. Appl. Phys.*, **64**, 877–887 (1988).
48. T. Martin, C. R. Whitehouse, and P. A. Lane, "Growth mechanisms studies in CBE/MOMBE," *J. Cryst. Growth*, **107**, 969–977 (1991).
49. A. S. Jordan and A. Robertson, "Thermodynamic analysis of AsH_3 and PH_3 decomposition including subhydrides," *J. Vac. Sci. Technol. A*, **12**, 204–215 (1994).
50. M. B. Panish and R. A. Hamm, "A mass spectrometric study of AsH_3 and PH_3 gas sources for molecular beam epitaxy," *J. Cryst. Growth*, **78**, 445–452 (1986).
51. M. Zahzouh, G. Hincelin, R. Mellet, and A. M. Pougnet, "Chemical beam epitaxy of high purity InP using tertiarybutylphosphine and 1,2 bis-phosphinoethane," *Mater. Sci. B*, **21**, 165–168 (1993).
52. D. Ritter, M. Wachter, M. Hurich, M. Keidler, B. Marheineke, and H. Heinecke, "Cracking behaviour of tertiarybutylphosphine and phosphine in MOMBE," presented at VI-EWMOVPE, Gent, Belgium, 1995.
53. E. A. Beam III, T. S. Henderson, A. C. Seabaugh, and J. Y. Yang, "The use of tertiarybutylphosphine and tertiarybutylarsine for the metalorganic molecular beam epitaxy of the $In_{0.53}Ga_{0.47}As/InP$ and $In_{0.48}Ga_{0.52}P/GaAs$ material systems," *J. Cryst. Growth*, **116**, 436–446 (1992).
54. D. Ritter, M. B. Panish, R. A. Hamm, D. Gershoni, and I. Brener, "Metalorganic molecular beam epitaxy of InP, $Ga_{0.47}In_{0.53}As$ and GaAs with tertiarybutylarsine and tertiarybutylphosphine," *Appl. Phys. Lett.*, **56**(15), 1448–1450 (1990).
55. S. H. Li, N. I. Buchan, C. A. Larsen, and G. B. Stringfellow, "OMVPE growth mechanism for GaP using tertiarybutylphosphine and trimethylgallium," *J. Cryst. Growth*, **96**, 906–914 (1989).
56. T. P. Fehlner, "A mass spectrometric investigation of the low-pressure pyrolysis of diphosphane-4," *J. Am. Chem. Soc.*, **89**(25), 6477–6482 (1967).
57. H. Heinecke, "Potentiality and challenge of MOMBE (CBE)," *Mater. Sci. Eng. B*, **9**, 83–91 (1991).
58. H. Heinecke, B. Baur, R. Höger, B. Jobst, and A. Miklis, "Effect of surface orientation on GaInAsP material composition in MOMBE (CBE)," *J. Cryst. Growth*, **124**, 170–175 (1992).

59. H. Baumeister, E. Veuhoff, M. Popp, and H. Heinecke, "GRINSCH GaInAsP MQW laser structures grown by MOMBE," *J. Cryst. Growth*, **188**, 266–274 (1998).
60. T. Isu, K. Morishita, S. Goto, Y. Nomura, and Y. Katayama, "Real-time observations of III-V-growth on patterned substrates by μ-RHEED," *J. Cryst. Growth*, **127**, 942–948 (1993).
61. E. Tokumitsu, Y. Kudou, M. Konagai, and K. Takahashi, "Molecular beam epitaxial growth of GaAs using trimethylgallium as a gas source," *J. Appl. Phys.*, **55**, 3163–3165 (1984).
62. E. Tokumitsu, Y. Kudou, M. Konagai, and K. Takahashi, "Metalorganic molecular-beam epitaxial growth and characterization of GaAs using trimethyl- and triethylgallium sources," *Jpn. J. Appl. Phys.*, **24**, 1189–1192 (1986).
63. H. Heinecke, A. Brauers, F. Grafahrend, C. Plass, N. Pütz, K. Werner, M. Weyers, H. Lüth, and P. Balk, "Selective growth of GaAs in the MOMBE and MOCVD systems," *J. Cryst. Growth*, **77**, 303–309 (1986).
64. G. J. Davies, W. J. Duncan, P. J. Skevington, C. L. French, and J. S. Foord, "Selective area growth for opto-electronic integrated circuits (OEICs)," *Mater. Sci. Eng. B*, **9**, 93–100 (1991).
65. Y. Kawaguchi, H. Asahi, and H. Nagai, "Gas source MBE growth of high-quality $In_{x}Ga_{1-x}As$," *Proc. 18th Int. Conf. Solid State Devices and Materials*, Tokyo, 1986, pp. 619–622.
66. D. A. Andrews, M. A. Z. Rejman-Green, B. Wakefield, and G. J. Davies, "Selective area growth of InP/InGaAs multiple quantum well lasers structures by metalorganic molecular beam epitaxy," *Appl. Phys. Lett.*, **53**, 97–98 (1988).
67. O. Kayser, "Selective growth of InP/GaInAs in LP-MOVPE and MOMBE/CBE," *J. Cryst. Growth*, **107**, 989–998 (1991).
68. H. Heinecke, B. Baur, R. Schimpe, R. Matz, C. Cremer, R. Höger, and A. Miklis, "Selective area epitaxy of InP/GaInAsP heterostructures by MOMBE," *J. Cryst. Growth*, **120**, 376–381 (1992).
69. R. Matz, H. Heinecke, B. Baur, R. Primig, and C. Cremer, "Facet growth in selective aea epitaxy of InP by metalorganic MBE," *J. Cryst. Growth*, **127**, 230–236 (1993).
70. Y. Chen, J. E. Zucker, T. H. Chiu, J. L. Marshall, and K. L. Jones, "Quantum well electroabsorption modulators at 1.55 μm using single-step selective area chemical beam epitaxial growth," *Appl. Phys. Lett.*, **61**, 10–12 (1992).
71. H. Sugiura, T. Nishida, R. Iga, T. Yamada, and T. Tamamura, "Facet growth of InP/InGaAs layers on SiO_2-masked InP by chemical beam epitaxy," *J. Cryst. Growth*, **121**, 579–586 (1992).
72. M. Gotoda, T. Isu, S. Maruno, and Y. Nomura, "Selective embedded growth of InGaAs/InP double-heterostructures by chemical beam epitaxy," *J. Cryst. Growth*, **125**, 502–508 (1992).
73. H. Heinecke, "Surface selective growth of GaInAsP heterostructures by metalorganic MBE," *J. Cryst. Growth*, **127**, 126–135 (1993).
74. H. Heinecke, B. Baur, A. Miklis, R. Matz, C. Cremer, and R. Höger, "Evidence for vertical superlattices grown by surface selective growth in MOMBE (CBE)," *J. Cryst. Growth*, **124**, 186–191 (1992).
75. T. H. Chiu, Y. Chen, J. L. Zucker, J. L. Marshall, S. Shunk, and S. N. G. Chu, "Selective

area growth of InGaAsP/InP waveguide modulator structures by chemical beam epitaxy," *J. Cryst. Growth*, **127,** 169–174 (1993).

76. H. Heinecke and E. Veuhoff, "Evaluation of III-V growth technologies for optoelectronic applications," *Mater. Sci. Eng. B*, **21,** 120–129 (1993).
77. H. Heinecke, E. Veuhoff, N. Pütz, M. Heyen, and P. Balk, "Kinetics of GaAs growth by low pressure MOCVD," *J. Electron. Mater.*, **13,** 815–830 (1984).
78. H. Heinecke, A. Milde, R. Matz, B. Baur, and R. Primig, "Novel III-V heterostructures fabricated by metalorganic molecular beam epitaxy," *Phys. Scr.*, **T55,** 14–19 (1994).
79. E. Veuhoff, H. Heinecke, J. Rieger, H. Baumeister, R. Schimpe, and S. Pröhl, "Improvements in selective area growth of InP by metalorganic vapor phase epitaxy," in *Proc. 4th Int. Conf. InP and Related Materials*, Newport, R.I., 1992, pp. 210–213.
80. A. Usui, "Atomic layer epitaxy," *Mater. Res. Soc. Symp. Proc.*, **198,** pp. 183–193 (1990).
81. D. B. Gladden, W. D. Goodhue, C. A. Wang, and G. A. Lincoln, "Electrical and structural characterization of GaAs vertical-sidewall epilayers grown by atomic layer epitaxy," *J. Electron. Mater.*, **21,** 109–114 (1992).
82. S. Ando, S. S. Chang, and T. Fukui, "Selective area epitaxy of GaAs/AlGaAs on (111)B substrates by MOCVD and applications to nanometer structures," *J. Cryst. Growth*, **115,** 69–73 (1991).
83. H. Heinecke, M. Wachter, and U. Schöffel, "Facet formation and characterization of III-V structures grown on patterned surfaces," *Microelectron. J.*, **28,** 803–816 (1997).
84. M. Wachter, C. Menke, and H. Heinecke, "Anisotropic surface diffusion at crystal facet transitions during localized Ga-In-As-P growth by MOMBE," *Microelectron. J.*, **28,** 841–848 (1997).
85. A. Iberl, H. Göbel, and H. Heinecke, "Characterization of III/V-heterostructures grown by selective area epitaxy using double-crystal x-ray diffractometry with high lateral resolution," *J. Phys. D Appl. Phys.*, **28,** A172–A178 (1995).
86. M. Ganeau, R. Chaplain, A. Rupert, A. Le Corre, M. Salvi, H. L'Haridon, and D. Lecrosnier, "Oxygen complexes in III-V compounds as determined by secondary-ion mass spectrometry under cesium bombardment," *J. Appl. Phys.*, **66,** 2241–2247 (1989).
87. J. L. Benchimol. F. Alaoui, Y. Gao, G. Le Roux, E. V. K. Rao, and F. Alexandre, "Chemical beam epitaxy of indium phosphide," *J. Cryst. Growth*, **105,** 135–142 (1990).
88. T. K. Uchida, T. Uchida, K. Mise, F. Koyamu, and K. Iga, "Extremely high Be doped $Ga_{0.47}In_{0.53}As$ growth by chemical beam epitaxy," *J. Cryst. Growth*, **105,** 366–370 (1990).
89. W. T. Tsang, F. S. Choa, and N. T. Ha, "Zinc-doping of InP during chemical beam epitaxy using diethylzinc," *J. Electron. Mater.*, **20,** 541–544 (1991).
90. R. A. Hamm, D. Ritter, H. Temkin, M. B. Panish, and M. Geva, "p-type doping of InP and $Ga_{0.47}In_{0.53}As$ using diethylzinc during metalorganic molecular beam epitaxy," *Appl. Phys. Lett.*, **58**(21), 2378–2380 (1991).
91. W. T. Tsang, F. S. Choa, R. A. Logan, T. Tanbun-Ek, and A. M. Sergent, "All-gaseous doping during chemical-beam epitaxial growth of InGaAs/InGaAsP multi-quantum-well lasers," *Appl. Phys. Lett.*, **59**(9), 1008–1010 (1991).
92. T. Sudersena Rao, C. Lacelle, S. J. Rolfe, M. Dion, J. Thompson, P. Marshall, P. Chow-Chong, D. Ross, M. Davies, and A. P. Roth, "Chemical beam epitaxy growth of 1. 3 μm

InGaAsP/InP double heterostructure lasers using all gas source doping," *Appl. Phys. Lett.*, **65**(8), 1015–1017 (1994).

93. C. W. Snyder, J. W. Lee, R. Huss, and R. A. Logan, "Catastrophic degradation lines at the facet of InGaAsP/InP lasers investigated by transmission electron microscopy," *Appl. Phys. Lett.*, **67**, 488–490 (1995).

94. S. L. Jackson, M. T. Fresina, J. E. Baker, and G. E. Stillman, "High-efficiency silicon doping in InP and $In_{0.53}Ga_{0.47}As$ in gas source and metalorganic molecular beam epitaxy using silicon tetrabromide," *Appl. Phys. Lett.*, **64**, 2867–2869 (1994).

95. D. Ritter, R. A. Hamm, M. B. Panish, and M. Geva, "Beryllium δ-doping studies in InP and $Ga_{0.47}In_{0.53}As$ during metalorganic molecular beam epitaxy," *Appl. Phys. Lett.*, **63**, 1543–1545 (1993).

96. E. Veuhoff, H. Baumeister, R. Treichler, M. Popp, and H. Heinecke, "Zn doping of InP/GaInAsP device structures in metalorganic molecular beam epitaxy," *J. Cryst. Growth*, **164**, 402–408 (1996).

97. H. Heinecke, K. Werner, M. Weyers, H. Lüth, and P. Balk, "Doping of GaAs in metalorganic MBE using gaseous sources," *J. Cryst. Growth*, **81**, 270–275 (1987).

98. H. Ando, N. Okamoto, A. Sandhu, and T. Fuji, "Gas-source MBE growth of n-type InP using TEI, PH_3 and Si_2H_6," *Jpn. J. Appl. Phys.*, **30**, L1696-L1698 (1991).

99. K. Kimura, S. Horiguchi, K. Kamon, M. Shimazu, M. Mashita, M. Mihara, and M. Ishii, "Silicon doping from disilane in gas source MBE of GaAs," *J. Cryst. Growth*, **81**, 276–280 (1987).

100. E. Veuhoff, H. Baumeister, J. Rieger, M. Gorgel, and R. Treichler, "Comparison on Zn and Mg incorporation in MOVPE InP/GaInAsP laser structures," *J. Electron. Mater.*, **20**, 1037–1041 (1991).

101. B. Marheineke, E. Veuhoff, and H. Heinecke, "Dopant incorporation behaviour during MOMBE growth of InP on (100), {111} and nonplanar surfaces," *J. Cryst. Growth*, **188**, 183–190 (1998).

102. W. Walukiewicz, J. Lagowski, L. Jastrzebski, P. Rava, M. Lichtensteiger, C. H. Gatos, and H. C. Gatos, "Electron mobility and free-carrier absorption in InP; determination of the compensation ratio," *J. Appl. Phys.*, **51**(5), 2659–2668 (1980).

103. J. J. Yang, R. P. Ruth, and H. M. Manasevit, "Electrical properties of epitaxial indium phosphide films grown by metalorganic chemical vapor deposition," *J. Appl. Phys.*, **52**(11), 6729–6734 (1981).

104. R. Bhat, C. Caneau, C. E. Zah, M. A. Koza, W. A. Bonner, D. M. Hwang, S. A. Schwarz, S. G. Menocal, and F. G. Favire, "Orientation dependence of S, Zn, Si, Te and Sn doping in OMCVD growth of InP and GaAs: application to DH lasers and lateral p-n junction arrays grown on non-polar substrates," *J. Cryst. Growth*, **107**, 772–778 (1991).

105. M. Kondo, C. Anayama, T. Tanahashi, and S. Yamazaki, "Crystal orientation dependence of impurity dopant incorporation in MOVPE-grown III-V materials," *J. Cryst. Growth*, **124**, 449–456 (1992).

106. T. Imori, T. Takayuki, K. Ushikubo, K. Kondo, and K. Nakamura, "High-purity InP layer grown by metalorganic chemical vapor deposition using tertiarybutylphosphine," *Appl. Phys. Lett.*, **59**(22), 2862–2864 (1991).

107. T. Sudersena Rao, C. Lacelle, R. Benzaquen, S. J. Rolfe, S. Charbonneau, P. D. Berger, A. P. Roth, T. Steiner, and M. L. W. Thewalt, "Electrical, optical properties, and surface

morphology of high purity InP grown by chemical beam epitaxy," *J. Appl. Phys.*, **76**(9), 5300–5308 (1994).

108. E. J. Thrush, C. G. Cureton, and A. T. R. Briggs, "MOCVD grown InP/InGaAs structures for optical receivers," *J. Cryst. Growth*, **93**, 870–876 (1988).

109. H. Protzmann, Z. Spika, B. Spill, G. Zimmermann, W. Stolz, E. O. Göbel, P. Gimmich, and J. Lorberth, "MOVPE growth and characterization of InP using the novel P-source ditertiarybutylphosphine," presented at VI-EWMOVPE, Gent, Belgium, 1995, contribution A16.

110. M. Keidler, D. Ritter, H. Baumeister, M. Druminski, and H. Heinecke, "Planar selective area growth of DH laser structures using hydride, tertiarybutyl and ditertiarybutyl group V precursors in MOMBE," *J. Cryst. Growth*, **188**, 168–175 (1998).

111. H. Ando, S. Yamamura, and T. Fujii, "Recent progress in multi-wafer CBE system," *J. Cryst. Growth*, **164**, 1–15 (1996).

112. T. Tanbun-Ek, R. A. Logan, H. Temkin, K. Berthold, A. F. J. Levi, and S. N. G. Chu, "Very low threshold InGaAs/InGaAsP graded index separate confinement heterostructure quantum well lasers grown by atmospheric pressure metalorganic vapor phase epitaxy," *Appl. Phys. Lett.*, **55**, 2283–2285 (1989).

113. H. Temkin, N. K. Dutta, T. Tanbun-Ek, R. A. Logan, and A. M. Sergent, "InGaAs/InP quantum well lasers with sub-mA threshold current," *Appl. Phys. Lett.*, **57**, 1610–1612 (1990).

114. R. M. Redstall, A. P. Skeats, D. M. Cooper, and A. L. Burness, "Preliminary reliability studies of 1.55mm graded index separate confinement buried heterostructure (GRIN-SCH) multiple quantum well (MQW) lasers grown by metalorganic vapor phase epitaxy (MOVPE)," *Electron. Lett.*, **26**(15), 1132–1133 (1990).

115. B. Borchert, B. Stegmüller, H. Baumeister, J. Rieger, E. Veuhoff, H. Hedrich, and H. Lang, "High performance 1. 55 μm quantum-well metal-clad ridge-waveguide distributed feedback lasers," *Jpn. J. Appl. Phys.*, **30**, L1650-L1652 (1991).

116. N. Carr, A. K. Wood, J. Thompson, N. Maung, R. M. Ash, and A. J. Moseley, "The metal-organic vapor phase epitaxy growth of GaInAsP and GaAlInAs based graded refractive index seperate confinement heterostructure multiple quantum well lasers incorporating linearly graded confinement layers," *Mater. Sci. Eng. B*, **9**, 355–360 (1991).

117. N. Carr, J. Thompson, A. K. Wood, R. M. Ash, D. J. Robbins, A. J. Moseley, and T. Reid, "A comparison of MOVPE grown strained and unstrained MQW lasers incorporating continuously graded or single composition confinement layers," *J. Cryst. Growth*, **124**, 723–729 (1992).

118. G. L. Price, "Critical thickness and growth-mode transitions in highly strained $In_xGa_{1-x}As$ films," *Phys. Rev. Lett.*, **66**, 469–472 (1991).

119. H. Sugiura, "MOMBE growth of highly tensile-strained InGaAsP MQWs and their applications to 1.3 μm wavelength low threshold current lasers," *J. Cryst. Growth*, **175/176**, 1205–1209 (1997).

120. Y. Imajo, A. Kasukawa, T. Namegaya, and T. Kikuta, "1.3μm $InAs_yP_{1-y}$/InP strained-layer quantum well laser diodes grown by metalorganic chemical vapor deposition," *Appl. Phys. Lett.*, **61**, 2506–2508 (1992).

121. M. Yamamoto, N. Yamamoto, and J. Nakano, "MOVPE growth of strained InAsP/InGaAsP quantum-well structures for low-threshold 1.3 μm Lasers," *IEEE Trans. Quantum Electron.*, **30**, pp. 554–561 (1994).

122. J. F. Carlin, A. V. Syrbu, C. A. Bersetz, J. Behrend, A. Rudra, and E. Kapon, "Low threshold 1.55 μm wavelength InAsP/InGaAsP strained multiquantum well laser diode grown by chemical beam epitaxy," *Appl. Phys. Lett.*, **71**, 13–15 (1997).
123. R. G. Walker, "Simple and accurate loss measurement technique for semiconductor optical waveguides," *Electron. Lett.*, **21**(13), 581–583 (1985).
124. H. Künzel, P. Albrecht, R. Gibis, M. Hamacher, and S. Schelhase, "MOMBE growth of high quality GaInAsP (λ_g = 1.05 μm) for waveguide applications," *J. Cryst. Growth*, **164**, 449–453 (1996).
125. J. H. Angenent, M. Erman, J. M. Auger, R. Gamonal, and P. J. A. Thijs, "Extremely low loss InP/GaInAsP rib waveguides," *Electron. Lett.*, **25**(10), 628–629 (1989).
126. H. Künzel, P. Albrecht, S. Ebert, R. Gibis, P. Harde, R. Kaiser, H. Kizuki, and S. Malchow, "MOMBE grown GaInAsP (λ_g = 1.05/1.15 μm) waveguide for laser integrated photonic ICs," *J. Cryst. Growth*, **188**, 281–287 (1998).

CHAPTER EIGHT

Physics and Technological Control of Surfaces and Interfaces of InP-Based Materials

HIDEKI HASEGAWA
Hokkaido University

8.1	Introduction	248
8.2	Semiconductor Heterointerfaces of InP-Based Materials	250
	8.2.1 General Properties of Heterointerfaces	250
	8.2.2 Band Offsets and Their Measurement	251
	8.2.3 Overview on Theory of Band Lineup and Its Application to InP-Based Heterointerfaces	253
	8.2.4 Band Lineup Data for InP-Based Heterointerfaces	257
	8.2.5 Possible Artificial Control of Band Lineup	260
8.3	Metal–Semiconductor Interfaces of InP-Based Materials	260
	8.3.1 Schottky Barrier Formation and Current Transport	260
	8.3.2 Mechanism of Fermi-Level Pinning	263
	8.3.3 Properties of Metal–Semiconductor Interfaces of InP-Based Materials and Attempts at SBH Control	266
8.4	Surfaces and Surface Passivation of InP and Related Materials	269
	8.4.1 Properties of Clean and Real Surfaces	269
	8.4.2 Surface Passivation and Properties of Interface States	270

InP-Based Materials and Devices: Physics and Technology, Edited by Osamu Wada and Hideki Hasegawa.
ISBN 0-471-18191-9 © 1999 John Wiley & Sons, Inc.

8.4.3 Formation of Insulator Films on InP-Based Materials 272
8.4.4 Recent Approaches to Surface Passivation Using
Interface Control Layers 273
References 274

8.1 INTRODUCTION

Semiconductor–semiconductor (S-S) interfaces, metal–semiconductor (M-S) interfaces, and insulator–semiconductor (I-S) interfaces are basic constituent elements of all semiconductor devices, including those based on InP-based materials. As an example, a typical structure of an InP-based high-electron-mobility transistor is shown in Fig. 8.1. One clearly sees here that such an advanced device consists of multitudes of interfaces. In this chapter, properties, physical understanding, and technological control of S-S, M-S, and I-S interfaces of InP-based materials are reviewed and discussed. Before going to the specific details of various kinds of InP-related interfaces, some general introductory remarks are given below.

The basic motivation of interface formation from the viewpoint of device physics is to create potential-energy differences at the interface, as shown schematically in Fig. 8.2a and b, and utilize them for realization of various device operations. In present-day devices such as that shown in Fig. 8.1, with feature sizes down to the submicron range, interfaces are used to confine carries near the interface and to control their average number, movement direction, average velocity, and/or recombination–generation statistics. Mathematically speaking, interfaces are utilized to provide potential-energy boundary conditions for semiclassical macroscopic semiconductor equations that describe ensemble average motions of carriers treated as classical particles.

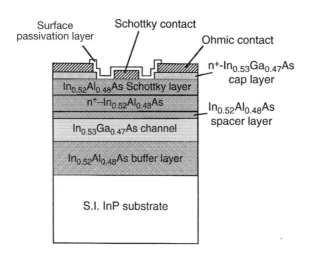

FIGURE 8.1 InP-based high-electron-mobility transistor (HEMT).

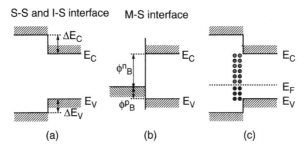

FIGURE 8.2 (*a*) and (*b*) Ideal and (*c*) real interfaces.

For future devices with feature sizes in the decameter–nanometer range or below, quantum nanoelectronics based on various quantum devices, including quantum wire transistors, quantum interference devices, single-electron devices, quantum dot lasers, and so on, have emerged and they are currently being studied intensively. In this quantum regime, S-S, I-S, and M-S interfaces will still be utilized. However, their role will be much more delicate in nature, and will be to control the quantum mechanical motion of individual electrons in artificial quantum structures whose boundaries are defined by interfaces. Examples of quantum structures include quantum wells, quantum wires, quantum dots, tunneling and resonant tunneling barriers, superlattices, and so on. In these structures, interfaces are used to impose potential-energy boundary conditions directly to the Schrödinger equation, so as to control not the ensemble average motion of semiclassical particles, but the quantum mechanical wave-particle motion of individual electrons. Thus we see that interfaces already play extremely important roles in present-day devices, but they will become even more important in the future and will require atomic-level perfection so as to manipulate wavefunctions.

Interfaces are also used to protect devices. Surface passivation is such an interface technology, and its objectives are (1) to terminate and passivate active dangling bonds at the semiconductor surface, and (2) to protect devices from environmental influences. Passivation of III-V compound semiconductors is known to be difficult, due to the appearance of a high density of interface states at I-S interfaces. More generally, although S-S, M-S, and I-S interfaces are formed to produce desired potential profiles for electrons and holes that are required for device operation or device protection, formation of real interfaces, particularly interface formation between dissimilar materials, very often results in generation of interface states within the energy gap, as shown in Fig. 8.2c, and they tend to fix the position of the Fermi level at a certain equilibrium position within the energy gap, since even a slight deviation of the Fermi level from the equilibrium position produces a large amount of electrostatic charge, resulting from an occupancy change of interface states. This phenomenon, called *Fermi-level pinning,* causes various unwanted phenomena in devices. Thus realization of a desired interface potential profile without introducing interface states is a key point for interface formation.

8.2 SEMICONDUCTOR HETEROINTERFACES OF InP-BASED MATERIALS

8.2.1 General Properties of Heterointerfaces

The relationships between energy gap E_g and a lattice constant a are summarized in Fig. 8.3 for various binary and ternary III-V material systems. Ternary, quaternary, or more complicated alloy semiconductors that have lattice parameters in the vicinity of that of InP and can therefore be grown on InP substrates with acceptable quality may loosely be defined as InP-based materials, including InP itself. It can be seen from Fig. 8.3 that a large variety of heterointerfaces can be formed from InP-based materials.

Generally, heterointerfaces can be classified into lattice-matched interfaces and lattice-mismatched interfaces. The latter can further be classified into (1) coherently strained or pseudomorphic interfaces free of misfit dislocations, and (2) relaxed heterointerfaces where strain is partially or fully relaxed with introduction of misfit dislocations. The maximum thickness of the coherently strained layer one can reach without introduction of misfit dislocations, called the *critical layer thickness,* can be calculated from the theory by Matthews and Blakeslee [1], or from the theory by People and Bean [2], provided that the elastic constants of the materials are known. Elastic properties of InP-based materials can be found in the literature [3–5].

Extensive efforts during the past 20 to 30 years have led to tremendous progress in the epitaxial growth of binary, ternary, and quaternary III-V alloy materials on InP substrates by MBE and MOVPE. This has made it possible to realize, in repro-

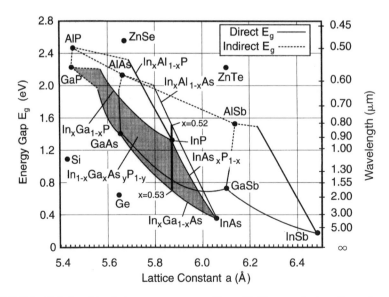

FIGURE 8.3 Relationships between energy gap, E_g, and lattice constant, a, for various III-V materials.

ducible and controllable fashion, atomically abrupt heterointerfaces that are free of interface defects. However, the conditions to realize such high-quality interfaces are severe. For example, heterointerfaces can be formed either by a continuous growth process or by a discontinuous process involving growth interruption and regrowth. Obviously, the latter has a greater degree of freedom for device fabrication. However, the quality of regrown interfaces is generally known to be inferior to that of continuously grown interfaces, even if the growth interruption is made in ultrahigh vacuum [6–8]. It is known empirically that only lattice-matched interfaces, or pseudomorphic interfaces satisfying the condition of critical layer thickness, that are carefully prepared by optimized continuous growth procedures can be nearly ideal, being free of interface defects. In others, there is a strong tendency that trapping states are introduced at the interface, and these interface states may cause Fermi-level pinning as well as various other unwanted phenomena for device operation.

The thickness condition necessary to obtain high-quality strained epitaxial layers is a severe one and limits the compositions of various alloys to lattice parameter ranges that are close to that of the InP substrate. Thus accessible composition ranges are strong functions of the desired thickness. One approach to relax this condition is the *strain-compensated structures,* where multiple interfaces with tensile and compressive strains are repeated in alternating fashion [9–12]. More recently, a concept of a novel compliant universal (CU) substrate has been proposed and partially verified [13], and this may relax the foregoing limitation and widen accessible composition ranges of semiconductor alloys of various compounds in future. In the CU substrate, a thin layer of high-density screw dislocation network is produced on a thick substrate by twist wafer bonding so as to increase lattice flexibility.

The availability of varieties of high-quality heterointerfaces on InP substrates has led to the current flourishing of InP-based advanced electronic and optoelectronic devices based on heterointerfaces where potential-energy profiles for electrons and holes within semiconductor structures can be tailored almost arbitrarily. The actual potential profile at the heterointerface naturally depends on the abruptness of the atomic profile of the interface region. When the atomic profile is perfectly abrupt in the atomic scale, the resultant potential profile becomes also abrupt to monolayer levels, as has been confirmed through experiments such as optical emission from narrow quantum wells. On the other hand, one can also tailor and create various graded potential profiles by designing and forming various graded atomic profiles.

8.2.2 Band Offsets and Their Measurement

Types of Band Offsets The most important fundamental property of an abrupt semiconductor heterointerface is the mutual alignment or lineup of conduction and valence bands at the interface. The differences in the band edges at interface, ΔE_v and ΔE_c, are called valence and conduction *band offsets,* respectively. They are also called *band discontinuities.* As is well known, there are two types of band lineup, type I and type II, as shown in Fig. 8.4. In type I, the bandgap region of the smaller bandgap material is included within that of the larger bandgap material at the inter-

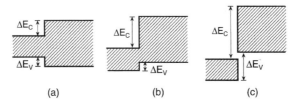

FIGURE 8.4 Types of heterointerfaces: (*a*) type I; (*b*) type II (staggered type); (*c*) type II (broken gap type).

face, and in type II this is not the case. In the type I lineup, the following relation obviously holds:

$$\Delta E_v + \Delta E_c = \Delta E_g \tag{1}$$

where ΔE_g is the difference in the bandgap energy. The fractions, defined by

$$Q_v = \frac{\Delta E_v}{\Delta E_g} \quad \text{and} \quad Q_c = \frac{\Delta E_c}{\Delta E_g} \tag{2}$$

are often regarded as phenomenological characteristic quantities of the heterointerface, since there often exists an empirical tendency that they are independent of the alloy composition. On the other hand, the type II lineup is further classified into staggered and broken gap alignments, as shown in Fig. 8.4*b* and *c*, respectively. In both cases, the following relation holds:

$$|\Delta E_v - \Delta E_c| = \Delta E_g \tag{3}$$

Measurements of Band Offset The magnitudes of band offset can be determined [14] by (1) electrical measurements, (2) optical measurements, and (3) ultraviolet and x-ray photoelectron spectroscopy (XPS) measurements. Electrical methods include thermionic emission measurements over the heterointerface energy barrier, internal photoemission measurements over the barrier, admittance spectroscopy, measurements of C–V carrier density profiles of the heterointerface region, measurements of $(1/C)^2$–V characteristics of the heterointerface region, and Hall measurements of two-dimensional electron gas (2DEG) density for various spacer thicknesses. Optical methods utilize quantum-well samples and measure their optical absorption spectra, photoluminescence spectra, or photoinelastic scattering spectra. In band offset measurement using x-ray photoelectron spectroscopy, a heterostructure with a sufficiently thin overlayer that allows the escape of photoelectrons from the inner material is used. Then, by measuring the energy depth of a particular core level from the valence band edge for each material, as well as the energy difference between the two core levels, the valence band offset can be calculated in a straightforward fashion.

It should be noted that each method has its own source of measurement errors. For example, C–V methods are sensitive to extra charge due to filling of interface states. This was indeed shown to be a source of discrepancy between different measurements for the determination of band offset in the InGaAs/InP interface [15]. In optical measurements using quantum wells, one should recall the fact that the confined quantum states are sensitive to the actual potential profile at the interface. In the case of photoelectron spectroscopy measurements, band bending due to Fermi-level pinning can be a cause of error [16]. From the viewpoint of sample preparations, an unintentional interface layer is often formed by interface reactions such as As–P exchange, and this can give rise to erroneous results.

Properties of Band Offset Previous extensive experimental works on GaAs-based lattice-matched heterointerfaces such as AlGaAs/GaAs have indicated strongly that within typical measurement accuracies of ±10 to 20 meV, the valence and conduction band offsets of a strain-free heterointerface are characteristic quantities of the heterointerface, being determined by the material combination forming the interface itself. Thus they are independent (1) of the external electric or magnetic field, (2) of conduction types and doping levels of the materials forming the interface, (3) of the crystalline direction of the interface plane, and (4) of the growth sequence. In addition, when three different kinds of interfaces are formed from three different materials A, B, and C, the following transitivity rule holds:

$$E_i(A/B) + E_i(B/C) + E_i(C/A) = 0 \quad (4)$$

where $E_i(A/B)$ denotes the conduction ($i = c$) or valence band ($i = v$) offset at the interface between the materials A and B. However, so far, the validity of these properties has not been well established for InP-based lattice-matched heterointerfaces. For example, a large growth sequence dependence of band offsets was reported for InGaAs/InP interface grown by MOMBE and characterized by XPS [17].

When the interface is under strain, the behavior of the heterointerface becomes far more complex. Although the band offsets probably do not depend on external fields, conduction type, and doping, they do depend on crystalline direction and growth sequence, since strain distribution near the interface obviously depends on them. The transitivity relation is also lost.

8.2.3 Overview on Theory of Band Lineup and Its Application to InP-Based Heterointerfaces

Approaches for Band Lineup Calculation Since the band offset is an important quantity for analysis and design of heterostructure devices, a theory that can predict its magnitude to a desired accuracy will be extremely useful. However, the present status of theories of band lineup is still not predictive enough, although they can provide fairly good estimates in certain well-studied cases. There also exists a conceptual disagreement concerning the basic lineup mechanism, as explained below.

There are two basic approaches to the theory of band lineup. One is an abinitio type of self-consistent numerical calculation of the band structure of a crystal, including a heterointerface, using a supercell method. This is a rigorous calculation, but it is a difficult and tedious one that has to be repeated for specific material combinations and crystal orientations. Several authors attempted this type of calculation for InP-based heterointerfaces [18–21].

The other approach is more intuitive. It attempts to connect bulk band structures at interface on the basis of plausible arguments. This type of approach is based on the experimentally observed validity of the transitivity rule in Eq. (4), which suggests that the band offset is determined by the difference in some characteristic energy of each material. If this approach works well, it is very practical and useful particularly for device design, since it allows simple and straightforward determination of band offsets of heterointerfaces with arbitrary combinations of materials and alloy compositions by utilizing a vast database covering the bulk band structures. However, there exist two conflicting views for this approach: one that we might call natural band lineup with a negligible interface dipole, and a second that stresses the importance of an interface dipole that leads to lineup at the charge neutrality level. It is not clear at present which view is more appropriate.

Natural Band Lineup In this view one assumes that there exists a common energy reference for band lineup and that bands align naturally with respect to this common energy reference after interface formation. In other words, it ignores possible interface charge transfer and creation of interface dipole during interface formation. The electron affinity rule [22], the lineup theory of Frenseley and Kroemer [23], the LCAO theory of natural band lineup of Harrison [24], and the model solid theory of Van de Walle and Martin [25] fall in this category.

The classical electron affinity rule takes the vacuum level as the energy reference. The difficulty with this rule arose from recognition that the measured values of the electron affinity include an unknown contribution from atomic rearrangements at the surface, such as surface reconstructions and relaxations, which are different from atomic arrangements at the heterointerfaces. On the other hand, theoretical determination of band edge location with respect to vacuum is not an easy task. This is because most of the bulk band structure calculations assume infinitely large crystals without surface. An exception is LCAO theory by Harrison [24], where atomic term energies are given with respect to vacuum. On a more empirical basis, Langer and Heinrich [26] proposed that transition metal (TM) deep levels serve as the energy reference for band lineup. However, an observation that is in conflict with the general validity of this idea and supportive of lineup at the charge neutrality level has been presented on the basis of data compilation [27].

The most accurate theory based on the spirit of natural band lineup is the model solid theory of Van de Walle and Martin [25], in which the average potential values of various materials are calculated with respect to a common energy reference, assuming semi-infinite hypothetical model solids without surface reconstruction and relaxation. The band lineup can then be determined by placing bulk bands according to the average potential values. The values calculated for the average potential for valence and conduction band edges are tabulated in Ref. 28. The values for mixed alloys may be obtained by linear interpolation. The band offsets can be obtained as the difference in the average potential with corrections for the difference in spin-orbit splitting energy.

Band Lineup at Charge Neutrality Level This view is based on the idea that interface charge transfer and the resulting interface dipole are the main factors that determine the band lineup. According to this view, the interface dipole due to interface charge transfer is generated at the heterointerface so as to screen the difference in the average potential on both sides of the interface by the dielectric screening mechanism. It is further assumed that each material has a characteristic charge neutrality energy level E_0, where virtual gap states in the energy gap change its character from donor type to acceptor type. The position of E_0 is determined by the bulk band structure. Then the bands align in such a way that the charge neutrality level itself aligns at the interface within an error of $\Delta E_0/\varepsilon$, due to dielectric screening. Here ε is the semiconductor permittivity and ΔE_0 is the difference in E_0. The band offset between materials 1 and 2 is given approximately by

$$\Delta E_v = |(E_{v1} - E_{v0}) - (E_{v2} - E_0)|$$
$$\Delta E_c = |(E_{c1} - E_0) - (E_{c2} - E_0)|$$
(5)

E_0 was shown to exist for a one-dimensional system by Flores and Tejedor [29]. It was redefined by Tersoff [30] for three-dimensional systems as midgap energy, using a cell-averaged Green"s function. In a tight binding context, it corresponds to the average sp^3 hybrid orbital energy, according to Hasegawa et al [31,32] and Harrison and Tersoff [33]. As a similar and alternative band lineup energy, Cardona and Christensen [34] introduced the dielectric midgap energy (DME) level, based on fitting the band structure to Penn's two-band model. Mönch [35] has attempted to determine the location of E_0 using sp^3s^* theory from the viewpoint of band lineup at the M-S interface, as discussed below. The location of E_0 for InP-based materials is summarized in Table 8.1. Using this table, one can make a rough estimate of the band offset of the heterointerface without strain. The values for mixed alloys may be obtained by linear interpolation, as was also done in Table 8.1.

The two views described above are seemingly very different concerning the role of interface charge transfer, although both satisfy the transitivity rule. However, the actual calculated values are not so different. Their agreement with measured values are fair but not good enough. There was a theoretical effort to reconcile the two views above by first calculating the bulk valence band edge with respect to a common reference level using atomic sphere approximation similarly to the method used in the model solid theory, then making an interface dipole correction using an

TABLE 8.1 Locations of Charge Neutrality Level E_0 for Various III-V Materials Measured from the Valence Band Maximum

Material	Theory (eV)			Experiment (eV)
	Ref. 30	Ref. 34	Ref. 35	Ref. 32
InP	0.76	0.87	0.86	0.90
GaAs	0.70	0.55	0.52	0.45
InAs	0.50	0.62	0.50	0.50
AlAs	1.05	0.92	0.92	0.95
GaP	0.81	0.73	0.83	0.90
$In_{0.53}Ga_{0.47}As$	0.59	0.59	0.51	0.47
$In_{0.52}Al_{0.48}As$	0.76	0.76	0.70	0.71

interface bond polarity (IBP) model [36]. It was found that after this procedure, the transitivity rule of Eq. (4) is lost by 10 to 70 meV.

Effects of Strain Finally, the effect of strain is discussed briefly. With the presence of strain, a complication arises from the fact that the bandgap energy and band offsets are not only a function of the material combination and alloy composition, but also a function of the strain distribution near the interface. Note that the bandgap energies in Fig. 8.3 are unstrained values, and if the material becomes strained, they change after heterointerface formation. Coherent strain at the heterointerface can be decomposed into two components. One is the hydrostatic component, which changes the bandgap energy together with a corresponding change in the band offsets. The other is the in-plane uniaxial strain component, which splits the degenerate valence bands at $k = 0$, which under appropriate conditions causes the light-hole band to lie higher than the heavy-hole band. As an example, a schematic illustration of strain effects is given in Fig. 8.5 for an InP-based strained heterointerface of $In_xGa_{1-x}As$/InP for various alloy compositions. Thus coherent strain can give rise to greater freedom of potential-energy design for devices.

The most frequently used theory to estimate the band offset of a strained heterointerface is the model solid theory of Van de Walle [28]. According to this theory, the band offsets are obtained in the following way. First, the offset value for the hypothetical unstrained heterointerface is obtained. Then the strain distribution near the heterointerface is obtained from the sample structure and elastic constants. Finally, unstrained band offsets are corrected, with the energy shifts of valence and conduction band edges as well as splitting of the valence band edges calculated with respect to the unstrained case by using the calculated strain components and the appropriate deformation potentials. For this, one needs to know the conduction and valence band hydrostatic deformation potentials, a_c and a_v, and the valence band shear deformation potentials: b for (100) uniaxial strain and d for (111) uniaxial strain. If the conduction band is indirect and degenerate at the minima, one also has to correct for splitting due to uniaxial strain, using the values of conduction band

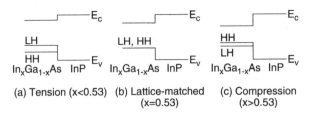

FIGURE 8.5 Effect of strain at $In_xGa_{1-x}As/InP$ heterointerface with coherently strained $In_xGa_{1-x}As/InP$.

shear deformation potentials. Values of these deformation potentials are tabulated in Ref. 28.

On the other hand, Cardona and Christensen also discussed the effects of strain on the band offsets from the viewpoint of band lineup at the dielectric midgap energy (DME) [34]. Here, the deformation potential calculated for the DME itself was included, in addition to the deformation potentials for strain correction of the unstrained band offset values. Similarly, a tight-binding theory of strained heterointerfaces was developed by Anderson and Jones from the viewpoint of lineup by the interface dipole [37].

Thus, for a discussion of band offsets at strained interfaces, one needs knowledge of individual band edge deformation potentials, which are more difficult to measure than conventional bandgap hydrostatic deformation potentials. Measuring the pressure dependence of the 1.2-eV PL emission from a type II $In_xAl_{1-x}As/InP$ interface, Yeh et al. [38] found the valence band hydrostatic deformation potentials of $In_xAl_{1-x}As(0 \leq x \leq 0.52)$ to follow $-1.9x + 0.2$ (eV), which agrees with DME calculation rather than that of model solid theory.

As a more specific theoretical calculation on InP-based strained interfaces, Wang and Stringfellow [39] performed a theoretical analysis for the $In_xGa_{1-x}As/InP$ system based on the model solid theory. Band lineups for InGaAsP/InP and InGaAlAs/InP systems were calculated by Ishikawa and Bowers [40] by applying deformation potential corrections to the unstrained offsets based on Harrison"s natural band lineup theory.

8.2.4 Band Lineup Data for InP-Based Heterointerfaces

Lattice-Matched Heterointerfaces The most widely used InP-based lattice-matched heterointerfaces for electronic and optoelectronic devices are shown in Fig. 8.6 with typical measured values of band offsets. The lattice-matched $In_{0.53}Ga_{0.47}As/InP$ and $In_{0.52}Al_{0.48}As/In_{0.53}Ga_{0.47}As$ interfaces exhibit type I lineup, whereas the lattice-matched $In_{0.52}Al_{0.48}As/InP$ interface is of type II staggered lineup. Values of band offsets of these interfaces have been reviewed critically in various data books [3,4,40] and are not repeated here except for the following brief survey.

Even for the most studied $In_{0.53}Ga_{0.47}As/InP$ heterointerface, there still exist cer-

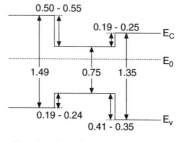

FIGURE 8.6 Band lineup for $In_{0.52}Al_{0.48}As/In_{0.53}Ga_{0.47}As/InP$. Units are eV.

tain ambiguities in the literature as to the exact values of the band offsets, although recent measurements seem to agree that $\Delta E_c \approx 250$ meV, $\Delta E_v \approx 350$ meV, and $Q_c/Q_v \approx 4:6$. In the case of the $In_{0.52}Al_{0.48}As/In_{0.53}Ga_{0.47}As$ system, data suggest that $\Delta E_c \approx 500$ meV, $\Delta E_v \approx 200$ meV, and $Q_c/Q_v \approx 7:3$. A slight extension of this system is the lattice-matched $In_{0.52}(Al_xGa_{1-x})_{0.48}As/InGaAs$ system, for which it was found [42] that ΔE_c is in direct proportion to x such that

$$\Delta E_c = 530x \text{ (meV)} \tag{6}$$

with $Q_c/Q_v \approx 7:3$ being independent of x.

As for the $In_{0.52}Al_{0.48}As/InP$ interface, not much work on direct determination of band offset has been done, although it is generally agreed that it is a staggered lineup [43–47]. The transitivity rule, if it holds, suggests that $\Delta E_c \approx 250$ meV and $\Delta E_v \approx 150$ meV, and this is in agreement with XPS measurement [43]. However, a PL measurement gave $\Delta E_c \approx 390$ meV and $\Delta E_v \approx 290$ meV [44]. A slight extension of this system is the lattice-matched $In_{1-x-y}Ga_xAl_yAs/InP$ system with $x + y = 0.47$ [48]. For this system, the following formula was obtained by PL measurements [49]:

$$\Delta E_c = 0.245 - 1.179y - 0.3y^2 \tag{7}$$

As can be seen from Eq. (7) and others [50], this system goes to type II at around $y = 20\%$.

According to Adachi [41], the quaternary alloy system $In_{1-x}Ga_xAs_yP_{1-y}$ frequently used in optical device can be lattice matched to InP by satisfying the following conditions:

$$x = \frac{0.189y}{0.4176 - 0.0125y} \approx \frac{y}{2.2} \tag{8}$$

Various authors measured the band lineup in this system, and they are also reviewed

8.2 SEMICONDUCTOR HETEROINTERFACES OF InP-BASED MATERIALS

in data books mentioned above. Adachi [41] gave the following formulas on the basis of the work by Langer and Heinrich [26], where transition metal impurity levels were used as the reference for band lineup:

$$\Delta E_v(y) = 502y - 152y^2 \text{ (meV)} \quad (9a)$$

$$\Delta E_c(y) = 286y + 3y^2 \text{ (meV)} \quad (9b)$$

Strained Heterointerface As an example showing the effect of strain, measured data [51] for InAs/AlAs for cases of thin InAs grown on AlAs substrate and thin AlAs grown on InAs substrate are shown in Fig. 8.7. We see that strain indeed has an extremely large effect on the band lineup.

InP-based strained heterointerfaces that are very often utilized in electronic and optoelectronic devices are $In_xGa_{1-x}As/InP$ and $In_xGa_{1-x}As/In_{0.52}Al_{0.48}As$ systems. For example, InP-based HEMTs with an In-rich pseudomorphic $In_xGa_{1-x}As$ channel ($x > 0.53$) has demonstrated superb high-frequency performance due to reduced effective mass and increased conduction band offset [52–54]. Optical properties of strained quantum wells (QWs) using these systems were reviewed by Gershoni and Temkin [55]. As for the band lineup of the $In_xGa_{1-x}As/InP$ system, Cavicchi et al. [56] obtained ΔE_c values of 175, 210, and 315 meV for $x = 0.37$, 0.53, and 0.69, respectively. Wang and Stringfellow [39] performed PL measurements on single QWs and compared the results favorably with an theoretical analysis based on the model solid theory. For an $In_xGa_{1-x}As/In_{0.52}Al_{0.48}As$ system [57–60], Huang et al. [57] obtained the following formula using I–V–T and C–V measurement:

$$\Delta E_c = 0.384 + 0.254x \quad (x < 0.54) \quad (10a)$$

$$= 0.344 + 0.487x \quad (x > 0.58) \quad (10b)$$

For an $In_{0.53}Ga_{0.47}As/In_xAl_{1-x}As$ system, Lee et al. determined ΔE_c by C–V profiling [61]. There is some interest also in strained $In_xGa_{1-x}As/In_yGa_{1-y}As$ interfaces for electronic and optoelectronic applications, and some measurements have been taken [62–64]. A natural extension of the lattice-matched $In_{1-x}Ga_xAs_yP_{1-y}/InP$ sys-

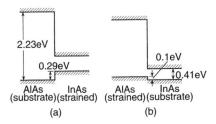

FIGURE 8.7 Strain effect on the band lineup of InAs/AlAs heterointerfaces.

tem is to introduce strain by changing compositions, and some data and theory are available for this system [40,65–67].

Considerable interest has been raised in using strained InAsP for use as the well layer in long-wavelength optoelectronic devices instead of lattice-matched or strained InGaAs or InGaAsP. The advantages include larger ΔE_c, smaller in-plane hole effective mass, less serious carrier overflow at interfaces, reduced alloy scattering than in quaternary alloys, and so on. Various compressively strained and strain-compensated multi-quantum wells using InAsP/InP [68–71], InAsP/InGaAsP [72–77], and InAsP/InGaP [77–83] heterointerfaces have been studied, mainly from the viewpoint of device application. Limited reports have indicated that $\Delta E_v = 0.25\Delta E_g$ for $InAs_xP_{1-x}$/InP with $x = 0.65$ to 0.67 [70,71] and that the $InAs_xP_{1-x}$/$In_{0.53}Ga_{0.47}As_{1-y}P_y$ superlattice shows type I alignment for $x > 0.58$ and $y < 0.21$ and type II alignment for $x < 0.58$ and $y > 0.21$ [73].

8.2.5 Possible Artificial Control of Band Lineup

Since the band offset is a quantity of prime importance for a heterointerface, it is very useful for device design if it can be modified and tuned artificially by a suitable means. The feasibility of such band offset change by insertion of a group IV interface monolayer into a III-V heterointerface was proposed theoretically [84,85]. It relies on the creation of an intrinsic interface dipole due to polarization of the interlayer. Its experimental verification by XPS measurements then followed for the MBE-grown AlGaAs/Si/GaAs [86,87]. However, a more detailed XPS investigation on MBE-grown GaAs/Si/AlAs and InGaAs/Si/InAlAs interfaces [88,89] indicated that the observed XPS core-level shifts are not due to band offset change but to sharp band bending within the escape depth of photoelectrons caused by surface Fermi-level pinning at midgap and Si delta doping. Another approach is to utilize the dipole created by delta doping near the interface, although the potential profile inevitably becomes less steep. This was first tried for a GaAs-based system [90] and then extended to an InGaAs/InP system [91].

8.3 METAL–SEMICONDUCTOR INTERFACES OF InP-BASED MATERIALS

8.3.1 Schottky Barrier Formation and Current Transport

To consider Schottky barrier formation at M-S interface, Fermi-level positions of various metals and the positions of the semiconductor energy bands before interface formation are shown in Fig. 8.8a with reference to the vacuum level. As shown in Fig. 8.8a, the Fermi-level position of the metal changes over more than 4 eV, depending on the metal species.

The simplest mutual alignment of the bands after M-S interface formation is to align the bands to each other naturally, keeping the energy distance of each band from the vacuum level unchanged at the interface. This is usually referred to as the *ideal Schottky limit,* although the same idea was reached earlier by Mott [92]. Be-

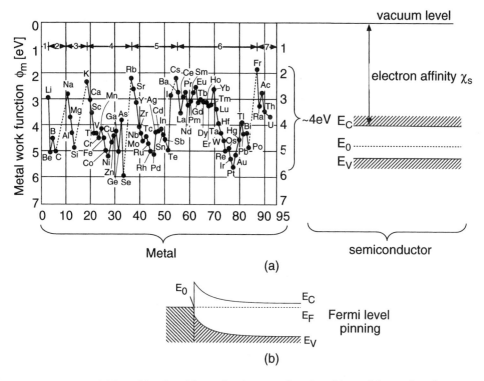

FIGURE 8.8 (*a*) Fermi-level positions of various metals and positions of the semiconductor energy bands before formation of a metal–semiconductor (M-S) interface; (*b*) a real M-S interface.

cause of this, it is hereafter referred to as the *Mott–Schottky limit*. The Schottky barrier height (SBH) in this Mott–Schottky limit, denoted as ϕ^n_{BMS} and ϕ^p_{BMS} for n- and p-type materials, are given, respectively, by

$$\phi^n_{BMS} = \phi_m - \chi_s \quad \text{and} \quad \phi^p_{BMS} = \chi_s + E_g - \phi_m \tag{11}$$

where ϕ_m is the metal work function, and χ_s and E_g are the electron affinity and bandgap energy of the semiconductor, respectively. Thus SBHs are linearly dependent on the metal work function in the Mott–Schottky limit. Furthermore, for certain work function values, Schottky barriers should disappear, and ideal ohmic contacts for electrons or holes should be formed.

However, in reality, there is a certain characteristic energy E_0 for a given semiconductor, and a strong tendency exists that the metal Fermi level aligns in the vicinity of E_0, as shown in Fig. 8.8*b*. It was found by Tersoff [93] that the physical meaning of E_0 is the charge neutrality level discussed in Section 8.2.3. E_0 lies in most cases in the energy gap, but an important exception is the case of InAs, where it lies within the conduction band. Thus there is a strong tendency that large differences in work functions or electronegatives of metals are screened after M-S interface formation. This is called *Fermi-level pinning* at the M-S interface. Thus the SBH value becomes almost entirely independent of or only weakly dependent on metal work function. How strongly SBH depends on the metal work function is usually expressed phenomenologically in terms of the interface index S, defined by $S = d\phi_B^n/d\phi_m$. Using S, SBHs of real interfaces, ϕ_B^n and ϕ_B^p for n- and p-type can be expressed as

$$\phi_B^n = S\phi_{BMS}^n + (1-S)\phi_{BB}^n \quad \text{and} \quad \phi_B^p = S\phi_{BMS}^p + (1-S)\phi_{BB}^p \tag{12}$$

where ϕ_{BB}^n and ϕ_{BB}^p are the values of n- and p-type SBHs in the strongly pinned limit or so-called *Bardeen limit*, given respectively by

$$\phi_{BB}^n = E_c - E_0 \quad \text{and} \quad \phi_{BB}^p = E_0 - E_v \tag{13}$$

$S = 1$ gives the Mott–Schottky limit and $S = 0$ gives the Bardeen limit. Experimentally observed values of S are known to show chemical trends for different semiconductors. Namely, S is very small for elemental and III-V semiconductors, but takes larger values in II-VI semiconductors [94,95]. Equation (12) implies the following relation:

$$\phi_B^n + \phi_B^p = E_g \tag{14}$$

This relationship is known to hold empirically within a scatter of experimental data of about ±0.05 eV.

Given a Schottky barrier height ϕ_B, the current density J over the barrier can be expressed empirically as

$$J = J_0 \exp\left(\frac{qV}{nkT}\right)\left[1 - \exp\left(\frac{-qV}{kT}\right)\right] \tag{15}$$

where J_0 is the reverse saturation current density and n is the ideality factor. $n = 1$ corresponds to the ideal thermionic emission. J_0 is given by

$$J_0 = A^*T^2 \exp\left[\frac{-(\phi_B - \Delta\phi_B)}{kT}\right] \tag{16a}$$

with

$$A^* = \frac{4\pi m^* q k^2}{h^3} = 1.2 \times 10^6 \left(\frac{m^*}{m_0}\right)[\text{A m}^{-2}\text{ K}^{-2}] \tag{16b}$$

where A^* is the Richardson constant, $\Delta\phi_B$ the barrier height reduction by the image force, and m^* and m_0 are the carrier effective mass and the free electron mass, respectively. The effective value of the Richardson constant encountered in real M-S interfaces further includes various factors, such as quantum reflection of carriers at the interface, phonon scattering, and tunneling probability through an interfacial layer if it exists [96], and is called the *effective Richardson constant* A^{**}.

Equation (15) gives highly rectifying behavior and provides a basis for various device applications. Two important figures of merit for Schottky barriers are the barrier height ϕ_B and the ideality factor n. Although a simple transport model based on thermionic emission gives $n = 1$, there are various reasons for n becoming larger than unity, including image force barrier lowering, the presence of an interfacial layer, recombination in the depletion region, and charging and discharging of the interface state, as discussed in detail by Rhoderick and Williams [92]. More recently, Tung has shown that inhomogeneities cause $n > 1$ [97]. The measurement techniques of Schottky barrier heights [92] include the log J versus V plot, $J/(1 - \exp(-qv/kT)$ versus V plot, $\log(J_0/T^2)$ versus T^{-1} plot (Richardson plot), C^{-2} versus V plot, internal photoemission, surface photovoltage spectroscopy, photoemission spectroscopy, scanning tunneling spectroscopy (STS), and ballistic electron emission microscopy (BEEM).

Metal–semiconductor interfaces are also used to realize nonrectifying metal contacts to devices in order to supply power and to move electrical signals in and out. Successful formation of these contacts, called *ohmic contacts,* is supremely important in device fabrication technology. The merit of an ohmic contact is expressed by the specific contact resistance, which is the resistance through the interface having a unit area cross section, usually expressed in units of $\Omega \cdot cm^2$. Since Schottky barriers are formed at most metal–semiconductor interfaces, a standard way of realizing ohmic contacts is to form a thin and sufficiently highly doped semiconductor layer just underneath the metal so that the resulting semiconductor depletion layer width is sufficiently narrow for the carriers to tunnel freely through a field emission or thermionic field emission mechanism [92]. Practically, this is accomplished by depositing one or more metal films containing donors or acceptors and subsequently alloying the contact at a high temperature for a short time so that a thin, highly doped layer is formed through a recrystallization process. On the other hand, if a Schottky barrier is not formed or its height is vanishingly low, one can obviously realize ohmic contact without alloying. If they can be formed, such nonalloyed ohmic contacts are superior to alloyed contacts because alloying requires detailed empirical optimization, tends to be irreproducible, may degrade device performance, may interfere with other processing steps, and tends to produce rough surface morphology with irregular boundaries.

8.3.2 Mechanism of Fermi-Level Pinning

The mechanism that is responsible for Fermi-level pinning has been one of the central issues in the study of the M-S interface over 50 years, and numerous models have been proposed and discussed. A famous example is the interfacial layer theory of Cowley and Sze [98], where the presence of a thin insulating layer with interface states at the I-S interface explains the pinning behavior. However, extensive study in recent years of semiconductor surfaces and interfaces has revealed that Fermi-level pinning also exists on surfaces, at I-S interfaces and unoptimized S-S interfaces, as well as on intimate M-S interfaces without interfacial layers. Furthermore, it has been found that a strong correlation exists among surfaces and interfaces concerning the energy position of the pinning. Well-known examples include midgap pinning on clean MBE surfaces, air-exposed surfaces, and I-S, M-S, and regrown S-S interfaces of GaAs.

Recent models attempt to account for these correlations in explaining the origin of Fermi-level pinning. The models include the unified defect model of Spicer et al. [99], MIGS (metal-induced gap state model of Heine [100] and later Tersoff [93], DIGS (disorder-induced gap state) model of Hasegawa and Ohno [32,101,102], and the effective workfunction (EWF) model of Woodall and Freeouf [103]. Basic ideas for these models are outlined below, and their essence is summarized in Table 8.2.

TABLE 8.2 Major Models for Fermi-Level Pinning

Model	Origin of Pinning	N_{ss} Distribution and Pinning Position	Applicable Interfaces[a]	Nature of Pinning
Unified defect model [99]	Deep level related to stoichiometry, especially AsGa		V-S, S-S, I-S, M-S	Extrinsic
MIGS model [93,100]	Penetration of metal wave function into semiconductor	Midgap energy	M-S	Intrinsic
DIGS model [32,101,102]	Loss of two-dimensional periodicity, due to disorder of bonds at interface	Mean hybrid orbital energy	V-S, S-S, I-S, M-S	Extrinsic
Effective workfunction model [103]	Precipitation of As and P cluster at interface	Pinned at E_F of metallic cluster	V-S, S-S, I-S, M-S	Extrinsic

[a]V-S, Vacuum–semiconductor interface; S-S, semiconductor–semiconductor interface; I-S, insulator–semiconductor interface; M-S, metal–semiconductor interface.

Despite its long history in research, the dominant mechanism for Fermi-level pinning on semiconductor surfaces and interfaces has not yet been established sufficiently well.

Unified Defect Model This model is based on a series of photoemission measurements on clean surfaces having submonolayer- to monolayer-level coverage of various metals and oxygen. According to this model, Fermi-level pinning is due to discrete deep levels of stoichiometry-related defects that are formed during a high-energy interface formation process. Thus the pinning position is the energy level of the defects. For GaAs, EL2-related donor and acceptor levels originating from arsenic antisites are supposed to be such defects [104]. For InP, In and P vacancies were suggested to be such defects [104], but no well-established data exist for the energy levels of these defects. More recent work on hydrogenated metal contacts on GaAs and InP emphasizes the importance of the amphoteric nature of stoichiometry-related native defects [105].

MIGS Model This model asserts that intimate contact between a metal and a semiconductor results in intrinsic pinning of the Fermi level, due to the gap-state continuum formed by penetration of the evanescent wave portion of the metal wavefunction into a semiconductor. This gap-state continuum has a charge neutrality level E_0, whose meaning was discussed in Section 8.2.4. As mentioned previously, the position of the change neutrality level is determined by the bulk band structure of semiconductor and has nothing to do with the metal workfunctions. How strongly the MIGS continuum pins the Fermi level determines on the value of S and on the penetration length of MIGS into the semiconductor. Mönch [106] has derived an empirical formula for chemical trends in S by summarizing supercell-type theoretical calculations.

DIGS Model This model asserts that deposition of a metal or insulator on semiconductor surfaces disturbs the crystalline perfection of the semiconductor surfaces and forms a disorder-induced gap state (DIGS) continuum much like gap states in amorphous semiconductors. The DIGS continuum has a continuous energy and spatial distribution of gap states of acceptor and donor type whose energy boundary is given by the charge neutrality level E_0, discussed in Section 8.2.4. In this model, chemical trends of S were shown to be related exponentially to the energy distance, $E_0 - E_v$ [102], which can be interpreted as a measure of energy required to bring valence band states into the gap through disordering.

According to this model, Fermi-level pinning is extrinsic and process dependent. When the degree of disorder is high, the Fermi level is firmly pinned at the same position, E_0, as in the case of the MIGS model. However, when an intimate M-S interface is formed without causing disorder, the ideal Mott–Schottky limit should be realized.

EWF Model This model is based on the correlation between III-V Schottky barrier heights and the Fermi-level pinning positions of defective III-V heterointerfaces with interface states. According to this model, clusters of elemental As or P atoms are created at interface during interface formation, and the effective workfunctions of these clusters determines the Fermi-level pinning position. The same model was also used to account for the semi-insulating property of low-temperature LT-grown GaAs by MBE [107].

8.3.3 Properties of Metal–Semiconductor Interfaces of InP-Based Materials and Attempts at SBH Control

Properties of Conventional Schottky Barriers and Ohmic Contacts on InP-Based Materials Research on Schottky barriers has been done both from the viewpoint of basic science to clarify barrier formation mechanisms and from the viewpoint of device applications. Realizability, performance, and reliability of devices such as metal–semiconductor field-effect transistors (MESFETs), high-electron-mobility transistors (HEMTs), and M-S-M integrated photodetectors depend sensitively on the properties of Schottky barriers. Unfortunately, Schottky barrier heights (SBHs) of metal contacts produced by conventional metal deposition techniques on n-type InP-based materials are too low for use in these devices.

For example, SBH values of n-InP are typically in the range 0.4 to 0.5 eV for metals such as Au, Ag, and Cu, with somewhat smaller values of about 0.3 eV for Al and In [108]. These values are in reasonable agreement with the Fermi pinning positions at low metal coverage determined by photoemission spectroscopy and surface photovoltage spectroscopy techniques [108]. The value of S obtained by least-squares fitting is below 0.1, indicating strong pinning.

As for the SBH on $In_xGa_{1-x}As$, its alloy composition dependence was measured by Kajiyama et al. [109], which indicated that $(E_g - \phi_B^n)$ is independent of x, obeying the common anion rule [110]. However, later photoemission study by Brillson et al. [111] did not follow the common anion rule. SBH values for the lattice-matched $In_{0.53}Ga_{0.47}As$ were found to be typically 0.2 eV for Au, Ni, and Al [112].

For $In_xAl_{1-x}As$, the alloy composition dependence of the SBH was measured by several groups of workers [113–117]. The results indicated that the common anion rule does not hold. For the lattice-matched composition of $In_{0.52}Al_{0.48}As$, the SBH is typically within the range 0.6 to 0.8 eV for Au and Al. Furthermore, Mönch [118] compiled alloy composition dependencies of the SBH for $In_xGa_{1-x}As$ and $In_xAl_{1-x}As$ and tried to interpret the results in terms of the MIGS model.

Attempts to increase the above-mentioned low SBH values on n-type InP-based materials by depositing metals with large metal work functions by any of standard process so far have failed generally, due to strong Fermi-level pinning.

The fact that SBHs are low in n-type InP-based materials implies that it is easier to form good ohmic contacts on n-type InP-based materials than on p-type InP-based materials. This is indeed so. Standard technologies for ohmic contact formation on InP-based materials have been reviewed [3,4] and are not repeated here. Since InAs and $In_xGa_{1-x}As$ with x near unity have the unique property that the charge neutrality level E_0 lies in the conduction band, as one can see from Table 8.1, one can utilize them to realize n-type nonalloyed ohmic contacts by putting these layers on top after composition grading or some other suitable heterostructure design so as not to form barriers and not to introduce dislocations.

Attempts to Control the SBH by Improved Interface Formation Processes
The common feature of the models for Fermi-level pinning explained in Section 8.3.2 is that some kinds of interface states, including states in metallic clusters, are the origin of Fermi-level pinning. Thus, whether Fermi-level pinning can be removed depends on whether these states can be removed. For example, if MIGS is the major pinning mechanism in Schottky barriers in InP-based materials, there exists little hope for controlling the SBH in these materials, because the MIGS model asserts that the Fermi-level pinning is an intrinsic property of an intimate M-S interface. On the other hand, the three other models assert that Fermi-level pinning is due to extrinsic interface states and should then be removed by removing defects, interface disorder, or group V clusters, respectively. Thus various attempts to unpin the Fermi level have been made up to now to remove interface states on the basis of various ideas and processing methods. Another approach to controlling the SBH is to accept the presence of Fermi-level pinning and to try to modify the SBH by inserting an interlayer that gives rise to an increase in the SBH. This interlayer can be either an insulator or a semiconductor layer. However, the two seemingly different approaches described above cannot be separated clearly from each other in reality. This is because any kind of surface treatment that attempts to reduce interface states may lead to the formation of some kind of interlayer with a dipole, or conversely, artificial formation of an interlayer may reduce the interface states. Major approaches made on InP-based materials to control the SBH are summarized below.

Sulfur and Selenium Surface Treatments Since the discovery by Sandroff et al. [119] of PL intensity enhancement on GaAs surface by sulfur treatment, considerable work has been done on sulfur and selenium passivation of GaAs surfaces. When such treatments were made prior to Schottky barrier formation, SBH values were found to become more dependent on the metal work function, indicating phenomenologically the reduction of Fermi-level pinning [120,121]. As an example, $S = 0.53$ is reported for $(NH_4)_2S_x$ treatment of GaAs [121]. Following these works, a limited amount of work using $(NH_4)S_x$, H_2S, and Se was done on InP-based materials [122–124], giving rise to similar results. However, although it is well established that such treatments reduce native oxide components and may be beneficial for epitaxial regrowth of III-V and II-VI semiconductors, their usefulness for SBH control has not been well established, due to its unclear mechanism and problems in reproducibility and long-term stability.

Low-Temperature Deposition Deposition of Pd on n-InP at cryogenic temperature was reported to give an extremely high room-temperature SBH value of 0.96 eV [125]. Formation of insulating phosphides was suggested to be responsible for increased SBH values.

Plasma Surface Treatments Certain plasma surface treatments were reported to be effective for SBH control. SBH increase was observed on Au Schottky contacts formed on n-InP exposed to hydrogen plasma [126]. High SBH values of 0.7 and 0.86 eV were realized for Au Schottky contacts on n-$In_{0.53}Ga_{0.47}As$ and n-$In_{0.52}Al_{0.48}As$, respectively, by surface phosphidization using phosphine plasma [127,128].

Electrochemical Deposition Recently, substantially increased SBH values of 0.89, 0.50, and 0.89 eV were realized [129,130] on n-type InP, $In_{0.53}Ga_{0.47}As$, and $In_{0.52}Al_{0.48}As$, respectively, by depositing Pt using an in situ electrochemical process. These SBH values are 300 to 450 meV higher than conventional values. Stable InP MESFETs were also fabricated using this process [131]. The process consisted of controlled anodic etching of semiconductor followed by a subsequent cathodic deposition of metal in situ in the same electrolyte using the pulsed mode. The resulting Schottky diodes showed nearly ideal thermionic emission characteristics with excellent agreement between the C–V and I–V SBH values. The result was explained [132] in terms of the DIGS model [32]. Namely, an electrochemical process with an extremely low processing energy (several hundred millivolts) at room temperature was found to produce intimate contact of a semiconductor with nanometer-sized metal particles. Thus the process does not cause disorder on the semiconductor surface and realizes a nearly pinning-free interface with low DIGS density.

Formation of an Insulator Interlayer This technique has a long history. For n-InP, various chemical, photochemical, and thermal oxidation processes [133–137] were tried, and SBH values of 0.8 to 0.9 eV were achieved. Formation of a thin CdO_x layer by metal adsorption and oxidation resulted in a SBH value of 0.7 eV and led to stable InP MESFET operation [138]. Deposition of a thin PN_x film by plasma processes realized an effective SBH value of 0.8 eV [139]. Similar approaches were also made for n-InGaAs, realizing SBH values of 0.4 to 0.5 eV [140–142]. The difficulty with this approach is the controllability and reproducibility of the insulator interlayer. Care has also to be paid to the fact that the forward current is reduced by the tunneling factor, effectively reducing the Richardson constant. Charging and discharging of interface states at insulator–semiconductor interfaces sometimes cause hysteresis behavior.

Formation of a Semiconductor Interlayer An obvious and simple form of semiconductor interlayer is a thin counterdoped layer with doping density N, permittivity ε, and thickness d. The change of SBH is then given by

$$\Delta SBH = \frac{qNd^2}{2\varepsilon} \quad (17)$$

provided that the modified SBH is below E_g. This approach was tried on n-InP [143–145] and n-InGaAs [146–148]. The key question in this approach is how to form a well-defined, thin, highly doped, fully depleted p-type layer without causing in-diffusion of impurities and/or amorphousization of the highly doped layer.

Instead of counterdoping into the same semiconductor, an attempt was made to grow a doped ultrathin MBE-grown Si interface control layer (Si ICL) on the surface of III-V material, where ΔSBH in Eq. (17) can be realized in a separate Si ICL which can be doped to a high level. An ΔSBH value of 300 to 400 meV was realized in n-GaAs and n-InP [149,150].

Another approach to the control of SBH values is to grow a semiconductor layer with a larger bandgap. For n-type material, ΔSBH is given approximately by

$$\Delta\text{SBH} \approx \Delta E_c \tag{18}$$

provided that the interlayer is thick enough to prevent tunneling. Many attempts have been made, particularly on $In_{0.53}Ga_{0.47}As$, using various materials, such as lattice-mismatched GaAs [148,151,152], AlGaAs [153], and InGaP [148,154] and lattice-matched InP [148,155] and InAlAs [156,157], with considerable success.

8.4 SURFACES AND SURFACE PASSIVATION OF InP AND RELATED MATERIALS

8.4.1 Properties of Clean and Real Surfaces

Semiconductor surfaces are traditionally classified into clean surfaces and real surfaces. *Clean surfaces* are those with a negligible amount of impurity atoms adsorbed on surfaces. They can be obtained by cleaving semiconductor crystals in ultrahigh vacuum (UHV). Fresh crystal surfaces obtained by molecular beam epitaxy (MBE) may also be regarded as clean surfaces. On the other hand, *real surfaces* are those in the real world, such as air-exposed or chemically etched surfaces. Present-day semiconductor devices are usually fabricated starting from wafers having real surfaces, although UHV-based in situ processing may become important in the future.

On clean surfaces, dangling bonds of surface atoms interact with each other and generally form surface structures with surface relaxation and reconstruction where atom positions and symmetry of atom arrangements differ from those expected from the bulk structure. These microscopic atom arrangements are generally known to satisfy the electron counting model [158]. Reflecting the two-dimensional periodicity on the surface, two-dimensional surface state bands having in-plane wave vectors are generally formed that either lie in resonance with bulk bands or overlap the energy gap region.

On cleaved (110) surfaces of GaAs, InP, and other III-V materials, a particular type of surface relaxation is known to take place in such a way as to push cations out and pull anions in so as to push surface state bands out of the energy gap region [159]. On the other hand, characteristic 2 × 4 structures are formed by a regular array of group V dimers, and missing dimers appear on As-stabilized GaAs [160] and

P-stabilized InP MBE surfaces [161]. Originally, the array was reported to consist of three dimers and one missing dimer, but later, an array of two dimers and two missing dimers was found to be observed more frequently [162]. Although the dimer structures are similar in cases of GaAs and InP, the depth of the dimer trench seems to be different [163]. According to theoretical calculations, these 2×4 surfaces are free of surface states within the gap [164]. However, pinning of the Fermi level is usually observed in the measurements. For GaAs, kink defects in the 2×4 missing dimer structures were proposed to be deep acceptors responsible for pinning based on observation of a strong correlation between the kink density and Si doping [165]. However, such correlation does not seem to exist on InP despite the existence of strong pinning [163], and doubt has been raised as to the validity of the kink-deep acceptor model for GaAs [166].

On the other hand, real surfaces possess greater or lesser amounts of surface oxides, and their behavior is similar to that at the I-S interface. Usually, no well-defined surface relaxation or reconstruction is observed. A high density of surface states generally exists and their density of states usually forms a continuum in the energy gap. However, they are not extended states with in-plane wavevectors but highly localized states such as deep levels with a capture cross section of 10^{-14} to 10^{-16} cm^2. Their dynamic behavior is usually described by Shockley–Read–Hall statistics.

8.4.2 Surface Passivation and Properties of Interface States

Surface passivation is an interface technology whose objective is to terminate and passivate active dangling bonds at the semiconductor surface and to protect devices from environmental influences. The standard technique for surface passivation is to cover the semiconductor surface with a suitable insulating film. By having an insulator with a large energy gap, the resultant thick, high potential barrier at the I-S interface is expected to protect the delicate carrier motions in the device well from any influence of the outside world. A more active use of such a film is in metal–insulator–semiconductor field-effect transistors (MISFETs).

Since the main objective of surface passivation is to passivate sensitive surface dangling bonds, the most important criterion for selecting a passivation film is the density of electron interface states at the I-S interface. Other practically important criteria include film density or impermeability against outside impurity atoms and ions, adherence to the semiconductor surface, thermal stability, chemical stability, mechanical stress at the interface, and compatibility with other device processing technologies. If the I-S interface is placed under high fields, as in the case of applications to a gate dielectric for MISFETs and insulated gate HEMTs, and to surface passivation of MESFET and HEMT devices, two electrical properties are also important: electrical resistivity and electrical breakdown field strength.

Despite numerous possible combinations of semiconducting and insulating materials, and despite the well-advanced status of the semiconductor electronics, nearly ideal passivation satisfying all the criteria above is available only in a combination of silicon and silicon dioxide where a minimum interface state density N_{ssmin}

8.4 SURFACES AND SURFACE PASSIVATION OF InP AND RELATED MATERIALS

down in the range $10^9 \text{cm}^{-2}\text{ eV}^{-1}$ can be achieved. Passivation of compound semiconductors, including InP-based materials, is still a key technological issue that has not been completely solved.

The major difficulty comes from the density of electronic interface states at the I-S interface. Insulator formation on InP-based materials either by various oxidation processes or by various insulator deposition processes in general results in an I-S interface with high-density interface states (surface states) within the energy gap. As already mentioned, this interface state is a highly localized state at the interface with typical capture cross-section values, in the range 10^{-14} to 10^{-16} cm². The interface state density generally shows a U-shaped continuous distribution in energy. In the case of InP, the minimum value of 10^{11} to 10^{13} cm^{-2} eV^{-1} takes place at 0.4 to 0.5 eV below E_c [32], as shown schematically in Fig. 8.9a. These interface states tend to cause Fermi-level pinning at this minimum N_{ss} position, E_0, being largely independent of the insulator species and formation process, the conduction type, and the doping level of InP. A similar situation takes place in GaAs with a deeper E_0 position, about 0.9 eV below E_c, and a higher N_{ssmin} value of 10^{12} to 10^{13} cm^{-2} eV^{-1}, as shown in Fig. 8.9b. This difference in the position of E_0 makes surface passivation of n-type InP somewhat easier than passivation of n-type GaAs. The physical meaning of E_0 was interpreted as the sp^3 hybrid orbital charge neutrality level by Hasegawa and Ohno [32], as discussed in Section 8.2.3, and this led to the proposed DIGS model.

Interface states at the pinned I-S interface cause various unwanted phenomena in devices. Major adverse effects due to interface states in III-V devices include (1) irrealizability or poor gate control and instability in MISFETs, (2) high source resistance in planar MESFETs due to surface depletion, (3) frequency dispersion of transconductance in MESFETs [167], (4) side gating in MESFET integrated circuits [168], (5) gain reduction in HBTs due to surface recombination [169], (6) quantum efficiency reduction in optoelectronic devices and solar cells due to sur-

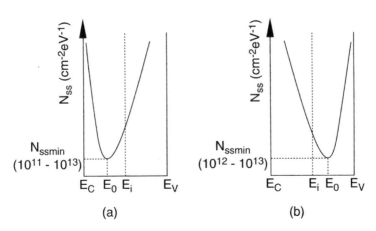

FIGURE 8.9 Interface state distributions for (a) InP and (b) GaAs.

face recombination, (7) dark currents in photodetector due to surface generation, (8) catastrophic mirror damage in lasers due to recombination enhanced interface reaction, (9) recombination–generation noise in electronic and optoelectronic devices, and (10) long-term instability phenomena in electronic and optoelectronic devices.

Recent experiments in GaAs and InGaAs near-surface quantum wells [170,171] have shown that the photoluminescence intensity of the quantum well is drastically reduced exponentially as the well-to-surface distance is reduced below 100 Å, due to interaction between confined quantum states and surface states. This indicates that surface passivation will become a critical technological issue for future nanostructure devices in the quantum regime.

8.4.3 Formation of Insulator Films on InP-Based Materials

Following the examples of the Si–SiO_2 system, initial attempts at passivation of InP-based materials were made by growing native oxide films by various oxidation techniques. Thermal, plasma, and anodic native oxides of InP consist of In_2O_3, P_2O_5, and $InPO_4$ [3]. They tend to separate in the oxide bulk rather than making a uniform amorphous mixture throughout the oxide bulk. Similarly, thermal and anodic oxide of InGaAs is a mixture of Ga_2O_3, In_2O_3, and As_2O_3, with a tendency for the As_2O_3 to escape in high-temperature processing, due to its volatile nature [4]. In these systems, the In_2O_3 component is semiconducting and hygroscopic and tends to cause deterioration of electrical properties and the chemical and thermal stability of the native oxides. In fact, an early model [172] for drain current drift in InP MISFETs attributed the instability to this component. Under certain conditions, $In(PO_3)_x$ (with x in the vicinity of 3) is formed [173], and the presence of this component at the interface was reported beneficial in reducing interface state densities. As compared with GaAs MIS systems, the interface state density in InP MIS systems having native oxides can be at least one order of magnitude lower, being in the range of 10^{11} cm^{-2} eV^{-1} or even lower, and this led to successful realization of InP MISFETs, unlike the case of GaAs. However, problems of electrical, chemical, and thermal instability have not yet been completely solved.

To realize native oxide-free and stable InP and InGaAs MISFETs, considerable interest has been paid to direct deposition of SiO_2 [174–182], Si_3N_4 [174,181–185], and P_xN_y films [186–189] by thermal, plasma-enhanced, and photo-enhanced chemical vapor deposition (CVD) processes. Expected benefits are better insulating properties and improved thermal stability. Prior to deposition, natural oxides were removed using such surface treatments as sulfur treatment, PCl_3/H_2 etching, and H_2 plasma cleaning. Although these efforts resulted in interface state densities in the range 10^{11} cm^{-2} eV^{-1}, the instability issues could not be removed entirely. Other films, such as As_2O_3 [190], $InPS_4$ [191], diamond carbon [192], and SrF_2 [193], were also investigated.

Thus, previous efforts based on conventional oxidation and insulator deposition processes, targeted primarily toward achieving stable InP and InGaAs MISFETs, did not reach practical exploitation, although some of the InP MISFET showed impressive high-frequency and power performance [194–196]. Not only MISFETs, but also other devices, such as HEMTs, HBTs, and optoelectronic devices, require insulator films for surface passivation. For these, empirically optimized silicon-based insulators such as SiO_2, Si_3N_4, and SiO_xN_y produced by various CVD processes are presently used in practical devices [197–204].

8.4.4 Recent Approaches to Surface Passivation Using Interface Control Layers

Extensive effort in the past has shown that Fermi-level pinning at I-S interfaces of InP-based materials cannot be removed completely using only standard approaches of direct insulator formation by various oxidation and insulator deposition techniques. Thus more recent work on surface passivation utilizes various kinds of interface control layers (ICLs). The basic surface passivation structure including an interface control layer (ICL) is shown in Fig. 8.10. An ICL is inserted at the interface of two different materials. The concept of an ICL was first proposed by Hasegawa et al. for GaAs and InGaAs where the ICL material was an MBE-grown ultrathin Si layer [205,206]. The role of the ICL is to remove interface states by providing a coherent and smooth transition of bonds from the semiconductor to the outer dielectric layer. Then, other practical important properties of the passivation film can be satisfied by choosing suitable outer dielectric materials, such as SiO_2 and Si_3N_4. Thus the ICL should provide suitable bond matching to either side of the interface and should act as a reaction and diffusion barrier during interface formation to realize a sharp, well-defined interface.

The surface passivation structures on InP-based materials including ICLs tested up to the present are summarized below together with references.

1. Oxide or insulator ICL
   $Al_2O_3/In(PO_3)_3/InP$ [207–209]
   SiO_2/ECR oxide/InP   [210]
   SiN_x/anodic oxide/InP   [211]
   SiN_x/PN/sulfur-treated InP   [212]

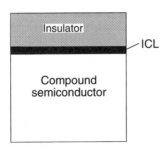

FIGURE 8.10 Basic passivation structure, including interface control layer (ICL).

2. ICL containing sulfur

SiO_2/S/InP	[213]
SiO_2/SiS_2/InP	[214]
Si_3N_4/polysulfide/InP	[215]
SiO_2/CdS/InP	[216,217]
SiN_x/InS/InP	[218–220]

3. ICL utilizing an ultrathin Si layer

SiO_2/Si/InGaAs	[205,206]
Si_3N_4/Si/InGaAs	[221,222]
SiO_2/Si_3N_4/Si/InGaAs	[223]
SiO_2/Si/InP	[224,225]
$Si_3N_4SiN_x$/Si/InP	[226]

Of the above, the Si ICL-based passivation scheme seems very promising, achieving a minimum interface state density in the range 10^{10} cm^{-2} eV^{-1} for InGaAs [223] and realizing very stable, high-gain InGaAs MISFETs [222,227,228] and InP-based insulated gate HEMTs [229]. It was also used for surface passivation of GaAs- and InP-based near-surface quantum structures [230,231], enhancing the PL intensity by a factor of 500 to 1000.

Application of the Si ICL technique to wide-gap InP was limited due to the very narrow energy gap of a strained Si ICL. However, recent theoretical calculation and experiment [226] have indicated that the conduction and valence band states of an Si ICL can be pushed out of the InP energy gap due to the quantum confinement effect in a surface Si quantum well.

Another apparently promising approach is ICL with CdS and InS. The latter has realized stable InP-based HIGFET operating up to 10 GHz [219,220].

As seen above, the trend in passivation technology is to understand and control the interface structure at the monolayer level, and ultimately, progress in this direction will lead to a successful and practical surface passivation technology.

REFERENCES

1. J. W. Matthews and A. E. Blakeslee, "Defects in epitaxial multilayers," *J. Cryst. Growth*, **27**, 118–125 (1974).
2. R. People and J. C. Bean, "Calculation of critical layer thickness versus lattice mismatch for Ge_xSi_{1-x}/Si strained-layer heterostructures," *Appl. Phys. Lett.*, **47**(3), 322–324 (1975).
3. *Properties of Indium Phosphide,* EMIS Data Rev. Ser., No. 6, INSPEC, London, 1991.
4. P. Bhattacharya, ed., *Properties of Lattice-Matched and Strained Indium Gallium Arsenide,* EMIS Data Rev. Ser., No. 8, INSPEC, London, 1993.
5. P. Bhattacharya, ed., *Properties of III-V Quantum Wells and Superlattices,* EMIS Data Rev., Ser., No. 15, INSPEC, London, 1996.

6. D. L. Miller, R. T. Chen, K. Elliot, and S. P. Kowalczyk, "Molecular-beam-epitaxy GaAs regrowth with clean interfaces by arsenic passivation," *J. Appl. Phys.*, **57**, 1922–1927 (1985).
7. E. Ikeda, H. Hasegawa, S. Ohtsuka, and H. Ohno, "Electronic properties and modeling of lattice-mismatched and regrown GaAs interfaces prepared by metalorganic vapor phase epitaxy," *Jpn. J. Appl. Phys.*, **27**, 180–187 (1988).
8. T. Saitoh, H. Tomozawa, T. Nakagawa, H. Takeuchi, and H. Hasegawa, "In-situ photoluminescence and capacitance-voltage characterization of InAlAs/InGaAs regrown heterointerfaces by molecular beam epitaxy," *J. Cryst. Growth*, **150**, 96–100 (1995).
9. J. Y. Tsao and W. Dodson, "Excess stress and the stability of strained heterostructures," *Appl. Phys. Lett.*, **53**, 848–850 (1988).
10. D. C. Houghton, "Strain relaxation kinetics in $Si_{1-x}Ge_x/Si$ heterostructures," *J. Appl. Phys.*, **70**, 2136–2151 (1991).
11. G. A. Vawter and D. R. Myers, "Useful design relationships for the engineering of thermodynamically stable strained-layer structures," *J. Appl. Phys.*, **65**, 4769–4773 (1989).
12. D. C. Houghton, M. Davies, and M. Dion, "Design criteria for structurally stable, highly strained multiple quantum well devices," *Appl. Phys. Lett.*, **64**(4), 505–507 (1994).
13. F. E. Ejeckam, Y. H. Lo, M. Seaford, H. Q. Hou, and B. E. Hammons, "A lattice engineered compliant universal (CU) substrate using a (twist) wafer bonding technology," *IEEF LEOS Meet. PDG*, **25** (1996); F. E. Ejeckam, Y. H. Lo, S. Subramanian, H. Q. Hou, and B. E. Hammons, "Lattice engineered compliant substrate for defect-free heteroepitaxy growth," *Appl. Phys. Lett.*, **70**, 1685–1687 (1997).
14. E. T. Yu, J. O. McClain, and T. C. McGill, "Band offsets in semiconductor heterojunctions," *Solid State Phys.*, **46**, 1–146 (1992).
15. S. R. Forrest, P. H. Schmit, R. B. Wilson, and M. L. Kaplan, "Measurement of the conduction band discontinuities of InGaAsP/InP heterojunctions using capacitance–voltage analysis," *J. Vac. Sci. Technol. B*, **4**, 37–44 (1986).
16. H. Ohno, H. Ishii, K. Matsuzaki, and H. Hasegawa, "Absence of growth sequence dependence of AlAs/GaAs heterojunction band discontinuity determined by x-ray photoelectron spectroscopy," *J. Cryst. Growth*, **95**, 367–370 (1989).
17. J. P. Landeman, J. C. Garcia, J. Massies, P. Maurel, G. Jezequel, J. P. Hirtz, and P. Alnot, "Direct determination of the valence-band offsets at $Ga_{0.47}In_{0.53}As/InP$ and $InP/Ga_{0.47}In_{0.53}As$ heterostructures by ultraviolet photoemission spectroscopy," *Appl. Phys. Lett.*, **60**, 1241–1243 (1992).
18. M. S. Hybertsen, "Role of interface strain in a lattice-matched heterostructure," *Phys. Rev. Lett.*, **64**, 555–558 (1990).
19. M. Peressi, S. Baroni, A. Baldereschi, and R. Resta, "Electronic structure of $InP/Ga_{0.47}In_{0.53}As$ interfaces," *Phys. Rev. B*, **41**, 106–110 (1990).
20. M. S. Hybertsen, "Interface strain at the lattice-matched $In_{0.53}Ga_{0.47}As/InP$ (001) heterointerface," *J. Vac. Sci. Technol. B*, **8**, 773–778 (1990).
21. M. S. Hybertsen, "Band offset transitivity at the InGaAs/InAlAs/InP(001) heterointerfaces," *Appl. Phys. Lett.*, **58**, 1759–1761 (1991).
22. R. L. Anderson, "Experiments on Ge–GaAs heterojunctions," *Solid-State Electron.*, **5**, 341–351 (1962).
23. W. R. Frensley and H. Kroemer, "Theory of the energy-band lineup at an abrupt semiconductor heterojunction," *Phys. Rev. B*, **16**, 2642–2652 (1977).

24. W. A. Harrison, "Elementary theory of heterojunctions," *J. Vac. Sci. Technol.*, **14**, 1016–1021 (1977).
25. C. G. Van de Walle and R. M. Martin, "Theoretical study of band offsets at semiconductor interfaces," *Phys. Rev. B*, **35**, 8154–8165 (1987).
26. J. M. Langer and H. Heinrich, "Deep-level impurities: a possible guide to prediction of band-edge discontinuities in semiconductor heterojunctions," *Phys. Rev. Lett.*, **55**, 1414–1417 (1985).
27. H. Hasegawa, "Universal alignment of transition metal impurity levels in III-V and II-VI compound semiconductors," *Solid State Commun.*, **58**, 157–160 (1986).
28. C. G. Van de Walle, "Band lineups and deformation potentials in the model-solid theory," *Phys. Rev. B*, **39**, 1871–1883 (1989).
29. F. Flores and C. Tejedor, "Energy barriers and interface states at heterojunctions," *J. Phys. C Solid State Phys.*, **12**, 731–749 (1979).
30. J. Tersoff, "Theory of semiconductor heterojunctions: the role of quantum dipoles," *Phys. Rev. B*, **30**, 4874–4877 (1984).
31. H. Hasegawa, H. Ohno, and T. Sawada, "Hybrid orbital energy for heterojunction band lineup," *Jpn. J. Appl. Phys.*, **25**, L265–L268 (1986).
32. H. Hasegawa and H. Ohno, "Unified disorder induced gap state model for insulator–semiconductor and metal–semiconductor interfaces," *J. Vac. Sci. Technol. B*, **4**, 1130–1138 (1986).
33. W. A. Harrison and J. Tersoff, "Tight-binding theory of heterojunction band lineups and interface dipoles," *J. Vac. Sci. Technol. B*, **4**, 1068–1073 (1986).
34. M. Cardona and N. E. Christensen, "Acoustic deformation potentials and heterostructure band offsets in semiconductors," *Phys. Rev. B*, **35**, 6182–6194 (1987).
35. W. Mönch, "Empirical tight-binding calculation of the branch-point energy of the continuum of interface-induced gap states," *J. Appl. Phys.*, **80**, 5076–5083 (1996).
36. W. R. L. Lambrecht and B. Segall, "Interface-bond-polarity model for semiconductor heterojunction band offsets," *Phys. Rev. B*, **41**, 2832–2848 (1990).
37. N. G. Anderson and S. D. Jones, "Optimized tight-binding valence bands and heterojunction offsets in strained III–V semiconductors," *J. Appl. Phys.*, **70**, 4342–4356 (1991).
38. C. N. Yeh, L. E. McNeil, R. E. Nahory, and R. Bhat, "Measurement of the $In_{0.52}Al_{0.48}As$ valence-band hydrostatic deformation potential and the hydrostatic-pressure dependence of the $In_{0.52}Al_{0.48}As/InP$ valence-band offset," *Phys. Rev. B*, **52**, 682–687 (1995).
39. T. Y. Wang and G. B. Stringfellow, "Strain effects on $Ga_xIn_{1-x}As/InP$ single quantum wells grown by organometallic vapor-phase epitaxy with $0 < x < 1$," *J. Appl. Phys.*, **67**(1), 344–352 (1990).
40. T. Ishikawa and J. E. Bowers, "Band lineup and In-plane effective mass of InGaAsP or InGaAlAs on InP strained-layer quantum well," *IEEE J. Quantum Electron.*, **30**, 562–570 (1994).
41. S. Adachi, *Physical Properties of III-V Semiconductor Compounds: InP, InAs, GaAs, GaP, InGaAs, and InGaAsP*, Wiley-Interscience, New York, 1992.
42. Y. Sugiyama, T. Inata, T. Fujii, Y. Nakata, S. Muto, and S. Hiyamizu, "Conduction band edge discontinuity of $In_{0.52}Ga_{0.48}As/In_{0.52}(Ga_{1-x}Al_x)_{0.48}As$ ($0 \leq x \leq 1$) heterostructures," *Jpn. J. Appl. Phys.*, **25**, L648–L650 (1986).

43. J. R. Waldrop, E. A. Kraut, C. W. Farley, and R. W. Grant, "Measurement of AlAs/InP and InP/In$_{0.52}$Al$_{0.48}$As heterojunction band offsets by x-ray photoemission spectroscopy," *J. Vac. Sci. Technol. B*, **8**, 768–772 (1990).
44. L. Aina, M. Mattingly, and L. Stecker, "Photoluminescence from AlInAs/InP quantum wells grown by organometallic vapor phase epitaxy," *Appl. Phys. Lett.*, **53**, 1620–1622 (1988).
45. C. Klingelhöfer, S. K. Krawczyk, M. Sacilotti, P. Abraham, and Y. Monteil, "Mapping of the band offset at InAlAs/InP interfaces using room temperature spectrally resolved scanning photoluminescence," *Proc. 5th Int. Conf. InP and Related Materials*, Paris, 1993, pp. 195–198.
46. M. A. Garcia Perez, T. Benyattou, A. Tabata, G. Guillot, M. Sacilotti, P. Abraham, Y. Monteil, and J. Tardy, "Optical characterization of InAlAs/InP single and double heterostructures," *Proc. 5th Int. Conf. InP and Related Materials*, Paris, 1993, pp. 155–158.
47. Y. Hakone, Y. Kawamura, K. Yoshimatsu, H. Iwamura, T. Ito, and N. Inoue, "Bistability of electroluminescence in InAlAs/InP type II MQW diodes," *Appl. Surf. Sci.*, **117/118**, 725–728 (1997).
48. M. Allovon and M. Quillec, "Interest in AlGaInAs on InP for optoelectronic applications," *IEE Proc. J*, **139**, 148–152 (1992).
49. J. Böhrer, A. Krost, and D. B. Bimberg, "Composition dependence of band gap and type of lineup in In$_{1-x-y}$Ga$_x$Al$_y$As/InP heterostructures," *Appl. Phys. Lett.*, **63**, 1918–1920 (1993).
50. M. Sacilotti, F. Motisuke, Y. Monteil, P. Abraham, F. Iikawa, C. Montes, M. Furtado, L. Horiuchi, R. Landers, J. Morais, L. Cardoso, J. Decobert, and B. Waldman, "Growth and characterization of type-II/type-Ii AlGaInAs/InP interfaces," *J. Cryst. Growth*, **124**, 589–595 (1992).
51. C. Ohler, J. Moers, A. Förster, and H. Lüth, "Strain dependence of the valence-band offset in arsenide compound heterojunctions determined by photoelectron spectroscopy," *J. Vac. Sci. Technol. B*, **13**, 1728–1735 (1995).
52. J. M. Kuo, T.-Y. Chang, and B. Lalevic, "Ga$_{0.4}$In$_{0.6}$As/Al$_{0.55}$In$_{0.45}$As pseudomorphic modulation-doped field-effect transistors," *IEEE Electron Device Lett.*, **8**, 380–382 (1987).
53. G.-I. Ng, D. Pavlidis, M. Jaffe, J. Singh, and H.-F. Chau, "Design and experimental characteristics of strained In$_{0.52}$Al$_{0.48}$As/In$_x$Ga$_{1-x}$As ($x > 0.53$) HEMT's," *IEEE Trans. Electron Devices*, **36**, 2249–2259 (1989).
54. M. Wojtowicz, R. Lai, D. C. Streit, G. I. Ng, T. R. Block, K. L. Tan, P. H. Liu, A. K. Freudenthal, and R. M. Dia, "0.10µm graded InGaAs channel InP HEMT with 305 GHz f_T and 340 GHz f_{max}," *IEEE Electron Device Lett.*, **15**, 477–479 (1994).
55. D. Gershoni and H. Temkin, "Optical properties of III-V strained-layer quantum wells," *J. Luminesc.*, **44**, 381–398 (1989).
56. R. E. Cavicchi, D. V. Lang, D. Gershoni, A. M. Sergent, J. M. Vandenberg, S. N. G. Chu, and M. B. Panish, "Admittance spectroscopy measurement of band offsets in strained layers of In$_x$Ga$_{1-x}$As grown on InP," *Appl. Phys. Lett.*, **54**(8), 739–741 (1989).
57. J.-H. Huang, T. Y. Chang, and B. Lalevic, "Measurement of the conduction-band discontinuity in pseudomorphic In$_x$Ga$_{1-x}$As/In$_{0.52}$Al$_{0.48}$As heterostructures," *Appl. Phys. Lett.*, **60**(6), 733–735 (1992).

58. M. A. Garcia Perez, T. Benyattou, A. Tabata, G. Guillot, M. Gendry, V. Drouot, and G. Hollinger, "Control of the interface roughness in highly strained $In_xGa_{1-x}As/In_{0.52}Al_{0.48}As$ heterostructures," *Proc. 6th Int. Conf. InP and Related Materials,* Santa Barbara, Calif., 1994, pp. 571–574.

59. S. Hanatani, M. Shishikura, S. Tanaka, H. Kitano, T. Miyazaki, and H. Nakamura, "A strained InAlAs/InGaAs superlattice avalanche photodiode for operation at an IC-power-supply voltage," Proc. 7th Int. Conf. InP and Related Materials, Sapporo, Japan, 1995, pp. 369–372.

60. S. Ababou, G. Guillot, S. Clark, G. Halkias, A. Georgakilas, K. Zekentes, D. Stievenard, X. Letartre, and M. Lannoo, "Admittance spectroscopy measurements on AlInAs/GaInAs/AlInAs quantum well structures: evidence of a deep level assisted tunneling," *Proc. 5th Int. Conf. InP and Related Materials,* Paris, 1993, pp. 151–154.

61. P. Z. Lee, C. L. Lin, J. C. Ho, L. G. Meiners, and H. H. Wieder, "Conduction-band discontinuities of $In_xAl_{1-x}As/In_{0.53}Ga_{0.47}As$ n-isotype heterojunctions," *J. Appl. Phys.,* **67**(9), 4377–4379 (1990).

62. A. Tabata, T. Benyattou, G. Guillot, M. Gendry, G. Hollinger, and P. Viktorovitch, "Optical and structural characterizations of $In_{0.53}Ga_{0.47}As/In_xGa_{1-x}As$ strained quantum wells on InP substrate," *Proc. 4th Int. Conf. InP and Related Materials,* Newport, R.I., 1992, pp. 140–143.

63. R. Schwedler, F. Brüggemann, K. Wolter, Ch. Jaekel, R. Kersting, A. Kohl, K. Leo, and H. Kurz, "Optical properties of shallow $In_{1-x}Ga_xAs/In_{1-y}Ga_yAs$ superlattices for optoelectronic applications," *Proc. 5th Int. Conf. InP and Related Materials,* Paris, 1993, pp. 656–659.

64. A. Godefroy, A. Lecorre, F. Clérot, J. Caulet, A. Poudoulec, S. Salaün, and D. Lecrosnier, "$In_{1-x}Ga_xAs/In_{.53}Ga_{.47}As$ strained superlattices grown by gas source molecular beam epitaxy," *Proc. 5th Int. Conf. InP and Related Materials,* Paris, 1993, pp. 143–146.

65. X. P. Jiang, P. Thiagarajan, G. A. Patrizi, G. Y. Robinson, H. Temkin, J. M. Vandenberg, D. Coblentz, and R. A. Logan, "Optical transitions in strained quantum wells of $In_xGa_{1-x}As_yP_{1-y}$ with InGaAsP barriers," *Proc. 6th Int. Conf. InP and Related Materials,* Santa Barbara, Calif., 1994, pp. 463–465.

66. T. K. Woodward, T.-H. Chiu, and T. Sizer II, "Multiple quantum well light modulators for the 1.06 μm range on InP substrates: $In_xGa_{1-x}As_yP_{1-y}$/InP, $InAs_yP_{1-y}$/InP, and coherently strained $InAs_yP_{1-y}/In_xGa_{1-x}P$," *Appl. Phys. Lett.,* **60**(23), 2846–2848 (1992).

67. K. W. Goossen, J. E. Cunningham, M. B. Santos, and W. Y. Jan, "Measurement of modulation saturation intensity in strain-balanced, undefected InGaAs/GaAsP modulators operating at 1.064 μm," *Appl. Phys. Lett.,* **63**(4), 515–517 (1993).

68. A. Kasukawa, T. Namegaya, T. Fukushima, N. Iwai, and T. Kikuta, "1.3 μm $InAs_yP_{1-y}$-InP strained-layer quantum-well laser diodes grown by metalorganic chemical vapor deposition," *IEEE J. Quantum Electron.,* **29**, 1528–1535 (1993).

69. P. N. Stavrinou, S. K. Haywood, L. Hart, M. Hopkinson, J. P. R. David, and G. Hill, "Strain effects in InAsP/InP MQW modulators for 1.06 μm operation," *Proc. 7th Int. Conf. InP and Related Materials,* Sapporo, Japan, 1995, pp. 536–539.

70. R. P. Schneider, Jr. and B. W. Wessels, "Photoluminescence excitation spectroscopy of $InAs_{0.67}P_{0.33}$/InP strained single quantum wells," *J. Electron. Mater.,* **20**, 1117–1123 (1991).

71. P. Stavrinou, S. K. Haywood, L. Hart, X. Zhang, M. Hopkinson, J. P. R. David, and G. Hill, "Varying strain in InAs$_{1-x}$P$_x$/InP multiple quantum well device structures," *Proc. 5th Int. Conf. InP and Related Materials,* Paris, 1993, pp. 652–655.

72. M. Yamamoto, N. Yamamoto, and J. Nakano, "MOVPE growth of strained InAsP/InGaAsP quantum-well structures for low-threshold 1.3-μm lasers," *IEEE J. Quantum Electron.,* **30,** 554–561 (1994).

73. C. Francis, P. Boucaud, F. H. Julien, J. Y. Emery, and L. Goldstein, "Photoluminescence study of band-gap alignment of intermixed InAsP/InGaAsP superlattices," *J. Appl. Phys.,* **78**(3), 1944–1947 (1995).

74. G. Patriarche, A. Ougazzaden, and F. Glas, "Inhibition of thickness variations during growth of InAsP/InGaP and InAsP/InGaAsP multiquantum wells with high compensated strains," *Appl. Phys. Lett.,* **69**(15), 2279–2281 (1996).

75. H. Sugiura, M. Ogasawara, M. Mitsuhara, H. Oohashi, and T. Amano, "Metalorganic molecular beam epitaxy of strain-compensated InAsP/InGaAsP multi-quantum-well lasers," *J. Appl. Phys.,* **79**(3), 1233–1237 (1996).

76. J. F. Carlin, A. V. Syrbu, C. A. Berseth, J. Behrend, A. Rudra, and E. Kapon, "InAsP quantum wells for low threshold and high efficiency multi-quantum well laser diodes emitting at 1.55 μm," *Proc. 9th Int. Conf. InP and Related Materials,* Hyannis, Mass., 1997, pp. 563–566.

77. H. Uenohara, M. R. Gokhale, J. C. Dries, and S. R. Forrest, "Low threshold InAsP/InGaP/InGaAsP/InP strain-compensated and compressively-strained 1.3 μm lasers grown by GSMBE," *Proc. 9th Int. Conf. InP and Related Materials,* Hyannis, Mass., 1997, pp. 555–558.

78. X. S. Jiang and P. K. L. Yu, "Strain compensated InAsP/InP/InGaP multiple quantum well for 1.5 μm wavelength," *Appl. Phys. Lett.,* **65**(20), 2536–2538 (1994).

79. X. B. Mei, K. K. Loi, H. H. Wieder, W. S. C. Chang, and C. W. Tu, "Strain-compensated InAsP/GaInP multiple quantum wells for 1.3 μm waveguide modulators," *Appl. Phys. Lett.,* **68**(1), 90–92 (1996).

80. X. B. Mei, W. G. Bi, C. W. Tu, L. J. Chou, and K. C. Hsieh, "Quantum confined Stark effect near 1.5 μm wavelength in InAs$_{0.53}$P$_{0.47}$/Ga$_y$In$_{1-y}$P strain-balanced quantum wells," *J. Vac. Sci. Technol. B,* **14**(3), 2327–2329 (1996).

81. N. Yokouchi, N. Yamanaka, N. Iwai, T. Matsuda, and A. Kasukawa, "InAsP/InGaP all ternary strain-compensated multiple quantum wells and its application to long wavelength lasers," *Proc. 7th Int. Conf. InP and Related Materials,* Sapporo, Japan, 1995, pp. 57–60.

82. A. Ougazzaden, F. Devaux, E. V. K. Rao, L. Silvestre, and G. Patriarche, "1.3 μm strain-compensated InAsP/InGaP electroabsorption modulator structure grown by atmospheric pressure metal-organic vapor epitaxy," *Appl. Phys. Lett.,* **70**(1), 96–98 (1997).

83. X. B. Mei, K. K. Loi, W. S. C. Chang, and C. W. Tu, "Benefits and limitations in barrier design in InAsP/GaInP strain-balanced MQWS for improving the 1.3 μm waveguide modulator performance," *Proc. 9th Int. Conf. InP and Related Materials,* Hyannis, Mass., 1997, pp. 437–439.

84. A. Muñoz, N. Chetty, and M. R. Martin, "Modification of heterojunction band offsets by thin layers at interface: role of the interface dipole," *Phys. Rev. B,* **41,** 2976–2981 (1990).

85. M. Peressi, S. Baroni, R. Resta, and A. Baldereschi, "Tuning band offsets at semiconductor interfaces by interlayer deposition," *Phys. Rev. B*, **43**, 7347–7350 (1991).
86. L. Sorba, G. Bratina, G. Ceccone, A. Antonini, F. J. Walker, M. Micovic, and A. Franciosi, "Tuning AlAs–GaAs band discontinuities and the role of Si-induced local interface dipoles," *Phys. Rev. B*, **4**, 2450–2453 (1991).
87. G. Bratina, L. Sorba, A. Antonini, G. Biasiol, and A. Franciosi, "AlAs–GaAs heterojunction engineering by means of group-IV elemental interface layers," *Phys. Rev. B*, **45**, 4528–4531 (1992).
88. M. Akazawa, H. Hasegawa, H. Tomozawa, and H. Fujikura, "Reappraisal of Si-interlayer-induced change of band discontinuity at GaAs–AlAs heterointerface taking account of delta-doping," *Jpn. J. Appl. Phys.*, **31**, L1012–L1014 (1992).
89. M. Akazawa, H. Hasegawa, H. Tomozawa, and H. Fujikura, "Investigation of valence band offset modification at GaAs–AlAs and InGaAs–InAlAs heterointerfaces induced by a Si interlayer," *Inst. Phys. Conf. Ser.*, No. 129, 1992, pp. 253–256.
90. T.-H. Shen, M. Elliot, R. H. Williams, and D. Westwood, "Effective barrier height, conduction band offset, and the influence of p-type d doping at heterojunction interfaces: the case of the InAs/GaAs interface," *Appl. Phys. Lett.*, **58**, 842–844 (1991).
91. J. Almeida, T. dell"Orto, C. Coluzza, A. Fassò, A. Baldereschi, G. Margaritondo, A. Rudra, H. J. Buhlmann, and M. Ilegems, "Inhomogeneous and temperature-dependent p-InGaAs/n-InP band offset modification by silicon δ doping: an internal photoemission study," *J. Appl. Phys.*, **78**, 3258–3261 (1995).
92. E. H. Rhoderick and R. H. Williams, *Metal–Semiconductor Contacts,* 2nd ed., Oxford Science Publications, Oxford, 1988.
93. J. Tersoff, "Schottky barrier heights and the continuum of gap states," *Phys. Rev. Lett.*, **52**, 465–468 (1984).
94. S. Kurtin, T. C. McGill, and C. A. Mead, "Fundamental transition in the electronic nature of solids," *Phys. Rev. Lett.*, **22**, 1433–1436 (1969).
95. L. J. Brillson, "Transition in Schottky barrier formation with chemical reactivity," *Phys. Rev. Lett.*, **40**, 260–263 (1978).
96. E. H. Rhoderick and R. H. Williams, *Metal–Semiconductor Contacts,* 2nd ed., Oxford Science Publications, Oxford, 1988, p. 96.
97. R. T. Tung, "Electron transport at metal–semiconductor interfaces: General theory," *Phys. Rev. B*, **45**, 13509–13523 (1992).
98. S. M. Sze, *Physics of Semiconductor Devices,* 2nd ed., Wiley, New York, 1981, p. 270.
99. W. E. Spicer, P. W. Chye, P. R. Skeath, C. Y. Su, and I. Lindau, "New and unified model for Schottky barrier and III-V insulator interface states formation," *J. Vac. Sci. Technol.*, **16**, 1422–1433 (1979).
100. V. Heine, "Theory of surface states," *Phys. Rev.*, **138**, A1689–A1696 (1965).
101. H. Hasegawa, "Theory of Schottky barrier formation based on unified disorder induced gap state model," *Proc. 18th Int. Conf. Physics of Semiconductors,* 1986, Vol. 1, pp. 291–294.
102. H. Hasegawa, "Control of compound semiconductor Schottky barriers by interface control layer," in *Metal–Semiconductor Interfaces,* ed. A. Hiraki, IOS Press, Tokyo, 1995, p. 280.
103. J. M. Woodall and J. L. Freeouf, "GaAs metallization: some problems and trends," *J. Vac. Sci. Technol.*, **19**, 794–798 (1981).

104. W. E. Spicer, R. Cao, K. Miyano, T. Kendelewicz, I. Lindau, E. Weber, Z. L. Weber, and N. Newman, "From synchrotron radiation to I-V measurements of GaAs Schottky barrier formation," *Appl. Surf. Sci.*, **41/42**, 1–16 (1989).
105. R. L. Van Meirhaeghe, W. H. Laflere, and F. Cardon, "Influence of defect passivation by hydrogen on the Schottky barrier height of GaAs and InP contacts," *J. Appl. Phys.*, **76**, 403–406 (1994).
106. W. Mönch, "Chemical trends of barrier heights in metal–semiconductor contacts: on the theory of the slope parameter," *Appl. Surf. Sci.*, **92**, 367–371 (1996).
107. A. C. Warren, J. M. Woodall, J. L. Freeouf, D. Grischknowsky, D. T. McInturff, M. R. Mellochan, and N. Otsuka, "Arsenic precipitates and the semi-insulating proper ties of GaAs buffer layers grown by low-temperature molecular beam epitaxy," *Appl. Phys. Lett.*, **57**, 1331–1333 (1990).
108. *Properties of Indium Phosphide*, EMIS Data Rev. Ser., No. 6, INSPEC, London, 1991, p. 325.
109. K. Kajiyama, Y. Mizushima, and S. Sakata, "Schottky barrier height of n-$In_xGa_{1-x}As$ diodes," *Appl. Phys. Lett.*, **23**, 458–459 (1973).
110. O. McCaldin, T. C. McGill, and C. A. Mead, "Correlation for III-V and II-VI semiconductors of the Au Schottky barrier energy with anion electronegativity," *Phys. Rev. Lett.*, **36**, 56–58 (1976).
111. L. J. Brillson, M. L. Slade, R. E. Vittura, M. K. Kelly, W. Tache, G. Margaritondo, J. M. Woodall, P. D. Kirchner, G. D. Pettit, and S. L. Wright, "Fermi level pinning and chemical interactions at metal–$In_xGa_{1-x}As$(100) interfaces," *J. Vac. Sci. Technol. B*, **4**, 919–923 (1986).
112. P. Bhattacharya, ed., *Properties of Lattice-Matched and Strained Indium Gallium Arsenide*, EMIS Data Rev. Ser., No. 8, INSPEC, London, 1993, p. 146.
113. C. L. Lin, P. Chu, A. L. Kellner, H. H. Wieder, and E. A. Rezek, "Composition dependence of Au/$In_xAl_{1-x}As$ Schottky barrier heights," *Appl. Phys. Lett.*, **49**, 1593–1595 (1986).
114. P. Chu, C. L. Lin, and H. H. Wieder, "Schottky barrier height of $In_xAl_{1-x}As$ epitaxial and strained layers," *Appl. Phys. Lett.*, **53**, 2423–2425 (1988).
115. L. P. Sadwick, C. W. Kim, K. L. Tan, and D. C. Streit, "Schottky barrier heights of n-type and p-type $Al_{0.48}In_{0.52}As$," *IEEE Electron Device Lett.*, **12**, 626–628 (1991).
116. F. Fueissaz, M. Gaihanou, R. Houdré, and M. Ilegems, "Measurements of Al–AlInAs Schottky barriers prepared in situ by molecular beam epitaxy," *Appl. Phys. Lett.*, **60**, 1099–1101 (1992).
117. J.-I. Chyi, J.-L. Shieh, R.-J. Lin, J.-W. Pan, and R.-M. Lin, "Schottky barrier heights of $In_xAl_{1-x}As$ ($0 \le x \le 0.35$) epilayers on GaAs," *J. Appl. Phys.*, **77**, 1813–1815 (1995).
118. W. MWumonch, "Schottky contacts on ternary compound semiconductors: compositional variations of barrier heights," *Appl. Phys. Lett.*, **67**, 2209–2211 (1995).
119. C. J. Sandroff, R. N. Nottenburg, J. C. Bischoff, and R. Bhat, "Dramatic enhancement in the gain of a GaAs/AlGaAs heterostructure bipolar transistor by surface chemical passivation," *Appl. Phys. Lett.*, **51**, 33–35 (1987).
120. M. S. Carpenter, M. R. Melloch, and T. E. Dungan, "Schottky barrier formation on $(NH_4)_2S$-treated n- and p-type (100)GaAs," *Appl. Phys. Lett.*, **53**, 66–68 (1988).
121. J.-F. Fan, H. Oigawa, and Y. Nannichi, "Metal-dependent Schottky barrier height with the $(NH_4)_2S_x$-treated GaAs," *Jpn. J. Appl. Phys.*, **27**, 2125–2127 (1988).

122. Y. Nannichi and H. Oigawa, "The effect of sulfur on the surface of III-V compound semiconductors," *Extended Abstr. 22nd Conf. Solid State Devices and Materials,* Sendai, Japan, 1990, pp. 453–456.

123. T. Sugino, K. Goda, H. Okitani, and J. Shirafuji, "Surface passivation of InP by in-situ dry process using PH_3-plasma and Se vapor," *Proc. 4th Int. Conf. InP and Related Materials,* Newport, R.I., 1992, pp. 290–293.

124. S. Habibi, M. Totsuka, J. Tanaka, and S. Matsumoto, "Improvement in Schottky diode characteristics of metal–$In_{0.52}Al_{0.48}As$ contact using an in situ photochemical etching and surface passivation process," *Proc. 7th Int. Conf. InP and Related Materials,* Sapporo, Japan, 1995, pp. 821–824.

125. Z. Q. Shi, R. L. Wallace, and W. A. Anderson, "High-barrier height Schottky diodes on N–InP by deposition on cooled substrates," *Appl. Phys. Lett.,* **59,** 446–448 (1991).

126. T. Sugino, H. Yamamoto, T. Yamada, H. Ninomiya, Y. Sakamoto, K. Matsuda, and J. Shirafuji, "Characteristics of Au/n-InP Schottky junctions formed on H_2- and PH_3-plasma treated surfaces," *J. Electron. Mater.,* **20,** 1001–1006 (1991).

127. T. Sugino, Y. Sakamoto, and J. Shirafuji, "Schottky barrier height of phosphidized InGaAs," *Jpn. J. Appl. Phys.,* **32,** L239–L242 (1993).

128. T. Sugino, I. Yamamura, A. Furukawa, K. Matsuda, and J. Shirafuji, "Improved barrier height of Schottky junctions formed on phosphidized AlInAs," *Proc. 6th Int. Conf. InP and Related Materials,* Santa Barbara, Calif., 1994, pp. 632–635.

129. N.-J. Wu, T. Hashizume, H. Hasegawa, and Y. Amemiya, "Schottky contacts on n-InP with high barrier heights and reduced Fermi-level pinning by a novel in situ electrochemical process," *Jpn. J. Appl. Phys.,* **34,** 1162–1167 (1995).

130. T. Sato, S. Uno, T. Hashizume, and H. Hasegawa, "Large Schottky barrier heights on indium phosphide-based materials realized by in-situ electrochemical process," *Jpn. J. Appl. Phys.,* **36,** 1811–1817 (1997).

131. S. Uno, T. Hashizume, S. Kasai, N.-J. Wu, and H. Hasegawa, "0.86 eV platinum Schottky barrier on indium phosphide by in situ electrochemical process and its application to MESFETs," *Jpn. J. Appl. Phys.,* **35,** 1258–1263 (1996).

132. H. Hasegawa, T. Sato and T. Hashizume, "Evolution mechanism of nearly-pinning platinum/n-type indium phosphide interface with a high Schottky barrier height by in-situ electrochemical process," *J. Vac. Sci. Technol. B,* **15,** 1227–1235 (1997).

133. O. Wada and A. Majerfeld, "Low leakage nearly ideal Schottky barriers to n-InP," *Electron. Lett.,* **14,** 125–126 (1978).

134. O. Wada, A. Majerfeld, and R. N. Robson, "InP Schottky contacts with increased barrier height," *Solid-State Electron.,* **25,** 381–387 (1982).

135. H. Yamagishi, "Schottky contacts on n-InP surface treated by plasma-induced oxygen radicals," *Jpn. J. Appl. Phys.,* **25,** 1691–1696 (1986).

136. Y. S. Lee and W. A. Anderson, "High-barrier height metal-indulator-semiconductor diodes on n-InP," *J. Appl. Phys.,* **65,** 4051–4056 (1989).

137. Z. Q. Shi and W. A. Anderson, "MIS diodes on n-InP having an improved interface," *Solid-State Electron.,* **34,** 285–289 (1991).

138. H. Sawatari, M. Oyake, K. Kainosho, H. Okazaki, and O. Oda, "Study of InP MESFETs based on CdO_x interfacial layers grown by the adsorption/oxidation method," *Proc. 4th Int. Conf. InP and Related Materials,* Newport, R.I., 1992, pp. 297–300.

139. T. Sugino, Y. Sakamoto, T. Miyazaki, K. Kousaka, K. Matsuda, and J. Shirafuji, "Char-

acterization of InP Schottky junctions formed by in-situ remote plasma process," *Proc. 8th Int. Conf. InP and Related Materials*, Schwäbisch-Gmünd, Germany, 1996, pp. 685–688.
140. D. V. Morgan, and J. Frey, "Increasing the effective barrier height of Schottky contacts to n-In$_x$Ga$_{1-x}$As," *Electron. Lett.*, **14**, 737–738 (1978).
141. K. C. Hwang, S. S. Li, C. Park, and T. J. Anderson, "Schottky barrier height enhancement of n-In$_{0.53}$Ga$_{0.47}$As by a novel chemical passivation technique," *J. Appl. Phys.*, **67**, 6571–6573 (1990).
142. T. J. Licata, M. T. Schmidt, D. V. Podlenik, V. Liberman, R. M. Osgood, Jr., W. K. Chan, and R. Bhat, "The formation of elevated barrier height Schottky diodes to InP and In$_{0.53}$Ga$_{0.47}$As using thin, excimer laser deposited Cd interlayers," *J. Electron. Mater.*, **19**, 1239–1246 (1990).
143. G. P. Schwartz and G. J. Gualtieri, "Metal/p+/n enhanced Schottky barrier structures on (100) InP," *J. Electrochem. Soc.*, **133**, 1021–1025 (1986).
144. Z. Abid, A. Gopinath, F. Williamson, and M. I. Nathan, "Direct Schottky contact InP MESFET," *IEEE Electron Device Lett.*, **12**, 279–280 (1991).
145. R. Tyagi, T. P. Chow, J. M. Borrego, and K. A. Pisarczyk, "Improved Al/InP Schottky barriers by coimplantation of Be/P," *Proc. 5th Int. Conf. InP and Related Materials*, Paris, 1993, pp. 349–352.
146. C. Y. Chen, A. Y. Cho, K. Y. Cheng, and P. A. Garbinski, "Quasi-Schottky using a fully depleted p+-GaInAs layer grown by molecular beam epitaxy," *Appl. Phys. Lett.*, **40**, 401–403 (1982).
147. J. H. Kim, S. S. Li, L. Figueroa, T. F. Carruthers, and R. S. Wagner, "A high-speed InP-based In$_x$Ga$_{1-x}$As Schottky barrier infrared photodiode for fiber-optic communications," *J. Appl. Phys.*, **64**, 6536–6540 (1988).
148. P. Korodos, M. Marso, R. Meyer, and H. Lüth, "Enhanced Schottky barriers on n-In$_{.53}$Ga$_{.47}$As using p-InGaAs, GaAs, InP and InGaP surface layers," *Proc. 4th Int. Conf. InP and Related Materials*, Newport, R.I., 1992, pp. 230–233.
149. K. Koyanagi, S. Kasai, and H. Hasegawa, "Control of GaAs Schottky barrier height by ultrathin molecular beam epitaxy Si interface control layer," *Jpn. J. Appl. Phys.*, **32**, 502–510 (1993).
150. S. Kasai and H. Hasegawa, "Optimization of interface control layer for InP Schottky barriers," *Proc. 6th Int. Conf. InP and Related Materials*, Santa Barbara, Calif., 1994, pp. 220–223.
151. C. Y. Chen, S. N. G. Chu, and A. Y. Cho, "GaAs/Ga$_{0.47}$In$_{0.53}$As lattice-mismatched Schottky barrier gates: influence of misfit dislocations on reverse leakage current," *Appl. Phys. Lett.*, **46**, 1145–1147 (1985).
152. J. Selders, P. Roentgen, and H. Beneking, "Ga$_{0.47}$In$_{0.53}$As JFETs and MESFETs with OM-VPE-grown GaAs surface layers," *Electron. Lett.*, **22**, 14–16 (1986).
153. T. Kikuchi, H. Ohno, and H. Hasegawa, "Ga$_{0.47}$In$_{0.53}$As metal–semiconductor–metal photodiodes using a lattice mismatched Al$_{0.4}$Ga$_{0.6}$As Schottky assist layer," *Electron. Lett.*, **24**, 1208–1210 (1988).
154. A. H. Terani, D. Decoster, J. P. Vilcot, and M. Razeghi, "Monolithic integrated photoreceiver for 1.3–1.55-µm wavelengths: association of a Schottky photo diode and a field-effect transistor on GaInP–GaInAs heteroepitaxy," *J. Appl. Phys.*, **64**, 2215–2218 (1988).
155. D. Kuhl, F. Hieronymi, E. M. Böttcher, T. Wolf, A. Krost, and D. Bimberg, "Very high-

speed metal–semiconductor–metal InGaAs:Fe photodetectors with InP:Fe barrier enhancement layer grown by low pressure metalorganic chemical vapor deposition," *Electron. Lett.*, **26**, 2107–2109 (1990).

156. H. Ohno, J. Barnard, C. E. C. Wood, and L. F. Eastman, "Double heterostructure $Ga_{0.47}In_{0.53}As$ MESFETs by MBE," *IEEE Electron Device Lett.*, **1**, 154–155 (1980).

157. S. R. Bahl, M. H. Leary, and J. A. del Alamo, "Mesa-sidewall gate leakage in InAlAs/InGaAs heterostructure field-effect transistors," *IEEE Trans. Electron Devices*, **39**, 2037–2043 (1992).

158. M. D. Pashley, "Electron counting model and its application to island structures on molecular-beam epitaxy grown GaAs(001) and ZnSe(001)," *Phys. Rev. B*, **40**, 10481–10487 (1989).

159. R. M. Feenstra and A. P. Fein, "Surface morphology of GaAs(110) by scanning tunneling microscope," *Phys. Rev. B*, **32**, 1394–1396 (1985).

160. M. D. Pashley, K. W. Haberern, W. Friday, J. M. Woodall, and P. D. Kirchner, "Surface of GaAs(001) (2 × 4)-c(2 × 8) determined by scanning tunneling microscopy," *Phys. Rev. Lett.*, **60**, 2176–2179 (1988).

161. H. Yamaguchi and Y. Horikoshi, "Unified model for first-order transition and electrical properties of InAs (001) surfaces based on atom-resolved scanning tunneling microscopy imaging," *J. Cryst. Growth*, **150**, 148–151 (1995).

162. T. Hashizume, Q. K. Xue, J. Zhou, A. Ichimiya, and T. Sakurai, "Structures of As-rich GaAs(001)-(2 × 4) reconstructions," *Phys. Rev. Lett.*, **73**, 2208–2211 (1994).

163. Y. Ishikawa, T. Fukui, and H. Hasegawa, "Missing-dimer structures and their kink defects on molecular beam epitaxially grown (2 × 4) reconstructed (001) InP and GaAs surfaces studied by ultrahigh-vacuum scanning tunneling microscopy," *Jpn. J. Appl. Phys.*, **36**, 1749–1755 (1997).

164. D. J. Chadi, "Atomic structure of GaAs(100)-(2 × 1) and (2 × 4) reconstructed surfaces," *J. Vac. Sci. Technol. A*, **5**, 834–837 (1987).

165. M. D. Pashley and K. W. Haberern, "Compensating surface defects induced by Si doping of GaAs," *Phys. Rev. Lett.*, **67**, 2697–2700 (1991).

166. Y. Ishikawa, T. Fukui, and H. Hasegawa, "Kink defects and Fermi level pinning on (2 × 4) reconstructed molecular beam epitaxially grown surfaces of GaAs and InP studied by ultrahigh-vacuum scanning tunneling microscopy and x-ray photoelectron spectroscopy," *J. Vac. Sci. Technol. B*, **15**, 1163–1172 (1997).

167. J. Graffeuil, Z. Hadjoub, J. P. Fortea, and M. Pouysegur, "Analysis of capacitance and transconductance frequency dispersion in MESFETs for surface characterization," *Solid-State Electron.*, **29**, 1087–1097 (1986).

168. H. Hasegawa, T. Kitagawa, T. Sawada, and H. Ohno, "A new side-gating model for GaAs MESFETs based on surface avalanche breakdown," *Inst. Phys. Conf. Ser.*, No. 74, 1984, pp. 521–526.

169. C. J. Sandroff, R. N. Nottenburg, J.-C. Bischoff, and R. Bhat, "Dramatic enhancement in the gain of a GaAs/AlGaAs heterostructure bipolar transistor by surface chemical passivation," *Appl. Phys. Lett.*, **51**, 33–35 (1987).

170. J. Moisson, K. Elcess, F. Houzay, J. Y. Marzin, J. M. Gérard, F. Barthe, and M. Bensoussan, "Near-surface GaAs/$Ga_{0.7}Al_{0.3}As$ quantum well: interaction with the surface states," *Phys. Rev. B*, **41**, 12945–12948 (1990).

171. Z. Sobiesierski, D. I. Westwood, D. A. Woolf, T. Fukui, and H. Hasegawa, "Photolumi-

nescence spectroscopy of near-surface quantum wells: electronics coupling between quantized energy levels and the sample surface," *J. Vac. Sci. Technol. B,* **11,** 1723–1726 (1993).

172. J. F. Wagner, K. M. Geib, C. W. Wilmen, and L. L. Kazmerski, "Native oxide formation and electrical instabilities at the insulator/InP interface," *J. Vac. Sci. Technol. B,* **1,** 778–781 (1983).

173. G. Hollinger, E. Bergignat, J. Joseph, and Y. Robach, "On the nature of oxides on InP surfaces," *J. Vac. Sci. Technol. A,* **3,** 2082–2088 (1985).

174. L. G. Meiners, "Electrical properties of SiO_2 and Si_3N_4 dielectric layers on InP," *J. Vac. Sci. Technol.,* **19,** 373–379 (1981).

175. P. Viktorovitch, J. Vojtek, I. Thomas, K. Choujaa, J. Tardy, R. Blanchet, B. Agius, F. Plais, and A. Chovet, "Frequency, energy and spatially resolved characterization of interface traps in metal–insulator–InP transistors based on noise and current drift measurements," *Proc. 4th Int. Conf. InP and Related Materials,* Newport, R.I., 1992, pp. 218–220.

176. J. Tardy, I. Thomas, P. Viktorovitch, V. Drouot, B. Agius, F. Plais, J. P. Dupin, and D. Barbier, "InP MISFETs with SiO_2 gate insulator deposited by distributed electron cyclotron resonance plasma deposition," *Proc. 4th Int. Conf. InP and Related Materials,* Newport, R.I., 1992, pp. 306–309.

177. F. Kiel, W. Kulisch, and R. Kassing, "In situ characterization of remote plasma treated and passivated InP by integral photoluminescence," *Proc. 4th Int. Conf. InP and Related Materials,* Newport, R.I., 1992, pp. 347–350.

178. M. Ochiai, R. Iyer, B. Bollig, and D. Lile, "Analysis of SiO_2/InP interfaces using gated photoluminescence and Raman spectroscopy," *Proc. 4th Int. Conf. InP and Related Materials,* Newport, R.I., 1992, pp. 658–660.

179. F. Schulte, Ch. Werres, J. Splettstöber, D. Schmitz, R. Tuzinski, K. Heime, and H. Beneking, "Analog and digital performance of MISFETs on p- and n-GaInAs," *Proc. 4th Int. Conf. InP and Related Materials,* Newport, R.I., 1992, pp. 503–506.

180. A. Fathimulla, D. Gutierrez, and H. Hier, "A novel insulated-gene InP/InAlAs MODFET," *Proc. 5th Int. Conf. InP and Related Materials,* Paris, 1993, pp. 428–431.

181. A. Fathimulla and D. Gutierrez, "ECR-plasma-deposited gate dielectrics for InP MISFETs," *Proc. 5th Int. Conf. InP and Related Materials,* Paris, 1993, pp. 447–450.

182. L. S. How Kee Chun, J. L. Courant, P. Ossart, and G. Post, "In-situ surface preparation of InP-based semiconductors prior to direct UVCVD silicon nitride deposition for passivation purposes," *Proc. 8th Int. Conf. InP and Related Materials,* Schwäbisch-Gmünd, Germany, 1996, pp. 412–415.

183. M. Yoshimoto, K. Takubo, T. Saito, T. Ohtsuki, M. Komoda, and H. Matsunami, "Direct photo-chemical vapor deposition of silicon nitride and its application to MIS structure," *IEICE Trans. Electron.,* **E75-C,** 1019–1024 (1992).

184. B. Lescaut, Y. I. Nissim, and J. F. Brese, "Passivation of InGaAs surfaces with an integrated process including an ammonia DECR plasma," *Proc. 8th Int. Conf. InP and Related Materials,* Schwäbisch-Gmünd, Germany, 1996, pp. 319–322.

185. Y.-H. Jeong, S. Takagi, F. Arai, and T. Sugano, "Effects on InP surface trap states of in situ etching and phosphorus-nitride deposition," *J. Appl. Phys.,* **62,** 2370–2375 (1987).

186. A. Astito, A. Foucaran, G. Bastide, M. Rouzeyre, J. L. Leclercq, and J. Durand, "Prepa-

ration electrical properties and interface studies of plasma nitride layers on n-type InP," *J. Appl. Phys.,* **70,** 2584–2588 (1991).

187. Y. Sakamoto, T. Sugino, T. Miyazaki, and J. Shirafuji, "Formation of PN_x/InP structure by in-situ remote plasma processes," *Proc. 7th Int. Conf. InP and Related Materials,* Sapporo, Japan, 1995, pp. 605–608.

188. Y. Matsumoto, T. Suzuki, K. Haga, H. Sasaki, M. Sakuma, T. Hanajiri, T. Sugano, and T. Katoda, "Improvement of resistivity of phosphorous nitride films deposited at 100°C on InP substrate," *Proc. 7th Int. Conf. InP and Related Materials,* Sapporo, Japan, 1995, pp. 609–611.

189. T. Sugino, Y. Sakamoto, T. Miyazaki, and J. Shirafuji, "Interface properties of PN_x/InP structures by in-situ remote plasma processes," *Appl. Surf. Sci.,* **104,** 428–433 (1996).

190. T. Kobayashi and Y. Shinoda, "Metal–insulator–semiconductor diodes fabricated on InP, InGaAsP, and InGaAs," *J. Appl. Phys.,* **53,** 3339–3341 (1982).

191. P. Klopfenstein, G. Bastide, M. Rouzeyre, M. Gendry, and J. Durand, "Interface studies and electrical properties of plasma sulfide layers on n-type InP," *J. Appl. Phys.,* **63,** 150–158 (1988).

192. V. J. Kapoor, M. J. Mirtich, and B. A. Banks, "Diamondlike carbon films on semiconductors for insulated-gate technology," *J. Vac. Sci. Technol. A,* **4**(3), 1013–1017 (1986).

193. S. Heun, M. Sugiyama, S. Maeyama, Y. Watanabe, and M. Oshima, "Electronic and structural properties of thin SrF_2 films on InP," *Proc. 7th Int. Conf. InP and Related Materials,* Sapporo, Japan, 1995, pp. 269–272.

194. T. Itoh and K. Ohata, "X-band self-aligned gate enhancement-mode InP MISFET's," *IEEE Trans. Electron Devices,* **30,** 811–815 (1983).

195. P. D. Gardner, S. Y. Narayan, S. G. Liu, D. Bechtle, T. Bibby, D. R. Capewell, and S. D. Colvin, "InP depletion-mode microwave MISFET's," *IEEE Electron Device Lett.,* **8,** 45–47 (1987).

196. L. Messick, D. A. Collins, R. Nguyen, A. R. Clawson, and G. E. McWilliams, "High-power high-efficiency stable indium phosphide MISFETs," Int. Electron Device Meeting, Washington, D.C., *IEDM Tech. Dig.,* 1988, pp. 767–770.

197. K. C. Hwang, P. Ho, M. Y. Kao, S. T. Fu, J. Liu, P. C. Chao, P. M. Smith, and A. W. Swanson, "W-band high power passivated 0.15 μm InAlAs/InGaAs HEMT device," *Proc. 6th Int. Conf. InP and Related Materials,* Santa Barbara, Calif., 1994, pp. 18–20.

198. K. C. Hwang, A. R. Reisinger, K. H. G. Duh, M. Y. Kao, P. C. Chao, P. Ho, and A. W. Swanson, "A reliable ECR passivation technique on the 0.1 μm InAlAs/InGaAs HEMT device," *Proc. 6th Int. Conf. InP and Related Materials,* Santa Barbara, Calif., 1994, pp. 624–627.

199. T. Hashizume and H. Hasegawa, "Surface damage on InP induced by photo- and plasma-assisted chemical vapor deposition of passivation films," *Proc. 7th Int. Conf. InP and Related Materials,* Sapporo, Japan, 1995, pp. 573–576.

200. D. Sachelarie, J. L. Pelouard, and J. L. Benchimol, "Orientation effect on Si_3N_4 passivated InP/InGaAs heterojunction bipolar transistors," *Proc. 8th Int. Conf. InP and Related Materials,* Schwäbish-Gmünd, Germany, 1996, pp. 149–151.

201. M. Arps, H.-G. Bach, W. Passenberg, A. Umbach, and W. Schlaak, "Influence of SiN_x passivation on the surface potential of GaInAs and AlInAs in HEMT layer structures," *Proc. 8th Int. Conf. InP and Related Materials,* Schwäbisch-Gmünd, Germany, 1996, pp. 308–311.

202. D. Caffin, L. Bricard, J. L. Courant, L. S. How Kee Chun, B. Lescaut, A. M. Duchenois, M. Meghelli, J. L. Benchimol, and P. Launay, "Passivation of InP-based HBT's for high bit rate circuit applications," *Proc. 9th Int. Conf. InP and Related Materials,* Hyannis, Mass., 1997, pp. 637–640.

203. Y. Umeda, T. Enoki, K. Arai, and Y. Ishii, "Silicon nitride passivated ultra low noise In-AlAs/InGaAs HEMT's with n+-InGaAs/n+-InAlAs cap layer," *IEICE Trans. Electron.,* **E75-C,** 649–655 (1992).

204. O. Schuler, H. Fourré, R. Fauquembergue, and A. Cappy, "Influence of parasitic capacitances on the performance of passivated InAlAs/InGaAs HEMTs in the millimeter wave range," *Proc. 8th Int. Conf. InP and Related Materials,* Schwäbisch-Gmünd, Germany, 1996, pp. 646–649.

205. H. Hasegawa, M. Akazawa, K. Matsuzaki, H. Ishii, and H. Ohno, "GaAs and $In_{0.53}Ga_{0.47}As$ MIS structures having an ultrathin pseudomorphic interface control layer of Si prepared by MBE," *Jpn. J. Appl. Phys.,* **27,** 2265–2267 (1988).

206. H. Hasegawa, M. Akazawa, H. Ishii, A. Uraie, H. Iwadate, and E. Ohue, "Characterization of InGaAs surface passivation structure having an ultrathin Si interface control layer," *J. Vac. Sci. Technol. B,* **8,** 867–873 (1990).

207. Y. Robach, J. Joseph, E. Bergignat, B. Commere, G. Holinger, and P. Viktorovitch, "New native oxide of InP with improved electrical interface properties," *Appl. Phys. Lett.,* **49,** 1281–1283 (1986).

208. T. Sawada, S. Itagaki, H. Hasegawa, and H. Ohno, "InP MISFET's with Al_2O_3/native oxide double-layer gate insulators," *IEEE Trans. Electron Devices,* **31,** 1038–1043 (1984).

209. H. Ishii, H. Hasegawa, A. Ishii, and H. Ohno, "X-ray photo-electron spectroscopy analysis of InP insulator–semiconductor structures prepared by anodic oxidation," *Appl. Surf. Sci.,* **41/42,** 390–394 (1989).

210. Y. Z. Hu, M. Li, Y. Wang, E. Irene, M. Rowe, and H. C. Casey, Jr., "Electron cyclotron resonance plasma process for InP passivation," *Appl. Phys. Lett.,* **63,** 1113–1115 (1993).

211. V. Devnath, K. N. Bhat, and P. R. S. Rao, "Stable InP MIS device using silicon nitride/anodic oxide double-layer dielectric," *IEEE Electron Device Lett.,* **18,** 114–116 (1997).

212. H. Ishimura, K. Sasaki, and H. Tokuda, "An extremely low interface state density of SiN_x with thin pn interface control layer on $(NH_4)_2S_x$ treated InP," *Inst. Phys. Conf. Ser.,* No. 106, 1990, pp. 405–410.

213. R. Iyer, R. R. Chang, A. Dubey, and D. L. Lile, "The effect of phosphorous and sulfur treatment on the surface properties of InP," *J. Vac. Sci. Technol. B,* **6,** 1174–1179 (1988).

214. R. Iyer, M. Ochiai, C. W. Chen, and D. L. Lile, "Passivating SiS_2 films on InP," *Proc. 4th Int. Conf. InP and Related Materials,* Newport, R.I., 1992, pp. 340–342.

215. R. W. M. Kwok, W. M. Lau, D. Landheer, and S. Ingrey, "Electrical and chemical stability of Al/SiN_x/InP-metal–insulator–semiconductor diodes with gas phase polysulfide exposure on InP," *J. Vac. Sci. Technol. A,* **11,** 990–995 (1993).

216. K. Vaccaro, H. M. Dauplaise, A. Davis, S. M. Spaziani, and J. P. Lorenzo, "Indium phosphide passivation using thin layers of cadmium sulfide," *Appl. Phys. Lett.,* **67,** 527–529 (1995).

217. K. Vaccaro, A. Davis, H. M. Dauplaise, and J. P. Lorenzo, "InP-based MISFETs using

CdS passivation," *Proc. 8th Int. Conf. InP and Related Materials,* Schwäbisch-Gmünd, Germany, 1996, pp. 693–696.

218. C. S. Sundararaman, P. Milhelich, R. A. Masut, and J. F. Currie, "Conductance study of silicon nitride/InP capacitors with an In_2S_3 interface control layer," *Appl. Phys. Lett.,* **64,** 2279–2281 (1994).

219. C. S. Sundararaman and J. F. Currie, "Performance of interface engineered SiN_x/ICL/InP/$In_{0.53}Ga_{0.47}As$/InP doped channel HIGFET's," *IEEE Electron. Device Lett.,* **16,** 554–556 (1995).

220. C. S. Sundararaman, M. Tazlauanu, P. Mihelich, A. Bensaada, and R. A. Masut, "A 1–10 GHZ interface engineered SiN_x/InP/InGaAs HIGFET technology," *Proc. 8th Int. Conf. InP and Related Materials,* Schwäbisch-Gmünd, Germany, 1996, pp. 697–700.

221. Z. Wang, D. S. L. Mui, A. L. Demirel, D. Biswas, J. Reed, and H. Morkoç, "Gate quality Si_3N_4/Si/n-$In_{0.53}Ga_{0.47}As$ metal–insulator–semiconductor capacitors," *Appl. Phys. Lett.,* **61,** 1826–1828 (1992).

222. D. S. L. Mui, Z. Wang, D. Biswas, A. L. Demirel, N. Teraguchi, J. Reed, and H. Morkoç, "Si_3N_4/Si/n-$In_{0.53}Ga_{0.47}As$ depletion-mode metal–insulator–semiconductor field-effect transistors with improved stability," *Appl. Phys. Lett.,* **62,** 3291–3293 (1993).

223. S. Kodama, S. Koyanagi, T. Hashizume, and H. Hasegawa, "Novel surface passivation scheme for compound semiconductor using silicon interface control layer and its application to near-surface quantum wells," *Jpn. J. Appl. Phys.,* **34,** 1143–1148 (1995).

224. H. M. Dauplaise, J. P. Lorenzo, E. A. Martin, K. Vaccaro, and G. O. Ramseyer, "Evaporated thin silicon interlayers for indium phosphide device applications," *Proc. 4th Int. Conf. InP and Related Materials,* Newport, R.I., 1992, pp. 293–296.

225. M. Shokrani and V. J. Kapoor, "Silicon dioxide with a silicon interfacial layer as an insulating gate for highly stable indium phosphide metal–insulator–semiconductor field effect transistors," *J. Electrochem. Soc.,* **138,** 1788–1794 (1991).

226. T. Takahashi, T. Hashizume, and H. Hasegawa, "Novel InP metal–insulator–semiconductor structure having an ultrathin silicon interface control layer," *Appl. Surf. Sci.,* **123/124,** 335–338 (1998).

227. M. Akazawa, H. Hasegawa, and E. Ohue, "$In_{0.53}Ga_{0.47}As$ MISFETs having an ultrathin MBE Si interface control layer and photo-CVD SiO_2 insulator," *Jpn. J. Appl. Phys.,* **28,** L2095–L2097 (1989).

228. S. Suzuki, S. Kodama, and H. Hasegawa, "A novel passivation technology of InGaAs surfaces using Si interface control layer and its application to field effect transistor," *Solid-State Electron.,* **38,** 1679–1683 (1995).

229. S. Suzuki, Y. Dohmae, and H. Hasegawa, "Fabrication and electrical characterization of InP-based insulated gate power HEMTs using ultrathin Si interface control layer," *Solid-State Electron.,* **41,** 1641–1646 (1997).

230. S. Kodama, S. Koyanagi, T. Hashizume, and H. Hasegawa, "Silicon interlayer based surface passivation of near-surface quantum wells," *J. Vac. Sci. Technol. B,* **13,** 1794–1800 (1995).

231. H. Fujikura, S. Kodama, T. Hashizume, and H. Hasegawa, "Surface passivation of $In_{0.53}Ga_{0.47}As$ ridge quantum wires using silicon interface control layers," *J. Vac. Sci. Technol. B,* **14,** 2888–2894 (1996).

CHAPTER NINE

Dry Process Technique for InP-Based Materials

KIYOSHI ASAKAWA
Femtosecond Technology Research Association (FESTA)

9.1	Introduction	290
9.2	Plasma Chemistry and Equipment	292
	9.2.1 Plasma Chemistry for III-V Compound Semiconductors	292
	9.2.2 Plasma Equipment	294
9.3	Smooth and Vertical Etching	295
	9.3.1 Chlorine-Based Etching	295
	9.3.2 Hydrocarbon-Based Etching	301
	9.3.3 Laser-Assisted Selective Gas Etching	303
9.4	Depth-Monitoring Technique	309
9.5	Device Applications	310
	9.5.1 Recent Progress in Damage Reduction	310
	9.5.2 Photonic Device Application	312
	9.5.3 Electronic Device Application	323
9.6	Estimation of Nonradiative Surface Recombination	325
	9.6.1 Time-Resolved Photoluminescence	326
	9.6.2 In Situ Characterization of Etching/Regrowth Interface	327
9.7	Summary	331
	References	331

InP-Based Materials and Devices: Physics and Technology, Edited by Osamu Wada and Hideki Hasegawa.
ISBN 0-471-18191-9 © 1999 John Wiley & Sons, Inc.

9.1 INTRODUCTION

The comprehensive development of dry etching for III-V compound semiconductors began with GaAs/AlGaAs-based materials in the early 1980s as an essential microfabrication technique for electronic GaAs integrated circuits (ICs) and optoelectronic ICs (OEICs) [1–3]. In the late 1980s, dry etching of III-V materials was extended to InP-based alloy semiconductors for the microfabrication of photonic devices required for 1.3- to 1.5-μm-wavelength optical communication systems [4–6]. Recent dry etching studies have focused on II-VI- and GaN-based materials for use in blue/green laser diode (LD)/light-emitting diode (LED) devices [7–10].

In the course of development, several problems with this etching technique have been solved, as shown by the overview of dry etching steps in Fig. 9.1. In the figure, each step has an advantage or disadvantage compared with conventional wet etching. The initial stage of etching study began with a search for an adequate etching gas to realize smooth etched surfaces. Carbon-containing halogen gases such as CCl_4, CCl_3F, and CCl_2F_2 were mainly used primarily for reactive ion etching (RIE) [11–13]. However, such gases often caused surface roughening, due to residual carbon-containing materials, so mixing of additional gases to scavenge this residue was carried out to solve this problem.

The development of a reactive ion beam etching (RIBE) system, based on ultrahigh-vacuum (UHV) and electron cyclotron resonance (ECR)-plasma technologies

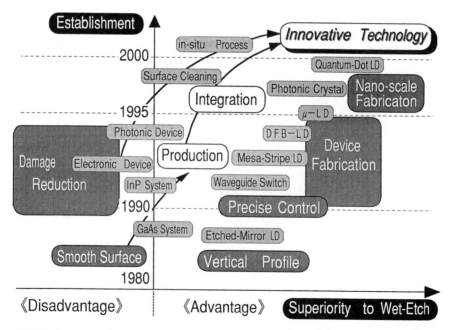

FIGURE 9.1 Overview of advancing dry etching steps in terms of advantages and disadvantages compared with conventional wet etching.

[14], enabled us to use a pure chlorine gas (Cl_2) to easily achieve mirrorlike smooth surfaces on the AlGaAs as well as the GaAs substrate [15]. This chlorine-based RIBE technique produced a high vertical profile [16] and depth-controlled [17] etching with the aid of a multilevel resist mask and endpoint monitoring techniques. These results could easily be applied to the mesa etching of several kinds of laser diodes [18,19], which involve the present notable vertical cavity surface emitting lasers (VCSELs) [20]. To solve the current problem of dry etching common to all III-V materials, we need a detailed study of dry-etching-induced damage and its influence on device performance [21–24]. To date, damage on the GaAs substrate has been investigated more intensively than that on the InP substrate. This is because the GaAs bulk or surface is more sensitive than the InP bulk or surface to the process-induced defect. In particular, in situ processing, which includes vacuum lithography [25,26], surface cleaning [27–31] and vacuum-through etching/overgrowth [32–38] using a chlorine-based gas, has been advanced primarily on the GaAs substrate as a promising technique to suppress or remove such damage and to fabricate nanostructures such as quantum wires or dots [34].

Although these chlorine-based etching techniques for GaAs systems were followed by etching of indium-containing materials such as InP, etched surfaces were often roughened, due to residues composed primarily of nonvolatile indium chlorides such as $InCl_3$. In recent years, however, this problem has been solved by two alternative methods. The first method is to increase the substrate temperature to improve the desorption rate of the indium chloride [39]. The second method is to use mixed gases such as methane (CH_4) [47], ethane (C_2H_6) [48], and H_2, which react with the indium to produce volatile metal-organic materials. These results are discussed in detail in Sections 9.2.1 and 9.3.

In this chapter we review recent dry process techniques. We focus on the fundamental characteristics of indium-containing compound materials (35–39) rather than on GaAs-based systems and their applications to photonic devices. InGaAs materials, which are important for 0.9-μm wavelength VCSELs and other strained-layer superlattice structures, are also included in the indium-containing materials shown here. In Section 9.2 we introduce basic plasma chemistry and equipment. In Section 9.3, a variety of etching techniques for the production of smooth and vertical surfaces are described. A large part of this section involves chlorine-based RIBE techniques reported recently. However, other promising techniques, such as a methane/ethane-based plasma etching [47,48] and a photo-assisted gas etching [93], are also described. An important practical depth-monitoring technique is discussed in Section 9.4. In Section 9.5 we describe device applications.

Section 9.6 deals with damage characterizations. We describe recent systematic studies on nonradiative surface recombination velocities [40,41] on mesa-etched sidewall surfaces using primarily GaAs-based materials. A time-resolved photoluminescence (PL) technique was used to study the nonradiative surface recombination velocities for quantum-well samples processed using different techniques [42]. In situ processing techniques were employed using UHV-based multichamber systems to provide the various structures [43]. Although these characterization techniques have not yet been applied to InP-based systems, important results were described here for reference to the InP case. Regarding GaAs-based dry etching, previous review articles should be referred to for details [44,45].

9.2 PLASMA CHEMISTRY AND EQUIPMENT

9.2.1 Plasma Chemistry for III-V Compound Semiconductors

In general, the elementary process of gas-surface chemistry in plasma-based dry etching is subject to the adsorption of reactive species such as neutral radicals and the formation of reaction products through chemical reaction with surface atoms and the desorption of reaction products [46]. The etching gas is usually selected so that the chemical reaction produces highly volatile reaction products. Halogen gases such as chlorine, fluorine, and bromine are preferable from the viewpoint of reaction product volatility for most compound semiconductors as well as for silicon and related materials. In particular, chlorine-based gases have often been used for GaAs etching because of their ability to produce highly volatile reaction products (such as $GaCl_3$), leading to high etching rates [11]. In contrast to this, InP-based materials have low-volatility indium chlorides (such as $InCl$, $InCl_2$, and $InCl_3$), which cause marked surface roughening at room temperature [39]. For these materials, two alternative methods are known to be useful; the first is to increase the substrate temperature in chlorine-based plasma etching to improve the desorption rate of indium chloride [39], and the second is to use mixed gases such as methane (CH_4) [47], ethane (C_2H_6) [48], and H_2, which react with the indium to produce volatile metal-organic materials such as $(CH_3)_3In$ and $(C_2H_5)_3In$.

Table 9.1 shows the substrate materials and etching gases reported to date. As described above, both chlorine- and methane/ethane-based gases are used for InP/InGaAs substrates. A comparison of the InP etching characteristics of these two gas systems is shown in Table 9.2. C_xH_y gas indicates CH_4 or C_2H_6 gas. The most prominent feature of C_xH_y-based etching is its ability to produce smooth etching and a significantly high etching rate (several hundred A/min) at room temperature. This is greatly superior to chlorine-based etching, in which an indium-containing substrate needs to be heated to more than 200°C to activate the desorption of the indium-containing nonvolatile reaction product, as discussed previously [35]. However, plasma-dissociated hydrocarbon materials often cause polymerization-related phenomena in the mask region. While deposition of a polymer on the mask positively contributes to an increase in the mask resistance, in particular, in deep etching, the eave-shaped lateral growth of a mask edge often prevents a vertical etching profile with a smooth sidewall surface [66]. Furthermore, carbon contamination on the sample surface or etching chamber wall is, in general, more serious than that with hydrocarbon-free gas plasma [66].

In contrast to C_xH_y-based etching, as discussed above, chlorine-based InP etching in early stages suffered from serious roughening of the etched surface. However, as will be discussed later, a smooth surface with an etching rate higher than that for C_xH_y-based etching has been achieved by controlling the substrate temperature

TABLE 9.1 Substrates, Equipment, and Etching Gases for Reported Dry Etching of Compound Semiconductors

Substrate	Equipment[a]	Etching Gas	Refs.
GaAs/AlGaAs	RIE	CCl_2F_2	11–13
	RIE	CCl_4, CCl_3F	13
	RIE	Cl_2	49
	RIE	PCl_3	50
	RIE	BCl_3	51
	RIE	HCl	50
	RIBE	Cl_2	15
	RIBE	CCl_4	52
	IBAE	Cl_2/Ar	53
	RIE	CH_4/H_2	54
	RIE	C_2H_6/H_2	55
	Gas etch	Cl_2	56, 57
GaP	RIE	CCl_4, Cl_2	13
GaSb	PE	H_2	58
InP/InGaAs	RIE	C_2H_6/H_2	48
	RIE	CH_4/H_2	47
	RIE	CH_3Cl	59
	RIE	CCl_2F_2	11
	RIE	CCl_3F	13
	RIE	CCl_4	60
	RIE	HCl	50
	RIE	Cl_2/Ar	3, 61
	RIBE	Cl_2	35, 36, 39, 68
	LAGE	HBr/F_2	93
GaN	RIBE	BCl_3/Ar	9
	RIE	$SiCl_4/Ar$	10
	RIE	$BCl_3/SiCl_4$	62
	RIBE	$Cl_2/H_2, CH_4/H_2/Ar$	63
	CAIBE	Cl_2	64
	Atom beam etch	Cl_2	65
ZnSe	RIE	Cl_2	7, 8
	RIE	CH_4/H_2	88, 89

[a] RIE, reactive ion etching; RIBE, reactive ion beam etching; CAIBE, chemically assisted ion beam etching; PE, plasma etching; LAGE, laser-assisted gas etching.

above 200°C [39]. Moreover, GaAs and AlGaAs are also smoothly etched under these conditions. This means that most compound alloy materials involving both nonvolatile In and easily oxidized Al elements can be etched smoothly only by chlorine-based etching. With regard to etching-induced damage, no comparison between these two gases has ever been reported. In Section 9.3, these characteristics are described in detail.

TABLE 9.2 Comparison of Etching Characteristics for C_xH_y- and Cl_2-Based Plasma

Characteristic	C_xH_y	Cl_2
Morphology	Good	Poor to good
Etch rate	Poor to good	Good
Carbon contamination: surface	Poor to good	Good
Carbon contamination: chamber	Poor	Good
Mask resistance	Good	Poor
Substrate temperature	Room temperature	$\geq 200°C$
Damage reduction	Good to ?	Good to ?

9.2.2 Plasma Equipment

Plasma conditions that control etching properties are characterized primarily by two factors: gas pressure and ion energy. Generally speaking, isotropic and anisotropic etching are carried out at higher and lower gas pressure, respectively, depending on the ion mean free path in the plasma. The ion energy, on the other hand, determines the penetration depth and degree of damage.

Figure 9.2 schematically shows how the RF and ECR plasmas often used in compound semiconductor dry etching are characterized on the gas pressure–ion energy map. The RF plasma often used in RIE is used at a gas pressure above 10^{-2} torr. Generally, parallel-plate-electrode RIE with a cathode configuration and without substrate bias provides us with a self-bias voltage V_{dc} ranging from several tens to

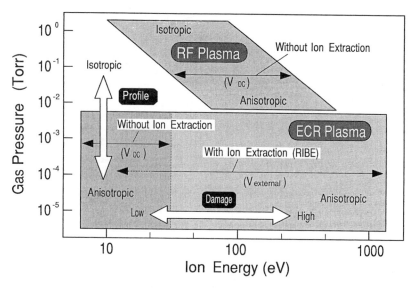

FIGURE 9.2 RF- and ECR-plasmas for compound semiconductors, characterized on the gas pressure–ion energy map.

several hundreds of volts. Since the V_{dc} value depends on the gas pressure, isotropic and anisotropic etching is only controlled by gas pressure. Although a negative substrate bias in the RIE is, to some extent, effective for anisotropic etching, the anisotropy is limited by the minimum gas pressure required to maintain a stable plasma discharge.

An ECR plasma, on the other hand, has the ability to stably discharge at a gas pressure ranging below 10^{-3} torr, as shown in Fig. 9.2. This means that its ion mean free path is longer than that in the RF plasma, which is favorable for high-aspect-ratio etching. ECR etching equipment without an ion extractor is subject to V_{dc} control only by gas pressure, similar to the RF plasma case. However, resulting from the high plasma density specific to ECR plasma, the V_{dc} value may be as low as 10 eV [14]. This means that this etching equipment can suppress damage extremely well, and it is thus particularly favorable for gate-recess processes in metal–semiconductor field-effect transistors (MESFETs) or high-electron-mobility transistors (HEMTs) [67]. ECR plasma equipment with an ion extractor is often called an RIBE. Because of the wide range of voltage controllabilities using external voltage, the RIBE is advantageous in realizing the high-aspect-ratio profiles that are needed for the mesa-stripe and mirror-facet etching in laser diodes, as shown in Section 9.4. In such a mesa-stripe structure, damage induced by ion bombardment is concentrated on the sidewall of the mesa stripe, so that in most cases it does not seriously deteriorate laser performance. This kind of damage influence is less than that in the above-mentioned gate-recess process, where most of the channel regions parallel to the etched bottom surface are markedly degraded, depending on the distance from the etched top surface. Characterization of the damage is discussed in detail in Sections 9.5.1 and 9.6. Another advantage of the ECR-RIBE is its potential to realize a high vacuum in the chamber, due to the discharge gas pressure being as low as 10^{-4} torr. Consequently, clean etching conditions such as those using pure chlorine gas, and compatibility with other ultrahigh-vacuum chambers, are possible. This advantage is discussed in Section 9.6.

9.3 SMOOTH AND VERTICAL ETCHING

In this section, smooth-surface and vertical-profile etching on InP-based material with a Cl_2-based ECR plasma, a hydrocarbon-based plasma, and laser assistance in a halogen gas atmosphere are reviewed separately. In Cl_2-based ECR-plasma etching, in particular, etching conditions to improve surface smoothness and sidewall profile verticality compared with the GaAs case are emphasized.

9.3.1 Chlorine-Based Etching

Temperature-Dependent Etching Contrary to GaAs etching, the etching of InP-based materials by chlorine-based plasma requires heating of the substrate as discussed above. Figure 9.3 shows the temperature dependence of the profile and etching rate for an InP substrate [68]. Etching was performed at a Cl_2 gas pressure of

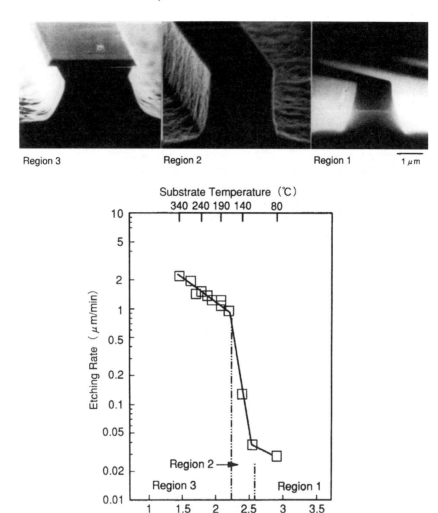

FIGURE 9.3 Temperature dependence of chlorine-based etching profile and etching rate for an InP substrate. (From Ref. 68.)

7×10^{-5} torr and an ion extraction voltage of 400 V. The resulting etching rates and profiles are categorized into three regions, as shown in Fig. 9.3. In the low-temperature region (region 1 in the figure), the etching rate is less than 0.05 μm/min, reflecting the low desorption rate for indium or indium/chloride-containing reaction products. The sidewall is tilted as it was for physical sputtering. In this region, a chemically enhanced etching rate common to reactive gas-plasma etching is barely achieved. In region 2, the rate increases abruptly, due to the increase in chemical enhancement as temperature increases. The profile is close to vertical. However, the bottom and sidewall surfaces have significant roughness. The rate-limiting process of this region is thought to be a chemical reaction. In the high-temperature region (region 3), on the other hand, the rate is saturated at more than 1 μm/min. This suggests that a rate-limiting process is the supply of etching species. It is notable that the sidewalls are composed of mesa and reverse-mesa profiles, which suggest crystallographic facets. The clear appearance of these facets is specific to crystallographic etching at high temperatures with gas molecules without plasma excitation (gas-etching mode) or with plasma-dissociated radicals (radical-etching mode). However, such a facet profile is not common in etching, including ion bombardment, as in this case. However, as shown in the next section, such crystallographic facets can be suppressed and vertical facets are made possible by increasing the ion voltage while maintaining the same temperature. Results on a vertical profile and a smooth surface produced with Cl_2-based InP RIBE are discussed in the following section.

Smooth-Surface and Vertical-Profile Etching The crystallographic sidewall profile in Fig. 9.3 was improved to vertical using a substrate temperature maintained at 230°C and an ion extraction voltage that was varied from 100 to 900 V. The results are shown in Fig. 9.4 [68]. It was found that the sidewall profile approaches vertical as the voltage increases. In addition to the profile, the grassy-roughened bottom surface at 500 V becomes smoother at 900 V. This abrupt change in profile and the surface roughness indicate that the effect of ion bombardment at 900 V plays two important roles. The first role is to provide a higher etching rate in the vertical direction along the bottom than the etching rate given in the lateral direction of the sidewall by radical species. Consequently, vertical etching is more dominant than crystallographic etching, even at high temperature. The second role of ion bombardment is to scavenge nonvolatile residues on the etched bottom, such as indium or indium–chloride droplets, which act as micromasks and cause grassy-roughened surfaces.

High-aspect-ratio etching for the InP substrate is also achieved using the RIBE conditions shown in Fig. 9.4. Figure 9.5 is a scanning electron microscope (SEM) photograph of InP vertical etching with 0.35-μm-wide line-and-space patterns [68]. The substrate temperature was 230°C, the ion voltage was 500V, and the Cl_2 gas pressure was 7×10^{-5} torr. The aspect ratio reaches 17 in this RIBE experiment.

The etched bottom smoothness at 900 V, shown in Fig. 9.4, was improved further by increasing the ion voltage to 1400 V and simultaneously tilting the sample at 15° from the horizontal plane and rotating it around a vertical axis. This sample movement was possible due to a specially designed manipulator in the RIBE apparatus. Figure 9.6 shows SEM photographs of etched samples with 4 μm and 0.35 μm stripe widths [68]. The sidewall profiles are vertical, and the bottom and sidewall surfaces are extremely smooth. The trenches at the bottom near the sidewall corners were formed by sputtering with ions reflected on the sidewall surface. Such trenches often appear when samples are tilted.

The surface roughness of the etched bottom, shown in Fig. 9.6, was measured with a scanning tunneling microscope (STM) technique. Figure 9.7 shows STM images of

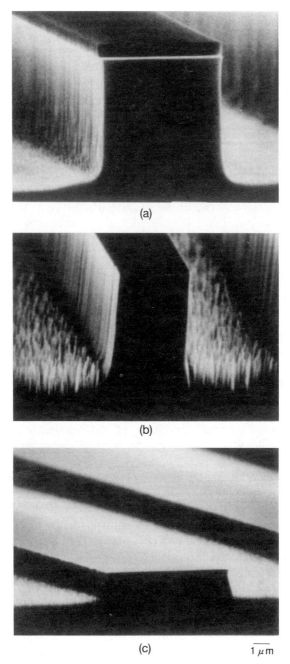

FIGURE 9.4 Ion energy dependence of chlorine-based etching profile and etched bottom morphology for an InP substrate: (*a*) 900 eV; (*b*) 500 eV; (*c*) 100 eV. (From Ref. 68.)

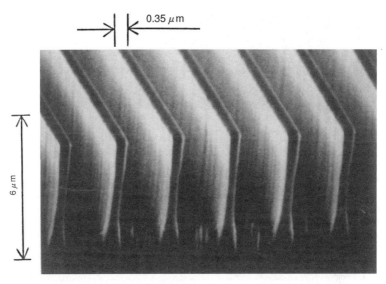

FIGURE 9.5 SEM photograph of InP vertical etching with 0.35-μm-wide line-and-space patterns. (From Ref. 95.)

microroughness on the InP substrate before and after RIBE [68], respectively. Before etching, the peak-to-peak roughness was 7 nm, which was presumed to be a result of surface pretreatment with wet etching. After 1-μm-deep RIBE, the surface roughness increased only by 4 nm (corresponding to 0.4% as the ratio of etching-induced roughness to etch depth). Such roughness can be ignored for most photonic device applications. We found that the Cl_2-based ECR-RIBE of InP substrates could provide us with a vertical profile and smooth bottom and sidewall surfaces if the RIBE was performed at a high ion voltage at a high substrate temperature.

Comparison of GaAs and InP Etching Optimum conditions for a vertical profile and a smooth surface in the Cl_2-based ECR-RIBE of InP substrates are summarized in Fig. 9.8. The figure shows the optimal flow for RIBE conditions starting from those for GaAs: (1) The etching rate was increased by heating the substrate from room temperature to 230°C. (2) Then the profile was improved from crystallographic to vertical by lowering the Cl_2 gas pressure from the 10^{-3} torr range to the 10^{-5} torr range. (3) Finally, the bottom surface was smoothed by increasing the ion energy from 400 to 500 eV to around 1 keV. With the GaAs substrate, on the other hand, both a smooth surface and a vertical profile were achieved over a wide range of gas pressures, ion voltages, and substrate temperatures. However, from the high-etch-rate point of view, optimal RIBE conditions for the GaAs are Cl_2 gas pressure in the range 10^{-3} torr, ion energy in the range 300 to 500 V, and a substrate temperature below 60°C, as shown in Fig. 9.8. Therefore, the optimal conditions for GaAs and InP substrates are different.

As mentioned in Section 9.1, difficulty during the initial stage of development in

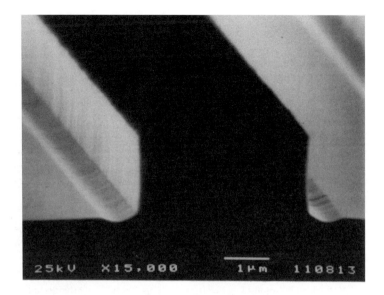

(a)

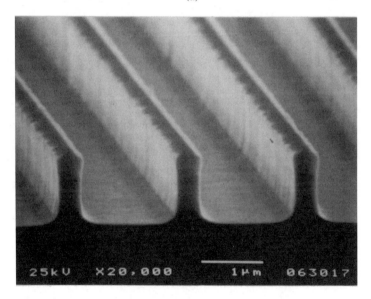

(b)

FIGURE 9.6 SEM photographs of etched samples with (a) 4-μm and (b) 0.35-μm stripe widths. (From Ref. 68.)

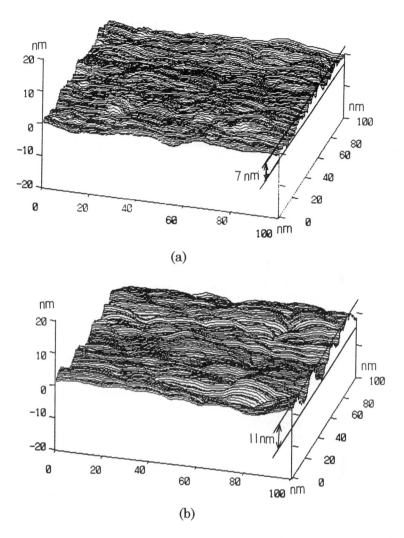

FIGURE 9.7 STM images of microroughness on the InP substrate: (a) roughness before RIBE; (b) roughness after RIBE. (From Ref. 68.)

dry etching of III-V compound materials for high-rate and smooth etching of aluminum- and indium-containing materials is due mainly to the easy oxidization of aluminum and the low volatility of indium. Results in Fig. 9.8, however, indicate that almost all the alloy compounds, including GaAs- and InP-based materials, can produce both vertical and smooth etching using Cl_2-based ECR-RIBE.

9.3.2 Hydrocarbon-Based Etching

Hydrocarbon-Based Plasma Chemistry   As described in Section 9.2.1, an ele-

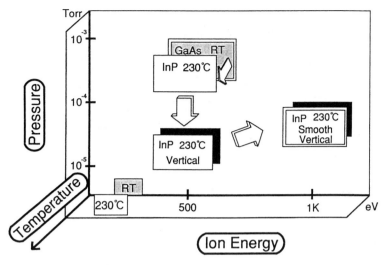

FIGURE 9.8 Summarized optimum conditions for a vertical profile and a smooth surface in the Cl_2-based ECR RIBE of InP substrates.

mentary process of dry etching for most semiconductors using reactive gas plasma is subject to the change of nonvolatile surface atoms into volatile reaction products with the aid of gas-surface chemistry between chemically reactive plasma species such as halogen radicals and substrate materials, thus enhancing desorption of the reaction products covering the surface. In the case of etching indium-containing materials, however, halogen gas plasma produces unfavorable products (i.e., nonvolatile indium–halogen compounds which cause surface roughening). Instead, hydrocarbon gas plasma such as CH_4 or C_2H_6 was found to be preferable for producing volatile, indium-containing metal-organic materials such as trimethyl indium [47]. It was also found that an excess supply of hydrocarbon was unfavorable because of polymer film formation on the etched surface [92]. Thus InP etching is currently being employed by introducing a deoxidizing gas such as hydrogen which can suppress the polymerization. A typical example is shown in Fig. 9.9, where a Ga-composition-dependent RIE rate of In_xGa_yAsP in a plasma of 20 vol % CH_4 in H_2 with no substrate heating is plotted at two different total gas pressures [47]. It is shown that compared with an InP etching rate, a GaAs etching rate is decreased by a factor of 2 to 3.

Smooth-Surface and Vertical-Profile Etching Matsui and his co-workers investigated the optimum RIE condition of InP, InGaAs, and GaAs substrates in detail using a mixture of C_2H_6 and H_2 as an etchant [90]. They achieved smooth and vertical etching for these three materials with 7% C_2H_6 concentration, 0.6 W/cm² RF power density, and 13.3 Pa total pressure, as shown by the SEM photographs in Fig. 9.10. They also found that surface morphologies are degraded at a higher C_2H_6 con-

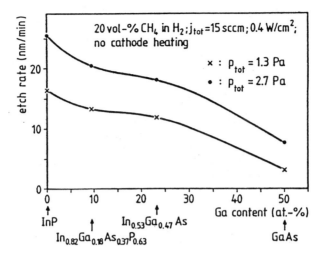

FIGURE 9.9 Ga-composition-dependent In_xGa_yAsP etching rate in a CH_4/H_2 RIE. (From Ref. 47.)

centration. Figure 9.11 shows SEM photographs of roughened InP surfaces for (a) 12% C_2H_6 and (b) 7% C_2H_6 concentrations at 16 Pa total pressure and 0.6 W/cm² RF power density [90]. Polymerization on the etched surface or on the sidewall due to the excess hydrocarbon supply was suppressed by adding O_2 gas intentionally to C_2H_6/H_2 gas, as shown in Fig. 9.12 [91]. It is found that etched-bottom roughnesses and eaves at the top of the sidewall originating from polymerization of the hydrocarbon without O_2 gas, as shown in Fig. 9.12a, disappear as the O_2 gas flow rate increases, as shown in Fig. 12b and c.

Surface Contamination Analysis Carbon contamination on the etched bottom and sidewall surfaces, predicted when a hydrocarbon plasma is used as an etchant, was investigated in detail using Auger electron spectroscopy. Figure 9.13 shows Auger spectra measured on the sidewall etched without O_2 and with O_2 at a 1-sccm flow rate and a spectrum of unetched InP substrate [92]. The results show that the ratio of the intensity of oxygen to that of indium increased from 0.37 to 0.53 with the addition of O_2, while the ratio of carbon to indium decreased dramatically, from 5.1 to 1.3, with the addition of O_2.

9.3.3 Laser-Assisted Selective Gas Etching

Instead of using a reactive gas plasma, photo-assisted gas-surface chemistry in a particular condition provides us with highly selective etching without significant damage, unlike the case of ion-bombardment-involving plasma etching. Takazawa and his co-workers succeeded in ArF-excimer-laser-assisted highly selective etching of InGaAs/InAlAs using an HBr and F_2 gas mixture [93]. In this technique, an HBr gas supply under the irradiation of ArF excimer laser light (193 nm) plays an impor-

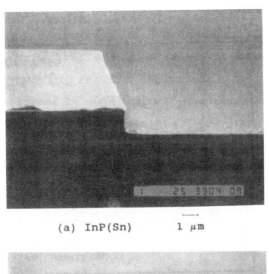

(a) InP(Sn) 1 μm

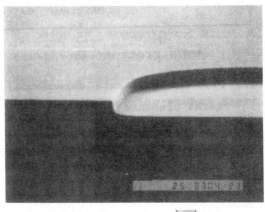

(b) GaInAs 1 μm

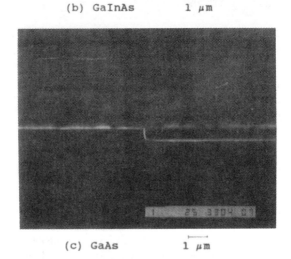

(c) GaAs 1 μm

FIGURE 9.10 SEM photographs of (*a*) InP, (*b*) GaInAs on InP, and (*c*) GaAs after etching with 7% C_2H_6 concentration, 0.6 W/cm^2 RF power density, and 13.3 Pa total pressure. (From Ref. 90.)

9.3 SMOOTH AND VERTICAL ETCHING

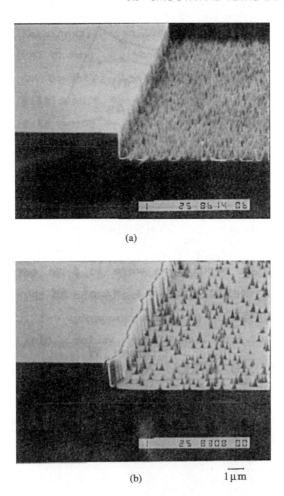

FIGURE 9.11 Surface morphologies on InP under etching conditions: (a) 12% C_2H_6 concentration, 13.3 Pa total pressure, and 0.6 W/cm² RF power density; (b) 7% C_2H_6 concentration, 16 Pa total pressure, and 0.6 W/cm² RF power density. (From Ref. 90.)

tant role in enhancing the desorption of nonvolatile indium bromide reaction products from InGaAs, while the addition of F_2 gas contributes significantly to reduction in the InAlAs etching rate resulting from the production of very nonvolatile AlF_x (probably $x = 3$) on top of the InAlAs surface. By virtue of these photochemical mechanisms, an etching rate ratio over 450 has been achieved.

As an example of experimental results, Fig. 9.14 shows InGaAs and InAlAs etching rates as a function of the incident laser fluence when using HBr gas at a partial pressure of 60 mtorr, with no F_2 gas and with F_2 gas at a pressure of 6 mtorr for a sample temperature of 110°C and a laser pulse frequency of 70 Hz [93]. The flu-

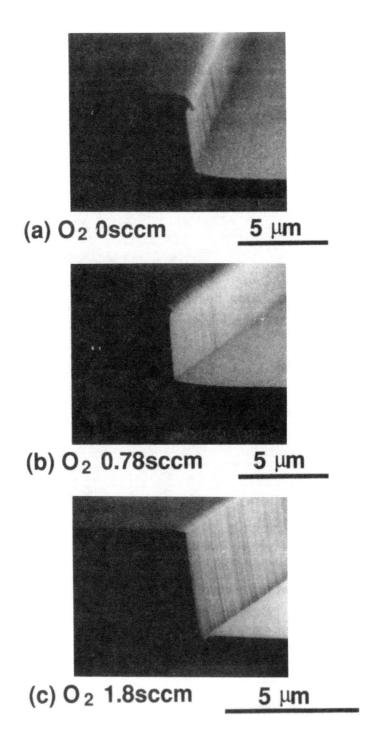

FIGURE 9.12 SEM photographs of InP substrates etched for 60 min using a C_2H_6/H_2 mixture: (*a*) without O_2 gas; (*b*) with O_2 gas at a 0.78-sccm flow rate; (*c*) with O_2 gas at a 1.8-sccm flow rate. The flow rates of C_2H_6 and H_2 were 2.9 and 52.9 sccm, respectively. The RF power density was 0.6 W/cm². (From Ref. 91.)

9.3 SMOOTH AND VERTICAL ETCHING   307

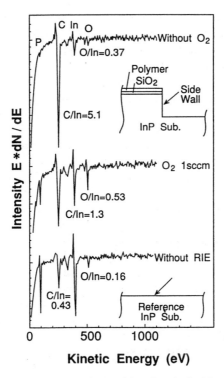

FIGURE 9.13 Auger spectra measured on a sidewall etched without O_2 and with O_2 at a 1-sccm flow rate, and a spectrum of unetched InP substrate. (From Ref. 92.)

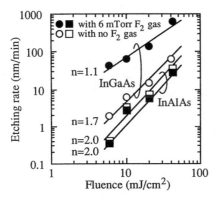

FIGURE 9.14 InGaAs and InAlAs etching rates, with no F_2 gas and with F_2 gas at a pressure of 6 mtorr, a sample temperature of 110°C, and a laser pulse frequency of 70 Hz. (From Ref. 93.)

ence dependence of the etching rate was fitted to a straight line in the log-log plot. An *n* value in the figure shows the power index. Due to the difference in the power index, the InGaAs/InAlAs etching rate ratio increases as the laser fluence decreases when the F_2 partial pressure is 6 mtorr.

Since the target material is not subject to ion bombardment in principle, this etching technique is suitable for device application where extremely low-damage etching is required, as in the case of gate-recess etching of InGaAs/InAlAs field-effect transistors. Damage was evaluated by the sheet carrier density and the carrier mobility in the InGaAs/InAlAs structure for Hall effect measurement. Figure 9.15 shows these parameters as a function of etching time [94]. An InGaAs cap layer 10 nm thick was removed in about 11 s, which resulted in exposure of the surface of the InAlAs layer. The dashed line in the figure is the calculated sheet carrier density. Reference 94 should be referred to for details on the calculation. A high carrier mobility is preserved after 4 min of etching. These results show that no damage is induced in the InGaAs channel layer during the etching.

Figure 9.16 is an SEM micrograph of a cross section of the HEMT structure after gate recess etching using the technique mentioned above [94]. The etching was stopped at the InAlAs layer surface. Since the etching is isotropic, lateral etching under the SiO_2 mask can be observed. This lateral etching results in electrical isolation between a gate electrode and the InGaAs layer forming source and drain regions. Good drain and gate *I–V* characteristics were reported for a 0.7- by 18-μm InAlAs/InGaAs HEMT, although Ref. 94 should be referred to for details on device performance.

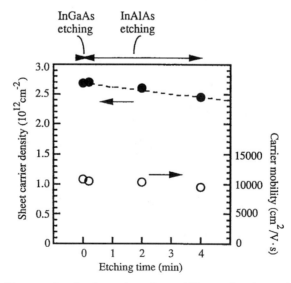

FIGURE 9.15 Sheet carrier density and carrier mobility as functions of etching time. (From Ref. 94.)

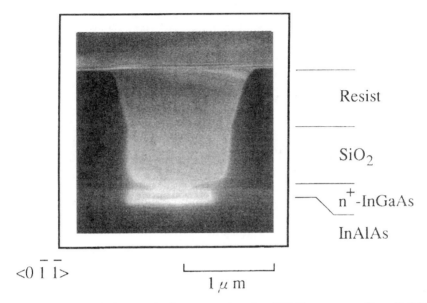

FIGURE 9.16 SEM photograph of a cross section of an HEMT gate recess. (From Ref. 94.)

9.4 DEPTH-MONITORING TECHNIQUE

The microfabrication of electronic/optoelectronic devices often requires precise control of etching depth in the range 10 nm. In a self-aligned structure (SAS) laser diode [69], for example, the microfabrication of a ridge waveguide above the active layer requires etching to a depth of 100 to 500 nm, leaving a 100- to 200-nm-thick cladding layer with an etch-depth precision of 10 to 20 nm just above the active layer. It is difficult to control such a precise depth by monitoring only the etching time. Thus a variety of endpoint monitoring methods for use during etching have been developed. The laser interferometry method is known to provide etch-depth control in the range 10 to 20 nm [17].

In this section we describe two examples of endpoint monitoring demonstrated using laser interferometry. The first is the monitoring of mesa-stripe RIBE at a 200- to 500-nm etch depth with ±20-nm precision on an MOVPE-grown GaInP/AlGaInP visible quantum-well laser wafer [69]. Figure 9.17a shows the layered structure of the visible laser wafer. The calculated sinusoidal curve, inserted in the layered structure, shows the reflected laser beam intensity. Figure 9.17b shows monitored sinusoidal curves of reflected laser intensity when the etching was performed on the top surface of the substrate. The measured result is close to the result calculated. Here the periodicity of the sinusoidal curve corresponds to 90 nm, depending on the reflective index of each layer. By applying this technique to an actual laser device, part of the top AlGaInP layer was etched to a depth of 200 to 500 ±20 nm.

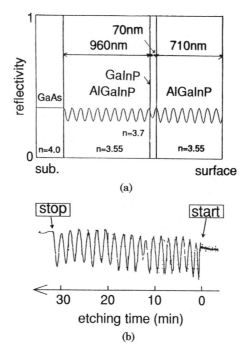

FIGURE 9.17 Explanation of endpoint monitoring using a laser interferometry for precise etching of the visible laser wafer: (*a*) layered structure; (*b*) monitored sinusoidal curves of reflected laser intensity. (From Ref. 69.)

The second demonstration of endpoint monitoring is done during in situ processing, where etching is followed by epitaxial overgrowth [43]. Here the etched region is buried by the overlayer, so that etching without endpoint monitoring results in great difficulty in controlling the etch depth. In a recent study on the in situ processing of an InGaAs/GaAs quantum-well wafer, the Cl_2 gas etching of a 20-nm-thick InGaAs quantum-well layer on a GaAs substrate was interrupted at an etching depth of 10 nm, and laser interferometry was used to monitor the 10-nm etch depth within ±2 nm [70]. The experiment was carried out in an ultrahigh-vacuum-based multichamber system that had MBE, RIBE, and Auger electron spectrometer chambers. The measured reflected laser intensity curve coincided well with the curve calculated. The remainder of the 10-nm-thick InGaAs layer was confirmed using an in situ Auger electron spectrometer. Details on the experimental results are described elsewhere [70].

9.5 DEVICE APPLICATIONS

9.5.1 Recent Progress in Damage Reduction

The most probable cause of damage is the bombardment of plasma-induced ions on the etched surface. Due to this unfavorable effect, almost all applications to electronic or photonic devices have suffered from deterioration in device properties. A typical example is the instability of threshold voltage V_{th} in GaAs MESFET or GaAs/AlGaAs HEMT devices whose gate electrodes were deposited on the dry-etched surface [71]. This is attributable to etching-induced electron traps near the gate electrode generated by ion bombardment. In recent years, many papers have reported electronic-damage behavior from results produced using a variety of characterization methods, such as C–V and Hall measurements. The damage has been reduced significantly using a high-plasma-density and therefore low-self-bias-voltage (V_{dc}) etching machine. Since an ECR plasma typically provides low V_{dc}, recent studies on the Hall effect in a delta-doped GaAs/AlGaAs two-dimensional electron gas (2DEG) structure using the ECR plasma resulted in excellent sheet carrier concentration and mobility performance [71]. The detailed results are described later in this section.

Another unfavorable effect of dry-etching-induced damage is deteriorated PL intensities caused by nonradiative recombination centers generated in the optically active layers by an attack of ions or plasma-induced contamination [23,72–74]. Although such a nonradiative defect is presumed to seriously degrade the photonic device properties in general, the degree of damage depends on two factors. The first is the relation between the length (or width) of the etched region and the carrier diffusion length. A typical example is a laser diode with an active-layer mesa-stripe patterned by dry etching. In this case, damage on the mesa-etched sidewall surface of an embedded stripe laser does not give rise to a serious problem in the threshold current. This is because mesa-stripe widths are usually larger than the diffusion length of injected carriers (i.e., typically more than 1 μm), so that most of the carriers recombine in the bulk region in the active layer radiatively before they diffuse to the carrier traps at the etched surface to recombine nonradiatively [41]. As a result, etching of the structure does not cause deterioration of the photoluminescence intensity of the entire active region. Typical demonstrated examples are shown in the next section.

The second factor is related to materials. As compared with GaAs-based systems, PL characteristics of the InP-based system are known to be less sensitive to the process-induced damage. Yablonovitch and his co-workers investigated in detail the difference of surface recombination velocities (SRVs) between InP/$In_{0.53}Ga_{0.47}As$ and GaAs/AlGaAs heterojunctions regrown on surfaces prepared chemically using several methods [40]. They showed that the SRV value of the InP-based material is more than an order of magnitude lower than that of the GaAs-based material for each processed surface.

Although quantitative comparisons between SRV values on the dry-etched GaAs and InP surfaces have not been reported to date, some papers report that a dry etching-induced damage layer on the InP-based system is so thin that it can be removed by chemical treatment after etching. For example, Fig. 9.18 shows PL spectra of nondoped n-type LEC-grown InP before and after C_2H_6/H_2 RIE [95]. The main peak at 1.37 eV is the donor–acceptor (DA) pair emission due to the residual Zn acceptors DA(Zn). The intensity of this peak is hardly decreased by RIE for up to 20 min. The new emission line at 1.38 eV appears after RIE for more than 20 min. This peak is considered to be the DA pair emission due to C acceptors that

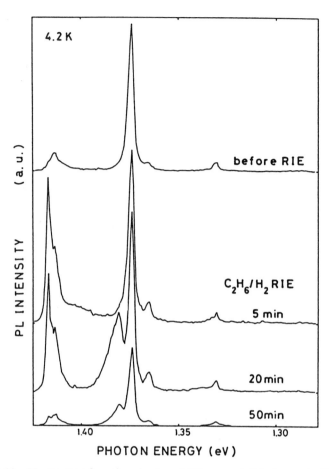

FIGURE 9.18 PL spectra of nondoped n-type LEC-grown InP before and after C_2H_6/H_2 RIE. Etching rate is 70 nm/min. (From Ref. 95.)

penetrated during RIE. Figure 9.19 shows the 1.37-eV emission intensity for various C_2H_6/H_2 RIE times as a function of a depth from the RIE-etched surface (i.e., the in-depth profile of $I_{DA(Zn)}$) [95]. The results show that the etching-induced damage seems to be restricted to the very surface region and can be removed by HF treatment. Along with the advancement in damage characterization on the etched surfaces, as discussed above, the application of dry etching to both photonic and electronic devices has progressed for several years. Some typical examples using both Cl_2-based RIBE and CH_4/C_2H_6-based RIE are reviewed in Sections 9.5.2 and 9.5.3.

9.5.2 Photonic Device Application

Figure 9.20 shows schematic cross-sectional views of photonic device structures that contain both devices with etched mesa stripes wider than 1 μm and devices un-

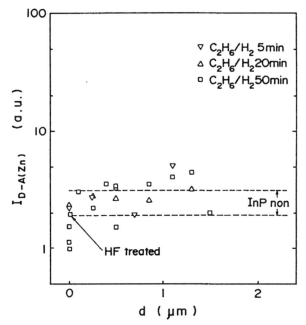

FIGURE 9.19 In-depth intensity profile of DA(Zn) emission peak (1.37 eV) for various C_2H_6/H_2 RIE times. The region between the dashed lines corresponds to PL intensity before the RIE. The square denoted by the arrow corresponds to a sample treated with HF after RIE. (From Ref. 95.)

der development with mesa stripes narrower than 1 μm. Each inset is arranged in order according to the minimum size of the mesa etching, as indicated by d. The insets shown at the upper level are structures whose active layers are etched off, and the insets at the lower level show structures in which etching is stopped just above the active layers. Sizes in the range 10 to 100 μm correspond to cavity lengths of the Fabry–Perot (FP) type etched mirror lasers [1–5,18,75], while sizes in the range 1 to 10 μm are the mesa-stripe widths of FP lasers and the air-post diameters of VC-SELs [20]. These devices have been confirmed to operate as well as the wet-etched lasers. Structures in the range 10 nm to 1 μm are under investigation and are promising for use in μ-cavities, photonic crystals, and quantum dots in the future. The experimental results of a distributed feedback (DFB)-LD, fabricated by a dry etching technique, are described below as a typical application in the range 100 nm in Fig. 9.20 [76].

Vertical and deep dry etching for fabricating mirror facets on an FP laser wafer, as one of the strongest motivations for the use of dry etching in photonic applications, has been demonstrated by many groups primarily using GaAs/AlGaAs materials, and excellent current-light output power characteristics similar to those for wet-etched lasers have been reported. These excellent results have also been confirmed by InP-based lasers. Figure 9.21 shows an SEM photograph and schematic view of a C_2H_6/H_2 RIE-ed facet on a GaInAsP/InP laser [48]. The total pressure of

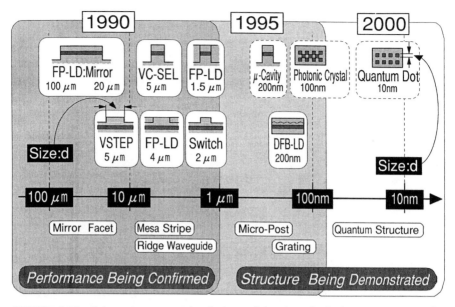

FIGURE 9.20 Schematic cross-sectional views of the photonic device structures containing both devices with mesa stripes wider than 1 μm and devices under development with mesa stripes narrower than 1 μm.

the mixed gas and the concentration of C_2H_6 were 13.3 Pa and 7 vol %, respectively. The total etched depth was 3.5 μm. In Fig. 9.22a, pulsed current-light output characteristics for (a) cleaved/cleaved, (b) etched/cleaved, and (c) etched/etched mirror configurations are shown, and typical lasing spectra of the laser with an etched/etched configuration are shown in Fig. 9.22b [48]. It was confirmed that these lasers have etched facets almost equivalent to the cleaved facets.

The precise size controllability of the dry etching feature without undercutting was put to practical use in the mesa-stripe etching of the lasers. Figure 9.23 shows a schematic cross-sectional view of a mesa-dry-etched GaInP/AlGaInP quantum-well visible laser [69]. Although this material is not categorized in the InP-based material, difficulty in dry etching is similar to that of the InP because of indium-containing materials, as discussed in Section 9.3.2. Etching was conducted in two steps; the first step was etching on a GaAs cap layer to a depth of about 300 nm at room temperature, and the second step was on a p-AlGaInP cladding layer to a depth of about 500 nm at 230°C. The stripe width was 4 μm. Etching depths were controlled by the laser monitoring technique discussed in Section 9.4. The threshold current of the dry-etched laser was smaller than that of the wet-etched laser. Figure 9.24a shows the current–light power characteristics of the dry-etched laser compared with those of the wet-etched laser shown in Fig. 9.24b. The threshold current of the dry-etched laser is lower than that of the wet-etched laser. It was observed from SEM images of the cross sections of mesa stripes that the mesa profile of the dry-etched stripe is symmetrical while that of the wet-etched stripe is asymmetrical, probably because

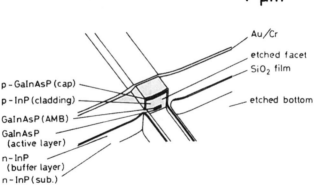

FIGURE 9.21 SEM photograph and schematic view of an RIE-etched facet. (From Ref. 48.)

of the use of a vicinal substrate. This difference in profiles may have resulted in different threshold currents, as shown in Fig. 9.24. Figure 9.25 shows the results of the aging test performed for dry- and wet-etched lasers, respectively, at an output power of 3 mW under an ambient temperature of 50°C. No significant change is observed in the operating current within the measured operating time (3300 h). Since the mesa dry etching is stopped just above the active layer, the etching-induced damage is considered not to reach the active region.

Figure 9.26 shows another example of RIBE application to photonic devices: microfabrication of air-post structures in InGaAs/GaAs VCSELs. In general, a VCSEL, whose cross-sectional shape in the horizontal plane is circular or square, leads to instability in the polarization direction of the emitted light because of a lack of in-plane unidirectionality in the light-emitting mechanism. Recently, however, sim-

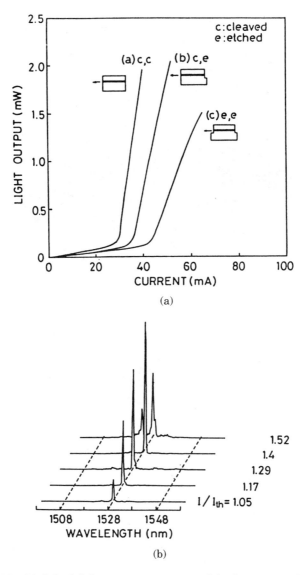

FIGURE 9.22 (a) Pulsed light output power versus injection current characteristics for cleaved–cleaved [curve (a)], etched–cleaved [curve (b)], and etched–etched [curve (c)] configurations; (b) typical lasing spectra of the laser with etched–etched configuration. (From Ref. 48.)

ply modified air-post structures were fabricated by two kinds of dry etching on a pair of parallel sidewalls and were found to contribute very effectively to the suppression of polarization instability [77]. The first air-post structure is an etching with a sawtoothed undulation on the vertical sidewall surface, shown in Fig. 9.26a. The second is an etching with a tilted profile, shown in Fig. 9.26b. As a result, con-

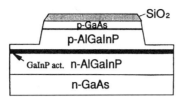

FIGURE 9.23 Schematic cross-sectional view of a mesa-dry-etched GaInP/AlGaInP quantum-well visible laser. (From Ref. 69.)

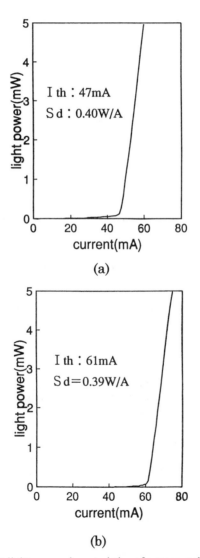

FIGURE 9.24 Current-light power characteristics of a mesa-etched GaInP/AlGaInP quantum-well visible laser: (*a*) dry-etched laser; (*b*) wet-etched laser. (From Ref. 69.)

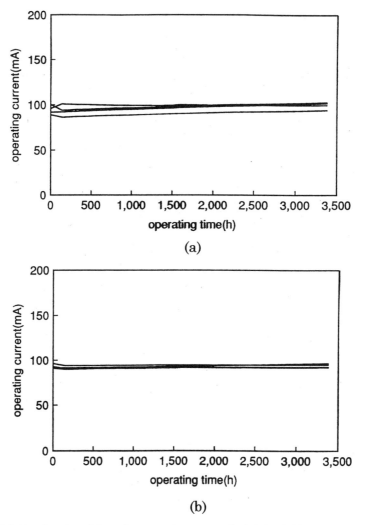

FIGURE 9.25 Results of the aging test of a mesa-etched GaInP/AlGaInP quantum-well visible laser: (*a*) dry-etched laser; (*b*) wet-etched laser. (From Ref. 69.)

trollability of the polarization direction in the range 30 to 50% for a conventional square column was well stabilized to 92%, as shown in the histogram in Fig. 9.27, where the sawtoothed undulation structure was adopted. Each bar in the histogram indicates a number of elements whose polarization directions show 0°, 90°, 45°, and switch. These structures are difficult to achieve by wet etching.

A high-aspect-ratio feature of dry etching can be used to prepare tall air-post structures, as shown in Fig. 9.28, where GaAs active layers are sandwiched by AlAs/AlGaAs-distributed Bragg reflector (DBR) layers to form a microcavity laser [96]. Although such a microcavity structure using the InP-based substrate has not

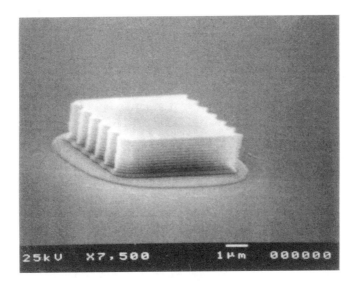

(a)

(b)

FIGURE 9.26 SEM photographs of air-post structures fabricated by RIBE: (*a*) etching with a sawtoothed undulation on the vertical sidewall surface; (*b*) etching with a tilted profile. (From Ref. 77.)

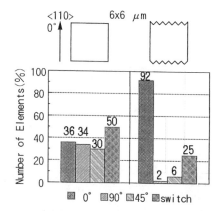

FIGURE 9.27 Histograms showing controllability of the polarization direction in dry-etched InGaAl/GaAs VCSELs compared for a conventional square column (left side) and a column having sawtoothed undulation on the vertical sidewall (right side). (From Ref. 77.)

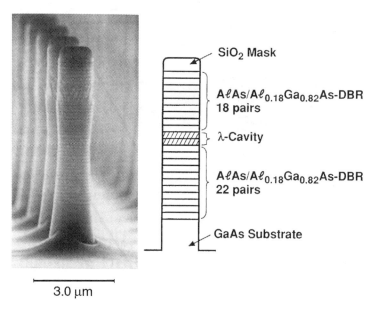

FIGURE 9.28 SEM photograph of pillar-type microcavities with diameters of about 1.3 μm defined by ECR RIBE and a schematic diagram showing the structure. The cavity region consists of GaAs SQW and $Al_{0.3}Ga_{0.7}As$ space layers. (From Ref. 96.)

been demonstrated up to now, a high-aspect-ratio characteristic of InP dry etching, shown in Fig. 9.5, will enable us to realize a similar result.

Most of the dry-etched sizes shown in Figs. 9.23 to 9.28 range from 1 to 5 μm, while the next example is etching with the pattern size in the submicrometer range. Figure 9.29 shows a schematic structure and band diagram of a novel InGaAsP/InP DFB laser [76]. Partially corrugated grating were fabricated using weighted-dose allocation variable-pitch electron beam (EB) lithography (termed WAVE) and Cl_2-plasma RIE. The gratings were etched on the InP cladding layer to a depth of 50 nm just about the InGaAsP MQW layer by heating the substrate to 200°C. Figure 9.30 shows SEM photographs of the cross sections for poly(methyl methacrylate) (PMMA) gratings before etching, dry-etched corrugation profiles on the substrate, and buried MQW substrate. Since the grating pitch was controlled precisely by this new WAVE method, precisely controlled multiple-wavelength λ/4-shifted DFB-LD arrays having excellent wavelength uniformity were achieved. These fine structures are difficult to fabricate in conventional wet etching and illustrate what dry etching can achieve.

The other advantages of dry etching are good uniformity with high precision over a wafer and reproducibility from wafer to wafer. However, when the etching mask is an insulating material such as SiO_2 and SiN_x, it is often a serious problem that there is positive charge buildup on the mask, which results in dielectric breakdown and ion deflection, leading to degradation of the verticality and uniformity of

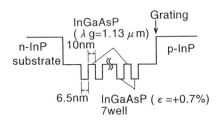

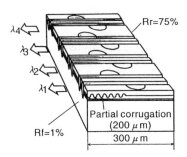

FIGURE 9.29 Schematic structure view and band diagram of a novel InGaAsP/InP DFB laser with partially corrugated gratings fabricated using weighted-dose allocation variable-pitch EB lithography (WAVE) and Cl_2-plasma RIE. (From Ref. 76.)

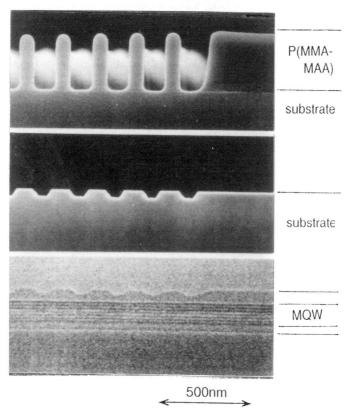

FIGURE 9.30 SEM photographs of cross sections of PMMA gratings before etching, dry-etched corrugation profiles on the substrate, and buried MQW substrate. (From Ref. 76.)

etching. To solve this problem, an electron shower was introduced in the etching chamber to neutralize the accumulated positive charge on the mask. Figure 9.31 shows the etched pattern on the InGaAsP-based vertical transmission optical amplifier (VTOA) wafer, indicating the effect of this electron shower [97]. Etching was performed by Cl_2-RIBE. In the case of etching without the electron shower (Fig. 9.31a), nonuniformity of the etching depth is observed as shown by the black pattern, while in the case of etching with an electron shower (Fig. 9.31b), no black stripe pattern is observed. Figure 9.32 shows a deep groove structure etched with an electron shower [97]. The 10-μm-deep groove with vertical sidewalls and smooth bottom surfaces was achieved by virtue of neutralization by the electron shower. This technique is important in practice from the full-wafer-technology point of view, not restricted to InP-based materials.

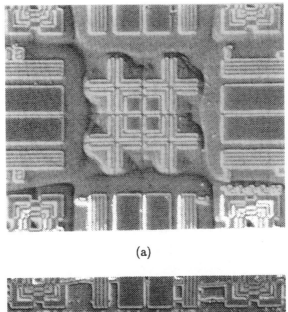

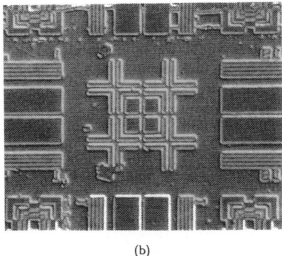

FIGURE 9.31 Effects of an electron shower: etching profile (*a*) without an electron shower and (*b*) with an electron shower. (From Ref. 97.)

9.5.3 Electronic Device Application

Dry etching is an indispensable technique for a practical production yield of fabricated short-channel MESFET and HEMT devices with gate widths of less than 0.25 μm. In this application, etching-induced damage, discussed in Section 9.5.1, has been improved considerably along with an improvement in surface smoothness over the past few years. Figures 9.33 to 9.35 show typical examples of a damage-re-

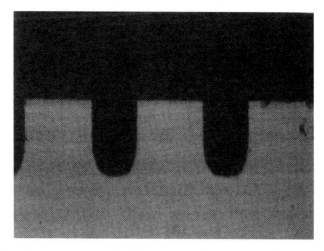

FIGURE 9.32 Groove structure etched by Cl_2 RIBE. The depth is about 10 μm. (From Ref. 97.)

duced, delta-doped 2DEG structure on a GaAs/AlGaAs wafer [71]. The layer structures with a delta-doped i-AlGaAs layer on an i-GaAs two-dimensional electron gas (2DEG) layer and a band diagram for this structure are depicted in Fig. 9.33. Figure 9.34 shows an SEM photograph of a cross section of a gate-recessed GaAs/AlGaAs wafer. The GaAs layer was recessed by selective dry etching using chlorine- and florine-mixed ECR plasma with a photoresist mask whose groove width was 800 nm. Etching was stopped at the top of the AlGaAs layer, due to the production of a

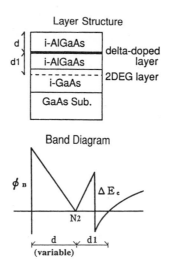

FIGURE 9.33 Layer structure with a delta-doped i-AlGaAs layer on an i-GaAs 2DEG layer and a band diagram for this structure. (From Ref. 71.)

nonvolatile aluminum fluoride on the AlGaAs surface. The etched surface is sufficiently smooth, as shown in Fig. 9.34. To study the effect of damage on the sheet carrier concentration and mobility, the position of the delta-doped layer was changed between 40 and 90 nm in the i-AlGaAs layer. Figure 9.35 shows the measured normalized sheet carrier concentration, N_{s1}/N_{s0}, and mobility, μ_1/μ_0, as a function of the depth of the delta-doped layer, where the microwave power was changed from 50 to 200 W. N_{s0} and μ_0 are the sheet carrier concentration and mobility, respectively, for the wet-etched sample; N_{s1} and μ_1 are those for the dry-etched sample. It was found that N_s and μ values for a microwave power of 50 W reveal almost no change in the range 40 to 90 nm. This favorable result was caused by the high plasma density specific to the ECR plasma (the machine was a "super-ECR" made by the Anelva Corporation) in which the self-bias voltage (V_{dc}) produced in the vicinity of the sample surface is considered to be as low as 10 V. Due to this low V_{dc} value specific to the ECR plasma, ion-bombardment-induced electron traps in the I-AlGaAs layer are rarely produced. Therefore, high-density ECR plasma etching is very promising for the fabrication of short-channel, high-speed devices.

9.6 ESTIMATION OF NONRADIATIVE SURFACE RECOMBINATION

As discussed in Section 9.5.1, it was suggested that although nonradiative recombination caused by dry-etching-induced damage generally degrades PL intensities [78–86], a degree of the damage depends on the size of the etched region. As will be shown later, for example, a dry-etched mesa stripe whose width is more than 1 μm

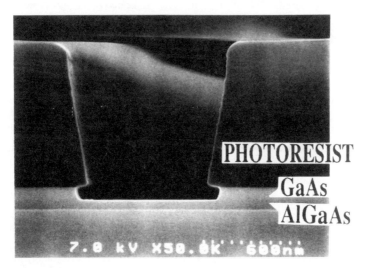

FIGURE 9.34 SEM photograph of a cross section of a gate-recessed GaAs/AlGaAs wafer. (From Ref. 71.)

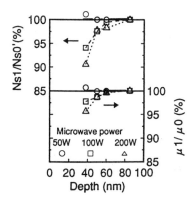

FIGURE 9.35 Measured normalized sheet carrier concentration N_{s1}/N_{s0} and mobility μ_1/μ_0 as a function of depth of the delta-doped layer for microwave power of 50 to 200 W. (From Ref. 71.)

does not significantly degrade the PL intensities. This is explained by the fact that the photo-excited or current-injected carriers in the wide area of the etched region mainly recombine radiatively in the middle bulk region before they diffuse to the carrier trap located at the etched-edge region to recombine nonradiatively. However, if the degree of the damage is quantitatively connected to material parameters such as nonradiative surface recombination velocity S_r and carrier lifetime τ, the device performance can be estimated more clearly in designing the device. In the following section, therefore, the surface recombination velocity S_r and carrier lifetime τ will be determined as a function of the size of the dry-etched portion by using time-resolved PL measurement. Analytical results to be discussed later can be applied to all the photonic compound semiconductors, irrespective of differences between GaAs- and InP-based alloy compositions.

9.6.1 Time-Resolved Photoluminescence

The carrier lifetime τ can be derived from time-resolved PL measurement in which photo carriers are pumped by a short optical pulse, and the decay time of PL intensities caused by the relaxation of the pulse-excited carriers is measured. Such carriers generally decay by the intrinsic lifetime τ_{int} specific to the bulk region if no etching is done [42]. However, if the photo-pumped region has been etched, the excited carriers nonradiatively decay at carrier traps generated by etching-induced damage faster than at the bulk region [41]. Such a carrier lifetime, τ_{eff}, which contains both an intrinsic term, τ_{int}, and a carrier trap–dependent term, R/S_r, where R is the width of the dry-etched region, has been analyzed in detail by our group by considering the carrier diffusion length D_r. For detailed procedures, see Ref. 41. The τ_{eff} versus R curve calculated is shown in Fig. 9.36. When the R value is large enough, as shown by region (a), the carrier decay is determined by the intrinsic lifetime, τ_{int}. When the R value is small enough, on the other hand, as shown by region (c), the carrier decay is limited by the surface recombination; that is, τ_{eff} is proportional to

9.6 ESTIMATION OF NONRADIATIVE SURFACE RECOMBINATION

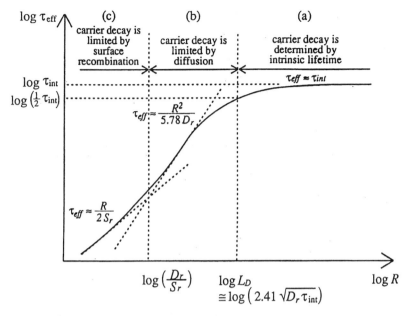

FIGURE 9.36 Calculated τ_{eff} versus R curve where, τ_{eff}, is the carrier lifetime and R is the width of the dry-etched region. (From Ref. 41.)

R/S_r. In the middle region, (b), the carrier decay is accelerated with decreasing R due to the surface recombination. However, the degree of decay is influenced by the diffusion. According to the curve shown here, τ_{eff} at the smaller value of R in region (c), is proportional to R, so that the parameter S_r is derived from the slope of the τ_{eff} versus R curve in region (c). Since the parameters τ_{eff} and R are given by the measurement, S_r can be determined in this manner. Furthermore, the value of R given approximately by the term $2.41 \sqrt{D_r \tau_{int}}$, which divides regions (a) and (b), provides us with a turning point where the dry etching will start to influence the degree of damage. Measured τ_{eff} versus R curves for several kinds of processed samples are shown in the next section.

9.6.2 In Situ Characterization of Etching/Regrowth Interface

As mentioned above, the degree of damage and contamination that appear at semiconductor surfaces depends strongly on the kinds of processing techniques used, such as growth interruption or etching. In this section, our recent study on the quantitative relationships between the carrier lifetime, surface recombination velocity, and structured size for GaAs/AlGaAs quantum-well samples is reviewed. Different degrees of lifetime deterioration are compared for different processing methods, such as growth interruption, air-exposure, plasma etching and gas etching, and for different sample structures, such as the mesa-etched stripe and flat quantum-well surface of GaAs substrates.

Figure 9.37 shows a schematic top view of the UHV-based multiple-chamber system used for the experiment [33]. The system is composed of MBE, RIBE, EB, Auger electron spectroscopy, and load-lock chambers. The background pressure of most chambers except for the RIBE and load-lock chambers was less than 5×10^{-10} torr. The vacuum of the RIBE chamber where plasma and gas etching were carried out was 2×10^{-8} torr. All the chambers except for the EB chamber were used in the experiment.

Figure 9.38 shows several sample-preparation methods [98]: Fig. 9.38a shows a mesa-striped structure that was patterned on the resist film by the conventional optical lithography and was RIBE-ed with the resist mask in the RIBE chamber. Sample 1 is RIBE-ed and sample 2 is wet etched to a thickness of about 200 nm on the sidewalls and bottom surfaces after RIBE [41]. Figure 9.38b shows a buried mesa structure that was RIE-ed with a resist mask and was accompanied by gas etching and regrowth in the RIBE and MBE chambers. The resist mask was removed before introducing the sample into the UHV chamber. The resulting sample is sample 3 [87]. Since these samples are mesa striped, the damage to be measured exists on the sidewall of the stripe. In Fig. 9.38c, on the other hand, a GaAs quantum well was MBE grown on the AlGaAs layer and the top surface of the layer was processed in the UHV chamber by several methods, as follows: Sample 4 was Cl_2 gas etched in the RIBE chamber and an AlGaAs cladding layer was regrown in the MBE chamber [43]. Sample 5 was growth interrupted and was held in the UHV tunnel for 15 h. After that, an AlGaAs layer was regrown in the MBE chamber [42]. Sample 6 was exposed to the air for 15 min and an AlGaAs layer was regrown in the MBE chamber [42]. Sample 7 is a continuously grown control sample for reference.

Figure 9.39 shows the measured results of relationships between carrier lifetime τ and pattern size d for the various samples shown in Fig. 9.38 [41–43,87,98]. The size d means a mesa width for mesa structures (i.e., samples 1, 2, and 3) and the

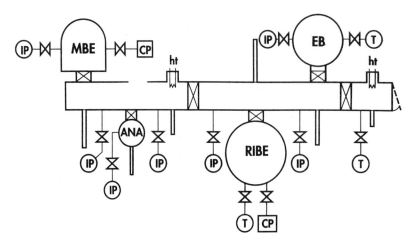

FIGURE 9.37 Schematic top view of the UHV-based multiple-chamber system used in the experiment. (From Ref. 33.)

9.6 ESTIMATION OF NONRADIATIVE SURFACE RECOMBINATION   329

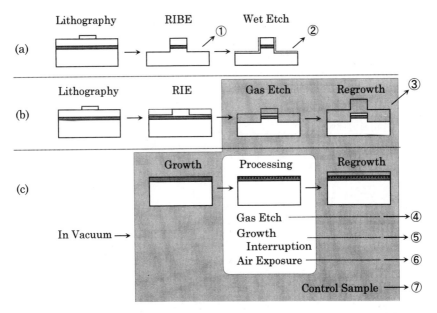

FIGURE 9.38 Several sample-preparation methods that contain lithography and dry mesa etching in an ex situ process and gas etching and MBE regrowth in an in situ process. (From Ref. 98.)

quantum-well thickness for plane heterostructures (samples 4, 5, 6, and 7). Each solid curve is a calculated relationship between τ versus d for the different combinations of τ_{int} and S_r shown in Fig. 9.36. Here the effect of the carrier diffusion is ignored, so that region (b) in Fig. 9.36 does not exist in Fig. 9.39. This approximation is reasonable when we view the change in carrier lifetime τ for the wide range of pattern size d. As a result, the S_r value for a continuously grown heterostructure (sample 7) is 1×10^2 cm/s, while S_r values for growth-interrupted (sample 5) and air-exposed (sample 6) structures are degraded to 7×10^2 and 3×10^3 cm/s, respectively. The S_r value for a gas-etched structure (sample 4), on the other hand, is 1×10^3 cm/s, between those for the growth-interrupted and air-exposed structures. This result is the lowest for gas-etched structures reported to date. However, S_r values for mesa-etched structures are further degraded in the range from 6×10^3 cm/s (sample 3) to 1×10^5 cm/s (samples 1 and 2).

Although samples 3 and 4 are both gas etched at the in situ steps, the S_r value for mesa structure 3 is larger than that for plane heterostructure 4. This is probably because the sidewall surface of the mesa structure is more roughened and contaminated than the interface of the plane heterostructure, due to air exposure after lithography at the ex situ step. Furthermore, the S_r value for RIBE-ed sample 1 is increased significantly to 1.5×10^5 cm/s compared with gas-etched sample 3. This is mainly because the sidewall surface damage, contamination, and probably surface roughness are more serious for the RIBE than for the gas-etched sample. In sample 2, on

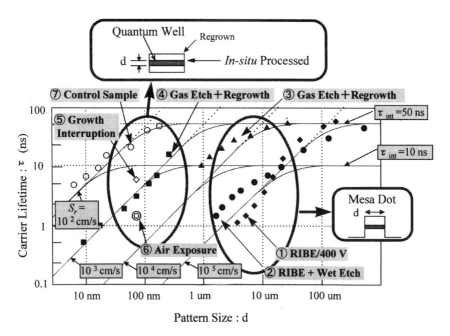

FIGURE 9.39 Measured results of relationships between carrier lifetime τ and pattern size d for various samples shown in Fig. 9.38. (From Ref. 98.)

the other hand, it is considered that the RIBE-induced damage layer has been almost removed by the wet etching to a thickness of about 200 to 400 nm on the mesa-structure surface. Nevertheless, the S_r value for sample 2 has not been improved significantly compared to that of sample 1, and it is still an order of magnitude larger than that for sample 3. This implies that contamination due to air exposure cannot be ignored, in addition to the effect of the mesa-etched surface being larger for the RIBE than for the gas etching.

In the RIBE-ed samples 1 and 2, the carrier lifetime τ_{eff} is decreased from 50 ns to almost 1 ns, along with the pattern size decrease to 1 μm. However, it has been reported that mesa-etched InGaAs/GaAs VCSELs with 5-μm-diameter air-post structures and GaInP/AlGaInP visible lasers with 4-μm-wide mesa stripes fabricated by RIBE exhibit no significant degradation in device performance, such as the threshold current, compared with those of wet-etched devices. This means that a decrease in the carrier lifetime down to nanosecond order caused by the dry etching, such as the RIBE and RIE described in this chapter, is likely to be ignored for photonic device performance. In other words, the degree of the effect of dry-etching-induced damage generally depends on pattern size, as discussed in Section 9.5.1. The experimental analyses shown above are based on 0.8-μm bandgap GaAs/AlGaAs heterojunctions. However, the analytical results discussed here can be applied to all photonic compound semiconductors.

9.7 SUMMARY

Cl_2- and CH_4/C_2H_6-based dry etching of III-V compound semiconductors for optoelectronic applications has been reviewed with a focus on indium-containing substrate materials such as InP. Plasma chemistry for the indium-containing material was first discussed compared with that for GaAs-based materials. A CH_4/C_2H_6-based plasma etches both GaAs- and InP-based substrates smoothly at room temperature. However, an excess supply of such a gas roughens the etched surface. On the other hand, a Cl_2-based plasma etches GaAs substrates at room temperature, while it requires heating the substrate for InP-based materials. Based on the plasma chemistry, recent experiments on smooth and vertical etching of InP-based system were discussed for Cl_2-based plasma etching, CH_4/C_2H_6-based plasma etching, and laser-assisted selective gas etching.

A variety of dry etching techniques have recently been applied to the InP-based photonic devices. The results exhibited excellent device performance for mesa-etched structures with an etched size of greater than 1 μm. For example, a GaInAsP/InP etched-mirror laser showed current versus light-power characteristics equivalent to those of a cleaved-mirror laser. InGaAs/GaAs VCSELS with about 5-μm diameters and sawtoothed-shaped air-post structures presented extremely low threshold currents and highly polarization-controllable characteristics. A buried dry-etched mesa-stripe GaInP/AlGaInP laser exhibited a lifetime greater than 3000 h.

The degree of dry-etching-induced damage was evaluated quantitatively in terms of the nonradiative surface recombination velocity S_r. The value of S_r for a Cl_2-gas-etched GaAs surface was as small as 1×10^3 cm/s when evaluated after the overgrowth of an AlGaAs layer. This value is equivalent to that for a growth-interrupted GaAs/AlGaAs interface and comes close to the $S_r = 1 \times 10^2$ cm/s for a continuously grown GaAs/AlGaAs interface. The result suggests that the gas-etching technique has the potential of fabricating nanostructures such as quantum dots by further reducing the value of S_r by an order of magnitude. Such an evaluation technique was emphasized as applicable to all photonic compound semiconductors, irrespective of differences between GaAs- and InP-based alloy compositions.

In conclusion, recent results on dry etching of III-V compound semiconductors, including InP-based materials, are encouraging and offer the promise of two possibilities: establishing a uniform, reproducible, and reliable full-wafer-microprocessing technique at the mass-production level, and realizing quantum effect devices having structures in the range from submicrometer to sub-100 nm with no associated degradation in electronic or optoelectronic properties.

REFERENCES

1. L. A. Coldren, K. Iga, B. I. Miller, and J. A. Rentschler, "GaInAsP/InP stripe-geometry laser with a reactive-ion-etched facet," *Appl. Phys. Lett.*, **37**, 681–683 (1980).
2. L. A. Coldren, B. I. Miller, K. Iga, and J. A. Rentschler, "Monolithic two-section

GaInAsP/InP active-optical-resonator devices formed by reactive ion etching," *Appl. Phys. Lett.*, **38**, 315–317 (1981).

3. L. A. Coldren and J. A. Rentschler, "Directional reactive-ion-etching of InP with Cl_2-containing gases," *J. Vac. Sci. Technol.*, **19**, 225–230 (1981).

4. O. Mikami, H. Akiya, T. Saitoh, and H. Nakagome, "CW operation of 1.5 μm GaInAsP/InP buried-heterostructure laser with a reactive-ion-etched facet," *Electron. Lett.*, **19**, 213–214 (1983).

5. H. Saitoh, Y. Noguchi, and H. Nagai, "CW operation of 1.5 μm InGaAsP/InP BH lasers with a reactive-ion-etched facet," *Electron. Lett.* **21**, 748–749 (1985).

6. N. Bouadma, J. F. Hogrel, J. Charil, and M. Carre, "Fabrication and characterization of ion beam etched cavity InP/InGaAsP BH lasers," *IEEE J. Quantum Electron.*, **23**, 909–914 (1987).

7. E. M. Clausen, Jr., G. Craighead, M. C. Tamargo, and J. L. De Moiguel, "Etching and cathodeluminescence studies of ZnSe," *Appl. Phys. Lett.*, **53**, 690–691 (1988).

8. T. Saitoh, T. Yokogawa, and T. Narusawa, "Reactive ion etching of ZnSe and ZnS epitaxial films using Cl_2 electron cyclotron resonance plasma," *Appl. Phys. Lett.*, **56**, 839–841 (1990).

9. S. J. Pearton, J. W. Lee, J. D. McKenzie, and C. R. Abernathy, "Dry etching damage in InN, InGaN, and InAlN," *Appl. Phys. Lett.*, **67**, 2329–2331 (1995).

10. I. Adesida, A. Mahajan, E. Andideh, M. A. Khan, D. T. Olsen, and J. N. Kuznia, "Reactive ion etching of gallium nitride in silicon tetrachloride plasma," *Appl. Phys. Lett.*, **63**, 2777–2779 (1993).

11. E. L. Hu and R. E. Howard, "Reactive-ion etching of GaAs and InP using $CCl_2F_2/Ar/O_2$," *Appl. Phys. Lett.*, **37**, 1022–1024 (1980).

12. R. E. Klinger and J. E. Green, "Reactive ion etching of GaAs in CCl_2F_2," *Appl. Phys. Lett.*, **38**, 620–622 (1981).

13. R. H. Burton and G. Smolinsky, "CCl_4 and Cl_2 plasma etching of III-V semiconductors and the role of added O_2", *J. Electrochem. Soc.*, **129**, 1599–1604 (1982).

14. S. Matsuo and Y. Adachi, "Reactive ion beam etching using a broad beam ECR ion source," *Jpn. J. Appl. Phys.*, **21**, L4–L6 (1982).

15. K. Asakawa and S. Sugata, "GaAs and AlGaAs equi-rate etching using a new reactive ion beam etching system," *Jpn. J. Appl. Phys.*, **22**, L653–L655 (1983).

16. K. Asakawa and S. Sugata, "GaAs and AlGaAs anisotropic fine pattern etching using a new reactive ion beam etching system," *J. Vac. Sci. Technol. B*, **3**, 402–405 (1985).

17. R. Müller, "In situ depth monitoring for reactive ion etching of InGaAs(P)/InP heterostructures by ellipsometry," *Appl. Phys. Lett.*, **57**, 1020–1021 (1990).

18. T. Yuasa, M. Mannou, K. Asakawa, K. Shinozaki, and M. Ishii, "Dry-etched-cavity pair-groove-substrate GaAs/AlGaAs multiquantum well lasers," *Appl. Phys. Lett.*, **48**, 748–750 (1986).

19. T. Yuasa, T. Yamada, K. Asakawa, S. Sugata, M. Ishii, and M. Uchida, "Short cavity GaAs/AlGaAs multiquantum well lasers by dry etching," *Appl. Phys. Lett.*, **49**, 1007–1009 (1986).

20. T. Numai, T. Kawakami, T. Yoshikawa, M. Sugimoto, Y. Sugimoto, H. Yokoyama, K. Kasahara, and K. Asakawa, "Record low threshold current in microcavity surface-emitting laser," *Jpn. J. Appl. Phys.*, **32**, L1533–L1534 (1993).

21. S. W. Pang, "Dry etching induced damage in Si and GaAs," *Solid State Technol,* April, 249–256 (1984).
22. S. Sugata and K. Asakawa, "Characterization of damage on GaAs in a reactive ion beam etching system using Schottky diodes," *J. Vac. Sci. Technol. B,* **6,** 876–879 (1988).
23. H. F. Wong, D. L. Green, T. Y. Liu, D. G. Lishan, M. Bellis, E. L. Hu, P. M. Petroff, P. O. Holtz, and J. L. Merz, "Investigation of reactive ion etching induced damage in GaAs–AlGaAs quantum well structures," *J. Vac. Sci. Technol. B,* **6,** 1906–1910 (1988).
24. Y. Ide, S. Kohmoto, and K. Asakawa, "Electrical characterization of very-low energy (0–30 eV) Cl-radical/ion-beam-etching induced damage using two-dimensional electron gas heterostructures," *J. Electron. Mater.,* **21,** 3–7 (1992).
25. Y. L. Wang, H. Temkin, L. R. Harriot, R. A. Logan, and T. Tanbun-Ek, "Buried-heterostructure laser fabricated by in situ processing techniques," *Appl. Phys. Lett.,* **57,** 1864–1866 (1990).
26. Y. Sugimoto, H. Kawanishi, and K. Akita, "Electron-beam-induced modification of GaAs oxide for in situ patterning of GaAs by Cl_2 gas etching," *Semicond. Sci. Technol.,* **7,** 160–163 (1992).
27. K. Asakawa, "GaAs and AlGaAs surface cleaning using an ECR radical beam gun," *Extended Abstr. 18th (1986 Int.) Conf. Solid State Devices and Materials,* Tokyo, 1986, pp. 129–132.
28. K. Asakawa and S. Sugata, "Damage and contamination-free GaAs and AlGaAs etching using a novel ultrahigh-vacuum reactive ion beam etching system with etched surface monitoring and cleaning method," *J. Vac. Sci. Technol. A,* **4,** 677–680 (1986).
29. S. Sugata and K. Asakawa, "GaAs cleaning with a hydrogen radical beam gun in an ultrahigh-vacuum system," *J. Vac. Sci. Technol. B,* **6,** 1087–1091 (1988).
30. S. Iwata and K. Asakawa, "Direct observation of GaAs surface cleaning process under hydrogen radical beam irradiation," *AVS Conf. Proc. 227 of 2nd Int. Conf. Advanced Processing and Characterization Technologies,* Fla., 1991, pp. 122–125.
31. Y. Ide and M. Yamada, "Role of Ga_2O in the removal of GaAs surface oxides induced by atomic hydrogen," *J. Vac. Sci. Technol. A,* **12,** 1858–1863 (1994).
32. N. Tanaka, I. Matsuyama, M. Lopez, and T. Ishikawa, "Photoluminescence study of GaAs/AlGaAs quantum wells regrown on in-situ Cl_2-etched GaAs buffer layers," *Extended Abstr. 1993 Int. Conf. Solid State Devices and Materials,* Makuhari, Japan, 1993, pp. 760–762.
33. S. Kohmoto, N. Takado, Y. Sugimoto, M. Ozaki, M. Sugimoto, and K. Asakawa, "In situ electron beam patterning for GaAs using electron-cyclotron-resonance plasma-formed oxide mask and Cl_2 gas etching," *Appl. Phys. Lett.,* **61,** 444–446 (1992).
34. T. Ishikawa, N. Tanaka, M. Lopez, and I. Matsuyama, "Nanometer-scale pattern formation of GaAs by in situ electron-beam lithography using surface oxide layer as a resist film," *J. Vac. Sci. Technol. B,* **13,** 2777–2780 (1995).
35. T. Tadokoro, F. Koyama, and K. Iga, "A study on etching parameters of a reactive ion beam etch for GaAs and InP," *Jpn. J. Appl. Phys.,* **27,** 389–392 (1988).
36. R. A. Barker, T. M. Mayer, and R. H. Burton, "Surface composition and etching of III-V semiconductors in Cl_2 ion beams," *Appl. Phys. Lett.,* **40,** 583–586 (1982).
37. T. Matsui, K. Ohtsuka, H. Sugimoto, Y. Abe, and T. Ohishi, "1.5 μm GaInAsP/InP buried-heterostructure laser diode fabricated by reactive ion etching using a mixture of ethane and hydrogen," *Appl. Phys. Lett.,* **56,** 1641–1642 (1990).

38. T. R. Hayes, M. A. Dreibach, P. M. Thomas, and W. C. Dautremont-Smith, "Reactive ion etching of InP using CH_4/H_2 mixtures: mechanisms of etching and anisotropy," *J. Vac. Sci. Technol. B*, **7**, 1130–1140 (1989).
39. T. Yoshikawa, S. Kohmoto, M. Ozaki, N. Hamao, Y. Sugimoto, M. Sugimoto, and K. Asakawa, "Smooth and vertical InP reactive ion beam etching with Cl_2 ECR plasma," *Jpn. J. Appl. Phys.*, **31**, L655–L657 (1992).
40. E. Yablonovitch, R. Bhat, C. E. Zah, T. J. Gmitter, and M. A. Koza, "Nearly ideal InP/$In_{0.53}Ga_{0.47}As$ heterojunction regrowth on chemically prepared $In_{0.53}Ga_{0.47}As$ surfaces," *Appl. Phys. Lett.*, **60**, 371–373 (1992).
41. Y. Nambu, H. Yokoyama, T. Yoshikawa, Y. Sugimoto, and K. Asakawa, "Time-resolved investigation of sidewall recombination and in-plane diffusivity in dry-etched InGaAs/GaAs air-post structures," *Appl. Phys. Lett.*, **65**, 481–483 (1994).
42. N. Hamao, Y. Sugimoto, M. Sugimoto, and H. Yokoyama, "Reduction in interfacial recombination velocity using in situ processing in AlGaAs/GaAs overgrown heterointerfaces," *Inst. Phys. Conf. Ser.*, No. 129, 609–614 (1992).
43. S. Kohmoto, Y. Nambu, K. Asakawa, and T. Ishikawa, "Reduced nonradiative recombination in etched/regrown AlGaAs/GaAs structure fabricated by in situ processing," *J. Vac. Sci. Technol. B*, **14**, 3646–3649 (1996).
44. K. Asakawa, N. Takadoh, M. Uchida, and T. Yuasa, "GaAs reactive ion beam etching and surface cleaning using enclosed ultrahigh-vacuum processing system," *NEC Res. Dev.*, **87**, 1–13 (1987).
45. Y. Sugimoto, T. Yoshikawa, N. Takado, S. Kohmoto, N. Hamao, M. Ozaki, M. Sugimoto, and K. Asakawa, "Dry etching process for GaAs and InP based devices," *NEC Res. Dev.*, **33**, 469–480 (1992).
46. J. W. Coburn and H. F. Winters, "Ion- and electron-assisted gas-surface chemistry: an important effect in plasma etching," *J. Appl. Phys.*, **50**, 3189–3196, (1979).
47. U. Niggebrügge, M. Klug, and G. Garus, "A novel process for reactive ion etching on InP using CH_4/H_2," *Inst. Phys. Conf. Ser.*, No. 79, 367–372 (1985).
48. T. Matsui, H. Sugimoto, T. Ohishi, Y. Abe, K. Ohtsuka, and H. Ogata, "GaInAsP/InP lasers with etched mirrors by reactive ion etching using a mixture of ethane and hydrogen," *Appl. Phys. Lett.*, **54**, 1193–1194 (1989).
49. V. M. Donnelly, D. L. Flamm, C. W. Tu, and D. E. Ibbotson, "Temperature dependence of InP and GaAs etching in a chlorine plasma," *J. Electrochem. Soc.*, **129**, 2533–2537 (1982).
50. G. Smolinsky, R. P. Chang, and T. M. Mayer, "Plasma etching of III–V compound semiconductor materials and their oxides," *J. Vac. Sci. Technol.*, **18**, 12–16 (1981).
51. H. Tamura and H. Kurihara, "GaAs and AlGaAs reactive ion etching in BCl_3–Cl_2 mixture," *Jpn. J. Appl. Phys.*, **23**, L731–L733 (1984).
52. R. A. Powell, "Reactive ion beam etching of GaAs in CCl_4", *Jpn. J. Appl. Phys.*, **21**, L170–L172 (1982).
53. M. W. Geis, G. A. Lincoln, N. Efremow, and W. J. Piacentini, "A novel anisotropic dry etching technique," *J. Vac. Sci. Technol.*, **19**, 1390–1393 (1981).
54. R. Cheung, S. Thomas, I. McIntyre, C. D. W. Wilkinson, and S. P. Beaumont, "Passivation of donors in electron beam lithographically defined nanostructures after methane/hydrogen reactive ion etching," *J. Vac. Sci. Technol. B*, **6**, 1911–1915 (1988).
55. S. J. Pearton, W. S. Hobson, F. A. Baiocchi, A. B. Emerson, and K. S. Jones, "Reactive

ion etching of InP, InGaAs, InAlAs: comparison of C_2H_6/H_2 with CCl_2F_2/O_2," *J. Vac. Sci. Technol. B,* **8,** 57–67 (1990).
56. N. Furuhata, H. Miyamoto, A. Okamoto, and K. Ohata, "Cl_2 chemical dry etching of GaAs under high vacuum conditions: crystallographic etching and its mechanism," *J. Electron. Mater.,* **19,** 201–208 (1989).
57. S. Kohmoto, Y. Ide, Y. Sugimoto, and K. Asakawa, "GaAs surface cleaning/etching using plasma-dissociated Cl radical," *Jpn. J. Appl. Phys.,* **32,** 5796–5800 (1993).
58. R. P. H. Chang and S. Darack, "Hydrogen plasma etching of GaAs oxide," *Appl. Phys. Lett.,* **38,** 898–900 (1981).
59. S. Semura and T. Hattori, *Dig. 51st Japan Applied Physics Society Meeting,* Vol. 3, 1990, p. 1150 (in Japanese).
60. R. H. Burton, H. Temkin, and V. G. Keramidas, "Plasma separation of InGaAsP/InP light-emitting diodes," *Appl. Phys. Lett.,* **37,** 411–412 (1980).
61. L. A. Coldren, K. Furuya, B. I. Miller, and J. A. Rentschler, "Etched mirror and groove-coupled GaInAsP/InP laser devices for integrated optics," *IEEE J. Quantum Electron.,* **18,** 1679–1688 (1982).
62. M. E. Lin, Z. Fan, Z. Ma, L. H. Allen, and H. Morkoç, "Reactive ion etching of GaN using BCl_3," *Appl. Phys. Lett.,* **64,** 887–888 (1994).
63. S. J. Pearton, C. R. Abernathy, and F. Ren, "Low bias electron cyclotron resonance plasma etching of GaN, AlN, and InN," *Appl. Phys. Lett.,* **64,** 2294–2296 (1994).
64. I. Adesida, A. T. Ping, C. Youtsey, T. Dow, M. A. Khan, D. T. Olsen, and J. N. Kuznia, "Characteristics of chemically assisted ion beam etching of gallium nitride," *Appl. Phys. Lett.,* **65,** 889–891 (1994).
65. S. J. Pearton, C. R. Abernathy, F. Ren, J. R. Lothian, P. W. Wisk, and A. Katz, "Dry and wet etching characteristics of InN, AlN, and GaN deposited by electron cyclotron resonance metalorganic molecular beam epitaxy," *J. Vac. Sci. Technol. A,* **11,** 1772–1775 (1993).
66. H. Sugimoto, T. Isu, H. Tada, T. Miura, T. Shiba, T. Kimura, and A. Takemoto, "Suppression of side-etching in $C_2H_6/H_2/O_2$ reactive ion etching for the fabrication of an InGaAsP/InP P-substrate buried-heterostructure laser diode," *J. Electrochem. Soc.,* **140,** 3615–3620 (1993).
67. K. Hikosaka, T. Mimura, and K. Joshi, "Selective dry etching of AlGaAs–GaAs heterojunction," *Jpn. J. Appl. Phys.,* **20,** L847–L850 (1981).
68. T. Yoshikawa, S. Kohmoto, M. Anan, N. Hamao, M. Baba, N. Takado, Y. Sugimoto, M. Sugimoto, and K. Asakawa, "Chlorine-based smooth reactive ion beam etching of indium-containing III-V compound semiconductor," *Jpn. J. Appl. Phys.,* **31,** 4381–4386 (1992).
69. T. Yoshikawa, Y. Sugimoto, H. Hotta, K. Tada, H. Kobayashi, H. Yoshii, H. Kawano, S. Kohmoto, and K. Asakawa, "GaInP/AlGaInP index-waveguide-type visible laser diodes with dry-etched mesa stripes," *Electron. Lett.,* **29,** 1690–1691 (1993).
70. T. Usui, S. Kohmoto, and K. Asakawa, submitted.
71. H. Oikawa, M. Kohno, and A. Mochizuki, "High selectivity and low damage dry etching of GaAs to AlGaAs using BCl_3/SF_6 gas in ECR plasma," *Report of Research Center of Ion Beam Technology,* Hosei University (*Proc. 13th Symp. Materials Science and Engineering,* Research Center of Ion Beam Technology, Hosei University), Suppl. 13, 1995, pp. 145–150.

72. M. Kawabe, N. Kanzaki, K. Masuda, and S. Namba, "Effects of ion etching on the properties of GaAs," *Appl. Opt.,* **17,** 2556–2561 (1978).
73. H. Yamada, H. Ito, and H. Inaba, "Reactive ion etching of GaAs and AlGaAs for integrated optics," *Proc. 5th Symp. Dry Processing,* Tokyo, 1984, pp. 73–78.
74. Y. Ide, N. Takado, and K. Asakawa, "Optical characterization of reactive ion beam etching induced damage using GaAs/AlGaAs quantum well structures," *Inst. Phys. Conf. Ser.,* No. 106, 495–500 (1989).
75. M. Uchida, S. Ishikawa, N. Takado, and K. Asakawa, "An AlGaAs laser with high-quality dry etched mirrors fabricated using an ultrahigh vacuum in situ dry etching and deposition processing system," *IEEE J. Quantum Electron.,* **24,** 2170–2177 (1988).
76. Y. Muroya, T. Nakamura, H. Yamada, and T. Torikai, "Precise wavelength control for DFB laser diodes by novel corrugation delineation method," *IEEE Photon. Technol. Lett.,* **9,** 288–290 (1997).
77. T. Yoshikawa, H. Kosaka, K. Kurihara, M. Kajita, Y. Sugimoto, and K. Kasahara, "Complete polarization control of 8 × 8 vertical-cavity surface-emitting laser matrix arrays," *Appl. Phys. Lett.,* **66,** 908–910 (1995).
78. J. P. Harbison, A. Scherer, D. M. Hwang, L. Nazar, and E. D. Beebe, "MBE regrowth of AlGaAs on ion etched GaAs/AlGaAs microstructures," *Mater. Res. Soc. Symp. Proc.,* **126,** 11–16 (1988).
79. B. E. Maile, A. Forchel, R. Germann, J. Straka, L. Korte, and C. Thanner, "Lateral quantization induced emission energy shift of buried GaAs/AlGaAs quantum wires," *App. Phys. Lett.,* **57,** 807–809 (1990).
80. A. Izrael, J. Y. Marzin, B. Sermage, L. Birotheau, D. Robein, R. Azouley, J. L. Benchimol, L. Henry, V. T. Mieg, F. R. Ladan, and L. Taylor, "Fabrication and luminescence of narrow reactive ion etched $In_{1-x}Ga_xAs/InP$ and $GaAs/Ga_{1-x}Al_xAs$ quantum wires," *Jpn. J. Appl. Phys.,* **30,** 3257–3260 (1991).
81. R. Bergmann, A. Menchig, G. Lehr, P. Kubler, J. Hommel, R. Rudeloff, B. Henle, F. Scholz, and H. Schweizer, "Improvement of the quality of InGaAs/InP quantum wires due to epitaxial overgrowth," *J. Vac. Sci. Technol. B,* **10,** 2893–2895 (1992).
82. E. M. Clausen, Jr., J. P. Harbison, L. T. Florez and B. Van de Gaag, "Assessing thermal Cl_2 etching and regrowth as methods for surface passivation," *J. Vac. Sci. Technol. B,* **8,** 1960–1965 (1990).
83. Y. Kadoya, H. Noge, H. Kano, H. Sakaki, N. Ikoma, and N. Nishiyama, "Molecular-beam-epitaxial growth of n-AlGaAs on clean Cl_2-gas etched GaAs surfaces and the formation of high mobility two-dimensional electron gas at the etch-regrown interfaces," *Appl. Phys. Lett.,* **61,** 1658–1660 (1992).
84. M. Hong, R. S. Freund, K. D. Choquette, H. S. Luftman, J. P. Mannaerts, and R. C. Wetzel, "Removal of GaAs surface contaminants using H_2 electron cyclotron resonance plasma treatment followed by Cl_2 chemical etching," *Appl. Phys. Lett.,* **62,** 2658–2660 (1993).
85. M. Hong, J. P. Mannaerts, L. Grober, S. N. G. Chu, H. S. Luftman, K. D. Choquette, and R. S. Freund, "Interfacial characteristics of AlGaAs after in situ electron cyclotron resonance plasma etching and molecular beam epitaxial regrowth," *J. Appl. Phys.,* **75,** 3105–3111 (1994).
86. T. A. Strand, B. J. Thibeault, D. S. L. Mui, L. A. Coldren, P. M. Petroff, and E. L. Hu, "Low regrowth-interface recombination rates in InGaAs–GaAs buried ridge lasers fabricated by in situ processing," *Appl. Phys. Lett.,* **66,** 1966–1968 (1995).

87. T. Ishikawa, I. Matsuyama, N. Tanaka, M. Lopez, M. Tamura, and Y. Nambu, "In situ fabrication of buried GaAs/AlGaAs quantum-well mesa-stripe structures with improved regrown interfaces," *Jpn. J. Appl. Phys.*, **34**, L1412–L1415 (1995).
88. M. A. Foad, A. P. Smart, M. Watt, C. M. Sotomayor-Torres, and C. D. W. Wilkinson, "Reactive ion etching of II-VI semiconductors using a mixture of methane and hydrogen," *Electron. Lett.*, **27**, 73–75 (1991).
89. M. A. Foad, M. Watt, A. P. Smart, C. M. Sotomayor-Torres, C. D. W. Wilkinson, W. Kuhn, H. P. Wagner, S. Bauer, H. Leiderer, and W. Gebhardt, "High-resolution dry etching of zinc telluride: characterization of etched surfaces by x-ray photoelectron spectroscopy, photoluminescence and Raman scattering," *Semicond. Sci. Technol.*, **6**, A115–A122 (1991).
90. T. Matsui, H. Sugimoto, T. Ohishi, and H. Ogawa, "Reactive ion etching of III-V compound using C_2H_6/H_2," *Proc. 1988 Dry Process Symp.*, Tokyo, 1988, pp. 119–124.
91. H. Sugimoto, K. Ohtsuka, and T. Isu, "Reactive ion etching of InP using C_2H_6/H_2O_2," *J. Appl. Phys.*, **72**, 3125–3128 (1992).
92. H. Sugimoto, T. Isu, H. Tada, T. Miura, T. Shiba, T. Kimura, and A. Takemoto, "Suppression of side-etching in C_2H_6/H_2O_2 reactive ion etching for the fabrication of an InGaAsP/InP p-substrate buried-heterostructure laser diode," *J. Electrochem. Soc.*, **140**, 3615–3620 (1993).
93. H. Takazawa, S. Takatani, and S. Yamamoto, "ArF-excimer-laser-assisted highly selective etching of InGaAs/InAlAs using HBr and F_2 gas mixture," *Jpn. J. Appl. Phys.*, **35**, L754–L756 (1996).
94. H. Takazawa, S. Takatani, K. Higuchi, and M. Kudo, "Fabrication of InAlAs/InGaAs high-electron-mobility-transistors using ArF-excimer-laser-assisted damage-free highly selective InGaAs/InAlAs etching," *Jpn. J. Appl. Phys.*, **35**, 6544–6548 (1996).
95. K. Ohtsuka, T. Ohishi, Y. Abe, H. Sugimoto, and T. Matsui, "Photoluminescence characterization of InP surface reactive ion etched by a gas mixture of ethane and hydrogen," *J. Appl. Phys.*, **70**, 2361–2365 (1991).
96. T. Tezuka, S. Nunoue, H. Yoshida, and T. Noda, "Spontaneous emission enhancement in pillar-type microcavities," *Jpn. J. Appl. Phys.*, **32**, L54–L57 (1993).
97. M. Ohashi, S. Nunoue, H. Hamasaki, M. Kushibe, G. Hatakoshi, N. Suzuki, and M. Nakamura, "Vertical transmission optical amplifiers for parallel optical buses", *Proc. 1997 Real World Computing Symp.*, Tokyo, 1997, pp. 575–582.
98. K. Asakawa, T. Yoshikawa, S. Kohmoto, Y. Nambu, and Y. Sugimoto, "Chlorine-based dry etching of III/V compound semiconductors for optoelectronic application," *Jpn. J. Appl. Phys.*, **35**, 373–387 (1998).

CHAPTER TEN

Heterostructure Field-Effect Transistors and Circuit Applications

JUERGEN DICKMANN
DaimlerChrysler AG

10.1	Introduction	340
10.2	Requirements Driven by Application	341
	10.2.1 Remote Sensing by Millimeter-Wave Sensors	341
	10.2.2 High-Resolution Radar Systems for Automotive Applications	342
10.3	Principle of Operation	343
	10.3.1 Basic Device Structure	343
	10.3.2 Two-Dimensional Electron Gas	345
10.4	Material Aspects	348
	10.4.1 Material Choice	348
	10.4.2 Strained Layers	348
	10.4.3 Electronic Band Structure	350
	10.4.4 Transport Properties Versus Material Composition	350
10.5	DC Operation	352
	10.5.1 Charge Control Model	353
	10.5.2 Discussion on Parasitic Effects	354
10.6	RF Operation	357
	10.6.1 Equivalent-Circuit Description	357
	10.6.2 Low Noise Operation	358

InP-Based Materials and Devices: Physics and Technology, Edited by Osamu Wada and Hideki Hasegawa.
ISBN 0-471-18191-9 © 1999 John Wiley & Sons, Inc.

10.6.3 Power Operation		363
10.7 Layer Structure Design		364
	10.7.1 Control of Channel Conductivity	364
	10.7.2 Control of Drain Current and Transconductance	365
10.8 Device Fabrication		367
	10.8.1 Epitaxial Growth	367
	10.8.2 Fabrication Process	369
10.9 Novel Processing and Materials		371
	10.9.1 Dry Process for InAlAs/InGaAs-Based Structure	371
	10.9.2 Alternative Materials	372
10.10 Status Quo HFETs and MMICs		374
	10.10.1 HFETs	374
	10.10.2 MMICs	377
10.11 Summary and Future Challenges		380
Acknowledgments		381
References		381

10.1 INTRODUCTION

The triumphal march of the high-electron-mobility transistor (HEMT) as the workhorse in modern high-speed electronic and optoelectronic applications began with the invention of the ancestor of all HEMTs, the $Al_xGa_{1-x}As/GaAs$ HEMT. The indium phosphide (InP)–based HEMT, in particular, the $In_yAl_{1-y}As/In_xGa_{1-x}As$ HEMT, is a natural extension of the GaAs-based HEMT. In recent years it has demonstrated its high-speed potential to push the performance limit beyond that of ancestors on GaAs. Thus it can improve present systems and enables the design of components for entirely new systems application. Over the last 10 years, rapid development in the field of HEMT research has taken place, which has led (as discussed later in this chapter) to a number of modifications of device structures. It would be appropriate to substitute for the HEMT a more general device, the heterostructure field-effect transistor (HFET).

The material quality and reproducibility, including those of the substrate, have been improved so that the advantages inherent in InP-based HFETs can be exploited fully. As a consequence, the highest cutoff frequencies for current gain f_T = 340 GHz [1], as well as for power gain f_{max} = 600 GHz [2], and the noise figure (e.g., $F_{min}(f = 94$ GHz$)$ = 1.2 dB [3]) lowest for all three terminal devices has been demonstrated by InP-based HFETs. Most of other key components for analog microwave and millimeter-wave applications have already been demonstrated on InP-based HFETs [10–13]. Other advantages include the fact that they show high gain at extremely low bias conditions, resulting in low power consumption and low-voltage power operation. They also operate well even at cryogenic temperatures with low noise, which is advantageous for earth observation systems. Moreover InP-based HFETs exhibit, at larger gate dimensions, performance comparable to those of GaAs-based counterparts. The rapid progress in material growth, especially the increased knowledge of strained layer epitaxy has shown possibilities of new devices

[4–6]. One of the most spectacular modifications has been driven by manufacturability-relevant reasons; the growth of InP-based devices (HFETs, HBTs, etc.) via metamorphic buffer layers on GaAs substrates [7–9]. This new technique enables circuit designers to consider the most appropriate devices for a certain electrical function on GaAs substrate. This will certainly revolutionize future concepts of millimeter-wave integrated circuits (MMICs) for many kinds of communication and sensor systems.

The aim of this chapter is to give a generic and comprehensive overview on InP-based HFETs and MMICs for present and future applications. In Section 10.2 we outline some key applications for InP-based HFETs. In Section 10.3 the principle of operation of InP-based HFETs is explained, and Section 10.4 covers some material aspects of device performance. In Sections 10.5 and 10.6 we describe basic features of dc- and RF operation, respectively. In Section 10.7 we give information on the influence of individual layers in a HFET layer structure on device performance, and Section 10.8 covers fabrication-relevant facts. Novel appoaches to processing and material composition for HFETs are discussed in Section 10.9, while Section 10.10 surveys the status of InP-based HFETs and MMICs. Section 10.11 is a summary and briefly touches future challanges in the field of InP-based HFETs and MMICs.

10.2 REQUIREMENTS DRIVEN BY APPLICATION

Research activities, especially in the industrial scene, are driven by system requirements, and hence by the demand for new electrical performance, which can presently not be met by any other technologies. To provide a comprehensive chain of ideas in this chapter, a few typical applications that require InP-based HFETs are reviewed.

10.2.1 Remote Sensing by Millimeter-Wave Sensors

Imaging radiometers are commonly used for remote sensing by millimeter waves. The applications of those sensors are in earth sensing and environmental climate and pollution observation [14]. They provide real data for studying global changes and processes in Earth's atmosphere. Other applications are in the field of radio astronomy, target seekers (electronic weapons), and medical thermography [15]. These imaging radiometers are arrays of several hundreds to thousands of pixels that are imaged simultaneously by arrays of sensors.

A typical block diagram of a superheterodyne receiver of one such sensor using MMICs is shown in Fig. 10.1. The low-noise amplifier (LNA) chip is used to provide low-noise amplification before downconversion to an IF frequency where it can be further amplified prior to signal processing. The requirement of the LNA is to provide the lowest noise figure possible and the highest gain simultaneously at extremely high frequencies. Typical frequencies of operation are in the W-band and above. To achieve the lowest noise figure possible, the millimeter-wave monolithic integrated circuits (MMICs) are cooled down to cryogenic temperatures.

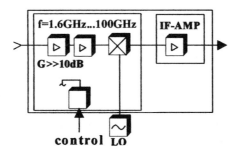

FIGURE 10.1 Block diagram of a superheterodyne receiver.

Hence operability at low temperature may be another requirement for InP-based HFETs.

10.2.2 High-Resolution Radar Systems for Automotive Applications

While communication systems in the low-GHz regime promote the use of gallium arsenide (GaAs) and silicon (Si) technologies, millimeter-wave radar systems also become feasible for civil applications with potentially high production volumes [16]. Such radar sensors can be used in cars for intelligent cruise control or in collision avoidance systems. In a mobile society in which the traffic density increases each year, there is an increasing demand for sensors that fulfill safety requirements. The key application of such systems is collision avoidance. Present system configurations provide information about the distance and relative speed within one antenna beam. This information would allow a safety-relevant assessment of the present traffic situation of a car and offer tracking and safety functions such as warning signs to initiate braking (in a later generation this would be done automatically). A limitation of present systems is the binary (yes/no) assessment; in other words, there is something/nothing in front. Thus classification of data is hard to perform, and this could lead to drastic misinterpretation. A higher resolution of the scene in front of the car could allow evaluation of the object identified by more appropriate classification of the reflected signals and would provide enough information to circumvent a direct collision. Such a radar system could be an imaging radar, shown in Fig. 10.2.

Due to administrative regulations in Europe, frequency bands around 77 GHz are considered for operation. In view of semiconductor development, such systems are challenging in two ways: high electrical performance at high millimeter-wave frequencies and at the same time, high-volume manufacturability with cost-constraint issues typical for the automotive industry. The technical requirements include low-noise, high-gain operation and power generation under a wide range of environmental constraints. To meet the cost requirement, the total number of MMICs and total wafer area per electrical function have to be minimized. Therefore, for example, a device technology would be of great interest that offers considerably higher gain per

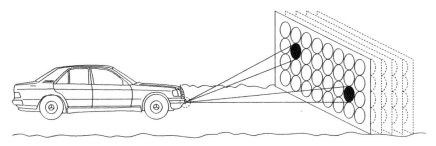

FIGURE 10.2 Imaging radar for automotive use.

stage (at least 2 dB) than that available with present techniques. It would be an additional advantage if this could be achieved at more relaxed gate dimensions.

For the applications described above and in Chapters 2 and 3, the technical targets of InP-based HFET technology can be summarized as follows: (1) systems components for operation at extremely high frequencies; (2) (very) low-noise, high-gain performance; (3) low-voltage operation; (4) low-temperature operation; and (5) design and fabrication of high-performance devices with relaxed gate dimensions to increase the yield and reduce the cost.

10.3 PRINCIPLE OF OPERATION

10.3.1 Basic Device Structure

A principal cross section of a basic configuration of an InP-based HFET is shown in Fig. 10.3a. The basic layer structure (Fig. 10.3b) is composed of a semi-insulating InP substrate followed by an InAlAs buffer layer, an InGaAs channel layer L_z, an InAlAs spacer layer d_{sp}, an InAlAs pulse-doped donor layer d_d, an InAlAs Schottky barrier layer d_u, and an InGaAs cap layer. The dark line in Fig. 10.3a marked "2DEG" (two-dimentional electron gas) depicts the flow of electrons in an HFET which forms the conductive path in the InGaAs channel layer. This conductive path is generated by the transfer of free electrons out of the doped InAlAs layer into the InGaAs channel, where they form a quasi-two-dimensional electron gas localized within an approximately 10-nm-thick layer at the heterointerface of the InAlAs and InGaAs layer. The situation is very similar to the inversion layer in an Si/SiO$_2$ material system [17]. Here the conduction band forms a quasi-triangular potential well. The difference is in that the carriers in the InGaAs are localized in the potential well not by inversion but by the electric field that is caused by the parent donors in the InAlAs and the mobile electrons in the InGaAs channel. Ideally, this very thin conductive layer in the quantum well is the only path under the gate electrode where mobile carriers flow. The InAlAs layers are entirely depleted from mobile carriers so that the gate modulates only the 2DEG.

The advantage of this device principle compared to those of MESFET and bipolar transistors is that the 2DEG is spatially separated from its parent donors and

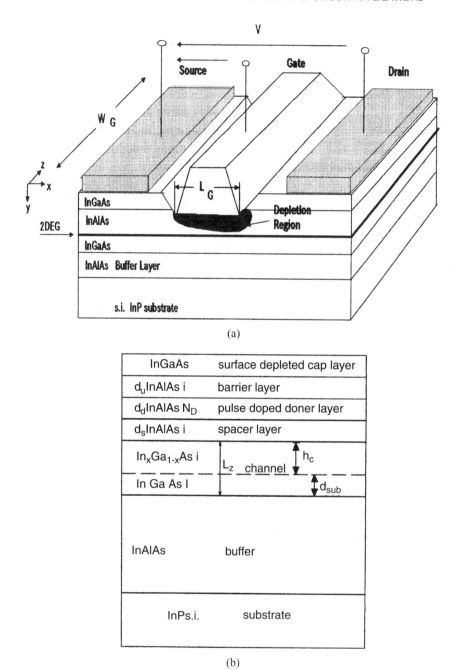

FIGURE 10.3 (*a*) Cross-sectional structure of InP-based HFET; (*b*) layer structure of pulse-doped InP-based HFET.

does not undergo ionized impurity (Coulomb) scattering and consequently exhibits much improved carrier mobility and velocity. Thus a very thin (about 10 nm) conductive path provides a high carrier sheet density (for electrons $>10^{12}$ cm^{-2}) with, simultaneously, excellent transport properties (mobility, velocity), even in proximity to the gate (15 to 30 nm). This feature translates into clear advantages for high-frequency operation.

10.3.2 Two-Dimensional Electron Gas

Charge Transfer Across a Heterojunction If two semiconductors with different bandgaps and electron affinities are brought in contact (grown on each other), a heterojunction is formed. The properties of the heterojunction are governed by the energy band lineup at the interface [18]. The heterojunction exhibits discontinuities in the conduction band (ΔE_C) and the valence band (ΔE_V). The standard material system used for InP-based HFETs is In$_y$Al$_{1-y}$As/In$_x$Ga$_{1-x}$As. The wide-bandgap semiconductor (InAlAs) has the smaller electron affinity, so that the conduction band of the narrow-bandgap semiconductor (InGaAs) lies lower in energy. The valence band of the narrow-bandgap material lies higher in energy than in the wide-bandgap material, in this heterojunction (type I heterojunction). The sum of ΔE_C and ΔE_V is equal to the bandgap difference E_G(InAlAs) $- E_G$(InGaAs). For conventional HFET application the wide-bandgap material is selectively n-doped, while the narrow-bandgap material is undoped. This doping principle is called *modulation doping* [19]. Due to the Fermi-level alignment, electrons are transferred from the InAlAs into the InGaAs. This charge transfer leads to a dipole layer formed by the positively charged depletion layer in the InAlAs and the electron accumulation (negative charge) in the InGaAs. Thus the conduction band of the InGaAs layer very close to the heterojunction forms a quasi-triangular quantum well with quantized density of states perpendicular to the heterojunction. The transferred electrons are localized in this very thin 10- to 12-nm quantum well and have two degrees of freedom for motion parallel to the heterojunction, for which they can be called a quasi-two-dimensional electron gas.

To describe the electron transfer across the heterojunction exactly, Schrödinger's and Poisson's equations have to be solved self-consistently using numerical means. However, analytical models well suited for determining the 2DEG concentration were reported for a modulation-doped AlGaAs/GaAs HFET [21,22]. Neglecting the interface states and assuming full depletion of the n-doped region and a triangular electron well with two populated subbands, the 2DEG concentration n_s can be calculated by equating the two expressions self-consistently for n_s deduced from both Poisson's equation (depleted wide-bandgap material $N_D d_d$) and Schrödinger's equation (maximum density of states in the quantum well). This procedure has been discussed extensively in the literature [21–26]. Here only the extract is summarized. The maximum sheet density n_s that can be transferred in the quantum well from the entirely depleted wide-bandgap material is given from Poisson's equation by [21]

$$n_s = \sqrt{\frac{\varepsilon N_D}{q}(\Delta E_C - E_{F2} - E_{Fi}) - N_D^2 d_{sp}^2 - N_D d_{sp}} \qquad (1)$$

where d_{sp} is the spacer layer, E_{F2} the separation of the conduction band edge and the Fermi level in the wide-bandgap material, E_{Fi} the Fermi level at the interface relative to the narrow-bandgap material's conduction band, N_D the donor density with 100% activation efficiency, $\varepsilon = \varepsilon_0 \varepsilon_r$ the permittivity, q the elemental charge, and ΔE_C the conduction band discontinuity.

From Schrödinger's equation the maximum sheet density in the quantum well with an assumption that two lowest subbands are populated is given by [21]

$$\begin{aligned} n_s &= \frac{DkT}{q} \sum_{j=0}^{\infty} \ln\left\{\left[1 + \exp\left(\frac{q(E_{Fi} - E_j)}{kT}\right)\right]\right\} \\ &\approx \frac{DkT}{q} \ln\left\{\left[1 + \exp\left(\frac{q(E_{Fi} - E_0)}{kT}\right)\right]\left[1 + \exp\left(\frac{q(E_{Fi} - E_1)}{kT}\right)\right]\right\} \\ &\approx \frac{DkT}{q}\left(\frac{q(E_{Fi} - E_0)}{kT} + \frac{q(E_{Fi} - E_1)}{kT}\right) = D(2E_{Fi} - E_0 - E_1) \quad \text{for } E_{Fi} \geq E_0, E_1 \quad (2)\end{aligned}$$

where D is the density of states, E_F the Fermi energy, and E_j the jth subband energy in the quantum well. The sheet density of the 2DEG is determined by equating the two expressions for n_s, Eqs. (1) and (2).

Influence of Layer Structure Parameters on Two-Dimensional Electron Gas Concentration From Eqs. (1) and (2), the most important dependence of n_s on the layer structure parameters can be summarized as follows:

$$n_s \propto \sqrt{N_D} \sqrt{\Delta E_C} \qquad (3)$$

Equation (3) reveals that n_s can be maximized by increasing the doping concentration N_D and the conduction band discontinuity ΔE_C. However, the square root in Eq. (3) also makes it obvious that n_s can increases monotonically, but shows a kind of saturation for large of N_D and ΔE_C. As explained in more detail in the next section, ΔE_C is a function of the InAs mole fraction (x,y) in the $In_yAl_{1-y}As/In_xGa_{1-x}As$ system and thus can be tailored by adjusting x and/or y. More rigorous numerical calculations were performed in the literature for this material system [6,25,26].

For a HFET layer structure the population of the first three subbands was calculated in [26] and verified experimentally by Shubnikov–de Haas measurements. The result is shown in Fig. 10.4. As can be seen, two first subbands carry nearly the entire 2DEG and thus verify the assumptions made in the analytical model for Eqs. (1) and (2). When compared with that for a comparable layer structure of pseudomorphic HFETs on GaAs, this value of n_s is about 30% higher. The spacer layer acts as a tunnel barrier and may reduce the electron transfer into the 2DEG. A practical compromise for device operation is to use d_{sp} = 2nm.

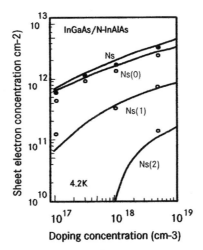

FIGURE 10.4 Subband population of 2DEG carriers in modulation-doped HFET layer structure. The figure x compares data measured with Shubnikov–deHaas (open circles) and calculated data (lines). Solid circle and line represent the data for total 2DEG concentration (N_s). $N_s(0)$, $N_s(1)$, and $N_s(2)$ are the 2DEG concentrations of the first, second and third populated subband in the quantum well, respectively; thus $N_s = N_s(0) + N_s(1) + N_s(2)$. (From Ref. 26.)

Another important parameter determining n_s is the channel thickness L_z. Figure 10.5 shows the influence of L_z on n_s for the lattice-matched case ($x = 0.53$, $y = 0.52$). This analysis has shown that for channel thicknesses of $L_z > 20$ nm, n_s behaves as in a layer structure without InAlAs buffer layer. Quantization and confinement effects are observed for smaller channel thicknesses. Due to the quantization effect, n_s starts to drop drastically for $L_z < 11$ nm. As a consequence, the channel thickness should be kept in the range 12 nm $< L_z <$ 20 nm for practical applications.

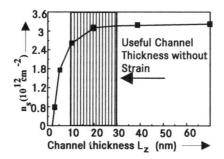

FIGURE 10.5 Dependence of 2DEG concentration on channel thickness L_z. The 2DEG concentration was calculated by self-consistently solving Schrödinger's and Poisson's equations as discussed in the text. (From Ref. 6)

10.4 MATERIAL ASPECTS

Since the InP-based HFET is a natural extension of the GaAs-based HFETs, the basic considerations made there can generally be adopted in the InP-based HFET [27,28]. Detailed discussion of InP-based materials and materials properties and growth techniques can be found in Chapters 4, 5, 7, and 8. In this section, material aspects most important for HFET operation are reviewed.

10.4.1 Material Choice

Material systems useful for HFET operation include $In_yAl_{1-y}As/In_xGa_{1-x}As$, $In_yAl_{1-y}As/InP$, $In_xGa_{1-x}As/InP$, and quaternary alloys such as InGaAsP in combination with InGaAs or InAlAs within a 1 to 2% lattice mismatch. The application of various materials are given in Section 10.9. One important aim of the layer structure design is to maximize the channel conductivity which is proportional to the product $n_s \times \mu$. As deduced in Section 10.3, the 2DEG concentration n_s is proportional to the square of the conduction band discontinuity. ΔE_C can be optimized by adjusting the composition of the materials forming the heterojunction. In an InAlAs/InGaAs system, this is done by increasing x and reducing y. That may cause a lattice constant mismatch between the InAlAs and InGaAs layers and with respect to the InP substrate. This lattice mismatch leads to strain incorporation in these layers, which has to be taken into account for HFET layer structure design.

10.4.2 Strained Layers

The concept of using strained layers to improve the performance of HFETs has been applied successfully to GaAs-HFETs and has generated a new type of HFET, the pseudomorphic HFET [29]. However, there exists a limitation in the applicability of strained layers for HFETs: Namely, the existence of a critical layer thickness up to which the lattice strain can be accommodated without deteriorating the carrier transport properties. Above this critical layer thickness it is energetically favorable for the strain to relax by forming dislocations.

To determine the critical layer thickness (h_c) with respect to the lattice mismatch, two theoretical models have been developed [30,31]. One model considers mechanical equilibrium conditions [30], and the other, energy balance [31]. The lowest values for h_c correspond to the mechanical equilibrium model. Most values of h_c adopted in HFET layer structures are close to this. However, growth conditions have been found to lead to variations in h_c even for nominally the same layer structures [4,32]. It is important to notice that the substrate determines the lattice constant, and the layers grown on top are enforced to match this lattice constant. In the case of InAlAs and InGaAs on InP, this leads to a tetragonal deformation from their normal cubic crystalline structures. A material with a lattice constant smaller than that of the substrate undergoes biaxial tension, while for the opposite case it undergoes biaxial compression. Up to the critical layer thickness h_c, the strain can be accommodated elastically. Such layers are perfect crystals and exhibit excellent transport properties. They are termed *pseudomorphic*.

The critical layer thicknesses calculated based on both models are shown in Fig. 10.6 for two materials most relevant for HFET applications: $In_yAl_{1-y}As$ and $In_xGa_{1-x}As$. To improve the device performance (also refer to Section 3) both parameters x and y can potentially be changed. From Fig. 10.6 it becomes obvious that h_c imposes stringent limitations in mole fraction, which can be utilized to improve the HFET performance. Especially for high InAs mole fractions x, the channel thickness L_z enters the quantization limited region of $L_z < 12$ nm (see also Fig. 10.5). A more detailed discussion of the impact of h_c on device performance is given in Section 10.7 and in Ref. 6.

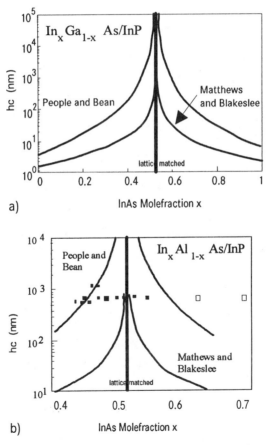

FIGURE 10.6 Critical layer thickness versus AlAs mole fraction for (*a*) InGaAs and (*b*) InAlAs grown on InP substrate. Lines are calculated results. Solid squares represent experimental data of pseudomorphic layers; open squares represent experimental data of layers assumed to have undergone relaxation. (From Refs. 30 and 31.)

10.4.3 Electronic Band Structure

The electronic band structure is another important issue in the design of HFETs. For $In_yAl_{1-y}As$ the electronic band structure imposes a maximum AlAs mole fraction ($1-y$) value of about 0.68. For higher ($1-y$) values, the material becomes an indirect transition type. Such a limitation does not exist for InGaAs. The Γ–L intervalley separation increases with the InAs-mole fraction x in $In_xGa_{1-x}As$. This property leads to higher peak velocities in bulk InGaAs and to higher average velocities in HFET channels.

10.4.4 Transport Properties Versus Material Composition

Electron Mobility Low field mobility can be interpreted as the capability of electron acceleration toward peak velocity. As discussed in the next section, even for HFETs with extremely short gate length there exists a mobility-limited region under the gate which has a direct impact on the average speed of the electrons. This influence becomes predominant for longer gate lengths. In Section 10.3 it was shown that a great stride ahead could be made by increasing the low field mobility by introducing the modulation-doped scheme. For such layer structures the ionized impurity scattering is greatly reduced, due to the spatial separation of electrons in the 2DEG from their parent donors and to the screening effect caused by extremely high 2DEG density.

Intensive investigations have been made to determine the limit of mobility in InP-based heterostructures, mostly on InAlAs/InGaAs [33–35]. Figure 10.7 shows the mobility calculated for a modulation-doped layer structure over the temperature range $T = 4$ to 300 K. Polar optical phonon scattering is the limiting mechanism from about 300 K down to 130 K. Below 130 K, the alloy disorder scattering becomes dominant. At room temperature the interaction with polar optical phonons limits the mobility to about 11,000 $cm^2/V \cdot s$. Alloy disorder scattering limits the mobility to values in the range 4 to 7×10^4 $cm^2/V \cdot s$. Other scattering mechanisms, such as ionized impurity scattering or acoustic phonon scattering, are relatively unimportant. Piezoelectric coupling contributes the least and is negligible in InAlAs/InGaAs modulation-doped structures [33–35].

Continuating our discussion of strained layers in Section 10.3, the dependence of electron mobility as a function of the InAs mole fraction (x) is important. Figure 10.8 shows the results calculated for mobility at 77 K. The circles are experimental values that support the general trend predicted from the calculations. As can be seen, the dependence of electron mobility is predominated by alloy disorder scattering. Mobility is minimum at $x = 0.3$ and increases with x. Therefore, for high-speed operation, the highest values of x possible with reference to h_c are desirable.

Electron Velocity For high speed, low-noise, power HFETs, the velocity–electric field characteristic of the channel material gives information about the ultimate performance. Along the gate length, the electric field ranges from a low (source) to an extremely high value (drain). In short-gate-length devices, an average

10.4 MATERIAL ASPECTS   351

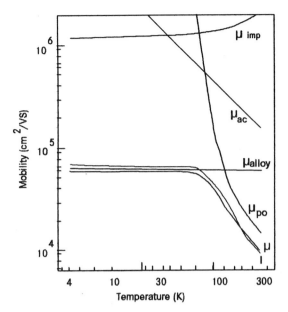

FIGURE 10.7 Calculated electron mobility in $In_{0.53}Ga_{0.47}As$ lattice matched with InP as a function of temperature. (From Ref. 33.)

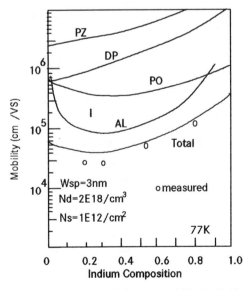

FIGURE 10.8 Calculated dependence of electron mobility in $In_xGa_{1-x}As$ as a function of InAs mole fraction. Circles show experimental data. (From Ref. 34.)

velocity is often deduced below the peak velocity [28] and emphasizes the importance of this field effect. The electron velocity-field characteristic for GaAs and InP calculated using the Monte Carlo method is described in Chapter 4. InGaAs offers higher velocities over a wider range of electric field than that of other semiconductors and thus distinguishes itself from others for high-speed operation. For power applications, InP offers high velocity at high electric field values (see Section 10.9.).

A question arises as to the potential of increasing electron velocity by using strained channel layers. Although numerous pseudomorphic InP-based HFETs have been reported so far, the transport properties in these devices have remained relatively unclear [36]. This point is important since compared with bulk material, quantum size effects and deformation of the lattice alter the band structure as well as the material parameters [32]. The $V(E)$ characteristic of InGaAs under strain was calculated by [36]. The most important result of this investigation is that the electron velocity (peak and saturation) does not improve monotonically with x but exhibits a maximum between 70 and 80% [6,36]. An increase in x from 53% to 78% leads an improvement in the peak velocity by about 20%, while the saturation velocity remains largely unaffected. The velocity field characteristic calculated is shown in Fig. 10.9. Besides the critical layer thickness, the strain effect imposes stringent limitations on the transport properties in HFET structures, which have to be dealt with in the design of HFET devices.

10.5 DC OPERATION

Models to describe charge transfer across a modulation-doped heterojunction have been investigated extensively to predict the I–V characteristics of HFETs [37–39].

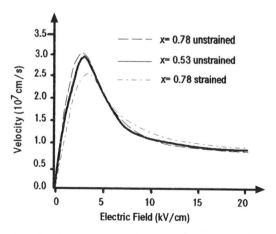

FIGURE 10.9 Calculated electron velocity–electric field characteristics of bulk $In_{0.53}Ga_{0.47}As$ and $In_{0.48}Ga_{0.22}As$ with and without lattice strain (From Ref. 36.)

Refer to Refs. 20 and 38 for overviews of this topic. For a rigorous description of device behavior, one-dimensional models have been proposed, while for shorter-gate-length devices more accurate two-dimensional numerical and seminumerical models are considered adequate [37]. In this section, key issues of device layout and layer structure are summarized, and parasitic effects and their origins are explained.

10.5.1 Charge Control Model

The steps involved in deriving an adequate description of the dc I–V characteristic are as follows:

1. Consider an intrinsic HFET, hence the part of the device channel under the gate. To describe the extrinsic HFET fully, the parasitic source and drain resistances have to be introduced.
2. Establish a submodel to describe the relationship of charge control of n_s by the Schottky gate (through an electric field perpendicular to the heterointerface).
3. Determine an approximation for the relationship of the velocity-field relation to be introduced in the saturation model (electron transport parallel to the heterojunction).
4. Combine the charge control model and the $V(E)$ approximation to derive I–V characteristics.
5. Refine this model by introducing additional models to describe parasitic effects, if necessary.

The very first model describing the I–V characteristics of an HFET was developed in [21] and extended by [23]. It is a one-dimensional model that is based on a linear dependence of the Fermi level on n_s. Although this model simplifies a real HFET, it is sufficiently accurate to yield fundamental relationships (equations) relevant to HFET operation. Ideal operation is characterized by the fact that the entire current is carried by the 2DEG. Consequently, this necessitates that the wide-bandgap layer (InAlAs) under the gate be fully depleted from mobile carriers. The layer structure therefore has to be designed to guarantee that the charges to deplete under the Schottky gate and the charges in the 2DEG match in the appropriate bias range. As discussed later, by adjusting the thickness–doping product for the doped InAlAs, the type of operation, enhancement or depletion, can be determined. In the ideal case where both depletion regions meet, the channel under the gate is considered to form a parallel-plate capacitance where the semiconductor is a dielectric and the gate and the 2DEG form conductive plates. This implies that for this range of gate bias (V_G), the gate modulates only the 2DEG, and a simple relationship of n_s versus V_G can be derived [21,23]:

$$n_s = \frac{1}{q} C_0 (V_G - V_{\text{th}}) = \frac{1}{8} \frac{\varepsilon}{d_{\text{eff}}} (V_G - V_{\text{th}}) \tag{4}$$

where $\varepsilon = \varepsilon_0\varepsilon_r$ is the dielectric constant of InAlAs, C_0 the 2DEG capacitance per unit gate area, V_{th} the threshold voltage, V_G the applied gate voltage, and d_{eff} the effective distance between the gate and the center of the 2DEG.

An intuitive aid in determining charge modulation is to consider the gate capacitance C_G as a function of V_G. There are three regions to be distinguished. Region I represents an area where the 2DEG starts to form. Region II represents the effective 2DEG modulation region. Notice that 2DEG concentration does not change over a certain bias range of V_G. In region III, a drastic rise in C_G is caused by parasitic charge modulation in the wide-band-gap InAlAs layer known as parallel conduction [40]. One difference in the basic assumptions of various HFET models made is the relationship of charge modulation in region II. The linear charge control model predicts a linear E_F-n_s relationship and fails to describe this region accurately, and thus the I-V characteristic in the threshold regime [22,23], hence other approaches have to be considered [37].

Once the charge control model is established, the I-V characteristic can be derived. The Schockley gradual channel approximation (GCA) holds at low fields and for relatively large gate lengths. This model assumes the saturation of I_{DS} if the pinch-off condition ($V_G = V_{th}$) is reached at $x = L_G$ [41]. An opposite situation is assumed in the velocity-saturation model (VSM) [42], which holds to some extent at high fields, hence for short gate lengths. It assumes that I_{DS} saturation occurs due to the velocity saturation in the entire channel region under the gate. For the GCA model a quadratic dependence of I_{DS}-V_G is predicted, while for the VSM model a linear I_{DS}-V_G relation is predicted. The maximum transconductance is in the VSM model [42],

$$g_{mVSM} = \frac{V_{sat}\varepsilon W_G}{d_{eff}} \quad (5)$$

In real devices, including those with very short gate lengths, it was reported [27] that a combination of GCM and VSM provides a better model. This is encountered in channel section models with pure GCM, VSM, and GCM–VSM transition sections [37]. In the channel section model, the channel is devided into sections with individual $V(E)$ characteristics. These $V(E)$ models imply that V_{sat} is set to the peak velocity in bulk material, which corresponds to the experience that the average velocity in the HFET channel is close to the peak velocity. However, there is no evidence that V_{sat} in $V(E)$ models equals to the peak velocity. Although the physical meaning of V_{sat} is not clear, it is very useful for device analysis.

10.5.2 Discussion on Parasitic Effects

The models described above, however, fail to describe parasitic effects present in real InP-based HFETs as are observed in a typical I-V characteristic of a high-performance lattice matched device with $L_G = 0.25$ μm, as shown in Fig. 10.10. In the following, pronounced parasitic effects are discussed.

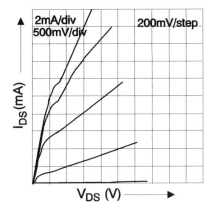

FIGURE 10.10 *I–V* characteristic of a 0.25-μm-gate lattice-matched InAlAs/InGaAs/InP-based HFET. (From Ref. 6.)

Transconductance Compression and Nonideal Drain Current Saturation
Transconductance compression with increasing V_G can be attributed to parallel conduction in wide-bandgap InAlAs layers [40]. Nonideal saturation behavior expressed by an increase in the output conductance (g_d) can be explained as follows:

1. The basic effect can be attributed to deconfinement reasons as known for GaAs-based HFETs [43]. At high electron energies caused by extremely high electric fields, most electrons at the drain side of the channel are three-dimensional in nature [36,44], which causes deconfinement of carriers from the heterointerface toward the buffer layer. This effect can be reduced by utilizing a wide-bandgap (InAlAs) buffer layer and a thin InGaAs channel layer.

2. A kink occurs in the dc *I–V* curve for V_{DS} in the saturation regime ($V_{DS} > 0.5$ V) and marks a drastic increase in the output conductance, known as a *kink effect*. This becomes even worse upon reducing the gate length and increasing the drain bias. The origin of the kink effect is still controversial, and there are at least three plausible mechanisms: real space transfer in the InAlAs buffer or barrier layer [45], impact ionization in the channel layer [46], and growth/material-related charge effects such as traps in InAlAs or in the buffer [47]. While the impact ionization has been proven to contribute to the kink [46], there is also experimental evidence that growth conditions have a dramatic impact on the kink effect [47–50]. HFETs manufactured in the same laboratory under the same conditions except for InAlAs growth conditions showed a drastic variation in kink effect [48].

The first important hint that the kink effect is trap related [47] demonstrated that the increase in g_d does not occur at RF frequencies. The frequency dispersion of g_d was found to fall in the range of trap time constants. A model that supports this assumption was developed [51]. The use of either a thin InGaAs [50] or an InAlAs/InGaAs superlattice buffer layer [6] showed that the kink effect can be reduced. Other experiments showed that a larger V/III ratio during InAlAs growth (low As overpressure) exhibited no kink effect [48]. Although there are individual successes in eliminating the kink effect, in general it is still a problem that has to be solved in InP-based HFETs.

Gate Leakage Current One drawback of InP-based HFETs compared with GaAs-based HFETs is higher gate leakage current, resulting in only moderate breakdown voltages for conventional layer structure. The origins are manifold:

1. For devices with low threshold voltages (e.g., enhancement mode) and with open channel bias ($V_G > 0$ V), parasitic gate current was observed to increase with V_{DS}. This gate current is believed to be caused by real space transfer of hot electrons whose energy is high enough to surpass the InAlAs barrier toward the gate and can be attributed to the low Schottky barrier height of the gate metal to InAlAs [52].

2. The reverse gate current can be explained by the low Schottky barrier height. In addition, this is due to the narrow bandgap of the InGaAs channel, which leads to enhanced impact ionization [46].

Breakdown Characteristics The power performance of InGaAs-containing HFETs is limited by low breakdown voltages and high gate leakage. This situation has stimulated investigations of the breakdown mechanisms in these devices [53–55]. According to the bias conditions relevant for power operation, recent investigations have classified the device breakdown into off-state [53] and on-state breakdown [54]. Although both relate to three-terminal operation conditions, the off-state condition is characterized by gate bias beyond threshold, $V_G > V_{th}$, and the on-state by open-channel conditions, $V_G \gg V_{th}$. From measurements on the temperature of the breakdown voltage in pseudomorphic InAlAs/In$_x$Ga$_{1-x}$As ($0.53 < x < 0.7$) HFETs, it was found that the impact ionization in the InGaAs channel is the dominant breakdown mechanism [54]. This is supported by experimental results [46] in which photo current was used to verify that the gate current is due to impact ionization in the open-channel regime. In the off-state breakdown, the situation is more complex. It was found [53–55] that the off-state breakdown is in general a two-step process of thermionic field emission followed by impact ionization at a potential step. As for bias dependence, thermionic field emission becomes dominant first, and impact ionization dominates at higher drain bias (breakdown regime). Using larger margins in HFET layout (larger geometries of L_G and L_{SD}, thicker barrier layers, double recess, etc.), breakdown voltages close to those of GaAs HFETs have been achieved in InP-based HFETs [53–55].

10.6 RF OPERATION

10.6.1 Equivalent-Circuit Description

For small-signal RF operation, the device can be described by a small-signal equivalent circuit (SEC) with a linear lumped-element description, as shown in Fig. 10.11. In the intrinsic FET, $C_{GS} + C_{GD}$ represents the total gate capacitance; C_{DS} models the interface to the buffer; R_i and g_{DS} represent the effect of channel resistance; g_m is the transconductance, which relates I_{DS} with the voltage swing across C_{GS}; R_G represents the gate resistance of the gate metal; R_s and R_D are the source and drain resistances; and τ represents the transit time. Knowledge of the network elements of SEC is necessary for device optimization and designing MMICs. One advantage of SEC is that on the assumption of accurately determined network elements, the RF performance can be predicted over a frequency range higher than the actual measurement range, so that it is possible to design the important figures of merit. One of the most important figures is the current-gain cutoff frequency f_T, which is given by the expression [56,57]

$$f_T = \frac{g_m}{2\pi\{(C_{GS} + C_{GD})[1 + g_d(R_s + R_d)] + C_{gd}g_m(R_s + R_d)\}}$$
$$\approx \frac{g_m}{2\pi(C_{GS} + C_{GD})} \approx \frac{g_m}{2\pi(C_{GS})} = \frac{V_{sat}}{2\pi L_{Geff}} \quad (6)$$

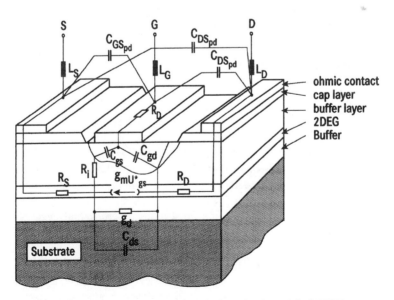

FIGURE 10.11 Small-signal equivalent-circuit model of HFET.

with L_{Geff} the electrically effective gate length (metal gate length plus depletion distance).

The other figure of merit of importance is the power-gain cutoff frequency (also called the maximum frequency of oscillation) f_{max} [56,57]:

$$f_{max} = \frac{f_T}{\sqrt{g_d(R_i + R_g + R_s) + 2\pi f_T C_{GD} R_g}} \quad (7)$$

$$f_{max} \approx \frac{f_T}{\sqrt{4g_d(R_i + R_g + R_s) + 2(C_{GD}/C_{GS})[(C_{GD}/C_{GS}) + g_m(R_i + R_s)]}} \quad (8)$$

Equation (7) is introduced as a general expression to describe f_{max}. Although Eq. (8) does not exactly predict f_{max} for the case that g_d becomes negligibly small (which is never the case in HFETs), it is one of the equations accepted by HFET designers as a practical approach to describing the influence of circuit elements.

It is emphasized that f_{max} depends strongly on direct input losses [the first term in the denominator of Eq. (8)] and input losses caused by feedback [the second term in the denominator of Eq. (8)]. The feedback capacitance C_{GD}, and hence the feedback impedance $(1/\omega C_{GD})$, become more important the higher the operation frequency is. The power-gain cutoff frequency is a more relevant figure of merit since it determines a frequency up to which the device can be used as an amplifier. Although f_T is an artificial parameter, it is useful for optimizing the layer structure since it is directly related to V_{sat} and L_G. The advantage of InP-based HFETs, especially of those fabricated according to the depleted cap layer concept [58], is extremely small feedback capacitance C_{GD}, which makes the intrinsic device more ideal (small Miller effect) and thus attractive for high-speed operation.

10.6.2 Low Noise Operation

The most attractive feature of InP-based HFETs is very low noise amplification with high gain, even in the millimeter-wave range. However, the low-frequency noise of InP-based HFETs is inferior to that in any bipolar technology and limits their use for oscillator applications. By improving the Schottky barrier height, HFETs can be used as mixers [11,59]. In this section we describe the physical origins of noise over various frequencies and show the potential and drawbacks of low-noise application of InP-based HFETs. See Refs. 60 and 61 for fundamental studies on this topic.

The general behavior of the noise figure as a function of frequency, which holds for unipolar and bipolar transistors, is shown in Fig. 10.12. For frequencies below f_c (typically, $< 10^8$ Hz), the frequency dependence is characterized by $1/f$ behavior and is termed low-frequency noise (LF noise). As the frequency increases above f_c, LF noise decreases and falls below the thermal noise level (white noise) associated with the HFET channel. For higher frequencies, the frequency behavior is characterized by linear behavior due to the capacitive coupling of the gate with the channel. The

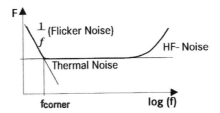

FIGURE 10.12 Dependence of noise figure as a function of frequency.

noise properties of any linear two-port element can be represented by a noise-free intrinsic element with external noise generators at the input and output ports. A schematic drawing of a generic circuit representation that holds for all frequencies is shown in Fig. 10.13. The noise generators of the intrinsic device are the noise-current generators (determined by the physical mechanisms) represented by the mean squares of $\overline{i_{ng}^2}$ and $\overline{i_{nd}^2}$. The noise voltage sources represent the thermal noise of the parasitic resistances, R_G, R_s, and R_D.

LF Noise In HFETs the low-frequency noise is caused by pure $1/f$ noise and generation–recombination (g-r) noise and is represented by the drain-current generator. While $1/f$ noise is a phenomenon that all amplifying devices have, g-r noise is due to technology- and material-related mechanisms [61]. The effect of gate leakage, which is often larger than in GaAs HFETs, can be represented by a local shot-noise gate source [62]. The gate leakage current is then expressed as $2q\,\Delta I_G$, which leads to a drain channel noise source. As a consequence, the gate leakage current governs the minimum noise figure, especially at low frequencies. The effect becomes less pronounced at higher frequencies.

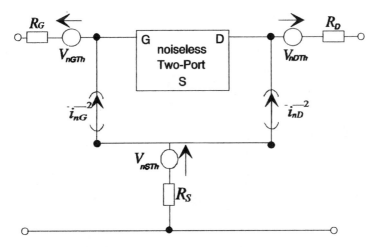

FIGURE 10.13 Equivalent-circuit representation of a noisy two port.

G-r noise due to traps manifests itself as bulges superimposed to the otherwise pure $1/f$ noise characteristic. Carrier density fluctuation due to g-r with discrete energy levels can be described by a Lorentzian spectral density and the corner frequency at which the spectrum crosses the thermal noise. G-r noise is thus frequency and temperature dependent, so temperature-dependent LF-noise measurements makes it possible to characterize traps [61]. Bias-dependent measurements can be made to get hints to the probably location of LF-noise sources in the HFET layer structure [61–63]. In InP-based HFETs, traps and g-r centers in InAlAs and at InAlAs/InGaAs interfaces are considered as major LF-noise sources (see Refs. 61–64 for details). It has also been observed that the noise level increases when the InAs mole fraction in the InGaAs channel increases. Since possible sources mentioned above are related to material (InAlAs), the device performance may be improved by using alternatives to InAlAs and by further optimizing the material growth technique.

The noise transition frequency f_c (corner frequency) is a figure of merit to classify the LF-noise performance of a device. While Si bipolar transistors and III/V HBTs show corner frequencies much less than 1 MHz, InAlAs/InGaAs HFETs typically show higher values in a range of some 100 MHz. Figure 10.14 shows the LF-noise behavior for some FET devices in comparison with an Al-free InP-based HFET [64]. It can clearly be seen that such devices are comparable to or slightly better than conventional MESFETs and GaAs-based HFETs. The use of InP as a cover layer for InAlAs may also be advantageous [65]. An HFET with an InP-etch stop layer showed slightly lower noise levels than that of a device with a conventional InAlAs-gate layer at a very low frequency of several 10 kHz. Nevertheless, the low-frequency noise levels are still too high for oscillator applications.

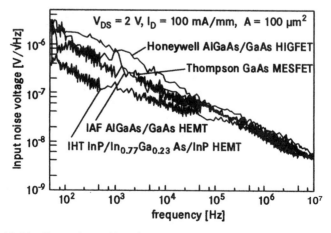

FIGURE 10.14 Comparison of low-frequency noise of GaAs-based HFETs with that of an aluminum-free InP-based HFET. (From Ref. 64.)

HF Noise High-frequency noise is related to the device channel and the capacitive coupling between the channel and the gate [60,66]. The gate noise is represented by a gate-current noise generator $\overline{i_{ng}^2}$ and is caused by charge fluctuation in the channel, which in turn induces the fluctuation of compensating charge on the gate electrode. This gate-noise current is proportional to f^2 in HFETs.

The channel noise is represented by a drain-current noise generator $\overline{i_{nd}^2}$ and is caused by various physical mechanisms driven by the electric field in the channel. In the linear region of the device channel, the GCA region, the channel noise is caused by the thermal noise (Johnson noise). A thermal noise voltage caused in the channel leads to a modulation of the channel resistance and causes a drain voltage fluctuation at the channel end (drain). The corresponding drain noise current is inversely proportional to $g_m I_{DS}$ [60,67]. In the high-field region, hot electron scattering [68], intervalley scattering [69], and high diffusion noise [67] contribute to the channel noise. The drain noise current is proportional to I_{DS}/V_{sat}^3 [67]. The superposition of both channel noise contributions yields the noise figure as a function of the drain current (see Fig. 10.15).

Another noise source is gate leakage. A new model that takes this effect into account by an additional parallel resistor to the gate capacitance and the resistor R_i has been proven to yield a good correlation between predicted and measured minimum

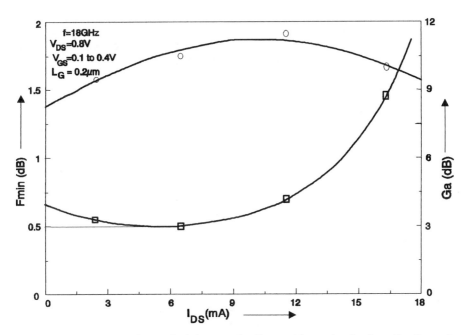

FIGURE 10.15 Dependence of minimum noise figure and associated gain as functions of drain–source current for 0.2-μm-gate InP-based HFET at room temperature. (From Ref. 123.)

noise figures even at low frequencies [71]. The negative influence of gate leakage on the noise figure vanishes at higher frequencies.

In InP-based HFETs, all noise sources relevant for channel noise are smaller than those in GaAs-based FETs. Reasons for this are: (1) thermal noise is inversely proportional to the transconductance, which is much higher in InP-based HFETs; (2) the larger energy separation between the Γ and L valleys reduces Γ–L intervalley scattering of electrons; (3) the larger electron velocity at the drain side of the gate leads to reduced diffusion noise; (4) the larger barrier height between the InGaAs channel and the InAlAs spacer and buffer layers reduces the hot electron transfer, hence the fluctuation of carrier velocity and eventually, the hot electron noise; and (5) higher transconductances reduce coupling between the channel and the gate electrode, which reduces the gate-current noise. These all lead to superior noise performance of InP-based HFETs over that of other transistor types, especially at higher frequencies. Figure 10.16 is a literature survey of the minimum noise figure as a function of the frequency in comparison with the standard GaAs HFET. The noise performance advantage vanishes at lower frequencies below about 10 GHz.

Another advantage of InP-based HFETs is that the impedance-matching conditions for maximum gain and minimum noise are very similar, hence allowing simultaneously high gain and low noise amplification at the same matching and bias condition (e.g., see Fig. 10.15). The noise resistance R_n of InP-based HFETs is smaller than for GaAs HFETs, which leads to less sensitivity of the minimum noise figure for bias variation. This smaller noise resistance R_n results from the higher transconductance. With reference to LNA application the higher gain allows the reduction of the second-stage noise contribution and thus the overall noise figure. Thus the minimum noise figure of HFETs exhibits the frequency dependence as [72]

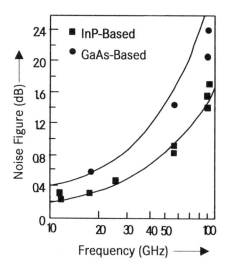

FIGURE 10.16 Noise performance of GaAs- and InP-based HFETs.

$$F_{\min} = 10\log(1 + cf). \tag{9}$$

In Refs. 72 and 73, values for $c = 8.6 \times 10^{-12}$ s, 7.4×10^{-12} s, and 4×10^{-12} s have been found experimentally for AlGaAs, AlGaAs/InGaAs, and InAlAs/InGaAs-InP HFETs, respectively, and this clearly demonstrates the advantage of InP-based HFETs in low-noise amplification (see also the solid lines in Fig. 10.16).

A further improvement in noise performance is achieved at cryogenic temperatures [74]. Noise temperatures under $T_n = 10$ K have been measured [5,75] for descrete devices at 18K ambient temperature. Receiver noise temperatures of 15 K at 40 to 50 GHz, 37 K at 60 GHz, and 50 K at 75 GHz have been measured [76]. These extremely good noise results are attributed to improved transport properties of the 2DEG (hence improved g_m and f_T), reduced coupling effect between $\overline{i_{ng}^2}$ and $\overline{i_{nd}^2}$, reduced thermally induced charge transfer to the buffer (improved g_d), and reduced gate leakage.

10.6.3 Power Operation

Power application of HFETs becomes reasonable at frequencies above 20 GHz. For lower frequencies, MESFETs and HBTs cover the performance requirements and cost restrictions. A device for power application at millimeter-wave frequencies should exhibit low leakage current; low knee voltage; high breakdown voltage; high open-channel current and high transconductance, which is synonymous with high f_T; and high and linear gain. In general, the power performance of InP-based HFETs is limited by the higher leakage current and lower breakdown voltage.

Recently, much effort has been made to improve gate leakage and breakdown characteristics in four ways: enhancement of the effective Schottky barrier height [77], reduction of peak electric field in the drift region [58], optimization of channel layer and spacer layer design [46], and improved process technology [78]. The Schottky barrier height was improved by using a high AlAs mole fraction in InAlAs layers or by introducing new materials, such as AlGaInP [79], as the Schottky layer. The reduction of peak electric field was achieved by using either double recess structures [80] or the surface-depleted cap-layer concept [58], which resulted in breakdown voltages close to those of GaAs-based HFETs. A composite InGaAs/InP [81] split-channel scheme and strained InAlAs spacer layers [46] improved the breakdown voltage and gate leakage behavior efficiently. A great stride ahead was made in gate leakage reduction by the introduction of mesa sidewall isolation [78], in which the channel layer was partially removed along the mesa slope, where the gate feed would otherwise get in contact with the InGaAs channel layer. These investigations have led to improvements in power performance of InP HFETs. Recent publications quote values for power-added efficiency which are comparable or even better than for GaAs HFETs. The performance achieved is listed in Table 10.1.

While InP-power HFETs are inferior below 60 GHz [84–86], they outperform GaAs-based HFETs at higher frequencies. The change in situation at higher frequencies may be due to the higher f_T value since power density remains constant up to $f = f_T$ and then drops drastically. The lower knee voltage and higher gain may also

TABLE 10.1 Power Performance of Power InP HFETs at Various Frequencies

P_{out}/W_G (mW·mm)	Power-Added Efficiency η (%)	Frequency f (GHz)	Ref.
0.35	30	60	82
0.35	41	60	5
0.26	26	94	5
0.3	33	94	83

be advantageous for the improvement of power-added efficiency. This is summarized in Fig. 10.17, which represents the status quo in the power performance of HFETs.

10.7 LAYER STRUCTURE DESIGN

10.7.1 Control of Channel Conductivity

In Section 10.6 the important influence of channel properties, μ, V_{peak}, and V_{sat}, on the device performance has been demonstrated. The channel conductivity is proportional to the product of n_s and μ. In this section the impact of individual layer structure design parameters on the channel conductivity is discussed.

The parameter terminology follows Fig. 10.3. The influence of the donor concentration N_D in the doping pulse d_d, the spacer layer thickness d_{sp}, and the channel thickness L_z on n_s has been discussed in Section 10.3. According to Eq. (1), n_s is proportional to $\sqrt{\Delta E_c}$, which can be increased by using strained channel layers. Due to the limit in layer thickness imposed by the critical layer thickness h_c, espe-

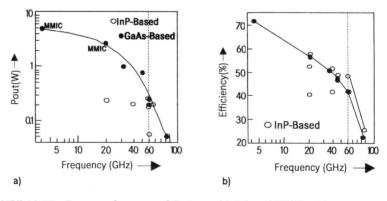

FIGURE 10.17 Power performance of GaAs- and InP-based HFETs: (*a*) output power versus frequency; (*b*) efficiency versus frequency.

cially for high values of InAs mole fraction x, L_z falls in the quantization-limited region. On the other hand, it has been shown that interface roughness of the back-side interface of the channel limits the quality of the transport properties for thicknesses $L_z < 30$ nm [87].

To relax these constraints the channel can be designed as a split channel, as done in Ref. 6. The approach is to subdivide the channel into a pseudomorphic part with a thickness h_c and a lattice-matched part with a thickness d_{sub} (see Fig. 10.3). To guarantee the utmost performance, the strained part is set to the critical layer thickness limit according to the mechanical equilibrium model [30]. The total channel thickness is then given by $L_z = h_c(x) + d_{sub}$ = constant = 40 nm. The sheet carrier concentration n_s as a function of the InAs mole fraction in the channel layer as calculated in Ref. 6 is shown in Fig. 10.18. The optimum value of n_s is 3.25×10^{12} cm^{-2} and is close to the maximum 2DEG concentration of a lattice-matched layer structure but at improved mobility (see Section 10.3). The reason that no higher values for n_s can be achieved is because the strain limits the maximum value of h_c. This also leads to the turning point in the n_s–x characteristic that maximizes n_s at $x = 0.75$.

10.7.2 Control of Drain Current and Transconductance

The impact of the total gate-to-channel separation on g_m and I_{DS} can be effectively analyzed by varying the barrier layer thickness d_u and the design of the doping pulse (N_D and d_d). The modeled influence of d_u on g_m and I_{DS} is shown in Fig. 10.19 [6]. The parameters used are $L_G = 0.25$ μm, $N_D d_d = 4 \times 10^{12}$ cm^{-2}, $R_D = 0.8$ Ω·mm, and $R_S = 0.4$ Ω·mm. The transconductance improves for reduced d_u and thus corresponds to the VSM [see Eq. (5)]. Another important effect is that the value of I_{DS} at the bias for maximum transconductance rises for reduced d_u. This parameter is very useful to design HFETs with a certain g_{mmax}–I_{DS} correlation at a certain bias point. It was also verified that a smaller value of d_d at a constant $N_d d_d$ product improves the transconductance, due to more efficient charge modulation [27]. For smaller

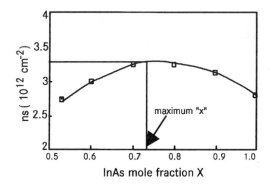

FIGURE 10.18 Calculated dependence of 2DEG carrier concentration as a function of InAs mole fraction in a In$_x$Ga$_{1-x}$As channel. (From Ref. 6.)

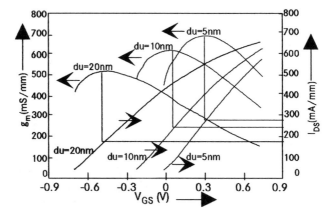

FIGURE 10.19 Calculated dependence of saturation current and transconductance as functions of gate bias with barrier layer thickness (d_u) as a parameter. (From Ref. 6.)

values of d_d, less voltage swing is necessary to modulate n_s [$g_m \sim (\Delta n_s/\Delta V_G)$]. Analysis that predicted the influence of the spacer layer thickness is straightforward. The larger the spacer layer d_{sp} becomes, the smaller the value of n_s and hence the lower the transconductance.

The way how the split channel mentioned in Section 10.7.1 should be designed in order to improve g_m and I_{DS} is shown in Fig. 10.20 [6]. Both g_m and I_{DS} do not improve monotonically with increased InAs mole fraction. They are now also functions of h_c and thus, in practice, functions of individual material growth conditions determining h_c. In most applications where a split-channel concept has been adopt-

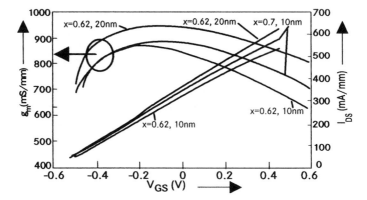

FIGURE 10.20 Calculated dependence of saturation current and transconductance as functions of gate bias for various InAs mole fractions in a HFET channel layer. There exists a maximum value for the InAs mole fraction up to which the HFET performance can be improved. (From Ref. 6.)

ed, values of $x \approx 0.7$ to 0.8 are found to be optimum for HFET application [1,6]. In Ref. 6 it was demonstrated that using the split-channel concept, an improvement in high-speed performance of about 40% can be achieved by using InAs molefractions of $0.53 < x < 0.7$.

To allow for effective modulation of the 2DEG by the gate and to prevent punchthrough of the drain potential to the source, the layer structure has to be designed so as to maintain Schockley's aspect ratio rule. This means that the ratio of the electrically effective gate length to the gate-to-2DEG separation (L_G/d_{eff}) should be much larger than unity. In practice, values of a factor of 3 and larger are acceptable. This becomes extremely problematic for gate lengths in the range 50 to 150 nm. Assuming an aspect ratio $L_G/d_{\text{eff}} = 5$ and a gate length of $L_G = 50$ nm, this requires an effective layer thickness of $d_{\text{eff}} = d_u + d_d + \Delta d = 10$ nm, which is difficult to control. For thicknesses of 10 nm and less, on the other hand, the gate-to-channel separation approaches the tunneling regime. Thus a gate length of 50 nm seems to be the lower limit in practical devices. Systematic investigations on this topic have been carried out [1,88].

The gate leakage (as a measure for breakdown) and output conductance characteristics could be improved drastically by using an undoped cap layer [89]. Since undoped cap-layer devices suffer from effective shielding of the 2DEG from changes in the surface potential and larger parasitic resistances, this approach was improved further by introducing the surface-depleted cap layer concept [58]. In this approach, the cap layer is entirely separated from mobile carriers by depletion due to surface states and the InAlAs barrier layer. The advantage is that the high-electric-field domain at the edges of the gate spreads out and moves toward the drain with increasing drain bias. The consequence is improved breakdown voltage, lower leakage current, and a drastically reduced feedback capacitance C_{GD}, in the range <0.1 pF/mm. This value is half that of a conventional layer structure. C_{GD} becomes a dominant feedback loss mechanism, limiting the power gain cutoff frequency, especially at high frequencies. The depleted-cap-layer concept therefore allows high f_{max}/f_T ratios [58,90]. However, one drawback of this approach is that f_T decreases due to a longer effective channel length. Although the advantage of this approach in achieving high values of f_{max} is huge, it has to be noted that the device with the higher f_T is better than another if the f_{max} values are the same.

10.8 DEVICE FABRICATION

10.8.1 Epitaxial Growth

Especially for short-gate ($L_G < 1$ μm) HFETs with small gate–channel distances, very thin layers with abrupt, smooth interfaces must be grown with precise and reproducible thickness control and good crystal quality. Furthermore, planar doping of the supply layer is a key to the support of optimum spacing between the carrier supply layer and the channel, on one hand, and between the supply layer and Schottky gate on the other.

Optimization of the Buffer Layer The electrical properties of structures grown on GaAs can be improved by the use of thick undoped GaAs buffer layers of high resistivity. Undoped lattice-matched InGaAs layers on InP suffer from low resistivity, caused by a high Si impurity level in the substrates, making InGaAs buffers useless for high-frequency devices. InAlAs buffer layers, especially those grown at reduced temperatures, have sufficiently high sheet resistance but degrade surface quality with increasing thickness. Optimum thickness can be evaluated by low-temperature 2DEG mobility measurements. Figure 10.21 shows the temperature dependence of 2DEG mobility in planar-doped HFET structures with InAlAs buffer layers between 50 and 100 nm thick [91]. Sheet carrier density is about 4×10^{12} cm^{-2} in all cases. At room temperature all devices have nearly identical mobilities of about 10,000 cm^2/V·s. The highest low-temperature mobility of 50,000 cm^2/V·s was achieved with a buffer layer thickness of 75 nm, which is about one order of magnitude smaller than for usual buffer layers for structures on GaAs. The advantages of such thin buffer layers are manifold. The total growth time compared to those of GaAs-based devices is reduced by nearly a factor of 2 making the postepitaxial costs for lattice-matched HFETs on InP comparable, despite the much higher substrate cost. Furthermore, mesa isolation by wet chemical etching can be done selectively down to the substrate with excellent reproducibility. Another advantage is that transmission lines and contact pads can be placed on a semi-insulating substrate while maintaining nearly planar surfaces.

Reproducibility and Material Quality The maintenance of a lattice-matching condition within a growth run, as well as run, to run is done more easily with gas-

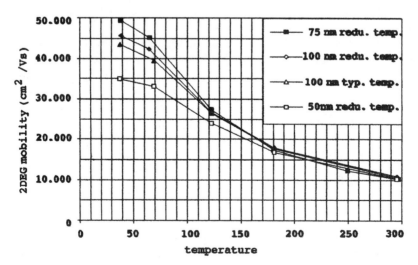

FIGURE 10.21 Temperature dependence of electron Hall mobility in pulse-doped HFET layer structures with different buffer layers. The abbreviation "redu.temp." represents a reduced growth temperature of the buffer layer, compared to typical growth temperatures. (From Ref. 91.)

source systems than with conventional MBE machines. On the other hand, the small thickness of HFET layers makes the structure insensitive to the lattice mismatch up to $\Delta a/a = 1$ to 2×10^{-3}. A slight deviation from the normal In content of about 53% and 52% in InGaAs and InAlAs, respectively, has nearly no effect on device performance. The sheet resistances of several standard HFET layers MBE grown on 2- and 3-in. InP substrates received from different suppliers are shown in Fig. 10.22 as an example. The sheet resistance of surface-depleted structures with a very thin doped cap layer is influenced primarily by the channel conductivity. The variation is less than 10% from a mean value of 115 Ω/square. Even a long duration of more than 6 months between runs has led to comparable results without recalibration.

10.8.2 Fabrication Process

The major fabrication technology of InP-based HFETs has been transferred from GaAs FET technology. Only a few process steps had to be refined to meet material inherent requirements. After material growth by either MBE or MOCVD, the process sequence consists of the following steps, which correspond to mask levels: (1) mesa isolation, (2) ohmic contact formation, (3) gate lithography (electron beam/mask), (4) metal resistors, (5) passivation, (6) air bridges, (7) wafer thinning, (8) via-hole formation, and (9) chip discretization. Process step 8 can be omitted when coplanar waveguide scheme is used (see also Section 10.10) [12,13,70,92].

Device isolation is usually performed by mesa etching. Using either wet chemical or dry etching processes, isolation grooves are etched down into the buffer or the semiconductor substrate. Implant isolation, which is easily done in GaAs, is difficult to apply to InGaAs. One particular problem for mesa isolation in InP-HFETs is the contact to be made between the gate feed and the InGaAs channel exposed along the mesa side walls. Due to the low Schottky barrier height of the

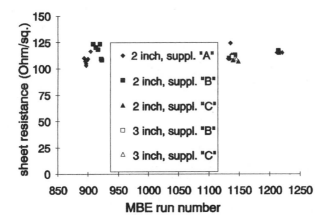

FIGURE 10.22 Reproducibility of the sheet resistance for surface-depleted cap-layer HFET layer structures. The growth run interruption represents 9 months. (After Ref. 91.)

gate metal to the InGaAs, a drastic increase occurs in the gate leakage current. Although the contact channel area L_zL_G is several hundred times smaller than the intrinsic gate area L_GW_G, it carries several times larger leakage current. To circumvent this problem, trench isolation [94] and mesa-sidewall etching have been developed [78].

Ohmic contacts are formed, alloyed or nonalloyed [93]. The formation of nonalloyed ohmic contacts is more feasible in InP-based HFETs since much higher doping levels can be achieved in InGaAs. For typical surface-depleted InAlAs/InGaAs HFETs, contact resistances are $R_K < 0.1$ Ω·mm. Practical benefits of nonalloyed ohmic contacts are the well-shaped edges of patterns and high reliability.

The most critical step in HFET technology is the gate process for high-speed operation. Lateral dimensions of state-of-the-art devices are in the range 50 to 100 nm [88]. To keep gate resistance small, a number of gate fabrication techniques have been invented [95,96]. They have in common a T- or diamond-shaped gate cross section, where the small footprint defines the gate length and the T-bar provides an increased metal cross section for lowering the gate resistance. The conventional way to realize such a structure is to use multilayer PMMA processes [95]. In recent years, especially for higher throughput, hybrid techniques such as dielectric-assisted gate fabrication have been developed [97,98]. A typical process flow is shown in Fig. 10.23. The process flow is as follows:

1. Deposition of a dielectric layer (e.g., polyimide). The material selection makes it possible to adjust its solubility against different solvents, which is helpful in the following process steps. After deposition, direct electron-beam exposure defines the gate foot print.
2. Transfer of the electron-beam pattern via oxygen plasma etching into the polyimide. The advantage of this process is that round profiles of the polyimide walls are achieved.
3. A second lithography step, to define the T-bar of the gate using a high-throughput wafer stepper.
4. Gate recessing.
5. Evaporation of gate metal.
6. Full passivation of the device surface.

The gate recess is the most critical step. It determines the L_G/d_{eff} ratio and hence the threshold voltage and transconductance. The typical recess depth is several tens of nanometers, with a precision of several nanometers. Material-selective etchants [52,96] and etch-stop-layer techniques [65] have been developed to control the recess depth precisely. HFETs are sensitive to surface effects, especially when the InAlAs is exposed to air. To improve device reliability, passivation is mandatory. Standard passivation is made by the deposition of dielectric films, typically Si_3N_4. Typical fabrication details for InP-based HFETs have been described in Ref. 96.

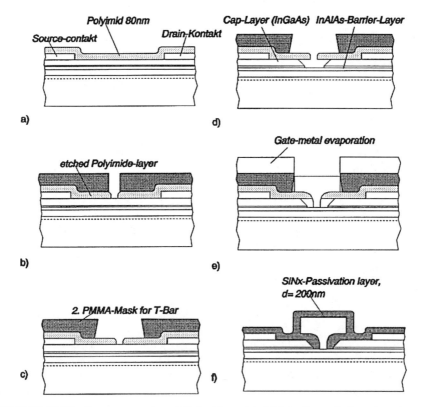

FIGURE 10.23 Process sequence of dielectric-assisted fabrication technique for sub-quarter-micron gates. (From Ref. 123.)

10.9 NOVEL PROCESSING AND MATERIALS

10.9.1 Dry Process for InAlAs/InGaAs-Based Structure

Wet chemical etchants are used for gate recessing. Typically, citric-based etchants have often been used, which oxidize the InAlAs with an etching selectivity of 25 to 30 [52,78,99]. To improve this further, dry etching techniques are utilized and yield selectivities up to 150 [100]. An elegant way to achieve excellent selectivity is the use of etch-stop layers in which separation between the gate and the center of the 2DEG is predetermined with the accuracy of epitaxy. A deviation in threshold voltage as small as 16 mV, reported for 0.1-μm InAlAs/InGaAs HFETs over 2-in. wafers [65], was sufficient for some digital applications [100]. The deviation in cutoff frequency was small; $f_{max} = 266 \pm 13$ GHz and $f_T = 189 \pm 6.1$ GHz. The split-channel concept described in Section 10.4 was used for high-speed performance.

10.9.2 Alternative Materials

InGaAs/InP and InGaAs/InGaAsP Heterojunctions An approach to improve power performance while maintaining high speed was proposed in Ref. 81 using an InGaAs/InP channel. As explained earlier, the 2DEG is three-dimensional in nature at the drain-side end of the gate, due to the high electric field [36,44], and it extends toward the buffer layer. In Section 10.3 it was shown that InP has a higher drift velocity at high electric fields. In addition, InP has a much lower impact ionization coefficient than that of InGaAs. Thus the split channel concept can combine the low-field transport property of InGaAs and the high-field property of InP to result in devices exhibiting higher breakdown and higher f_T values than those for conventional structures.

An extension of the split-channel concept is the use of a pure InP channel [101]. The electron mobility of 2500 cm^2/V·s with a 2DEG concentration of 3.6×10^{12} cm^{-2} has been realized in HFET layer structures. Even 5000 cm^2/V·s has been measured at room temperature and 27.000 cm^2/V·s at 77 K. A record power density of 1.45 W/mm at 30 GHz has been determined for a T-gate with a 0.3-μm gate length. The power-added efficiency was 24%, with a small signal gain of 6.2 dB. The cutoff frequencies were $f_{max} = 132$ GHz and $f_T = 80$ GHz. These results are very promising for power applications of InP-based HFETs.

An approach to capture all the benefits of a phosphor-containing channel is proposed in Ref. 102 by using the quarternary alloy (e.g., an In$_{0.73}$Ga$_{0.27}$As$_{0.6}$P$_{0.4}$ channel layer with a bandgap of 0.95 eV, being 0.2 eV larger than for the lattice-matched InGaAs). The larger bandgap is advantageous with respect to impact ionization since the threshold energy for impact ionization increases with the bandgap. On the other hand, the critical electric field necessary to achieve peak velocity is similar to that in InGaAs and thus provides comparably high mobilities. Carrier mobilities of 3900 and 10,000 cm^2/V·s were measured for a 2DEG concentration of $n_s = 1.5 \times 10^{12}$ cm^{-2} at 300 and 77 K, respectively. The channel breakdown voltage of 5 to 7 V was measured. The cutoff frequencies of $f_{max} = 105$ GHz and $f_T = 27$ GHz were quoted for a 1-μm-gate-length device.

Superlattice Channels An interesting approach to improving the electron mobility in the channel is the use of a (GaAs)$_m$(InAs)$_m$ superlattice channel [103,104], where m indicates the monolayer number. The use of a superlattice instead of bulk InGaAs should yield an effective reduction of alloy scattering, which was recognized in Section 10.3. A theoretical study of Ref. 103 recommended a superlattice period with $m \leq 4$, with a channel thickness of 10 to 20 nm. It was predicted that for small values of m, the superlattice band structure is essentially the same as for the lattice-matched alloy In$_{0.53}$Ga$_{0.47}$As, with negligible effect of lattice strain. With these small values of m, the channel has a perfect long-range order with no random potential fluctuation [103]. A $f_T = 28$ GHz value was measured for $L_G = 0.85$ μm gate length, which is encouraging [103]. The potential of this approach for low-noise application has yet to be proven.

Aluminium-Free HFETs As mentioned in Section 10.6, the use of InAlAs in the layer structure may cause problems. Roughness at the InGaAs/InAlAs interface and traps and defects in the InAlAs layers lead to an increase of low-frequency noise and are believed to contribute to the kink effect. A challenge is to achieve long-term stability of AlAs-containing materials, as has been anticipated in the AlGaAs system. Several groups have investigated alternatives to realize aluminum-free interfaces for InP-based HFETs. One approach using the modulation-doped principle realized an InP/InAs$_{0.6}$P$_{0.4}$ HFET [105]. In this structure InP works as the buffer and barrier layer and an InAsP layer is the channel. Electron mobilities of 6.500 and 55.300 cm^2/V·s for n_s = 2.2 × 10^{12} cm^{-2} were measured at 300 and 77 K, respectively. A value of f_{max} = 60 GHz was reported for an 0.5-μm-gate-length device, which shows the potential of such an approach. However, to achieve a reasonable Schottky barrier height for high device performance, an InAlAs layer was inserted in this experiment. Real Al-free HFETs can only be realized when the Schottky contact to the InP barrier layer is formed with junction-FET (JFET) like structures [64,94]. In Ref. 64 the Schottky layer was p-type doped. For an 0.25-μm-gate-length device with a p$^+$-InP/n$^-$-InP/InGaAs/n$^-$-InP structure [64], cutoff frequencies of f_T = 131 GHz and f_{max} = 152 GHz and a minimum noise figure of F_{min} = 0.9 dB were measured with an associated gain of G_a = 12 dB at 12 GHz for an unpassivated device. The two-terminal breakdown voltage was determined to be 16 V, which is extremely high for an InP-based device. Another approach to an Al-free JFET-like layer structure has been proposed [94]. The Schottky and spacer layers were InGaAs, and the Schottky layer was p-type doped. The channel layer was n-type InP. For a 0.6-μm-gate-length device, cutoff frequencies of f_T = 14.3 GHz and f_{max} = 37.5 GHz and two-terminal breakdown voltages in excess of 20 V were reported.

Metamorphic Approach Since InP substrates are more expensive than GaAs substrates, it is desirable to combine the advantages of heterostructures with the low cost and availability of 2- to 4-in. GaAs wafers [7]. Growing Ga$_{0.47}$In$_{0.53}$As directly on GaAs substrates instead of InP substrates results in a three-dimensional growth mode [106]. In this case the surface morphology and structural quality tend to be poor, and it therefore becomes necessary to design a buffer layer sequence to maintain two-dimensional growth. Epitaxy of strain-relaxed Al$_y$In$_{1-y}$As/Ga$_x$In$_{1-x}$As on GaAs with good material quality can be achieved by a stepwise or continuous change of In content (and thus the lattice constant) during the growth of a ternary Al$_y$In$_{1-y}$As or Ga$_x$In$_{1-x}$As buffer layer and by the use of different growth temperatures [107–109]. This relaxed buffer layer sequence represents a new substrate, and if the lattice constants of Al$_y$In$_{1-y}$As and Ga$_x$In$_{1-x}$As were the same, lattice-matched heterostructures with an arbitrarily chosen In content can be grown on this substrate. Thus lattice matching with the original substrate becomes less important.

A buffer layer concept to compensate the lattice misfit has been examined by varying the composition of Al$_x$Ga$_y$In$_{1-x-y}$As in step and linearly graded fashion [7], as shown in Fig. 10.24. Heterostructures were grown by MBE. The optimum growth temperature at a thermocouple readout was 600°C (pyrometer readout: 530°C) for

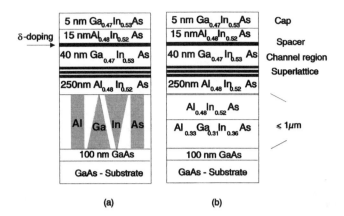

FIGURE 10.24 MBE-grown $Al_{0.48}In_{0.52}As/Ga_{0.47}In_{0.53}As$ HEMT structure with (*a*) linear-grading and (*b*) stepwise grading buffers. (Courtesy of K. Köhler, IAF.)

both lattice-matched growth on InP as well as lattice-relaxed growth on GaAs. Both types of buffers were grown at different growth temperatures (T_B) in the range 250 to 500°C (thermocouple readout). The step and linearly graded heterostructures both showed mirrorlike optical surface quality. Regarding microscopic surface morphology, cross hatching was observed for the linearly graded buffer structure, indicating the existence of misfit dislocations with a two-dimensional growth mode. The surface morphology of the step-graded buffer showed island formation and thus a rough surface due to a three-dimensional growth mode. The three-dimensional growth is caused by the quaternary $Al_{0.33}Ga_{0.31}In_{0.36}As$, having a lattice misfit greater than 2% with respect to GaAs [108].

The Hall mobility μ is plotted as a function of the buffer growth temperature T_B in Fig. 10.25. This shows higher-mobility μ_{77K} for linearly graded structures, with a maximum at $T_B = 450$°C. For step-graded structures a maximum at T_B of 400°C was measured. An electron concentration of 3.0×10^{12} cm^{-2} was measured for all samples. A decrease in the mobilities is observed for T_B values below 350°C and above 450°C for both buffer types. It is favorable to vary the lattice constant during buffer growth in steps as small as possible to achieve a continuous two-dimensional growth mode, thus providing good material and device properties [7,110]. The same HEMT structure grown lattice matched on InP yielded slightly better mobility than that of the linearly graded buffer. This difference may be attributed to the increase in interface roughness of the electron channel. Nevertheless, for a 0.3-μm gate length, values of $f_T = 90$ GHz and $f_{max} = 130$ GHz have been realized [110].

10.10 STATUS QUO HFETs AND MMICs

10.10.1 HFETs

InAlAs/InGaAs HFETs have shown the highest power gain cutoff frequency ($f_{max} = 600$ GHz, $x = 0.68$) and current gain cutoff frequency ($f_T = 340$ GHz, split channel,

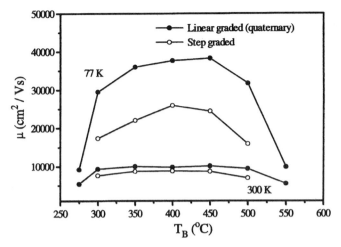

FIGURE 10.25 Hall mobilities at 300 and 77 K as functions of buffer-layer growth temperature. (Courtesy of K. Köhler, IAF.)

$x = 0.8$) among all types of transistors. It is interesting to note that these results have been achieved with pseudomorphic HFETs, which was also the method used for device improvement in GaAs HFETs. Figure 10.26 reviews the published cutoff frequencies for GaAs- and InP-based HFETs as a function of gate length. Comparison of the data clearly indicates two advantages of InP-based HFETs. First, InP-HFETs achieve higher cutoff frequencies for the same geometry and demonstrate the potential for applications over the full frequency band and beyond the 100-GHz border. Second, for a given application at a certain frequency, InP-based HFETs yield device performance comparable with that of GaAs devices but at larger gate geometries. This is advantageous for fabrication cost and yield, and hence for the overall cost constraints per device. This will become effective when the substrate cost decreases to a level comparable with that of GaAs.

To demonstrate the potential of InP-based HFETs at larger geometries, we have fabricated fully passivated lattice-matched, surface-depleted devices with 0.5-μm gate length and 80-μm gate width. The I_{DS}–V_{DS} curves are shown in Fig. 10.27. No kink effect is evident, and the two-terminal gate–drain breakdown voltage is determined to be 6 V, defined at $I_G = 1$ mA/mm. The device demonstrates a maximum transconductance of $g_m = 550$ mS/mm with associated $I_{DS} = 220$ mA/mm, which is comparable with production-relevant $L_G = 0.25$ μm GaAs HFETs. The mean values of cutoff frequencies over a 2-in. wafer are $f_T = 58$ GHz and $f_{max} = 140$ GHz. The minimum noise figure F_{min} is 0.7 dB at 12 GHz, with an associated gain of $G_a = 15$ dB. For a gate length of $L_G = 0.2$ μm, $g_m = 800$ mS/mm with an associated $I_{DS} = 240$ mA/mm, the two-terminal breakdown voltage is 5 V, $f_T = 130$ GHz, and $f_{max} = 250$ GHz. The minimum noise figure is $F_{min} = 0.5$ dB at 18 GHz with an associated gain of $G_a = 11$ dB. These are excellent values for fully passivated devices [123].

State-of-the-art devices at low-noise bias conditions offer an associated gain of 20, 9, and 7 dB at 18, 60, and 94 GHz, respectively. These values are typically 1 to 3 dB higher than for GaAs devices [2,72]. With increasing frequency, the associated

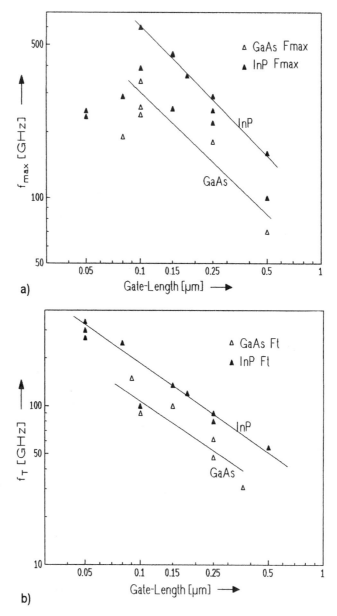

FIGURE 10.26 High-frequency performance of GaAs- and InP-based HFETs as a function of gate length. (*a*) power gain cutoff frequency f_{max}; (*b*) transit frequency f_T.

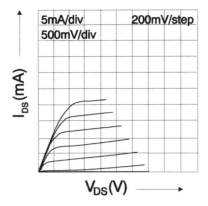

FIGURE 10.27 *I–V* characteristic of a 0.5-μm triangular gate InAlAs/InGaAs/InP HFET. (Courtesy of Daimler Benz.)

gain decreases and second-stage noise contribution in a multistage low-noise amplifier is no longer negligible. A figure of merit to clarify the benefit from higher associated gain is F_∞. F_∞ is the noise figure of a lossless infinite-stage amplifier, with each stage characterized by the noise figure and gain of the individual device [$F_\infty = 1 + (F - 1)/(1 - 1/G)$] [112].

10.10.2 MMICs

It is obvious from Table 10.2 that a higher associated gain leads to a lower F_∞ value and hence improved LNA noise. This proves that the InP HFET is best suited for low-noise applications. Thus the key analog circuit application of InP-based HFETs is for low-noise receiver front ends. Although InP-HFET-based MMIC technology is a relatively new technology, all relevant electrical functions to realize microwave and millimeter-wave T/R modules and front ends have been demonstrated. Among them are oscillators [113–115], power amplifiers [116], mixers [11,59,117], fre-

TABLE 10.2 Comparison of Noise Performance of GaAs and InP HFETs

Device Type	Frequency (GHz)	F_{min} (GHz)	G_a (dB)	F_∞ (dB)	Ref.
GaAs	18	0.50	15.1	0.52	111
GaAs	60	1.6	7.6	1.87	111
GaAs	94	2.4	5.4	3.09	111
InP	18	0.3	17.1	0.31	112
InP	60	0.9	8.7	1.03	98
InP	94	1.2	8.6	0.8	3
InP	94	1.3	7.2	1.39	4

quency modulators [11,59], circulators [12], and broadband and low-noise amplifiers [118–126].

Although microstrip line (MSL) is the conventional technique for the waveguide scheme, an increasing number of publications report the use of the coplanar waveguide (CPW) technique [12,13,70,92–93]. The attractive features and hence advantages over MSL are: (1) it is easier to fabricate (no substrate thinning and no backside processing required, thus less sensitive to process variations); (2) there is much higher fabrication yield and higher throughput; (3) source inductances are lower; (4) there is lower line coupling (better isolation); (5) radiation loss is lower; (6) circuits are generally more compact, due to better isolation performance; and (7) it is less sensitive to substrate thickness. There are, however, two disadvantages of CPW compared with MSL. First, the modeling for various patterns, including air-bridge crossover and edge effects, has not yet been well established. Second, the signal line and ground plane should be on the same plane, which means that the basic layout limitation for pattern density is tighter than for MSL.

High-performance low-noise amplifiers have been realized in MSL as well as in CPW. Noise figures range from 0.5 dB at 2 GHz [118] to 4.3 dB at 100 GHz [119]. Regardless of the waveguide scheme used, comparable noise figures can be realized. For example, at Q-band, a 2.4-dB noise figure is achieved for a two-stage CPW amplifier, compared with 2.3 dB for a MSL amplifier [70,120]. In both cases the gate length is 0.15 μm. For 0.1 μm for a MSL LNA, a minimum noise figure of F_{min} = 1.8 dB and an associated gain G_a = 26 dB were reported in Ref. 10. Extremely high frequency operation was achieved at 100 and 142 GHz [119,121]. At 100 GHz a minimum noise figure of 4.8 dB with 19 dB associated gain was reported for a three-stage amplifier using 0.1-μm-gate-length pseudomorphic (x = 0.6) HFETs.

Also noteworthy are the results for broadband amplifiers. A three-stage amplifier using 0.1-μm-gate-length pseudomorphic (x = 0.6) HFETs and the MSL technique showed a 3.3-dB minimum noise figure over 75 to 110 GHz with 11 dB asso-

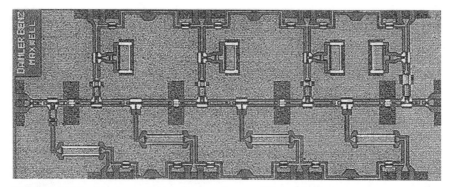

FIGURE 10.28 Four stage W-band low-noise amplifier with 0.25-μm-T-shaped gate InAlAs/InGaAs/InP HFET. Chip size is 1.3 × 3.4 mm². (From Ref. 123.)

ciated gain [122]. Using lattice-matched 0.25-µm-gate-length HFETs and the CPW technique, a four-stage amplifier was realized and a maximum gain of 15 dB with a minimum noise figure of less than 6dB was achieved. The 3-dB bandwidth ranged from 78 to 100 GHz (see Fig. 10.28). Beyond feedback amplifiers, lossy or reactively matched amplifiers and distributed amplifiers are most attractive for extremely broad bands. They typically show a large bandwidth and flat gain performance over the frequency band. Using a seven-stage distributed amplifier, a 5-dB gain was reported over a bandwidth of 5 to 100 GHz. This is the largest bandwidth for MMICs reported so far [124]. Using a feedback amplifier approach, a 17-dB gain over a bandwidth of 0.1 to 70 GHz was reported [125]. The noise figure was 5.8 dB (f = 8 to 16 GHz). The P_{1dB} value was measured to be 6 dB and the IP3 value was −23 dBc.

A comparable distributed amplifier using a cascode HFET configuration achieved 11 dB gain in 5 to 75 GHz, a 4.5-dB minimum noise figure and P_{1dB} = 10 dBm with IP3 = −27 dBc. The cascode configuration is a serial connection of two HFETs, one in common-source configuration followed by the second in common-gate configuration. The advantage of this configuration is in a drastic reduction of the feedback capacitance (Miller capacitance) and an improvement in the output resistance. At millimeter-wave frequencies the reverse isolation and output impedance are therefore improved. Thus the power gain is much higher than that of conventional common-source devices. For a balanced four-stage approach, a frequency band of 75 to 110 GHz, a gain of 23 dB, and a minimum noise figure of 6 dB at 94 GHz were reported [126].

A balanced power amplifier approach for high-efficiency Q-band operation was reported in Ref. 115. The amplifier performance reported was 33% power-added ef-

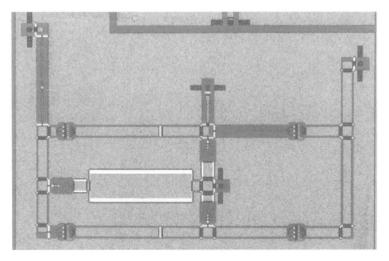

FIGURE 10.29 Active quasi-circulator using 0.25-µm-T-gate InAlAs/InGaAs/InP HFETs. Chip size is 1.75 × 1.9 mm². (From Ref. 12.)

ficiency with 26 dB maximum output power at 44 GHz. A CPW active quasi-circulator using InP-based HFETs as the active devices has been reported [12] for operation at 40 GHz (see Fig. 10.29). These MMICs are useful for signal separation of T/R signals. In this approach, HFETs switching transmission lines of different lengths allows either transmission or interference for different ports.

The advantage of HFETs in mixer applications includes higher conversion gain and lower noise figure compared with MESFETs and HBTs, for example. A potential drawback may be the higher low-frequency noise, which becomes disadvantageous for downconversion in the IF band. However, attractive mixer applications for up- and downconversion have been reported for W-band operation [59]. An active mixer has demonstrated 2.4-dB conversion gain with a 7.3-dB noise figure at 94-GHz RF and 85-GHz local oscillator (Lo) frequency. Complete downconverter, mixer, and LNA operations were realized at the same Lo and RF frequencies and showed a 3.6-dB noise figure and a 17.8-dB conversion gain. A balanced diode mixer for W-band operation using the same diode design as that used for HFETs was reported in Ref. 116, showing the advantage of integrating HFET amplifiers with HFET-like diode mixer circuits.

As explained earlier, InP-based HFETs suffer from their higher low-frequency noise. Therefore, spectrally clean (low phase noise) oscillators are difficult to realize. However, high-frequency baseband oscillators have been reported from several groups. A three-terminal transistor oscillator operating in the range 155 to 215 GHz has been reported [112]. A fully integrated D-band oscillator-doubler chain has been reported [113]. In the frequency band 130 to 132.8 GHz, an output power of 12 dB was measured. At Ka-band a dual-feedback HFET oscillator demonstrated 8.2 mW output power with a dc-to-RF conversion efficiency of 36%. InP-based HFETs are therefore applicable for oscillators operating at extremely high frequencies [114].

10.11 SUMMARY AND FUTURE CHALLENGES

In this chapter we have provided an overview of InP-based HFETs. It has been shown that InP-based HFETs are clearly the choice for low-noise, high-frequency applications. Performance advantages compared with those of GaAs-based HFETs have been shown by discrete devices and MMICs. Most of the electrical functions necessary for the realization of front ends for communication and sensor (radar) applications have already been demonstrated.

However, several problems remain to be solved. The most challenging is an improvement in RF output power. Others include the establishment of techniques to avoid the kink effect and to reduce low-frequency noise. To make InP-based HFET technology feasible for large-volume applications, useful techniques to reduce semiconductor wafer cost have to be developed. There is no doubt, however, that even today, InP-based HFETs are very attractive for applications where high performance is the key issue.

ACKNOWLEDGMENTS

The author would like to thank Dr. Koehler (Fraunhofer Institute for Applied Solid State Physics, Freiburg, IAF, Germany) for his contribution on methamorphic topics and Dr. Hachbarth (DaimlerChrysler Research Center, Ulm, Germany) for his contribution on material growth.

REFERENCES

1. L. D. Nguyen, A. Brown, M. A. Thomson and L. M. Jelloian, "50 nm InP high electron mobility transistors," *Microwave J.,* June 1993, pp. 96–101.
2. P. Smith, "InP based HEMTs for microwave and millimeterwave applications," *Proc. 7th Int. Conf. Indium Phosphide and Related Materials,* Sapporo, Japan, 1995, pp. 68–72.
3. K. H. G. Duh, P. C. Chao, S. M. J. Liu, P. Ho, M. Y. Kao, and J. M. Ballingall, "A super low-noise 0.1 μm T-gate InAlAs–InGaAs–InP HEMT", *IEEE Microwave Guided Wave Lett.,* **1**(5), 114–116 (1991).
4. D. C. Streit, K. L. Tan, P. H. Liu, and P. D. Chow, "MBE growth and characterization of lattice-matched and Pseudomorphic InGaAs/InAlAs/InP HEMTs." *Proc. 4th Int. Conf. Indium Phosphide and Related Materials,* Newport, R.I., 1992, pp. 682–684.
5. M. Y. Kao, P. M. Smith, P. C. Chao, and P. Ho, "Millimeter wave power performance of InAlAs/InGaAs/InP HEMTs," *Proc. IEEE/Cornell Conf. Advanced Concepts in High Speed Semiconductor Devices and Circuits,* 1991, pp. 469–477.
6. J. Dickmann, K. Riepe, A. Geyer, B. E. Maile, A. Schurr, M. Berg, and H. Daembkes, "$In_{0.52}Al_{0.48}As/In_xGa_{1-x}As$ ($0.53 < x < 1.0$) pseudomorphic high electron mobility transistors with high breakdown voltages: design and performances," *Jpn. J. Appl. Phys.,* **35**, 10–15(1996).
7. M. Haupt, K. Köhler, P. Ganser, S. Emminger, S. Müller, and W. Rothemund, "Growth of high quality $Al_{0.48}In_{0.52}As/Ga_{0.47}In_{0.53}As$ heterostructures using strain relaxed $Al_xGa_yIn_{1-x-y}As$ buffer layers on GaAs," *Appl. Phys. Lett.,* **69**(3), 412–414 (1996).
8. H. Rhodin, A. Nagy, V. Robbins, C. Su, C. Maddan, A. Wakita, J. Raggio, and J. Seeger, "Low-noise high-speed $Ga_{0.47}In_{0.53}As/Al_{0.48}In_{0.52}As$ 0.1 μm MODFETs and high-gain/bandwidth three-stage amplifier fabricated on GaAs," *Proc. 7th Int. Conf. Indium Phosphide and Related Materials,* Sapporo, Japan, 1995, pp. 73–76.
9. H. Masato, T. Matsuno, and K. Inoue, "$In_{0.5}Ga_{0.5}As/InAlAs$ modulation doped field effect transistors on GaAs substrates grown by low-temperature molecular beam epitaxy," *Jpn. J. Appl. Phys.,* **30**, 3850–3852 (1991).
10. L. Tran, R. Isobe, M. Delaney, R. Rhodes, D. Jang, J. Brown, L. Nguyen, M. Le, M. Thomson, and T. Liu, "High performance, high yield millimeter-wave MMIC LNAs using InP HEMTs," *Dig. IEEE Microwave Theory and Techniques Symp.,* 1996, pp. 9–12.
11. L. Tran, M. Delaney, R. Isobe, D. Jang, and J. Brown, "Frequency translation MMICs using InP HEMT technology." *Dig. IEEE Microwave Theory and Techniques Symp.,* 1996, pp. 261–264.
12. M. Berg, T. Hackbarth, B. E. Maile, S. Koßlowski, and J. Dickmann, "Active circulator

MMIC in CPW technology using quarter micron InAlAs/InGaAs/InP HFETs," *Proc. 8th Int. Conf. Indium Phosphide and Related Materials,* Schwäbisch Gmünd, Germany, 1996, pp. 68–71.

13. M. Berg, J. Dickmann, R. Guehl, and W. Bischof, "60 and 77-GHz monolithic amplifiers utilizing InP-based HEMTs and coplanar waveguides," *Microwave Opt. Technol. Lett.,* **11**(3), 139–145 (1996).

14. S. Weinreb, "Monolithic integrated circuit imaging radiometers," *Dig. IEEE Microwave Theory and Techniques Symp.,* 1991, pp. 405–407.

15. J. W. Waters, "Submillimeter heterodyne spectroscopy and remote sensing of the upper atmosphere," *Dig. IEEE Microwave Theory and Techniques Symp.,* 1991, pp. 391–393.

16. L. Raffaeli and E. Stewart, "Millimeter-wave monolithic components for automotive applications," *Microwave J.,* February 1992, pp. 22–32.

17. F. Stern and W. E. Howard, "Properties of semiconductor surface inversion layers in the electric quantum limit," *Phys. Rev. B,* **163,** 816–835 (1967).

18. L. Esaki, "Semiconductor superlattices and quantum wells," *Proc. 17th Int. Conf. Physics of Semiconductors,* Springer-Verlag, New York, 1984, p. 473.

19. R. Dingle, H. L. Störmer, A. C. Gossard, and W. Wiegmann, "Electron mobilities in modulation doped semiconductor heterojunction superlattices," *Appl. Phys. Lett.,* **33**(7), 665–667 (1987).

20. T. J. Drummond, W. T. Masselink, and H. Morkoç, "Modulation-doped GaAs/(Al,Ga)As heterojunction field-effect transistors: MODFETs," *Proc. IEEE,* **74**(6), 73–821 (1986).

21. D. Delagebeaudeuf and T. Linh, "Metal/(n)AlGaAs/GaAs two-dimensional electron gas FET," *IEEE Trans. Electron Devices,* **29,** 955–960 (1982).

22. T. J. Drummond, H. Morkoç, K. Lee, and M. Shur, "Model for modulation doped field effect transistors," *IEEE Electron Device Lett.,* **3,** 338–341 (1982).

23. K. Lee, M. Shur, T. J. Drummond, and H. Morkoç, "Current–voltage and capacitance–voltage characteristics of modulation-doped field-effect transistors," *IEEE Trans. Electron Devices,* **30,** 207–212 (1983).

24. B. Vinter, "Subbands and charge control in a two dimensional electron gas field-effect transistor," *Appl. Phys. Lett.,* **44**(3), 307–309 (1984).

25. K. S. Yoon, G. B. Stringfellow, and R. J. Huber, "Two-dimensional electron gas density calculation in $Ga_{0.47}In_{0.53}As/Al_{0.48}In_{0.52}As$, $Ga_{0.47}In_{0.53}As/InP$, and $Ga_{0.47}In_{0.53}As/InP/Al_{0.48}In_{0.52}As$ heterostructures," *J. Appl. Phys.,* **66**(12), 5915–5919 (1989).

26. Y. Nakata, S. Sasa, Y. Sugiyama, T. Fuji, and S. Hiyamizu, "Extremely high 2DEG concentration in selectively doped $Ga_{0.47}In_{0.53}As/n-Al_{0.48}In_{0.52}As$ heterostructures grown by MBE," *Jpn. J. Appl. Phys.,* **26**(1), L59–L61 (1987).

27. M. C. Foisy, P. J. Tasker, B. Hughes, and L. F. Eastman, "The role of inefficient charge modulation in limiting the current gain cut-off frequency of the MODFET," *IEEE Trans. Electron. Devices,* **35**(7), 871–877 (1988).

28. P. C. Chao, M. S. Shur, R. C. Tiberio, K. H. G. Duh, P. M. Smith, J. M. Ballingall, P. Ho, and A. A. Jabra, "Dc and microwave characteristics of sub-0.1 μm gate-length planar-doped pseudomorphic HEMTs," *IEEE Trans. Electron Devices,* **36**(3), 461–471 (1989).

29. H. Morkoç, T. Henderson, W. Kopp, and C. K. Peng, "High frequency noise of $Al_xGa_{1-x}As/In_yGa_{1-y}As$ MODFETs and comparison to $Al_xGa_{1-x}As/GaAs$ MODFETs," *Electron. Lett.,* **22**(11), 578–579 (1986).

30. J. W. Matthews and A. Blakeslee, "Defects in epitaxial multilayers," *J. Cryst. Growth*, **27,** 118–125 (1974).
31. R. People and J. C. Bean, "Calculation of critical layer thickness versus lattice mismatch for Ge_xSi_{1-x}/Si strained-layer heterostructures," *Appl. Phys. Lett.*, **47**(3), 322–324 (1985).
32. P. Chu and H. H. Wieder, "Properties of strained $In_xAl_{1-x}As$/InP heterostructures," *J. Vac. Sci. Technol. B*, **6**(4), 1369–1372 (1988).
33. D. Chattopadhyay, "Electron mobility in InGaAs quantum wells," *Phys. Rev. B*, **38**(18), 13429–13431 (1988).
34. K. Inoue, J. C. Harmand, and T. Matsuno, "High quality $In_xGa_{1-x}As$/InAlAs modulation doped heterostructures grown lattice mismatched on GaAs substrates," *J. Cryst. Growth*, **111,** 313–317 (1991).
35. T. Matsuoka, E. Kobayashi, K. Taniguichi, C. Hamaguchi, and S. Sasa, "Temperature dependence of electron mobility in InGaAs/InAlAs heterostructures," *Jpn. J. Appl. Phys.*, **29**(10), 2017–2025 (1990).
36. J. L. Thobel, L. Baudry, A. Cappy, P. Bourel, and R. Fauquembergue, "Electron transport of strained $In_xGa_{1-x}As$," *Appl. Phys. Lett.*, **56**(4), 346–348 (1990).
37. Y. Ando and T. Itoh, "DC, small-signal, and noise modeling for two-dimensional electron gas field-effect transistors based on accurate charge-control characteristics," *IEEE Trans. Electron Devices*, **37**(1), 67–75 (1990).
38. G. I. Ng, D. Pavlidis, M. Jaffe, J. Singh, and H. F. Chau, "Design and experimental characteristics of strained $In_{0.52}Al_{0.48}As/In_xGa_{1-x}As$ ($x > 0.53$) HEMTs," *IEEE Trans. Electron Devices*, **36**(10), 2249–2259 (1989).
39. I. C. Kizilyalli, K. Hess, J. L. Larson, and D. J. Widiger, "Scaling properties of high electron mobility transistors," *IEEE Trans. Electron Devices*, **33**(10), 1427–1433 (1986).
40. K. Lee, M. S. Shur, T. J. Drummond, and H. Morkoç, "Parasitic MESFET in (Al,Ga)As/GaAs modulation doped FETs and MODFET characterization," *IEEE Trans. Electron Devices*, **31**(1), 29–35 (1984).
41. W. Shockley, "A unipolar field-effect transistor," *Proc. IRE*, **40,** 1365–1376 (1952).
42. R. E. Williams and D. W. Shaw, "Graded channel FET's: improved linearity and noise figure," *IEEE Trans. Electron Devices*, **25,** 600–605 (1978).
43. L. F. Eastman, "III–V heterojunction field-effect transistor using indium alloys," *Tech. Dig. Int. Electron Device Meeting*, 1986, pp. 456–459.
44. J. L. Pelouard, R. Castagne, and P. Hesto, "Monte-Carlo study of ballistic transport in heterojunction bipolar transistors (HBTs) and in high electron mobility transistors (HEMTs)," in *Integrated Circuits, SPIE Proc.*, **795,** 41–57 (1987).
45. L. F. Palmateer, P. J. Tasker, W. J. Schaff, L. D. Nguyen, A. N. Lepore, and L. F. Eastman, "Observation of excess gate current due to hot electrons in 0.2µm gate length 100GHz f_T AlInAs/InGaAs/InP MODFETs," *Proc. Int. Symp. GaAs and Related Compounds,* Atlanta, Ga, 1987, Inst. Phys. Conf. Ser. No. 96, 1988, pp. 449–454.
46. C. Heedt, F. Buchali, W. Prost, W. Brockerhoff, D. Fritzsche, H. Nickel, R. Lösch, W. Schlapp, and F. J. Tegude, "Drastic reduction of gate leakage in InAlAs/InGaAs HEMTs using a pseudomorphic InAlAs hole barrier layer," *IEEE Trans. Electron Devices*, **41**(10), 1685–1690 (1994).
47. L. F. Palmateer, P. J. Tasker W. J. Schaff, L. D. Nguyen, and L. F. Eastman, "DC and rf

measurements of the kink effect in AlInAs/GaInAs/InP modulation doped field-effect transistors," *Appl. Phys. Lett.*, **54**(21), 2139–2141 (1989).

48. S. C. Palmateer, P. A. Maki, W. Katz, A. R. Calawa, J. C. M. Hwang, and L. F. Eastman, "The influence of V:III flux ratio on unintentional impurity incorporation during molecular beam epitaxial growth," *Proc. Int. Symp. GaAs and Related Compounds*, 1983, Inst. Phys. Conf. Ser. No. 7, 1984, pp. 217–222.

49. W. P. Hong, J.-E. Oh, P. K. Battacharya, and T. E. Tiwald, "Interface states in modulation doped $In_{0.52}Al_{0.48}As/In_{0.53}Ga_{0.47}As$ heterostructures," *IEEE Trans. Electron Devices*, **35**(10), 1585–1590 (1988).

50. A. S. Brown, U. K. Mishra, L. E. Larson, and S. E. Rosenbaum, "The elimination of dc *I–V* anomalies in GaInAs–AlInAs HEMT structures," *Proc. Int. Symp. GaAs and Related Compounds*, 1987, Inst. Phys. Conf. Ser. No. 96, 1988, pp. 445–448.

51. A. Brown and U. K. Mishra, "AlInAs–GaInAs HEMTs utilizing low-temperature alinas buffers grown by MBE," *IEEE Electron Device Lett.*, **10**(12), 565–567 (1989).

52. C. Heedt, "Fabrication and characterization of InAlAs/InGaAs/InP heterostructure field-effect transistors with low gate-leakage currents," dissertation, University of Duisburg, July 1995.

53. J. Dickmann, S. Schildberg, K. Riepe, B. E. Maile, A. Schurr, A. Geyer, and P. Narozny, "Breakdown mechanisms in pseudomorphic $InAlAs/In_xGa_{1-x}As$ high electron mobility transistors on InP:I:off-state," *Jpn. J. Appl. Phys.*, **34**, 66–71 (1995).

54. J. Dickmann, S. Schildberg, K. Riepe, B. E. Maile, A. Schurr, A. Geyer, and P. Narozny, "Breakdown mechanisms in pseudomorphic $InAlAs/In_xGa_{1-x}As$ high electron mobility transistors on InP:II:on-state," *Jpn. J. Appl. Phys.*, **34**, 1805–1808 (1995).

55. S. Bahl, J. del Alamo, J. Dickmann, and S. Schildberg, "Off-state breakdown in InAlAs/InGaAs MODFETs," *IEEE Trans. Electron Devices*, **42**(1), 15–22 (1995).

56. C. A. Liechti, "Microwave field-effect transistors–1976," *IEEE Trans. Microwave Theory Tech.*, **24**, 279–300 (1976).

57. M. B. Das, "Millimeter-wave performance of ultrasubmicrometer-gate field-effect-transistors: a comparision of MODFET, MESFET and PBT structures," *IEEE Trans. Electron Devices*, **34**(7), 1429–1440 (1987).

58. J. Dickmann, H. Dämbkes, H. Nickel, R. Lösch, W. Schlapp, J. Böttcher, and H. Künzel, "Influence of surface layers on the RF-performance of InAlAs/InGaAs HFETs," *IEEE Microwave Guided Wave Lett.*, **2**(2), 472–474 (1992).

59. P. D. Chow, K. Tan, D. Garske, P. Liu, and H. C. Yen, "Ultra low noise high gain W-band InP-based HEMT downconverter," *Dig. IEEE Microwave Theory and Techniques Symp.*, 1991, pp. 1041–1044.

60. A. van der Ziel, "Thermal noise in field-effect transistors," *Proc. IRE*, **44**, 811–818 (1956).

61. G. I. Ng, D. Pavlidis, M. Tutt, R. M. Weiss, and P. Marsh, "Low-frequency noise characteristics of lattice-matched ($x = 0.53$) and Strained ($x > 0.53$) $In_{0.52}Al_{0.48}As/In_xGa_{1-x}As$ HEMTs," *IEEE Trans. Electron Devices*, **39**(3), 523–532 (1992).

62. F. Danneville, G. Dambrine, H. Happy, and A. Cappy, "Influence of the gate leakage current on the noise performance of MESFETs and MODFETs," *Dig. IEEE Microwave Theory and Techniques Symp.*, 1993, pp. 373–376.

63. M. S. Thurairaj, M. B. Das, J. M. Ballingall, P. Ho, P. C. Chao, and Y. Kao, "Low-

frequency noise behaviour of 0.15μm gate-length lattice matched and lattice-mismatched MODFETs on InP substrates," *IEEE Electron Device Lett.*, **12**(8), 410–412 (1991).

64. M. Küsters and K. Heime, in minicourse text by J. Dickmann, "InP-based heterostructure FETs: design, manufacture, performance and applications," *Proc. 8th Int. Conf. Indium Phosphide and Related Materials,* Schwäbisch Gmünd, Germany, 1996.

65. T. Enoki, H. Ito, K. Ikuta, and Y. Ishii, "0.1 μm InAlAs/InGaAs HEMTs with an InP-recess-etch stopper grown by MOCVD," *Proc. 7th Int. Conf. Indium Phosphide and Related Materials,* Sapporo, Japan, 1995, pp. 81–84.

66. H. Statz, H. A. Haus, and R. A. Pucel, "Noise characteristics of gallium arsenide field-effect transistors," *IEEE Trans. Electron Devices,* **21,** 549–562 (1974).

67. C. Bergamachi, W. Patrick, and W. Bächtold, "Determination of the noise source parameters in InAlAs/InGaAs HEMT heterostructures based on measured noise temperature dependence on the electric field," *Proc. 6th Int. Conf. Indium Phosphide and Related Materials,* Paris, 1994, pp. 21–24.

68. F. Klaassen, "On the influence of hot carrier effects on the thermal noise of field effect transistors," *IEEE Trans. Electron Devices,* **17,** 858–862 (1970).

69. W. Bächtold, "Noise behaviour of GaAs field effect transistors with short gate-length," *IEEE Trans. Electron Devices,* **19,** 674–680 (1972).

70. Y. Umeda, T. Enoki, K. Arai, and Y. Ishii, "High-performance InAlAs/InGaAs HEMTs and their application to 40GHz monolithic amplifier," *Proc. Int. Conf. Solid State Devices and Materials,* Tsukuba, Japan, 1992, pp. 573–575.

71. R. Reuter, S. van Waasen, and F. J. Tegude, "A new model of HFETs with special emphasis on gate-leakage," *IEEE Electron Device Lett.,* **16**(2), 74–76 (1995).

72. P. M. Smith, "Status of InP HEMT technology for microwave receiver applications," *Dig. IEEE Microwave Theory and Techniques Symp.,* 1996, pp. 5–8.

73. K. H. G. Duh, S. M. Liu, L. F. Lester, P. C. Chao, P. M. Smith, M. B. Das, B. R. Lee, and J. M. Ballingall, "Ultra-low noise characteristics of millimeter-wave high electron mobility transistors," *IEEE Electron Device Lett.,* **9,** 521–524 (1988).

74. H. Mattes and M. Pilz, "Cryogenically coolable HEMT amplifiers in the frequency range 2–45GHz," *Proc. European Microwave Conf.,* 1994, pp. 1–6.

75. M. W. Pospieszalski, L. D. Nguyen, M. Liu, M. A. Thompson, and M. J. Delaney, "Very-low noise and low power operation of cryogenic AlInAs/GaInAs/InP HFETs," *Dig. IEEE Microwave Theory and Techniques Symp.,* 1994, pp. 1345–1347.

76. M. W. Pospieszalski, W. J. Lakatosh, L. D. Nguyen, M. Lui, T. Liu, M. A. Thompson, and M. J. Delaney, "Q- and E-band cryogenically coolable amplifiers using AlInAs/GaInAs/InP HEMTs," *Dig. IEEE Microwave Theory and Techniques Symp.,* 1995, pp. 1121–1124.

77. J. J. Brown, A. S. Brown, S. E. Rosenbaum, A. S. Schmitz, M. Matloubian, L. E. Larson, M. A. Melendes, and M. A. Thompson, "Study of the dependence of $Ga_{0.47}In_{0.53}As/Al_xIn_{1-x}As$ power HEMT breakdown voltage on Schottky layer design and device layout," presented at 51st Device Research Conf., June 1993, Santa Barbara, Calif.

78. S. R. Bahl, M. H. Leary, and J. A. del Alamo, "Mesa sidewall gate-leakage in InAlAs/InGaAs HFETs," *IEEE Trans. Electron Devices,* **39**(9), 2037–2043 (1992).

79. K. B. Chough, W. P. Hong, C. Caneau, J. I. Song, and J. R. Hayes, "OMCVD grown AlInAs/InGaAs HEMTs with AlGaInP Schottky layer," presented at 51st Device Research Conf., June 1993, Santa Barbara, Calif.
80. K. Y. Hur, R. A. McTaggert, B. W. LeBlanc, W. E. Hoke, P. J. Lemonias, A. B. Miller, T. E. Kazior, and L. M. Aucoin, "Double recess AlInAs/GaInAs/InP HEMTs with high breakdown voltages," *Proc. GaAs IC Symp.,* 1995, pp. 101–104.
81. T. Enoki, K. Arai, T. Akazaki, and Y. Ishii, "Novel channel structures for high frequency InP-based HFETs," *IEICE Trans. Electron.,* **E76-C**(9), 1402–1411 (1993).
82. M. Matloubian, L. M. Jelloian, A. S. Brown, L. D. Nguyen, L. E. Larson, M. J. Delaney, M. A. Thompson, R. A. Rhodes, and J. E. Pence, "V-band high-efficiency high-power AlInAs/InGaAs/InP HEMTs," *IEEE Trans. Microwave Theory Tech.,* **41**(12), 2206–2210 (1993).
83. P. M. Smith, S.-M. J. Liu, M. Y. Kao, P. Ho, S. C. Wang, K. H. G. Duh, S. T. Fu, and P. C. Chao, "W-band high efficiency InP-based power HEMT with 600 GHz f_{max}," *IEEE Microwave Guided Wave Lett.,* **5**(7), 230–232 (1995).
84. M. Matloubian, A. S. Brown, L. D. Nguyen, L. E. Larson, M. A. Melendes, and M. A. Thompson, "High power and high efficiency AlInAs/InGaAs on InP HEMTs," *Dig. IEEE Microwave Theory and Techniques Symp.,* 1991, pp. 721–724.
85. M. Matloubian, A. S. Brown, L. D. Nguyen, M. A. Melendes, L. E. Larson, M. J. Delaney, M. A. Thompson, R. A. Rhodes, and J. E. Pence, "20-GHz high efficiency AlInAs/InGaAs on InP power HEMT," *IEEE Microwave Guided Wave Lett.,* **3**(5), 142–144 (1993).
86. K. Y. Hur, R. A. McTaggert, M. P. Ventresca, R. Wohlert, L. M. Aucoin, and T. E. Kazior, "High performance millimeter wave AlInAs/InGaAs/InP HEMTs with individually grounded source finger bias," *Dig. IEEE Microwave Theory and Techniques Symp.,* 1995, pp. 465–468.
87. W. P. Hong, J. Singh, and P. K. Battacharya, "Interface roughness scattering in normal $In_{0.53}Ga_{0.47}As$–$In_{0.52}Al_{0.48}As$ modulation doped heterostructures," *IEEE Electron Device Lett.,* **L-7**(8), 480–482 (1986).
88. L. D. Nguyen, A. S. Brown, M. A. Thompson, and L. M. Jelloian, "50nm-self-aligned-gate pseudomorphic AlInAs/InGaAs high electron mobility transistors," *IEEE Trans. Electron Devices,* **39**(9), 2007–2014 (1992).
89. Y. C. Pao, C. K. Nishimoto, R. Majidi-Ahy, J. Archer, N. G. Bechtel, and J. S. Harris, "Characterization of surface-undoped $In_{0.52}Al_{0.48}As/In_{0.53}Ga_{0.47}As/InP$ high electron mobility transistors," *IEEE Trans. Electron Devices,* **37**(10), 2165–2169 (1990).
90. J. Dickmann, H. Haspeklo, A. Geyer, H. Dämbkes, and R. Lösch, "High performance fully passivated InAlAs/InGaAs/InP HFET," *Electron. Lett.,* **28**(7), 647–649 (1992).
91. T. Hackbarth, M. Berg, B. E. Maile, F.-J. Berlec, and J. Dickmann, "MBE growth of lattice matched HFETs on InP: material quality and reproducibility," *Proc. 8th Int. Conf. Indium Phosphide and Related Materials,* Schwäbisch-Gmünd, Germany, 1996, pp. 101–103.
92. J. Dickmann, S. Koßlowski, B. E. Maile, H. Haspeklo, S. Heuthe, A. Geyer, K. Riepe, A. Schurr, H. Daembkes, H. Künzel, and J. Böttcher, "High-gain 28 GHz coplanar waveguide monolithic amplifier on InP substrate," *Electron. Lett.,* **29**(5), 493–495 (1993).

93. A. Katz, *Indium Phosphide and Related Materials: Processing, Technology, and Devices,* Artech House, Norwood, Mass., 1992, pp. 307–335
94. M. M. Hashemi, J. B. Shealy, S. P. DenBaars, and U. K. Mishra, "High-Speed p$^+$ GaInAs-n InP heterojunction JFETs (HJFETs) grown by MOCVD," *IEEE Electron Device Lett.,* **14**(2), 60–62 (1993).
95. B. E. Maile, "Fabrication limits of nanometer T and Γ gates: theory and experiment," *J. Vac. Sci. Technol. B,* **11**(6), 2502–2508 (1993).
96. J. Dickmann, K. Riepe, H. Haspeklo, B. E. Maile, H. Dämbkes, H. Nickel, R. Lösch, and W. Schlapp, "Novel fabrication process for Si$_3$N$_4$ passivated InAlAs/InGaAs/InP HFETs," *Electron. Lett.,* **28**(19), 1849–1851 (1992).
97. K. Hosogi, N. Nakano, H. Minami, T. Katoh, K. Nishitani, M. Otsubo, M. Katsumata, and K. Nagahama, "Photo/EB hybrid exposure process for T-shaped gate super low-noise HEMTs," *Electron. Lett.,* **27**(22), 2011–2013 (1991).
98. K. Numila, M. Tong, A. A. Ketterson, and I. Adesida, "Fabrication of sub–100nm T-gates with SiN passivation layer," *J. Vac. Sci. Technol. B,* **9**(6), 2870–2874 (1991).
99. N. Yoshida, T. Kitano, Y. Yamamoto, K. Katoh, H. Minami, H. Takano, T. Sonoda, S. Takamiya, and S. Mitsui, "A super low noise V-band AlInAs/InGaAs HEMT processed by selective wet gate recess etching," *Dig. Microwave Theory and Techniques Symp.,* 1994, pp. 645–648.
100. Y. Ishii, "Fabrication technologies of InP-based digital ICs and MMICs," *Proc. 8th Int. Conf. Indium Phosphide and Related Materials,* Schwäbisch Gmünd, Germany, 1996, pp. 53–56.
101. O. Aina, M. Burgess, M. Mattingly, A. Meerschaert, J. M. O'Connor, M. Tong, A. Ketterson, and I. Adesida, "A 1.45W/mm, 30GHz InP-channel power HEMT," *IEEE Electron Device Lett.,* **13**(5), 300–302 (1992).
102. W. P. Hong, R. Bhat, J. R. Hayes, C. Nguyen, M. Koza, and G.-K. Chang, "High-breakdown, high gain InAlAs/InGaAsP quantum-well HEMT's," *IEEE Electron Device Lett.,* **12**(10), 559–561 (1991).
103. J. Singh, "A proposal for a high-speed In$_{0.52}$Al$_{0.48}$As/In$_{0.53}$Ga$_{0.47}$As MODFET with an (InAs)$_m$(GaAs)$_m$ superlattice channel," *IEEE Electron Device Lett.,* **7**(7), 436–439 (1986).
104. N. Nishiyama, H. Yano, S. Nakajima, and H. Hayashi, "n-AlInAs/(InAs)$_3$(GaAs)$_1$ superlattice modulation-doped field effect transistor grown by molecular beam epitaxy," *Electron. Lett.,* **26**(13), 885–887 (1990).
105. W. P. Hong, J. R. Hayes, R. Bhat, P. S. D. Lin, C. Nguyen, D. Yang, and P. K. Battacharya, "Novel strained InP/InAsP quantum-well HEMTs," *Proc. 4th Int. Conf. Indium Phosphide and Related Materials,* Newport, R.I., 1992, postdeadline paper.
106. T. Schweizer, K. Köhler, and P. Ganser, "Principal difference between the transport properties of normal AlGaAs/InGaAs/GaAs and inverted GaAs/InGaAs/AlGaAs modulation doped heterostructures," *Appl. Phys. Lett.,* **60**, 469–471 (1992).
107. H. Masato, T. Masuno and K. Inoue, "In$_{0.5}$Ga$_{0.5}$As/InAlAs modulation-doped field effect transistors on GaAs substrates grown by low-temperature molecular beam epitaxy," *Jpn. J. Appl. Phys.,* **30B**, 3850–3852 (1991).
108. J. Chen, J. M. Fernández, J. C. P. Chang, K. Kavanagh, and H. H. Wieder,"Modulation-doped In$_{0.3}$Ga$_{0.7}$As/In$_{0.29}$Al$_{0.71}$As heterostructures grown on GaAs by step grading," *Semicond. Sci. Technol.,* **7**, 601–603 (1992).

109. S. M. Lord, B. Pezeshki, S. D. Kim, and J. S. Harris, "1.3 μm exciton resonances in In-GaAs quantum wells grown by molecular beam epitaxy using a slowly graded buffer layer," *J. Cryst. Growth*, **127**, 759–764 (1993).

110. T. Fink, M. Haupt, G. Kaufel, K. Köhler, J. Braunstein, and H. Massler, "AlInAs/GaInAs/AlInAs MODFETs fabricated on InP and on GaAs with methamorphic buffer: a comparision," *Proc. 22nd Int. Conf. Compound Semiconductors*, Cheju Island, Korea, 1995, pp. 835–838.

111. K. H. G. Duh, P. C. Chao, P. Ho, A. Tessmar, S. M. J. Liu, M. Y. Kao, P. M. Smith, and J. M. Ballingall, "W-band InGaAs HEMT low noise amplifiers," *Dig. Microwave Theory and Techniques Symp.*, 1990, pp. 595–598.

112. P. C. Chao, A. J. Tessmar, K. H. Duh, P. Ho, M. Y. Kao, P. M. Smith, J. M. Ballingall, S.-M. Liu, and A. A. Jabra, "W-band low-noise InAlAs/InGaAs lattice matched HEMTs," *IEEE Electron Device Lett.*, **11**(1), 59–62 (1990).

113. G. M. Rebeiz et al., "The highest-frequency (155GHz and 215GHz) three-terminal transistor oscillator in the world reported," *IEEE Antennas Propagat.*, **36**(2), 36–38 (1994).

114. Y. Kwon, D. Pavlidis, P. Marsh, G. I. Ng, T. Brock G. Munns, and G. I. Haddad, "A fully integrated monolithic D-band oscillator-doubler chain using InP-based HEMTs, *Tech. Dig. GaAs IC Symp.*, 1992, pp. 51–54.

115. Y. Kwon, G. I. Ng, D. Pavlidis, R. Lai, and T. Brock, "High efficiency monolithic Ka-band oscillators using InAlAs/InGaAs HEMTs," *Tech. Dig. GaAs IC Symp.*, 1991, pp. 263–266.

116. A. Kurdoghlian, W. Lam, C. Chou, L. Jellian, A. Igawa, M. Matloubian, L. Larson, A. Brown, M. Thompson, and C. Ngo, "High-efficiency InP-Based HEMT MMIC power amplifier," *Tech. Dig. GaAs IC Symp.*, 1993, pp. 375–377.

117. Y. Kwon, D. Pavlidis, P. Marsh, G. I. Ng, and T. Brock, "A planar heterostructure diode W-band mixer using monolithic balanced integrated approach on InP," *Tech. Dig. GaAs IC Symp.*, 1992, pp. 67–69.

118. S. E. Rosenbaum, L. M. Jellion, L. E. Larson, U. K. Mishra, D. A. Pierson, M. S. Thompson, T. Liu, and A. S. Brown, "A 2-GHz three-stage AlInAs–GaInAs–InP HEMT MMIC low-noise amplifier," *IEEE Microwave Guided Wave Lett.*, **3**(8), 265–267 (1993).

119. H. Wang, R. Lai, T. H. Chen, P. D. Chow, J. Velebir, K. L. Tan, D. C. Streit, P. H. Liu, and G. Ponchak, "A monolithic W-band three-stage LNA using 0.1μm InAlAs/InGaAs/InP HEMT Technology," *Dig. IEEE Microwave Theory and Techniques Symp.*, 1993, pp. 519–523.

120. D. C. W. Lo, R. Lai, H. Wang, K. L. Tan, R. M. Dia, D. C. Streit, P. H. Liu, J. Velebir, B. Allen, and J. Berenz, "A high-performance monolithic Q-band InP-based HEMT low-noise amplifier," *IEEE Microwave Guided Wave Lett.*, **3**(9), 299–301 (19913).

121. H. Wang, R. Lai, D. C. W. Lo, D. C. Streit, M. W. Pospieszalski, and J. Berenz, "A 140GHz monolithic low-noise amplifier," *Dig. Int. Electron Device Meeting, IEDM'94*, 1994, pp. 933–934.

122. K. H. Duh et al., "Advanced millimeter-wave InP HEMT MMICs," *Proc. 5th Int. Conf. Indium Phosphide and Related Materials*, Paris, 1993, pp. 493–496.

123. M. Berg and J. Dickmann, unpublished results

124. R. Majidi-Ahy, M. Riaziat, C. Nishimoto, M. Glenn, S. Silvermann, S. Weng, Y. C. Pao,

G. Zdasiuk, S. Bandy, and Z. Tan, "5–100GHz InP CPW MMIC 7-section distributed amplifier," Dig. *IEEE Microwave and Millimeter Wave Monolithic Circuits Symp.,* 1990, pp. 31–34.

125. C. J. Maddan, R. L. van Tuyl, M. V. Le, and L. D. Nguyen, "A 17-dB Gain, 0.1–70GHz InP HEMT Amplifier IC," Tech. Dig. ISSCC, 1994, pp. 178–179.

126. H. Wang, R. Lai, S. T. Chen, and J. Berenz, "A monolithic 75–110 GHz balanced InP-based HEMT amplifier," *IEEE Microwave Guided Wave Lett.,* **3**(10), 381–383 (1993).

CHAPTER ELEVEN

Heterojunction Bipolar Transistors and Circuit Applications

HIN-FAI FRANK CHAU
EiC Corporation

WILLIAM LIU
Texas Instruments, Inc.

11.1	Introduction	392
11.2	HBT Operation and Modeling	392
	11.2.1 Drift-Diffusion Approach	396
	11.2.2 Ensemble Monte Carlo Approach	398
	11.2.3 Small-Signal Equivalent-Circuit Model	398
11.3	InP Heterojunction Bipolar Transistor Design	402
	11.3.1 Material Properties	402
	11.3.2 Single Heterojunction Bipolar Transistors	404
	11.3.3 Double Heterojunction Bipolar Transistors	406
11.4	Fabrication Technology	412
11.5	HBT Characteristics	415
	11.5.1 DC Characteristics	415
	11.5.2 Thermal Characteristics	419
	11.5.3 Thermal Instabilities	423
	11.5.4 Microwave Characteristics	425
	11.5.5 Power Performance	430

InP-Based Materials and Devices: Physics and Technology, Edited by Osamu Wada and Hideki Hasegawa.
ISBN 0-471-18191-9 © 1999 John Wiley & Sons, Inc.

11.6	Circuit Applications	432
	11.6.1 Digital Applications	432
	11.6.2 Analog Applications	433
	11.6.3 Mixed-Signal Applications	435
	11.6.4 Microwave Applications	435
	11.6.5 Optoelectronic and Telecommunication Applications	436
11.7	Summary and Future Challenges	438
	Acknowledgments	439
	References	439

11.1 INTRODUCTION

InP-based heterojunction bipolar transistor (HBT) technology has been progressing rapidly over the past few years. The current-gain cutoff frequency and maximum frequency oscillation have both broken the 200-GHz performance barrier [1–6]. There are a number of inherent advantages of these devices compared with the more widely used and more mature GaAs-based HBTs, including smaller turn-on voltage, lower power consumption, higher speed, and larger gain. These properties lend themselves well to the development of integrated circuits characterized by very high performance and low power consumption. To date, digital and analog integrated circuits consisting of nearly 1000 InP HBTs have been demonstrated. In addition, InP HBT offers material compatibility and therefore integrability with lightwave technology devices such as InP-based photodetectors and lasers operating in the wavelength range 1.3 to 1.55 μm, making it an important technology for telecommunications and signal processing.

The operation and modeling of HBTs are described in Section 11.2. In particular, the drift-diffusion model, ensemble Monte Carlo approach, and small-signal equivalent-circuit model are introduced. In Section 11.3, material parameters and band lineups important to the design and operation of InP HBTs are presented. Detailed design issues and trade-offs in single- and double-heterojunction bipolar transistors are then discussed. Conventional and advanced fabrication processes are compared in Section 11.4. The HBT dc, thermal, microwave, and power characteristics as well as thermal instability phenomenon are described in Section 11.5. The digital, analog, mixed-signal, microwave circuit, and telecommunication applications are reviewed in Section 11.6. Finally, the prospects and future challenges of InP HBTs are presented.

11.2 HBT OPERATION AND MODELING

The idea of using a heterojunction in a bipolar transistor was introduced by Shockley [7] and was later developed by Kroemer [8–10]. There are two basic types of HBTs. Single-heterojunction bipolar transistors [SHBTs] refers to those in which only the base–emitter junction is a heterojunction. When both the base–emitter and base–collector junctions are heterojunctions, the device is classified as a double-heterojunction bipolar transistor (DHBT). Table 11.1 compares the generic layer structures of these two types of HBTs as applied to InP-based materials system. We consider only Npn transistors here since Pnp InP HBTs are considerably slower than their Npn counterparts and are not widely used.

Starting from the InP substrate, which can be either semi-insulating or heavily doped n-type, the HBT structure typically consists of a highly doped n-type InGaAs subcollector layer for low collector resistance, a lightly doped n-type InGaAs (or InP) collector layer in SHBT (or DHBT) for small base–collector capacitance, a heavily doped p-type InGaAs base layer for low base resistance, an n-type InP or In-AlAs emitter layer, and then a highly doped n-type InGaAs cap layer for small emitter resistance. In addition to InP and InAlAs, other quaternary alloys, such as InGaAsP and InAlGaAs, can be used as the wide-bandgap material for the formation of a heterojunction with InGaAs. We do not consider InAlAs collector DHBTs here because of several drawbacks, which are discussed in Section 11.3.3. The transistor characteristics and operation are largely determined by the materials used in the emitter, base, and collector layers, as well as their thicknesses and doping concentrations. The cap and subcollector layers are for contacting purposes and are heavily doped to reduce access and contact resistances.

The energy band diagram of an Npn SHBT structure is shown in Fig. 11.1. For simplicity, the cap and subcollector layers are not shown in the figure. E_{FE}, E_{FB}, and E_{FC} indicate the Fermi levels in the emitter, base, and collector, respectively, at a base–emitter voltage V_{BE} and a collector–base voltage V_{CB}. ΔE_C and ΔE_V are the discontinuities in the conduction and valence bands, respectively, at the base–emitter heterojunction. qV_n and qV_p are the heights of the potential energy barriers seen by the electrons from the emitter and by the holes from the base, respectively, when injected across the base–emitter heterojunction.

Under normal operating conditions, electrons are injected from the emitter into the base, diffuse through the neutral base, and are collected by the collector. Since the base–collector homojunction is under reverse bias, and because the base doping level is several orders of magnitude higher than the collector doping, the depletion region lies almost entirely in the collector layer. Electrons reaching the base–collector junction are swept to the collector electrode by the high electric fields present in the collector depletion or space-charge region (SCR). The difference in energy

TABLE 11.1 Generic Layer Structures of InP SHBT and DHBT

Layer	Doping Type	SHBT	DHBT
Contact cap	n^+	InGaAs	InGaAs
Emitter cap	n^+	InP or InAlAs	InP or InAlAs
Emitter	n	InP or InAlAs	InP or InAlAs
Base	p^+	InGaAs	InGaAs
Collector	n^-	InGaAs	InP
Subcollector	n^+	InGaAs	InGaAs
Substrate	S.I. or n^+	InP	InP

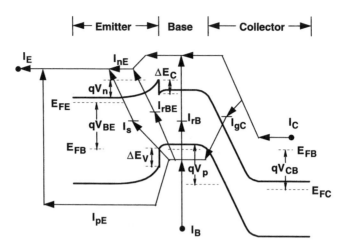

FIGURE 11.1 Energy band diagram of an Npn SHBT structure. The current components are indicated by heavy arrows.

bandgaps results in different energy barriers for electrons and holes at the heterojunction. For InP/InGaAs, InAlAs/InGaAs, InGaAsP/InGaAs, and InAlGaAs/InGaAs heterojunctions, the barrier for hole injection at the base–emitter heterointerface is larger than that for electron injection, leading to suppression of hole injection from the base into the emitter. This greatly enhanced emitter injection efficiency in an HBT significantly increases the current gain, and more important, allows the base doping concentration to be larger than that in the emitter without sacrificing current gain. A high base doping level reduces the base resistance, and a low emitter doping concentration decreases the base–emitter capacitance. Both the base resistance and base–emitter capacitance are important parameters in determining the microwave characteristics of an HBT.

As shown in Fig. 11.1, the three terminal currents of the HBT are given as follows:

Emitter current: $$I_E = I_{nE} + I_{pE} + I_s \quad (1)$$

Base current: $$I_B = I_{pE} + I_s + I_{rBE} + I_{rB} - I_{gC} \quad (2)$$

Collector current: $$I_C = I_{nE} - I_{rBE} - I_{rB} + I_{gC} \quad (3)$$

where I_{nE} is the electron current injected from the emitter into the base, I_{pE} the hole current injected from the base into the emitter, I_s the recombination current at the surface and base–emitter interface, I_{rBE} the recombination in the base–emitter depletion region, I_{rB} the recombination current in the neutral base region, and I_{gC} the generation current in the collector space-charge region.

The common-emitter current gain β is given by

$$\beta = \frac{I_C}{I_B} = \frac{I_{nE} - I_{rBE} - I_{rB} + I_{gC}}{I_{pE} + I_s + I_{rBE} + I_{rB} - I_{gC}} \quad (4)$$

When the generation current I_{gC} is negligible,

$$\beta < \frac{I_{nE}}{I_{pE}} = \beta_{\max}|_{I_{gC}=0} \propto \exp\left[\frac{q(V_p - V_n)}{kT}\right] \quad (5)$$

where $q(V_p - V_n)$ represents the difference between the energy barrier for holes injected from the base into the emitter and that for electrons injected from the emitter into the base. As shown in the energy band diagram of an abrupt heterojunction HBT in Fig. 11.1, $q(V_p - V_n) \approx \Delta E_V$ [9]. For graded heterojunction bipolar transistors, the conduction band discontinuity is graded to zero, and all the bandgap difference appears in the valence band. Therefore, the energy barrier difference $q(V_p - V_n) \approx \Delta E_g$ for graded HBTs [9].

The transconductance g_m is defined as

$$g_m = \frac{\partial I_C}{\partial V_{BE}} \quad (6)$$

which is an exponential function of V_{BE} since I_C is exponentially dependent on the base–emitter bias voltage. This exponential relationship is important in certain circuit applications and is not available in field-effect transistors (FETs) (see Chapter 10).

The turn-on voltage of an HBT is determined by the built-in or diffusion potential V_{bi} of the base–emitter heterojunction, which is expressed as

$$qV_{bi} = E_{gB} + \Delta E_C + kT \ln \frac{n_{no}}{N_{CE}} + kT \ln \frac{p_{po}}{N_{VB}} \quad (7)$$

$$\approx E_{gB} + \Delta E_C + kT \ln \frac{N_{DE}}{N_{CE}} + kT \ln \frac{N_{AB}}{N_{VB}} \quad (8)$$

where q is the electron charge, k is Boltzmann's constant, T is the temperature, E_{gB} the bandgap energy of the base, ΔE_C the conduction band discontinuity, n_{no} the electron density in the emitter, p_{po} the hole concentration in the base, N_{CE} the effective density of states in the emitter conduction band, N_{VB} the effective density of states in the base valence band, N_{DE} the emitter doping concentration, and N_{AB} the base doping concentration.

An essential element of a circuit technology is the need to control device characteristics precisely across an entire wafer. HBTs have an intrinsic advantage over FETs in that the turn-on voltage of the transistors depends predominantly on the bandgap energy of the base layer and the conduction band offset at the base–emitter heterojunction according to Eq. (7). Both the bandgap energy and the conduction band discontinuity are intrinsic properties of the materials and are very reproducible. Good uniformity of diode turn-on voltage is observed across an HBT wafer. In FETs, however, the threshold voltage is controlled by gate recess, and its variation depends largely on the process technology used. Overetching or underetching results in smaller- or larger-than-expected threshold voltage in a depletion-mode FET.

The carrier transport properties in an HBT can be obtained by solving the Boltzmann transport equation (BTE) [11]. However, because of the great complexity of the BTE, it cannot be solved analytically without simplifying assumptions. Next we describe two approaches for the modeling of HBTs.

11.2.1 Drift-Diffusion Approach

One popular approach to the BTE for bipolar transistors is the drift-diffusion approximation, which assumes that the electron temperature is in balance with the lattice and the temperature gradient is zero. The average electron velocity v is also assumed to adjust instantaneously to the electric field $\boldsymbol{F}$ through a linear relation $v = \mu \boldsymbol{F}$, where μ is the mobility. Approximate estimates of the behavior of HBTs can thus be obtained by solving the simplified transport, continuity, and Poisson equations simultaneously. Within the drift-diffusion approximation, these equations, when taking into account the position-dependent materials parameters, are expressed as follows:

Transport equations:
$$J_n = -q\mu_n n \frac{\partial \phi_n}{\partial x} = -q\mu_n n \frac{\partial}{\partial x}(\psi + \theta_n) + \mu_n kT \frac{\partial n}{\partial x} \tag{9}$$

$$J_p = -q\mu_p p \frac{\partial \phi_p}{\partial x} = -q\mu_p p \frac{\partial}{\partial x}(\psi - \theta_p) - \mu_p kT \frac{\partial p}{\partial x} \tag{10}$$

Continuity equations:
$$\frac{\partial n}{\partial t} = G - U + \frac{1}{q}\frac{\partial J_n}{\partial x} \tag{11}$$

$$\frac{\partial p}{\partial t} = G - U - \frac{1}{q}\frac{\partial J_p}{\partial x} \tag{12}$$

Poisson equation:
$$\frac{1}{\partial x}\left(\varepsilon \frac{\partial \psi}{\partial x}\right) = -q(p - n + N_D - N_A) \tag{13}$$

or
$$\frac{\partial^2 \psi}{\partial x^2} = -\frac{q}{\varepsilon}(p - n + N_D - N_A) - \frac{1}{\varepsilon}\frac{\partial \psi}{\partial x}\frac{\partial \varepsilon}{\partial x} \tag{14}$$

where J_n and J_p are the electron and hole current densities, respectively; μ_n and μ_p are the electron and hole mobilities, respectively; n and p are the electron and hole densities, respectively; ϕ_n and ϕ_p are quasi-Fermi potentials for electrons and holes, respectively; x is the position; ψ is the intrinsic Fermi potential; t is the time; G is the carrier generation rate; U is the recombination rate; ε is the permittivity; and N_D and N_A are the donor and acceptor densities, respectively.

The electron and hole densities are determined by

$$n = n_{io}\exp\left[\frac{q}{kT}(\psi - \phi_n + \theta_n)\right] \quad (15)$$

$$p = p_{io}\exp\left[\frac{q}{kT}(-\psi + \phi_p + \theta_p)\right] \quad (16)$$

and the quasi-Fermi potentials are given by

$$\phi_n = \psi + \theta_n - \frac{kT}{q}\ln\frac{n}{n_{io}} \quad (17)$$

$$\phi_p = \psi - \theta_p + \frac{kT}{q}\ln\frac{p}{n_{io}} \quad (18)$$

The position dependences are taken into account by the use of two factors, θ_n and θ_p, which are expressed as follows:

$$\theta_n = \frac{\chi}{q} + \frac{kT}{q}\ln\frac{N_C}{n_{io}} \quad (19)$$

$$\theta_p = -\frac{\chi + E_g}{q} + \frac{kT}{q}\ln\frac{N_V}{n_{io}} \quad (20)$$

where χ is the electron affinity; E_g is the bandgap energy; N_C and N_V are the effective densities of states in the conduction and valence bands, respectively; and n_{io} is the intrinsic carrier concentration of a reference material in the HBT.

The drift-diffusion formalism is an acceptable approximation as long as the device dimensions are large and the electric field is low, so that hot-electron effects are not dominant in determining the electron transport properties. In an abrupt junction HBT, however, electrons injected across the emitter–base heterointerface can have substantially higher electron temperatures than the lattice. In the collector space-charge region and in InP-based HBTs in particular, the high and sharply increased electric field at the base–collector junction heats up electrons rapidly and causes them to undergo a non-steady-state transition. In these situations, the electron velocity does not follow the electric field with a linear relation because of different energy and momentum relaxation times. The distances over which significant changes in carrier potential energy occur can be very short, and carriers are not in quasi-equilibrium. As a consequence, the accuracy of the drift-diffusion formulation is often inadequate for heterostructures. The preferred theoretical treatment of nonstationary transport and velocity overshoot effects is with the ensemble Monte Carlo approach [12].

11.2.2 Ensemble Monte Carlo Approach

In the ensemble Monte Carlo method, the trajectories of electrons in an Npn HBT are followed on a time-step basis. The motion of these electrons is influenced by external forces and various scattering mechanisms in the semiconductors. The external forces include applied electric fields, built-in electric field in a pn junction, built-in quasi-electric field due to compositional grading and impurity doping gradient, and so on. The electron drift process, or free flight, between two successive scattering events, is treated semiclassically. The durations of the free flights, the scattering mechanisms responsible for the interruption of the drift processes, and the final states of the scattered electrons are selected stochastically according to the given probabilities describing these microscope processes. The consequence of this is that any Monte Carlo method depends on the generation of a sequence of random numbers distributing with given probabilities describing the physical system under investigation. The Monte Carlo approach provides an exact solution to the Boltzmann transport equation as long as the energy band structures and various forms of scattering processes are known and is particularly suitable for studying non-steady-state transport. A large number of particles, on the order of 25,000 to 30,000, is typically used to represent all electrons in the HBT. Physical quantities such as the average velocity and energy as functions of space or time are obtained by averaging over the ensemble of electrons.

For a self-consistent HBT analysis, the motion of electrons and holes are solved simultaneously. Since the motion of holes is less important in an Npn HBT as far as the ballistic nature of it is concerned, a hybrid method can be employed; the motion of electrons is solved by the ensemble Monte Carlo method, while the motion of holes is solved by the drift-diffusion model. With the electron and hole distributions known from the Monte Carlo calculation and drift-diffusion approximation, respectively, Poisson equation can then be solved to give the updated potential and electric field profiles. This is coupled self-consistently with the solution of electron and hole transport and performed during the same time step. Solving the Poisson equation provides updated potential and electric field distributions for the Monte Carlo procedure. The technique thus provides an accurate and clear physical picture of electron transport in an Npn HBT. A detailed description of the ensemble Monte Carlo modeling of HBTs can be found in Ref. 13.

11.2.3 Small-Signal Equivalent-Circuit Model

The small-signal equivalent circuit of an HBT can be extracted from its measured S-parameters. Figure 11.2 shows the cross section of an InP HBT using triple-mesa and air-bridge technology. The figure indicates how the equivalent-circuit elements are connected and related to the physical parameters of the HBT, where R_E is the emitter resistance; R_{BE} is the emitter dynamic resistance; R_{B1} and R_{B2} are the intrin-

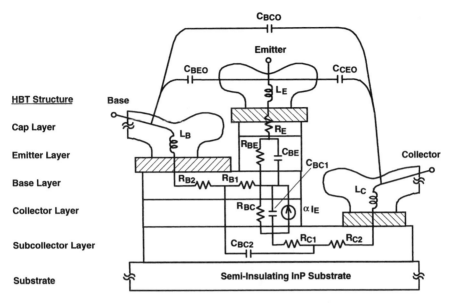

FIGURE 11.2 Cross section of an InP HBT using triple-mesa and air-bridge technology. The equivalent-circuit elements are related to the HBT physical parameters as shown.

sic and extrinsic base resistances, respectively; R_{C1} and R_{C2} are the intrinsic and extrinsic collector resistances, respectively; R_{BC} is the base–collector isolation resistance; C_{BE} is the base–emitter capacitance; C_{BC1} and C_{BC2} are the intrinsic and extrinsic base–collector capacitances, respectively; L_E, L_B, and L_C are the emitter, base, and collector inductances, respectively; C_{BEO}, C_{CEO}, and C_{BCO} are the coupling capacitances between base and emitter, collector and emitter, and base and collector, respectively; and α is the common-base current gain given by

$$\alpha = \frac{\alpha_0}{1+j\omega(\tau_b/1.2)} \frac{\sin\omega\tau_c}{\omega\tau_c} \exp\left[-j\omega\left(\frac{0.22\tau_b}{1.2}+\tau_c\right)\right], \tag{21}$$

where α_0 is the dc current gain, $\omega = 2\pi f$ and f is the frequency, τ_b is the base transit time, and τ_c is the transit time across the collector space-charge region involving more complex velocity transport effects.

Rearranging the equivalent circuit shown in Fig. 11.2 gives the T-type equivalent circuit frequently used in HBT small-signal modeling (Fig. 11.3). Alternatively, a π-equivalent circuit can also be used [14]. Unlike the T-equivalent circuit, however, the components of the π-equivalent circuit may not necessarily be related directly to the physical parameters of the HBT. The circuit elements of the π-equivalent circuit can, however, be converted from those of the T-equivalent circuit, and vice versa [15]. The equivalent circuit extraction can be conducted based on measured small-signal S-parameters of the HBT with the help of the optimization routines available in commercial microwave circuit simulators such as HP/EESOF Libra and Touchstone.

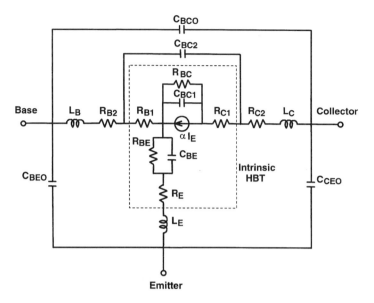

FIGURE 11.3 Small-signal equivalent-circuit model of an HBT.

The two major figures of merit for a microwave bipolar transistor are the current-gain cutoff frequency (f_T) and maximum frequency of oscillation (f_{max}), which are expressed as follows [16]:

$$f_T = \frac{1}{2\pi\tau_{ec}} = \frac{1}{2\pi(\tau_{ee} + \tau_b + \tau_c + \tau_{cc})} \tag{22}$$

$$f_{max} \approx \sqrt{\frac{f_T}{8\pi(R_B C_{BC})_{eff}}} \tag{23}$$

where τ_{ec} is the total HBT input response delay time, τ_{ee} the emitter–base charging time, τ_{cc} the collector charging time, and $(R_B C_{BC})_{eff}$ the product of the effective base resistance and base–collector capacitance. The four time constants τ_{ee}, τ_b, τ_c, and τ_{cc} can be expressed as [17]

$$\tau_{ee} = R_{BE}C_{BE} = \frac{nkT}{qI_E}C_{BE} \tag{24}$$

$$\tau_b \approx \frac{W_B^2}{2D_{nB}} + \frac{W_B}{v_e} \tag{25}$$

$$\tau_c = \frac{W_{SCR}}{2v_s} \tag{26}$$

11.2 HBT OPERATION AND MODELING

$$\tau_{cc} = (R_{BE} + R_E + R_C)C_{BE} = \left(\frac{nkT}{qI_E} + R_E + R_C\right)C_{BC} \qquad (27)$$

where C_{BC} is the total base–collector capacitance, I_E the emitter current, W_B the base width, D_{nB} the electron diffusion coefficient in the base, v_e the electron drift velocity in the base, W_{SCR} the width of the collector space-charge region, v_s the average electron velocity across the collector space-charge region, and R_C the total collector resistance.

The coefficient n in Eq. (24) takes into account the electrostatic potential distribution between the base and the emitter. In the complete depletion approximation, n is given by [18]

$$n = 1 + \frac{\varepsilon_E N_{DE}}{\varepsilon_B N_{AB}} \qquad (28)$$

The n factor is not related to any recombination centers and is due primarily to surface effects, which are small in InP/InGaAs HBTs. Tunnel currents and quantum mechanical reflections can also occur at the base–emitter heterojunction, leading to n values up to 2.

One of the basic requirements for the millimeter-wave bipolar structure is to reduce the carrier transit times by shrinking its intrinsic base and collector–base depletion widths, and yet in the same structure maintain a small time delay associated with charging the base–collector capacitance through the base resistance. Reducing the base layer thickness, however, increases the base resistance. In the design and fabrication of modern HBTs, the use of extremely high base doping (on the order of 10^{20} cm^{-3}) and self-aligned technology has greatly reduced the parasitics due to the base resistance. As the dimension of the base thickness shrinks, the primary intrinsic limitation of device speed has been attributed to collector transport. Over the past few years, great attention was paid to the reduction of collector transit time in GaAs HBTs by modifying the traditional n⁻ collector configuration to make use of the velocity overshoot phenomenon in the collector space-charge region [12, 19–22]. In InP HBTs, the velocity overshoot effects are much more significant and special collector designs are not needed to achieve f_T over 200 GHz [1,2].

If the goal of optimization is solely to achieve short collector transit time and therefore high-speed characteristics, one can of course reduce the thickness of the collector SCR alone without the need of modifying its design. In practice, however, the collector SCR thickness is limited by the base–collector breakdown. The breakdown voltage is inversely proportional to the collector doping through the equation $BV_{CBO} = \varepsilon_C E_m^2/2qN_C$. If the base–collector breakdown is limited by the reach-through effect (i.e., the entire low-doped collector is completely depleted), the breakdown voltage is lower and $BV_{CBO} = E_m W_C - qN_C W_C^2/2\varepsilon_C$. Because of the large voltage swing encountered at the output in a power HBT, a thick low-doped collector is commonly used. A thick collector, however, increases the distance over which the electrons need to travel. Breakdown voltages can also be increased by using collector materials with low-impact ionization coefficients or a large critical electric field for breakdown. Using a different collector material from the base may lead to current blocking due to the presence of the base–collector heterojunction. Overall, breakdown voltage and speed are the two most important figures that need to be compromised in most applications. Considerations of this type of compromise have been studied in detail elsewhere [23,24] and are discussed in Section 11.3.3.

11.3 InP HETEROJUNCTION BIPOLAR TRANSISTOR DESIGN

11.3.1 Material Properties

InP-based materials are attractive for high-speed-device applications because of their superior physical, electrical, and transport properties. Understanding the properties of these materials is important to an understanding of the design, operation, and characteristics of various types of InP HBTs. Table 11.2 compares the material properties of several InP-based materials, including InP, InGaAs, and InAlAs [25–28]. The indium mole fractions of the latter two are chosen such that they are lattice matched to InP substrates. This requires about 53% indium to be present in InGaAs and 52% indium in InAlAs. In addition to InP and InAlAs, other quaternary alloys, such as InGaAsP and InAlGaAs, can also be used as the wide-bandgap material for the formation of a heterojunction with InGaAs. InGaAs and InAlAs as well as their quaternaries can be grown by the traditional solid-source molecular beam epitaxy (MBE). Any phosphorus-containing layer including InP and InGaAsP can be grown by metal-organic molecular beam epitaxy (MOMBE), chemical beam epitaxy (CBE), metal-organic chemical vapor deposition (MOCVD), and more recently, solid-phosphorus-source MBE (SSMBE) [29].

First, InP has about 50% higher thermal conductivity than GaAs (0.68 W/cm·K versus 0.46 W/cm·K for GaAs) [28]. Devices fabricated on InP substrates can potentially operate at lower chip operating temperature than GaAs devices having the same power dissipation. Low junction temperature is important in extending the lifetime of a device. Second, since the dominant term in Eq. (7) is E_{gB}, use of the smaller-bandgap InGaAs base in InP HBTs results in lower turn-on voltage and

Table 11.2 Material Properties of InP, InGaAs, and InAlAs[a]

Material Property	InP	$In_{0.53}Ga_{0.47}As$	$In_{0.52}Al_{0.48}As$
Bandgap energy (eV)	1.35	0.75	1.46
Thermal conductivity (W/cm·K)	0.68	0.05	—
Γ–L valley separation (eV)	0.61	0.55	0.54
Electron effective mass m_0	0.078	0.041	0.074
Electron saturation velocity (cm/s)	1.5×10^7	7.0×10^6	6.0×10^6
Electron mobility at $N_d = 10^{17}$ cm^{-3} (cm^2/V·s)	3200	7000	900
Hole mobility at $N_a = 10^{17}$ cm^{-3} (cm^2/V·s)	150	300	180
Breakdown field (kV/cm) at $\alpha = 10^4$ cm^{-1}	480	220	520

[a]At 300 K.

smaller power consumption than in GaAs-based devices. This feature is particularly important for digital applications for which low power dissipation is essential. Furthermore, the smaller surface recombination velocity in InGaAs (10^3 cm/s versus 10^6 cm/s in GaAs [30]) leads to lower low-frequency noise and allows device scaling without suffering from degradation of the current gain. Being able to scale InP HBTs down to submicron dimensions with respectable dc current gain makes them particularly well suited for digital applications. The low-frequency or $1/f$ noise is important because it can be upconverted to RF frequencies in certain circuit applications, such as oscillators.

The electron transport properties of these materials are determined by their band structures. A large Γ–L valley separation $\Delta E_{\Gamma L}$ is important for high-speed transport because electrons gain more kinetic energy from the electric field and yet stay in the low-effective-mass Γ-valley without being scattered to the high-effective-mass satellite valleys. It allows better electron velocity overshoot and nonequilibrium transport over a larger distance in the collector, which in turn translate into higher speed characteristics. All InP-based materials have large Γ–L valley separation. In addition to $\Delta E_{\Gamma L}$, the electron effective mass m_e^* is also important because it determines how fast electrons can be excited to high velocities. Among the materials shown in Table 11.2, InGaAs has the smallest electron effective mass. While electron velocity overshoot in the collector determines the speed of an HBT at small collector–emitter V_{CE} biases such as in digital circuits, the electron saturation velocity v_{sat} plays a more important role in determining the collector transit time and cutoff frequency of a device at large V_{CE} biases or high fields, typically for power applications. Although InP has a larger electron effective mass than InGaAs, its electron saturation velocity is the highest among the three materials considered. InP, being a wide-bandgap material, is particularly attractive in InP power HBT designs because of its high breakdown field [31], high thermal conductivity, and high electron saturation velocity. Extended electron velocity overshoot and high electron saturation velocity in the collector reduce the transit times and increase the speed of the device.

In addition to bulk material properties, the particular band alignments of the heterojunctions are also important. Table 11.3 shows the band offsets of two commonly used InP-based heterostructures: InP/In$_{0.53}$Ga$_{0.47}$As and In$_{0.52}$Al$_{0.48}$As/In$_{0.53}$Ga$_{0.47}$As. The conduction band discontinuity ΔE_C is larger at the InAlAs/

TABLE 11.3 Band Discontinuities of InP/InGaAs and InAlAs/InGaAs Heterojunctions

Heterojunction	InP/ In$_{0.53}$Ga$_{0.47}$As	In$_{0.52}$Al$_{0.48}$As/ In$_{0.53}$Ga$_{0.47}$As
Conduction band discontinuity ΔE_C (eV)	0.23	0.5
Valence band discontinuity ΔE_V (eV)	0.37	0.21

InGaAs heterojunction and smaller in the InP/InGaAs system. The valence band offset ΔE_V is, however, smaller at the InAlAs/InGaAs heterointerface and larger at the InP/InGaAs heterojunction. The differences in band discontinuities play a key role in the resultant dc and transport characteristics of an HBT. The advantages and disadvantages of InAlAs and InP as the emitter and collector materials are explained in Sections 11.3.2 and 11.3.3.

11.3.2 Single Heterojunction Bipolar Transistors

The simplest InP-based HBT structures are those based on a single heterojunction design at the base–emitter junction. As shown in Table 11.1, both the base and collector layers are made of the same narrow-bandgap material (i.e., InGaAs) and a homojunction is formed at the base–collector junction. Wide-bandgap materials, which can be grown lattice matched on InP substrates, include InP, InAlAs, InAlGaAs, and InGaAsP. Since InAlGaAs and InGaAsP are quaternary materials, involving four elements in the alloys, they are considerably more difficult to be grown lattice matched to InP. Our discussions in this section, therefore, focus on the advantages and disadvantages of InP and InAlAs emitter HBTs. Quaternary emitters are described wherever appropriate. The characteristics of InP SHBTs are described in Section 11.5.

InP Versus InAlAs Emitters Both InP and InAlAs are wide-bandgap materials when compared to InGaAs and can be used as the emitter material. One of the key features that makes an HBT different from a homojunction bipolar transistor is the presence of a heterojunction at the base–emitter interface. As shown in Eq. (5), by having a wider bandgap in the emitter than in the base, the maximum current gain β_{max} is increased by a factor of $\exp(\Delta E_V/kT)$. From Table 11.3, the values of ΔE_V are 14 and 8 times the value of kT at room temperature for the InP/InGaAs and InAlAs/InGaAs heterostructures, respectively. The maximum current gain of an HBT can be orders-of-magnitude higher than the corresponding InGaAs homojunction bipolar transistor. (In practice, however, an HBT with an extremely high current gain of hundreds or thousands is less useful than a device with low base–emitter capacitance and small base resistance, which are important parameters for high-speed applications. For this reason, a low emitter doping concentration and a high base doping concentration are commonly used in an HBT.) In addition, the value of ΔE_V has an important influence on how the current gain of an HBT changes with temperature. Since ΔE_V is a very weak function of temperature, a base–emitter heterojunction with a small value of ΔE_V shows current gain degradation as temperature or power dissipation increases because more holes can overcome the additional hole barrier ΔE_V and be injected from the base into the emitter at elevated temperatures. A large ΔE_V value is therefore desirable to prevent hole injection from the base into the emitter and current gain degradation at elevated temperatures in an HBT. As shown in Table 11.3, the InP/InGaAs heterostructure has the larger ΔE_V between the two InP-based material systems.

In addition to the valence band offset, the conduction band discontinuity ΔE_C at the base–emitter heterojunction also plays an important role in both the dc and electron transport characteristics of an HBT. From the dc point of view, a large ΔE_C increases the turn-on voltage of the transistor according to Eq. (7). Since $(E_{gB} + \Delta E_C)$ is larger for the InAlAs/InGaAs heterojunction and smaller for the InP/InGaAs heterostructure, the turn-on voltage of the InAlAs/InGaAs heterojunction is greater than that of the InP/InGaAs. In the case when the heterojunction is compositionally graded, ΔE_C approaches 0 eV and the two InGaAs heterojunction systems have very similar turn-on voltages. Furthermore, because electrons injected across the abrupt base–emitter heterojunction are highly directional and are more or less perpendicular to the heterointerface, a large ΔE_C value such as that in the InAlAs/InGaAs system means that injected electrons are more confined to the intrinsic base region [32] and there is less chance for electrons to go to and recombine with holes at the extrinsic base surface, a property that improves the emitter injection efficiency and increases the current gain [33]. For the same reason, one would therefore expect slightly lower $1/f$ noise in InAlAs emitter than in InP emitter HBTs [34]. The resultant reduction in extrinsic base recombination also promotes lateral scaling for InAlAs emitter HBTs [35]. Although the InAlAs emitter HBTs are potentially more suitable for device scaling, recent experimental results showed that even small-area InP emitter HBTs with emitter dimensions of only 1×2 μm^2 can operate with a respectable dc current gain of 60 and a cutoff frequency f_T over 100 GHz at a collector current of only 0.6 mA [36]. The small additional reduction in extrinsic base recombination current in InAlAs emitter HBTs is thus not as significant and important as the reduction in surface recombination achieved solely by incorporating an InGaAs base, whose surface recombination velocity is three orders of magnitude lower than that in GaAs [30]. With an emitter size as small as 0.3 μm^2 in InAlAs emitter HBTs with a graded base–emitter junction, a dc current gain of 7 and an f_{max} of 99 GHz were achieved [37].

From the transport point of view, electrons in an abrupt base–emitter heterostructure enter the base with a very large energy (close to ΔE_C), and as a consequence, may have very high velocities if the ΔE_C is not too close to or larger than $\Delta E_{\Gamma L}$. These hot electrons injected from the emitter greatly reduce the base transit time, which would otherwise be determined by the slow electron diffusion process as in a homojunction bipolar transistor. Although a simplistic viewpoint states that most of the electrons injected from an InAlAs emitter into the InGaAs base in an abrupt InAlAs/InGaAs HBT do not suffer from intervalley scattering because the conduction band offset at the InAlAs/InGaAs base–emitter heterojunction is still smaller than Γ–L valley separation in the InGaAs base [35], recent Monte Carlo studies show that it is not necessarily the case. Because the InAlAs/InGaAs heterojunction has a conduction band offset of about 0.5 eV, which is so close to the Γ–L valley separation of 0.55 eV in the InGaAs base, there is a high possibility that electrons can be scattered to the high-electron-effective mass satellite valleys shortly after the injection across the base–emitter heterojunction [38]. Furthermore, if ΔE_C is too large and the base is too thin, electrons enter the collector region with large excess energy and initiate impact ionization after acquiring additional kinetic energy from the electric field [39,40]. For this viewpoint, it is therefore advantageous to use the

InP/InGaAs heterostructure or slightly decrease the conduction band discontinuity ΔE_C by replacing the InAlAs emitter with a smaller bandgap quaternary material such as $In_{0.52}(Al_xGa_{1-x})_{0.48}As$. In the latter case, the base transit time decreases initially with increasing x when x is small and the injection energy is low, but tends to saturate or increase only slightly at high injection energies, due to the large conduction band nonparabolicity of InGaAs [41]. The optimum aluminum mole fraction x from the transport point of view was found to be about 0.6 [38], ignoring some undesirable effects due to a reduction in the valence band discontinuity and less electron confinement to the intrinsic base region, as described earlier.

From the material growth point of view, InAlAs emitters can be grown in virtually all modern epitaxial growth techniques, such as MBE, MOMBE, CBE, and MOCVD. InP emitters, however, cannot be grown in the traditional MBE reactors but can be grown by MOMBE, CBE, MOCVD, and the more recent solid-phosphorus-source MBE technology (see Chapters 6 and 7 for details of epitaxial growth). From the processing point of view, however, InP emitters are more attractive because wet chemical etchants with extremely high selectivities are available for InP and InGaAs, thereby improving the uniformity and manufacturability of HBTs. The base resistance, for example, depends critically on where the base contact is deposited on the base layer. InP emitter HBTs have the additional advantage over those with InAlAs emitters in that the InGaAs cap and InP emitter layers can be removed selectively before the InGaAs base layer is exposed for base contact. The use of InP emitter HBTs and selective etchants for layer removal thus leads to better manufacturability and more uniform dc and RF characteristics across a wafer. If process variations can be minimized through selective etching, the remaining variations in transistor characteristics will come from layer thickness and doping uniformity during material growth.

11.3.3 Double Heterojunction Bipolar Transistors

Another important class of HBTs is the DHBT design. Although InP SHBTs are fast and simple in design and growth, their output conductance is high and breakdown voltage is low at high current densities. These are shown in Section 11.5.1 and explained in detail in Section 11.5.2. The poor output conductance is related to the small bandgap energy and poor thermal conductivity in the InGaAs collector. Since the bandgap of InGaAs in only 0.75 eV at room temperature, its intrinsic carrier concentration of 6.4×10^{11} cm^{-3} is 300,000 times higher than that in GaAs. Poor thermal conductivity in the InGaAs collector leads to a rapid increase in junction temperature with power dissipation, which further increases its intrinsic carrier concentration. At high power dissipation, the elevated junction temperature increases the base–collector reverse saturation or leakage current I_{CBO} by two to four orders of magnitude [42,43]. This base–collector leakage current does not go to the external base terminal but goes to the neutral base region of the transistor, similar to the impact ionization current component under avalanche breakdown at high collector bias. Because of this additional base current component I_{CBO}, the base–emitter junction becomes more forward biased, leading to a significant increase in the collector current. As a result, the "breakdown" voltage decreases rapidly with increasing power density.

Because of the high intrinsic reverse saturation current and poor thermal conductivity in the InGaAs collector, InP-based SHBTs are useful only for low-voltage or low-power-dissipation applications. To improve the output conductance and breakdown voltage at high power densities, the InGaAs collector in SHBTs needs to be replaced by a wider-bandgap material such as InP or InAlAs. Since the base is still InGaAs and both the emitter and collector are made of wider-bandgap materials, a double-heterojunction bipolar transistor is thus formed. The advantages and disadvantages of the two wide-bandgap collector materials are discussed next.

InP Versus InAlAs Collectors Unlike the use of InP and InAlAs emitters as described in Section 11.3.2, the selection of wide-bandgap collector material is more obvious and straightforward, for three reasons: First, while a larger conduction band offset ΔE_C at the base–emitter heterojunction gives rise to more energetic electron injection into the base, a conduction band offset at the base–collector heterojunction blocks the electron flow and prevents electrons from being collected in the collector. The height of this potential barrier for electrons is fixed and related only to the materials used to form the heterojunction. The effective barrier width, however, does change slightly with collector bias voltage. In certain abrupt heterojunction designs, the relative position of the top of the barrier to the bottom of the conduction band in the base can also be changed with collector bias. The overall effect is that a DHBT with an abrupt base–collector heterojunction of large ΔE_C may show poor transistor turn-on and saturation characteristics.

Second, although both InP and InAlAs have similar electron effective masses and Γ–L valley separations, the electron saturation velocity in InP is more than a factor of 2 times larger than that in InAlAs (Table 11.2). Since the collector space-charge region is a high-field region, the electron saturation velocity plays an important role in collector transit delay, which in turn affects the speed of the transistor.

Third, although there are limited data on the thermal conductivity in InAlAs, it is well known that when large numbers of foreign atoms are added to host lattice as in alloying, the thermal conductivity decreases significantly [25,28,44]. The thermal conductivity of a tertiary compound is usually lowest when the mole fraction of each constituent binary is within 30 to 70%, which is close to the In mole fraction x required for $In_xAl_{1-x}As$ to achieve lattice matching to InP substrates. This phenomenon was observed experimentally in AlGaAs, InGaAs, GaInP, InAsP, and GaAsP [45]. $In_{0.53}Ga_{0.47}As$ is a good example and has a thermal conductivity K_{th} of 0.05 W/cm·K, which is much smaller than the K_{th} values of 0.27 and 0.46 W/cm·K in InAs and GaAs, respectively [46]. $Ga_{0.51}In_{0.49}P$ is another good example and has a K_{th} value of 0.05 W/cm·K. InAlAs collector DHBTs are therefore expected to run at higher junction temperatures than InP collector transistors of the same dimensions, layer thicknesses, and power dissipation. For the same reason, although the use of quaternary alloys such as InGaAsP and InAlGaAs alleviates somewhat the current blocking due to the conduction band discontinuity at the base–collector heterojunction when compared to InAlAs/InGaAs heterojunction, the thermal conductivities of these materials are expected to be even worse. Because InAlAs has the large conduction band discontinuity of 0.5 eV when it forms a heterojunction with InGaAs, a low electron saturation velocity of 0.6×10^7 cm/s, and poor thermal conductivity, InP is the preferred collector material in InP DHBTs.

Base–Collector Heterojunction Design Several design techniques are available for the handling of base–collector heterojunction problems in an DHBT. They all serve one purpose: to reduce the current blocking effect due to the electron spike barrier in the conduction band at the base–collector heterojunction. The effectiveness of each method varies according to the value of ΔE_C, thickness of the collector layer, collector doping concentration, and operation current density. InP collector DHBTs are considered primarily since InP is the preferred material of choice for the collector.

To illustrate the complexity of the base–collector heterojunction design problem, let us first look at the simplest approach in which a thin undoped or lightly doped n⁻-InGaAs insertion layer is used between the p⁺-InGaAs base and n⁻-InP collector [47,48]. This type of collector layer design is sometimes known as composite collector structure. A thin, more highly doped n-InP layer is often introduced between the n⁻-InGaAs setback layer and n⁻-InP collector to further reduce the effective width of the spike barrier [49,50]. The reduction in barrier width allows more electron tunneling actions to occur, and electrons do not have to acquire energies greater than or equal to ΔE_C in order to transport through the potential barrier layer. Another way to look at it is as if the "effective" height of the barrier were reduced when tunneling phenomenon occurs.

Figure 11.4 shows energy band diagrams of an InP/InGaAs DHBT with a 4000-Å composite collector operated at low (5×10^4 A/cm²) and high (2×10^5 A/cm²) current density conditions, respectively. The band profiles were generated using the ensemble Monte Carlo simulator described in Ref. 12, and each dot represents 2.5×10^{15} electrons/cm³. The collector structure consisted of a 250-Å n⁻ InGaAs spacer (5×10^{16} cm⁻³) near the base, 2250-Å n⁻ InP (5×10^{16} cm⁻³), and 1500-Å i InP layers near the subcollector. The undoped InP layer is used to tailor the conduction band profile for high-speed transport similar to the undoped collector design described in Ref. 23 and to increase the breakdown voltage at high current densities [24]. We discuss the base–collector heterojunction design with this example by looking at three phenomena associated with the DHBT: current blocking, intervalley scattering, and breakdown.

Current Blocking By inserting an InGaAs spacer layer between the base and collector, one is able to reduce the influence of the conduction band barrier at the base–collector heterojunction on electron transport. The mechanism is explained as follows with the help of Fig. 11.4a. As long as the energy position of the peak of the spike barrier is lower than that of the bottom of the conduction band in the base layer, most electrons injected from the emitter and transported across the base layer through quasi-ballistic transport and/or diffusion mechanisms have sufficient kinetic energies to overcome the barrier and be collected efficiently in the collector. This

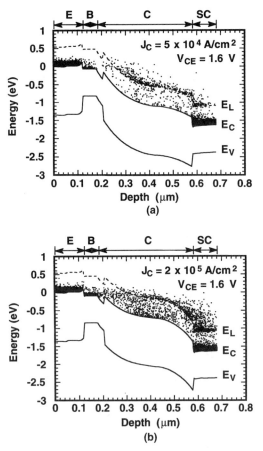

FIGURE 11.4 Energy band diagrams of an InP/InGaAs DHBT operating at (*a*) low (5×10^4 A/cm^2) and (*b*) high current density (2×10^5 A/cm^2) conditions. The band profiles were generated using a self-consistent ensemble Monte Carlo simulator. Each dot represents 2.5×10^{15} electrons/cm^3. The total collector thickness was 4000 Å and the V_{CE} bias was 1.6 V.

criterion can be met when the lightly doped InGaAs spacer layer is sufficiently thick, the electric field in that depleted layer is high and the device is under low current injection condition. As shown in Fig. 11.4a, the electric field, as indicated by the slope of the conduction band, is large near the base side of the collector, and the peak of the spike barrier is well below the bottom of conduction band in the base. Even slow electrons that rely on diffusion to travel across the base have sufficient energies to surmount the barrier. Very few electrons are scattered and trapped in the InGaAs spacer layer if the spacer layer is not too thick.

Under high current injection condition, the space charge in the collector space-charge region pulls the conduction band up with respect to the Fermi level (Fig. 11.4b). As soon as the energy peak of the spike barrier rises above the bottom of the conduction band in the base, electron transport is blocked. The blocking of electron flow reduces the collector current and results in further buildup of space charge, leading to larger base recombination current. Reduction in the collector current and increase in the base current result in rapid decrease in current gain as the injection level is increased further. Because electrons can no longer transport efficiently across the base–collector heterojunction, the base transit time increases and the operation speed of the device decreases. For high-speed transport at large current densities, the collector doping concentration near the base side of the collector has to be large enough to avoid this current blocking from happening. Figure 11.4b also shows that the energy peak of the spike barrier has increased and approached more closely the conduction band energy in the base but not yet exceeding the latter. The case represents the highest current density at which the device can operate without suffering from current blocking phenomenon. If the current density increases further, current blocking occurs. Electrons then accumulate and get trapped in the triangular well formed by the slope of the conduction band of the InGaAs spacer layer and the conduction band discontinuity ΔE_C of the InGaAs/InP heterojunction.

Intervalley Scattering The band profiles of the L-valleys are plotted in Fig. 11.4 as a function of depth at two different current densities. First, let us consider the case of low current density (Fig. 11.4a). As electrons traveled deeper into the collector, they gain more energies from the electric fields present in the collector space-charge region. If the electric fields are large or the slope of the conduction band is steep, they gain energies quickly over a short distance. Their energies are indicated by their energy positions relative to the conduction band E_C. When their energies are close to the Γ–L valley separation, a lot of electrons are scattered to the high effective mass L-valleys through intervalley scattering. As shown in Fig. 11.4a, because the electric fields near the base side of the collector are large under low current injection condition, a lot of electrons are scattered to high-effective-mass L-valleys at a relatively short distance of about 900 Å from the base. Note also that some electrons are scattered back to the Γ-valley at a distance of about 1600 Å from the base layer.

At high current densities, the energy band profile in the collector changes because of the presence of space charge (Fig. 11.4b). A comparison of Fig. 11.4a and b indicates that the electric field near the base side of the collector is lower, whereas that near the subcollector side of the collector is higher at higher current density [24]. This flatter conduction band profile aids electron transport because electrons can stay in the low-effective-mass Γ-valley over a larger distance of depth, resulting in velocity shoot effects over an extended distance. As shown in the figure, a great portion of the Γ-valley electrons can be found as deep as 2000 Å from the base in the collector space-charge region.

An optimum design window thus exists between high electric fields in the base side of the collector to prevent the energy peak of the spike barrier from rising above the conduction band energy in the base and triggering current blocking, and low electric fields in the same region to prevent electrons from gaining too much energy in the collector space-charge region and triggering intervalley scattering.

Breakdown Because electric fields build up at the collector–subcollector interface as current density increases [24], the subcollector side of the collector in the previous example is nonintentionally doped to maintain large breakdown voltage, especially at high current densities. A nonuniform doping profile is therefore not unusual in a typical DHBT design. To increase the breakdown voltage further, a thicker collector needs to be used. Depending on the desired operation voltage and current, a well-designed DHBT requires careful optimization of the spacer and collector layer thicknesses as well as the doping profile.

In general, an abrupt base–collector heterojunction with an InGaAs setback layer can be used only in thin collector designs (≤ 4000 Å) [47,48]. A graded heterojunction must be used for thicker collector DHBTs to avoid saturation problems or keep the "knee" voltage small in the I_C–V_{CE} characteristics. The design of the base–collector heterojunction becomes increasingly difficult as the collector gets thicker because the thick lightly doped part of the InP collector layer tends to lift the entire conduction band up with respect to the Fermi level in the collector region. The overall effect is that the energy position of the peak of the spike barrier gradually rises above the bottom part of the conduction band in the base, and current blocking occurs even at low injection levels as the collector gets thicker. A thick collector layer thus overcompensates the effects of the use of an InGaAs setback layer and doping spike near the base side of the collector. Further increase in the InGaAs spacer layer thickness and spike doping concentration may decrease the collector breakdown voltage significantly, and the benefits of the use of an InP collector may be gone.

In the InP/InGaAs heterojunction system, both step grading [51,52] and continuous compositional grading with InGaAsP quaternary materials can be used. For example, Yamahata et al. have recently achieved f_{max} and f_T values of 267 and 144 GHz, respectively, by applying a step-graded collector to thin collector structures (2700 Å) [4]. For power applications in which thick collector and high breakdown voltage are required, a continuous compositionally graded heterojunction with a proper doping profile needs to be used. The drawback of using such a compositional grading scheme, either step grading or continuous grading, is the difficulty in reproducing the composition of the InGaAsP grading layer and maintaining lattice matching. Alternatively, a chirped superlattice consisting of InGaAs and InP layers of variable thicknesses allows easier control of the grading [53,54]. Figure 11.5 illustrates the idea. Because the thickness of each constituent layer is very thin, an effective bandgap is formed across the superlattice through wavefunction interactions. When the thicknesses of each layer and period is varied in such a way that the effective bandgap is smallest near the base and largest near the collector side of the graded region, a continuous grading is achieved. Similar to the composite collector approach described earlier, the layer thicknesses and doping concentrations have to be designed properly to avoid significant changes in the electron transmission coefficient with collector bias voltage in the chirped superlattice. Our DHBTs designed in this way have shown a BV_{CBO} of 29 V defined at 1 μA leakage current, f_T of 69 GHz, and f_{max} of 166 GHz [42]. A similar type of DHBT power unit cell has demonstrated 2-W output power and 70% power-added efficiency at 9 GHz [55].

Although the transport and thermal properties in InAlAs are not as good as those

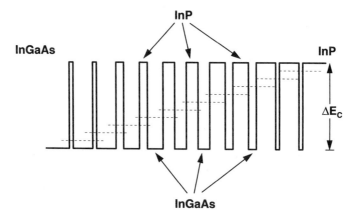

FIGURE 11.5 Schematic band diagram showing how a chirped superlattice consisting of InGaAs and InP layers of variable thicknesses can be used to achieve an effective continuous graded layer.

in InP, the biggest improvement in open-base breakdown voltage is expected from DHBTs with InAlAs collectors because the bandgap of InAlAs is largest among all the InP-based lattice-matched materials. The 0.5-eV ΔE_C present at the base–collector heterojunction represents the biggest challenge in InP-based base–collector heterojunction design [56]. Similar to the emitter design discussed in Section 11.3.2, a compromise solution can be made with a quaternary alloy such as InAlGaAs in the collector [57–59], which reduces the conduction band discontinuity when it forms a heterojunction with the InGaAs base and yet provides a better transport characteristics than InAlAs. Its thermal conductivity is, however, not known exactly but is expected to be worse than the binary compound InP.

11.4 FABRICATION TECHNOLOGY

The InP HBT fabrication process differs from that of GaAs HBTs in that deep isolation by ion implantation is not used because it does not result in similar high-resistivity layers as those obtainable in GaAs [60]. For shallow structures such as field-effect transistors (FETs), iron is very effective in producing high-resistivity InGaAs. However, since the atomic mass of iron is large, iron does not penetrate deep enough to isolate thick structures like HBTs. The triple-mesa technology is still by far the most commonly used process in high-performance InP-based heterojunction bipolar transistors (Fig. 11.2). To realize the HBT cross section shown in Fig. 11.2, a self-aligned InP HBT process can be used and is described as follows.

The emitter contact is first defined and deposited. A common metallization scheme used for the emitter contact is Ti/Pt/Au, which also acts as an etching mask during the emitter mesa etch. Refractory emitter contacts such as Wsi_x or TiW can also be used to improve contact stability. The emitter mesa etching steps are slightly different for InAlAs and InP emitter HBTs. In the case of InAlAs emitter HBTs, the InGaAs base is exposed by removal of the InGaAs cap and InAlAs emitter layers. This is typically done in a sulfuric acid–based (H_2SO_4:H_2O_2:H_2O) or a phosphoric acid-based etchant (H_3PO_4:H_2O_2:H_2O). Since the etching is nonselective between InAlAs and InGaAs layers, it is a timed etch and the surface breakdown voltage needs to be monitored during the etching to avoid overetching the base layer. In the case of InP emitter HBTs, the etching stops on the InP emitter layer. A hydrochloric acid–based etchant such as HCl:H_2O or HCl:H_3PO_4 is then used to remove the InP emitter layer. Since neither HCl etchant attacks InGaAs, the etching stops automatically on the top of the InGaAs base layer. The use of InP emitter thus leads to a more manufacturable process. However, the use of etchants with extreme selectivities means that a certain amount of overetching is needed to ensure complete removal of an upper epitaxial layer before the lower layer is etched with a different etchant. To avoid excessive undercutting and the resultant large separation between the base contact and emitter mesa, a reactive ion etch (RIE) process based on CH_4/H_2 plasma chemistry can be used to remove the InGaAs cap layer completely and the InP emitter layer partially. This plasma etch process is highly anisotropic and gives a nearly vertical etched profile [61]. Depending on the dc self-biased voltage of the plasma, the dry etch should be stopped at a distance that is sufficiently far away from the base to avoid damage to the base layer [62]. The RIE is then followed by a brief wet etch to remove the damaged etched surface and create the necessary emitter overhang for self-aligned base contact. By making use of a combination of RIE and selective wet chemical etching, a base–emitter separation of the order of 0.1 μm can be achieved [62].

Next, the base contact is defined. Owing to the emitter undercut formed in the emitter mesa etch, the Ti/Pt/Au base metal is self-aligned to the emitter contact without shorting the emitter metal (Fig. 11.2). An optional refractory metal layer such as W can first be deposited to improve contact stability [63]. This is followed by collector lithography in which the collector contact is defined. In the case when the collector layer is either InGaAs as in SHBTs or InAlAs as in DHBTs, the same sulfuric acid–or phosphoric acid–based etchant mentioned above can be used to expose the heavily doped InGaAs subcollector layer. If an InP collector layer is adopted, the base and collector layers are selectively removed by first etching the base in either H_2SO_4- or H_3PO_4-based etchant, and then the collector in an HCl-based etchant. If an InP/InGaAs superlattice grading structure is present, as in power DHBTs, a reactive ion etch process based on CH_4/H_2 plasma chemistry can first be used to etch through the superlattice grading layers before the HCl-based etchant is used.

After the subcollector layer is exposed, AuGe/Ni/Au collector metal is subsequently deposited (Fig. 11.2). Photoresist patterns are used to protect active device areas. Devices are then physically isolated from each other by removing the unwanted base, collector, and subcollector layers similar to the etching process used for the collector contact except for going one step further: by removing the InGaAs subcollector layer as well in the H_2SO_4- or H_3PO_4-based etchant.

There are several ways to connect the finished devices with other circuit elements. One is to deposit a layer of dielectric such as silicon nitride by the plasma-enhanced chemical vapor deposition (PECVD) technique to passivate the devices. The passivation step is followed by contact via hole etching to expose the emitter, base, and collector contacts, and then deposition of interconnect metal. If planarization is desired for multiple-level interconnects, the wafer can be coated with multiple spins of polyimide, followed by oxygen–argon plasma etch back to expose the emitter metal. Contact via holes are then etched to expose the base and collector contact, and interconnect metal is evaporated. At Texas Instruments, the process is similar to the first one except that electroplated air bridges are used instead of letting interconnect metal sit and cross over the edges of mesa islands (Fig. 11.2). The advantage of the air-bridge process is lower parasitic capacitances. The finished devices are then sintered in a furnace at 430°C for 1 min. Figure 11.6 shows the scanning electron microscopic (SEM) picture of a completed four-finger InP HBT using the air-bridge process.

Since extrinsic collector isolation implant is not a feasible technique to reduce the extrinsic base–collector capacitance in InP-based HBTs, novel techniques need to be used to reduce this parasitic capacitance. In InP DHBTs that employ InGaAs base, InP collector, and InGaAs subcollector layers, the collector can be undercut intentionally by excessive overetching of the InP collector layer [64]. More advanced technologies for other device types have also been demonstrated. For example, Song et al. demonstrated a novel InAlAs/InGaAs HBT structure with reduced parasitic base–collector capacitance achieved by incorporating a selectively grown

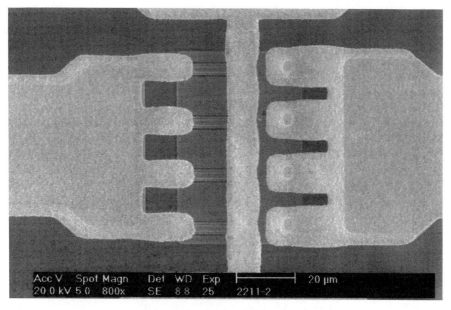

FIGURE 11.6 Scanning electron micrograph of a completed four-finger InP/InGaAs SHBT, with each finger having an emitter area of (2×20) μm^2.

buried InGaAs subcollector by chloride-transport vapor-phase epitaxy (VPE) and organometallic vapor-phase epitaxy (OMVPE) [65]. Matine et al. [66] and Bhattacharya et al. [67] investigated collector-up InP SHBTs with Schottky collector metal contacts for small base–collector capacitance. Also worth mentioning is a novel planarization process reported by Shigematsu et al. in which the sequence of emitter, base, and collector contact formation is reversed from the conventional process described above with reduced base–collector capacitance [68]. The extrinsic base resistance can be reduced by selective regrowth of heavily doped extrinsic base regions [69].

11.5 HBT CHARACTERISTICS

11.5.1 DC Characteristics

SHBTs: DC The common-emitter I_C–V_{CE} characteristics of a typical InP SHBT is shown in Fig. 11.7. The device had a 800-Å base doped at 3×10^{19} cm^{-3} and a 6000-Å collector doped at 2×10^{16} cm^{-3}. It had an actual emitter area of about 11 μm^2. The measured collector–emitter offset voltage $V_{CE, \text{offset}}$ was 110 mV. A V_{CE} saturation or knee voltage of about 0.5 V was measured at 7.5×10^4 A/cm^2. As shown, the transistor shows high output conductance and small breakdown voltage at large current levels. The reason for this is attributed to device self-heating, which significantly increases the base–collector reverse saturation or leakage current I_{CBO} [42,43], which is explained in detail in Section 11.5.2. Hence, even though the SHBT may exhibit a large collector–emitter breakdown voltage BV_{CEO} of over 7.5 V under

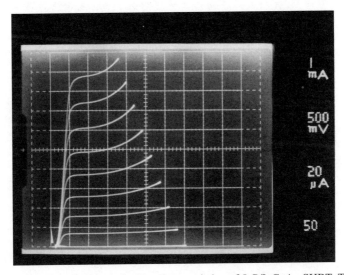

FIGURE 11.7 Common-emitter I_C–V_{CE} characteristics of InP/InGaAs SHBT. The actual emitter junction area was about 11 μm^2.

open-base conditions ($I_B = 0$ A), the breakdown voltage BV_{CE} ($I_B \neq 0$ A) is generally much smaller (< 3 V) at higher current densities ($J_C > 7 \times 10^4$ A/cm^2). BV_{CEO} is therefore not a good figure of merit for InP-based SHBTs because it is the BV_{CE} ($I_B \neq 0$ A) that limits the useful operating voltage of the devices. For this reason, InP-based SHBTs are useful only in low-voltage or low-power-dissipation applications.

Figure 11.8 shows the Gummel characteristics of three SHBTs with three different emitter materials: InP, $In_{0.52}Al_{0.48}As$, and $In_{0.52}(Al_{0.6}Ga_{0.4})_{0.48}As$. The ideality factors of the collector and base currents $n(I_C)$ and $n(I_B)$ were found to be 1.1 and 1.2, 1.3 and 1.4, and 1.0 and 1.1, respectively, for the InP/InGaAs, InAlAs/InGaAs, and InAlGaAs/InGaAs transistors. The high ideality factor in the collector current of the InAlAs/InGaAs SHBT is presumably due to electron tunneling through the conduction band spike barrier and quantum mechanical reflections at the abrupt base–emitter heterojunction. The Gummel plots also indicate the difference in turn-on voltages among these devices. Since the InAlAs/InGaAs SHBT has the largest ΔE_C value at the base–emitter heterojunction, it exhibits the largest turn-on voltage and the highest $n(I_C)$ value among the three transistors. Except for the small decrease in current at low current levels, all three devices showed negligible temperature dependences of current gain as the temperature increased 25 to 200°C[70]. The current gain degradation in all three InP-based transistors was much smaller than that of a typical AlGaAs/GaAs HBT [71] in the same temperature range.

DHBTs: DC The improvement in output characteristics and breakdown voltage of an InP/InGaAs/InP DHBT can be seen from the measured common-emitter I_C–V_{CE} characteristics shown in Fig. 11.9. The collector structure was the same as that described in Section 11.3.3. It composed of a 250-Å InGaAs spacer layer doped at 5×10^{16} cm^{-3}, a 2250-Å InP layer doped at 5×10^{16} cm^{-3}, and 1500-Å undoped InP. The base had a thickness of 800 Å and was doped at 3×10^{19} cm^{-3}. A 100-Å InGaAs spacer layer was inserted between the base and the InP emitter. A compari-

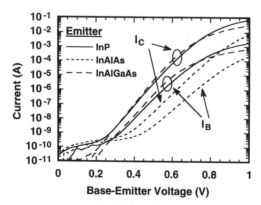

FIGURE 11.8 Gummel characteristics of InP/$In_{0.53}Ga_{0.47}As$, $In_{0.52}Al_{0.48}As$/$In_{0.53}Ga_{0.47}As$, and $In_{0.52}(Al_{0.6}Ga_{0.4})_{0.48}As$/$In_{0.53}Ga_{0.47}As$ SHBTs measured at $V_{CB} = 0$ V. The actual emitter junction area was 11 μm^2.

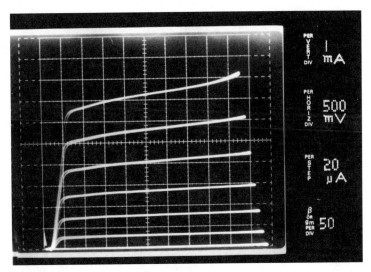

FIGURE 11.9 Measured common-emitter I_C–V_{CE} characteristics of a 11-μm^2 InP/InGaAs/InP DHBT. The total collector thickness was 4000 Å.

son between Figs. 11.7 and 11.9 clearly demonstrates the very much improved output conductance and breakdown voltage in the DHBT at high current densities. Even though both transistors had very similiar collector–emitter breakdown voltage BV_{CEO} of 7.5 V, the measured breakdown voltage BV_{CE} ($I_B \neq 0$ A) was less sensitive to the current density in the DHBT than in the SHBT. The common-emitter breakdown voltage of the DHBT remained high and was greater than 5.5 V at current densities as high as 7×10^4 A/cm^2. This value of BV_{CE} ($I_B \neq 0$ A) was significantly different from the SHBT in which a breakdown voltage of less than 3 V was measured at the same current density (Fig. 11.7). The measured collector–emitter offset voltage was 180 mV and the "knee" voltage was 0.7 V at 7.5×10^4 A/cm^2 current density in the double-heterojunction device.

Figure 11.10 shows the Gummel plots of the 11-μm^2 DHBT measured at $V_{CB} = 0$. It was found that the ideality factor was 1.15 and 1.30 for the collector and base currents, respectively. When a zero base–collector bias voltage was applied, a small hump appeared in the base current curve at $V_{BE} = 0.9$ V. It was due to a combination of base pushout and current blocking effects under large current injection conditions as described in Section 11.3.3. As the current density increases, the peak of the conduction band spike barrier at the base–collector heterojunction rises gradually with respect to the bottom of the base conduction band due to space-charge accumulation at the heterojunction. As soon as the peak of the spike rises above the bottom of the conduction band in the base, electrons injected or diffused from the base are blocked. It results in premature saturation of the collector current as indicated by the flat region of the collector current curve and sudden increase in base current in Fig. 11.10 for $V_{BE} > 0.9$ V. As a consequence, the current gain is compressed. Under such current injection conditions, a lot of electrons are trapped in the base since

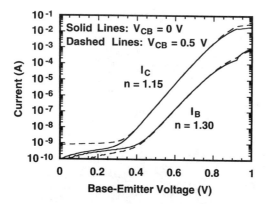

FIGURE 11.10 Gummel plots of a 11-μm^2 InP/InGaAs DHBT at $V_{CB} = 0$ V and $V_{CB} = 0.5$ V. The total collector thickness was 4000 Å.

they are no longer collected efficiently in the collector. The excess electrons result in a sudden increase in the recombination current and therefore the base current, as indicated by the hump in the base-current curve. The current blocking phenomenon can be suppressed easily in this case by applying a small base–collector reverse bias of 0.5 V. As indicated in Fig. 11.10, the hump in the base current disappears and the maximum collector current increases. The dc results indicate that the conduction band discontinuity at the base–collector heterojunction does not prohibit electron transport under normal operation conditions (at which the base–collector heterojunction is reverse biased) in the 4000-Å-thick collector InP/InGaAs/InP DHBT.

A power InP/InGaAs/InP DHBT was also fabricated. The MOMBE-grown epitaxial structure of the InP/InGaAs/InP power DHBT consists of a 3000-Å n$^+$ InGaAs subcollector (1×10^{19} cm^{-3}), 8000-Å n$^-$ InP collector (3×10^{16} cm^{-3}), 288-Å n$^-$ InGaAs/InP superlattice grading layer (of which 50% is InGaAs), 50-Å n$^-$ InGaAs spacer layer, 1000-Å p$^+$ InGaAs base (4×10^{19} cm^{-3}), 100-Å n InGaAs setback layer (1×10^{18} cm^{-3}), 900-Å n InP emitter (5×10^{17} cm^{-3}), and 100-Å n$^+$ InP, and 1700-Å n$^+$ InGaAs emitter contact layers (both 1×10^{19} cm^{-3}). A good power transistor should exhibit high breakdown voltage, small V_{CE} offset voltage, and small V_{CE} saturation voltage. In addition, it is also important to avoid current blocking in DHBTs at low current densities. The current blocking in DHBTs is accompanied by current gain compression, as described above. As shown in Fig. 11.11, the measured common-emitter I_C–V_{CE} characteristics of the completed 2×10 μm^2 power DHBTs show a very small V_{CE} offset voltage of only 85 mV and a V_{CE} saturation voltage of 1 V at 10 mA collector current (5×10^4 A/cm^2 current density). Current gain compression occurs at currents larger than 15 mA (7.5×10^4 A/cm^2 current density). The small dip in the I_C–V_{CE} characteristics is due to small changes in the electron transmission coefficient with collector–emitter voltage in the chirped superlattice and can be suppressed by reducing the thickness of the superlattice period and increasing the doping level in the superlattice grading layer. Despite the

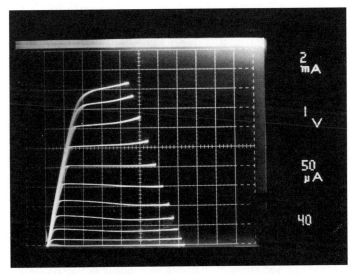

FIGURE 11.11 Measured common-emitter I_C–V_{CE} characteristics of a 2 × 10 μm^2 power InP/InGaAs/InP DHBT. The total collector thickness was 8288 Å.

use of a total of only 194-Å InGaAs in the collector, the V_{CE} saturation voltage in the power DHBT was small. The breakdown voltages BV_{CBO} and BV_{CEO} were 29 and 18 V, respectively. The good dc characteristics clearly demonstrate the effectiveness of chirped superlattice in thick collector DHBT designs. The RF and power performance of the power DHBTs are described in Sections 11.5.4 and 11.5.5, respectively.

11.5.2 Thermal Characteristics

As described in Section 11.3.1, InP is known to have higher thermal conductivity than GaAs (0.68 versus 0.45 W/cm·K at 300 K [25]). It is therefore common to assume that HBTs fabricated on InP substrates run at lower junction temperatures than in GaAs devices having the same geometry and power dissipation. This is, however, not necessarily true in the presence of an InGaAs subcollector layer (and also an InGaAs collector layer in SHBTs) because InGaAs has a thermal conductivity of only 0.05 W/cm·K at 300 K [25].

To better understand the influence of InGaAs on device thermal behavior, we have used a three-dimensional numerical thermal simulator developed at Texas Instruments to calculate the temperature distributions of one- and eight-finger InP SHBTs and DHBTs. The results were compared with the corresponding GaAs HBTs of similar layer thicknesses and dimensions. The InP HBTs were assumed to be mesa isolated and were therefore more nonplanar than the implant-isolated GaAs transistors. In case of GaAs HBTs, the collector, subcollector, and substrate materials were all made of GaAs. InGaAs and InP were taken to be the collector material in the InP SHBTs and DHBTs, respectively. For both types of InP transistors, the subcollector material was InGaAs, and its thickness on device junction temperature was investigated. Unless otherwise specified, the thicknesses of the collector layers in all devices were fixed at 0.8 μm and the finger size was 2 × 50 μm². The power dissipation per finger was 0.1 W and was assumed to be uniformly distributed in the intrinsic collector region under each finger. The latter assumption was sufficient for the purpose of illustrating the influence of InGaAs on device thermal resistance. No power dissipation was assumed in the base and emitter layers. 300-K thermal conductivity values were used and were assumed to be independent of temperature even though they all decrease with increasing temperature in practice.

Figure 11.12 compares the peak junction temperature rises as a function of subcollector layer thickness for one- and eight-finger GaAs HBTs and InP DHBTs and one-finger InP SHBTs. The junction temperature rise in GaAs HBTs is independent of the subcollector thickness because the subcollector layer is also made of GaAs and has a negligible thickness compared to the substrate thickness of 100 μm. In contrast, the junction temperature increases rapidly with increasing InGaAs subcollector thickness in all InP HBTs. The rises in one-finger InP SHBTs with a 0.8-μm InGaAs collector are far more significant than in multifinger GaAs HBTs and InP DHBTs of the same layer thicknesses. Even when the InGaAs subcollector layer thickness approaches 0 Å, the one-finger SHBT still has about the same peak junction temperature rise as that of an eight-finger InP DHBT with a 0.7-μm InGaAs subcollector. For the same power density per finger, multifinger transistor normally runs at a higher junction temperature than a transistor with fewer fingers because of thermal coupling among its fingers. In addition to the reliability concern, the high junction temperature in InP SHBTs also leads to a large reverse saturation current I_{CBO} in their base–collector diode characteristics. Consider, for example, a subcollector thickness of 3000 Å, a peak junction temperature rise of 139°C, and an ambient temperature of 25°C, which result in a peak junction temperature of about

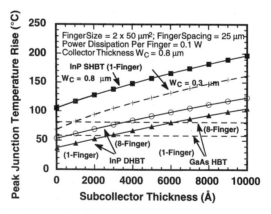

FIGURE 11.12 Dependence of junction temperature rise on subcollector layer thickness in one- and eight-finger GaAs HBTs and InP DHBTs and one-finger InP SHBTs.

164°C in the InP SHBT. Such a large temperature increase has been shown experimentally to give a two- to four-order-of-magnitude increase in base–collector saturation current I_{CBO} at small reverse-bias voltages [42,43], which is discussed in the latter part of this section.

When the InGaAs collector thickness is reduced to 0.3 μm, the one-finger SHBTs still run at a lot higher temperature than do eight-finger InP DHBTs with a 0.8-μm-InP collector, as shown in Fig. 11.12. A similar trend in peak junction temperature rise and high junction temperatures are observed when the subcollector layer thickness is fixed at 0.5 μm and the collector thickness varies in the SHBTs [72]. The results indicate that it is the total amount of InGaAs present in the collector and subcollector layer that is important in determining the junction temperature of the transistors. Comparison between the temperature distributions in GaAs HBTs and InP DHBTs of the same layer thicknesses shows that the latter operates at lower peak junction temperatures than do GaAs devices only when the InGaAs subcollector layer thickness is less than about 3500 Å (Fig. 11.12). This is an important consideration, especially for power applications, because the power performance of HBTs is thermally limited.

We have measured the average low-power thermal resistances R_{th} of 2×10 μm² InP SHBTs and DHBTs of different InGaAs thicknesses in the subcollector and/or collector from different wafer lots. To examine the first-order effect of the InGaAs layer on R_{th}, we applied the technique of Ref. 73 to the transistors and found that R_{th} increased substantially when more than 3000 Å of InGaAs was incorporated in the subcollector in DHBTs and was large in SHBTs. The results are plotted in Fig. 11.13.

In addition to the junction temperature considerations for reliable operation, the presence of InGaAs in the collector of an InP HBT has a significant influence on

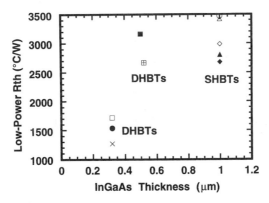

FIGURE 11.13 Measured average low-power thermal resistances of 2×10 μm² InP SHBTs and DHBTs of different InGaAs thicknesses in the subcollector and/or collector layers. Each data point represents a device from a different wafer lot. In the case of SHBTs, the InGaAs thickness includes both the 5000-Å InGaAs collector and the 5000-Å InGaAs subcollector. For DHBTs, the InGaAs subcollectors can be either 3200 or 5000 Å thick.

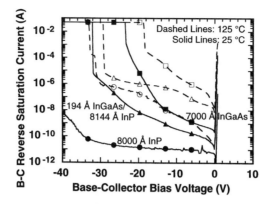

FIGURE 11.14 Measured base–collector reverse diode characteristics in three different collector structures at 25 and 125°C.

the base–collector reverse saturation current. Figure 11.14 compares the measured temperature dependences of reverse base–collector diode characteristics in three different collector structures: (1) 7000-Å n⁻-InGaAs, (2) 8000-Å n⁻-InP, and (3) 50-Å n⁻-InGaAs setback followed by 288-Å n⁻-InGaAs/InP superlattice grading (of which 50% is InGaAs) and 8000-Å n⁻-InP. As shown, by introducing only a total of 194 Å of InGaAs in the collector, the reverse saturation current increased substantially. Furthermore, an increase in substrate temperature by 100°C increased the saturation currents in all cases by two to four orders of magnitude. The reverse base–collector diode current includes the current component due to the intrinsic free carriers in the collector $I_{CBO,n}$, impact ionization $I_{CBO,i}$, and any process-related leakage current $I_{CBO,p}$ (such as that induced by dry etching). Impact ionization can be ignored at small base–collector reverse bias. If the process-related leakage is also negligible, the two- to four-order-of-magnitude increase in I_{CBO} at small reverse-bias voltage would be caused primarily by the increase in $I_{CBO,n}$. In the one-finger InP SHBT with a 0.8-μm collector and a 0.3-μm InGaAs subcollector layer, for example, the intrinsic carrier concentration in InGaAs increases from 6.4×10^{11} cm^{-3} at 25°C to 2.2×10^{14} cm^{-3} at 164°C peak junction temperature. Under the common-emitter operation configuration, the base terminal current is fixed. Therefore, this base–collector current component $I_{CBO,n}$ does not go to the external base terminal but goes to the neutral base and gets amplified through the current gain mechanism of the transistor, leading to an additional current component $\beta I_{CBO,n}$ in the collector current, where β is the current gain of the transistor. Based on this argument and ignoring thermal effects on impact ionization and other process-related leakages, InP HBTs with thick InGaAs subcollectors show worse output conductance and breakdown-like behavior than devices with thin subcollectors due to the increase in junction temperature and the corresponding exponential increase in intrinsic carrier concentration. If the impact ionization process is also involved in SHBTs, such as in cases when the base–collector reverse bias is large or when the local electric field in the collector is increased at high current densities [24], the output conductance in the SHBTs gets worse because of the anomalous increase in impact ionization rates in the InGaAs collector with increasing temperature [74]. The impact of this junction temperature dependence on the amount of InGaAs used in the collector and subcollector layers thus has particularly deleterious effects on InP SHBTs because of their high room-temperature diode saturation current and large-device thermal resistance. A transistor will suddenly blow up under high bias when I_{CBO} becomes excessive at elevated junction temperatures [75].

11.5.3 Thermal Instabilities

When two identical transistors are connected to common-emitter, common-base, and common-collector electrodes, each transistor is expected to conduct the same amount of collector current for any biases. However, as shown theoretically and demonstrated experimentally in Ref. 73, equal conduction occurs only when the power dissipation is low to moderate, such that the junction temperature rise above the ambient temperature is negligible. At high V_{CE} and/or high I_C operation where the transistor is operated at elevated temperatures, one transistor starts to conduct more current than the other. Eventually, one transistor conducts all the current while the other becomes electrically inactive. This phenomenon is known as *thermal instability*. The current domination results because of an intrinsic transistor property that as the junction temperature increases, the bias required to turn on some arbitrary current level decreases.

In a multifinger HBT, the transistor can be viewed as consisting of several identical sub-HBTs, with their respective emitter, base, and collector leads connected together. If one finger becomes slightly warmer than the others, its base–emitter junction turn-on voltage becomes slightly lower. Consequently, this particular finger conducts more current for a given fixed base–emitter voltage. This increased collector current, in turn, increases the power dissipation in the junction, raising the junction temperature even further. The thermal instability thus results in two phenomena. First, one finger of the transistor (the hot finger) hogs the input base current from the remaining fingers (the cold fingers), causing both the I_B and I_C of the hot finger to increase dramatically. The second phenomenon, which follows as a consequence of the first, is that the junction temperature in the hot finger increases significantly because of the increased intensity of power dissipation. Thermal instability does not normally occur in a single-finger HBT because the finger behaves as a single thermal unit. For a one-finger HBT with exceedingly long finger length, the thermal instability in itself due to extreme nonuniform temperature profile is, however, possible [76].

Figure 11.15 shows the transistor *I–V* characteristics of a two-finger InP DHBT, which was specially designed so that the collector current of each finger can be monitored individually [77]. The distance between the fingers is more than 100 μm, and the mutual thermal coupling can be neglected. Figure 11.16 illustrates the individual collector currents corresponding to the *I–V* characteristic of Fig. 11.15. For each set of two curves corresponding to a given total base current (I_{bT}), the dashed

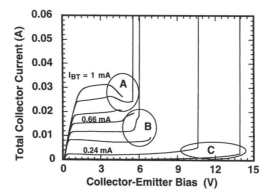

FIGURE 11.15 Measured I_C–V_{CE} characteristics of a two-finger InP DHBT, with each finger having an area of (2×10) μm^2. The plotted current is the sum of the two fingers' collector currents. The total base current applied to the DHBT varies from 0.07, 0.24, 0.44, 0.56, 0.66, 0.76, and 0.86 to 1 mA.

and solid curves denote the hot and cold finger currents, respectively. Although it is not clear from Fig. 11.15 whether thermal instability ever occurs, Fig. 11.16 clearly demonstrates that the transistor enters thermal instability—that one finger current increases, whereas the other decreases to zero. That is, when V_{CE} is to the left of the thermal instability loci, both fingers conduct relatively the same amount of current, as expected of the two identical fingers. Once the biases cross the instability loci, the two phenomena associated with thermal instability—that the current and the junction temperature increase in the hot finger—begin to surface. A study on the current gain variation in an InP DHBT is critical to an understanding of the I–V characteristic after thermal instability. Our measurements indicated that the current

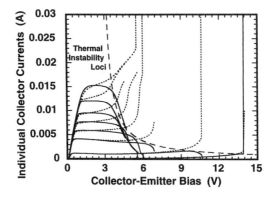

FIGURE 11.16 Measured individual collector currents corresponding to the I_C–V_{CE} of Fig. 11.15. The dashed and the solid curves denote the hot- and cold-finger currents, respectively.

gain β of one-finger InP DHBTs had negligible temperature dependence from 25 to 200°C, because of the large valence band discontinuity of 0.37 eV at the InP/InGaAs base–emitter heterojunction, but can either increase or decrease with I_C, depending on the present operating current level. β increases initially with I_C and then drops rapidly as I_C increases above 17 mA for the 2×10 μm^2 finger due to increased electron blocking effect at the base–collector heterojunction.

The familiar collapse of current gain characteristics in GaAs HBTs appears only at high current levels (region A in Fig. 11.15) [78,79]. While the gain collapse in GaAs HBT is due to its current gain decrease with temperature, the collapse in gain in InP DHBT is due to its current gain decrease at high currents. With further increase in V_{CE} in region A, the total collector current (I_{CT}) suddenly increases toward infinity, irreversibly shorting out the device itself. The failure mechanism is identical to that reported for GaAs HBT, determined to be due to the increasing I_{CBO} at elevated junction temperatures [75]. In region B, the current levels at which the device enters thermal instability is far below 17 mA; in this region, β increases as I_C increases. Therefore, when the hot finger current increases during thermal instability, the total current I_{CT} increases, as observed in Fig. 11.15, contrary to the collapse of current gain in region A. In region C, because the current levels are low, the corresponding V_{CE} values at which thermal instability takes place are larger than those in region B. Furthermore, since the externally applied base current is lower in region C, the threshold I_{CBO} (the amount of I_{CBO} required to trigger device failure) is smaller. Consequently, in region C it takes a smaller increase in I_{CT} past that at the thermal instability to result in a given power dissipation $I_{CT}V_{CE}$, which in turn results in enough temperature rise to generate the threshold I_{CBO}. The combined effects of lower I_{CBO} threshold and higher V_{CE} values during thermal instability in region C thus lead to sudden device failure, without displaying the noticeable trend in region B—that I_{CT} rapidly increases just prior to the failure.

In both regions A and B, the device survives as long as care is exercised and the applied collector–emitter bias does not increase too much beyond the aforementioned I_{CT} decrease or increase region. However, at low current levels (region C), I_{CT} increases toward infinity without giving a clear indication as to whether a thermal instability condition has occurred. In any case, care must be taken to avoid biasing devices near the thermal instability loci.

11.5.4 Microwave Characteristics

Impressive high-frequency characteristics have been reported for both InP SHBTs and DHBTs. Both the current-gain cutoff frequency and maximum frequency of oscillation have broken the 200-GHz performance barrier [1–6]. Next we describe typical RF characteristics of InP HBTs with different emitter and/or collector materials.

SHBTs: RF Impressive high-frequency characteristics were reported for both InP/InGaAs and InAlAs/InGaAs SHBTs lattice matched to InP substrates. Common-emitter InP/InGaAs and InAlAs/InGaAs SHBTs exhibited a record current-gain cutoff frequency f_T up to 209 GHz [2] and a record maximum oscillation frequency f_{max} as high as 236 GHz [3], respectively, at room temperature. Simultaneous achievement of both f_T and f_{max} above 160 GHz was also reported in common-emitter InP/InGaAs SHBTs [68].

The conduction band offset at the base–emitter heterojunction affects the base transport mechanisms, and an optimum ΔE_C exists in terms of good dc and RF performance, as discussed in Section 11.3.2. It is generally believed that hot-electron base transport in abrupt emitter HBTs greatly reduces the base transit time in thin base InP HBTs [80,81], but a recent study indicated that the reduced base transit time could simply be due to the high diffusion constant of thermalized electrons in the heavily doped base [82]. If the latter is correct, the choice of emitter material and the base–emitter heterojunction design play a less important role in base transport. The RF characteristics of an HBT depend largely on doping concentrations and base and collector layer thicknesses. For example, the record cutoff frequency f_T of 200 GHz in the CBE-grown InP/InGaAs SHBT was achieved through the use of an extremely thin base and collector layers of 420 and 1500 Å, respectively [1]. This magnitude of f_T is significantly higher than the best f_T value of 171 GHz reported in GaAs-based HBTs [22]. In fact, even though nonequilibrium transport mechanisms were involved in both devices, the former was achieved without the use of a special collector launcher structure. The electron velocity overshoot characteristics were more pronounced in InP SHBTs because of the smaller electron effective mass and larger Γ–L valley separation in InGaAs. In the 200-GHz f_T InP SHBTs reported, however, the breakdown voltage BV_{CEO} was only 4.2 V. At the current level at which the 200-GHz cutoff frequency was measured, the breakdown voltage BV_{CE} ($I_B \neq 0$ A) was only 1.2 V [1]. Also considering the 0.8-V knee voltage at the same current level, the peak-to-peak voltage swing available for circuit applications is limited to 0.4 V.

Comparison of maximum frequencies of oscillation, f_{max}, between InP and GaAs HBTs is, however, more complicated. In addition to the differences in intrinsic material properties described in Section 11.3.1, process technology also plays an important role in determining the ultimate parasitic resistances and capacitances. In the following discussions, the f_{max} performance of the less mature InP SHBTs fabricated with conventional triple-mesa technology is compared with that of the more sophisticated GaAs HBTs realized with implant isolation technology, graded InGaAs base design, and selective regrowth techniques [83].

The maximum frequency of oscillation, f_{max}, can be related to the current-gain cutoff frequency, f_T, through Eq. (23). The InP-based HBTs are intrinsically fast because of the small electron effective mass and large Γ–L valley separation in InGaAs, which allows highly nonequilibrium electron transport and extended velocity overshoot effects in the base and collector space-charge region. The cutoff frequencies, f_T, are thus larger in InP-based HBTs than those in GaAs-based devices of comparable structures and dimensions. For the same base doping concentration, the hole mobility in InGaAs is smaller than that in GaAs because of the presence of alloy scattering in the tertiary compound. Higher base sheet resistance is therefore expected in the InGaAs base layer than in GaAs base of the same thickness and doping level. The base contact resistance is, however, smaller on InGaAs than the corresponding GaAs base layer because of the smaller bandgap energy of InGaAs. Hence similar total base sheet and contact resistance is expected in both types of devices.

The base–collector capacitance depends on both the junction area and depletion width. The extrinsic base–collector junction area can be reduced in GaAs-based HBTs by ion implantation through the base layer into the extrinsic collector regions. Implant isolation, however, does not completely isolate InGaAs because of its small bandgap energy and large intrinsic carrier concentration. The resulting increase in base–collector capacitance partially offsets the speed improvement in InP-based SHBTs, but their maximum frequency of oscillation should, in principle, be comparable to the most sophisticated GaAs-based HBTs of similar structures and dimensions, without the need for complicated technologies.

The argument above was recently confirmed in InP-based SHBTs [3]. The extrinsic f_{max} of 236 GHz measured in a simple, conventional common-emitter InAlAs/InGaAs SHBT as shown in Fig. 11.17 compared well the best f_{max} value of 224 GHz obtained in a more sophisticated GaAs-based HBT, which utilized selective growth of heavily doped extrinsic base regions and compositionally graded InGaAs base structure [83]. Both device structures were grown by MBE and exhibited a current-gain cutoff frequency f_T value of 78 GHz. The cutoff frequency was high in the GaAs HBT because a thinner graded base layer (400 Å) and a thinner collector layer (5000 Å) were used. In contrast, the InP SHBT had a 800-Å base and a 6000-Å collector. The effective $(R_B C_{BC})_{eff}$ delay times as estimated using the simple approximate equation $(R_B C_{BC})_{eff} = f_T/(8\pi f_{max}^2)$ were very similar, being 0.056 and 0.062 ps, respectively, for the InP SHBT and GaAs HBT. Figure 11.18 summarizes recent microwave results of InP-based SHBTs plotted as a function of open-base breakdown voltage BV_{CEO} [1–3,67,84–89].

DHBTs: RF Although InP SHBTs are fast and simple in design and growth, their output conductance is high and breakdown voltage is low at high current densities

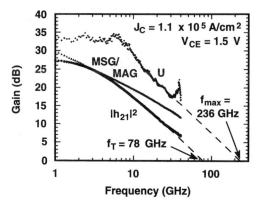

FIGURE 11.17 Measured RF performance of a common-emitter InAlAs/InGaAs SHBT at $V_{CE} = 1.5$ V and $I_C = 12.2$ mA. The emitter junction area was 11 μm^2.

428 HETEROJUNCTION BIPOLAR TRANSISTORS AND CIRCUIT APPLICATIONS

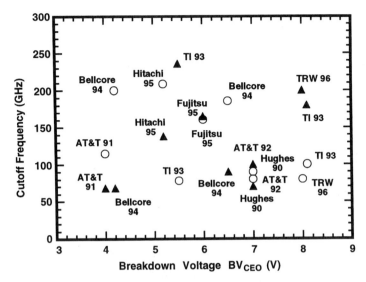

FIGURE 11.18 Recently reported microwave results of InP-based SHBTs plotted as a function of open-base breakdown voltage BV_{CEO}.

or large power dissipation, as explained in Section 11.5.2. The breakdown behavior can be improved by replacing the InGaAs collector in SHBTs with a wider bandgap material, such as InP as described in Section 11.3.3 and shown in Section 11.5.1. Because InP also has good transport properties, the microwave characteristics of DHBTs are comparable to those of SHBTs. Figure 11.19 shows the measured f_T and f_{max} of a 11-μm^2 InP/InGaAs/InP DHBTs with a 4000-Å collector plotted as a function of current density. The 4000-Å collector structure is the same as that described

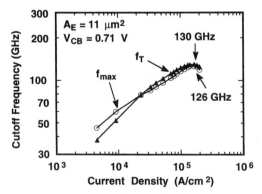

FIGURE 11.19 Measured f_T and f_{max} plotted as a function of current density for a 11-μm^2 common-emitter InP/InGaAs DHBT biased at $V_{CB} = 0.71$ V. The total collector thickness was 4000 Å.

in Sections 11.3.3 and 11.5.1. The V_{CB} bias voltage was fixed at 0.71 V. As shown in the figure, f_T and f_{max} increased from 38 and 46 GHz, respectively, at 4.5×10^3 A/cm², to peak values of 130 and 126 GHz at 1.6×10^5 A/cm². We have used the small-signal equivalent circuit shown in Fig. 11.3 to model the measured S-parameters under such peak f_T and f_{max} conditions. The extracted circuit parameters are given in Table 11.4. Under such bias conditions, the base transit time τ_b was found to be 0.47 ps, and the average electron velocity across the base was estimated to be about 1.7×10^7 cm/s, indicating that the electron transport across the base was not entirely due to the slow diffusion process but some nonequilibrium transport mechanism as well. The collector signal delay time τ_c was found to be about 0.44 ps, which leads to a collector transit time of $2\tau_c$ or 0.88 ps. The average electron velocity across the 2700-Å depleted collector space-charge region was thus estimated to be about 3.1×10^7 cm/s, which agrees well with the value predicted from Monte Carlo simulations [70]. The results indicate that even though a potential barrier is present at the abrupt base–collector heterojunction, nonequilibrium electron transport across the base and collector is possible in a properly designed base–collector heterojunction. At larger current densities, the current-gain cutoff frequency and thus the maximum oscillation frequencies decrease due to space-charge and electron-blocking effects.

In comparison, InAlAs/InGaAs/InAlAs DHBTs of the same dimensions with 4500-Å-thick collectors typically showed f_T and f_{max} values of 40 and 132 GHz, respectively. In this case, the base–collector heterojunction was linearly graded from InGaAs in the base to InAlAs in the collector with a 800-Å InAlGaAs grading layer. The base had a thickness of 800 Å and was doped at 5×10^{19} cm^{-3}. The collector doping was 2×10^{16} cm^{-3}. The f_T value of the InAlAs/InGaAs/InAlAs transistor was much smaller than that of the InP/InGaAs/InP DHBT described above. Since the collector of the former transistor was only 500 Å thicker than the latter one and both transistors had the same base thickness, the fact that the InAlAs/InGaAs/InAlAs

TABLE 11.4 Extracted Small-Signal Equivalent-Circuit Parameters of an InP/InGaAs DHBT Biased at V_{CB} = 0.71 V and J_C = 1.6 × 10⁵ A/cm²[a]

f_T (GHz)	130	f_{max} (GHz)	126
τ_b (ps)	0.444	τ_c (ps)	0.466
α_o	0.99	R_E (Ω)	4.76
C_{BE} (fF)	99	R_{B1} (Ω)	7.73
C_{BC1} (fF)	18	R_{B2} (Ω)	5.12
C_{BC2} (fF)	17	R_{C1} (Ω)	1.08
C_{BEO} (fF)	9	R_{C2} (Ω)	0.85
C_{CEO} (fF)	13	L_E (nH)	0.044
C_{BCO} (fF)	1	L_B (nH)	0.010
R_{BC} (kΩ)	90.9	L_C (nH)	0.030
R_{BE} (Ω)	1.45		

[a]The total collector thickness was 4000 Å.

DHBT had a much smaller f_T value indicates the poor electron transport characteristics expected in InAlAs. Their f_{max} values were, however, comparable, due to the reduced $(R_B C_{BC})_{eff}$ term in Eq. (23) in the InAlAs collector DHBT because the base doping concentration was much higher, and the collector was slightly thicker and more lightly doped.

For the 2×10 μm^2 InP/InGaAs/InP power DHBT with a 8288-Å-thick collector described in Section 11.5.1, the measured microwave performance is shown in Fig. 11.20. The extracted f_T and f_{max} values were 69 and 166 GHz, respectively, when they were biased at $V_{CE} = 3$ V and $I_C = 12.8$ mA. This f_{max} value is the highest in any InP DHBT that has a BV_{CEO} value greater than 8 V [72]. In comparison, our Ka-band GaAs HBTs typically have f_T and f_{max} values of 41 and 107 GHz, respectively, and BV_{CBO} and BV_{CEO} values of 24 and 17 V, respectively. The maximum available gain G_{max} value of the InP DHBTs at 30 GHz was 9.3 dB, which is also considerably higher than that in our conventional Ka-band GaAs power HBTs, which have a G_{max} value of about 6.9 dB at the same frequency. The results indicate that the InP HBT technology offers higher cutoff frequency, higher maximum frequency of oscillation, higher maximum available gain, and larger breakdown voltages than those of the corresponding GaAs HBTs.

The fastest InP-based DHBT reported to date was from an InP/InGaAs/InP DHBT structure with measured f_T and f_{max} values of 228 and 222 GHz, respectively, at the same bias conditions [6]. The highest f_{max} value extracted was 267 GHz, achieved in a 2700-Å-thick collector InP/InGaAs/InP DHBT with an emitter area of 0.8×5 μm^2 measured at an collector current of only 4 mA [4]. Figure 11.21 summarizes recent microwave results of InP-based DHBTs plotted as a function of open-base breakdown voltage BV_{CEO} [4,6,42,47,48,85,90–95].

11.5.5 Power Performance

Although limited power results have been reported on InP SHBTs [96], the limitations of SHBTs arise from their low breakdown voltage and high output conduc-

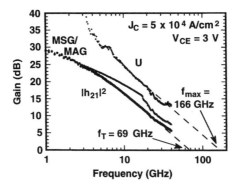

FIGURE 11.20 Measured microwave performance of a 2×10 μm^2 power InP/InGaAs/InP DHBT. The total collector thickness was 8288 Å. The measurement was made at $V_{CE} = 3$ V and $J_C = 5 \times 10^4$ A/cm^2.

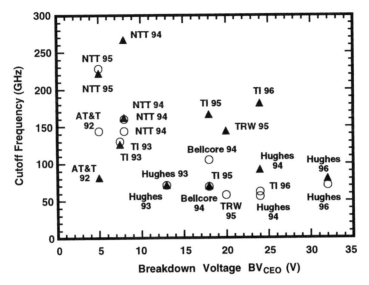

FIGURE 11.21 Recently reported microwave results of InP-based DHBTs plotted as a function of open-base breakdown voltage BV_{CEO}.

tance, which limit their usefulness to low-voltage and low-power applications. DHBTs offer better output conductance and breakdown characteristics, but require more careful base–collector heterojunction design. Too little InGaAs in the DHBT collector may lead to large saturation voltage and current blocking at low current densities. Too much InGaAs at the base–collector heterojunction results in high base–collector reverse saturation current and degradation of breakdown behavior [42].

We demonstrate here in Sections 11.5.1 and 11.5.4 that InP power DHBTs with small collector–emitter saturation voltage, large breakdown voltage, and high maximum frequency of oscillation can be achieved by putting as little as 194-Å InGaAs in the >8300-Å-thick collector by carefully controlling the use of InGaAs in the collector and subcollector layers. The measured common-emitter power performance of the power transistors at 30 GHz are given in Fig. 11.22. As shown, the common-emitter 2×10 μm^2 transistor delivered 19.1 mW CW output power (1.91 W/mm output power density), 5.3 dB associated gain, and 35.5% power-added efficiency (PAE) at 1 dB compression. The collector efficiency was 50.2%. The maximum output power density and the peak associated gain measured were 2.34 W/mm and 6.6 dB, respectively. The average collector current increased from 4 to 10 mA when the RF input power was increased from −2.5 to 9.5 dBm. The average collector current increased with RF input power because the transistor was biased in class AB operation. Despite the high breakdown voltage of the transistors, increasing the V_{CE} or V_{CB} bias voltage did not result in higher output power. As in the case of GaAs HBTs, the maximum output power appeared to be thermally limited. Some recently reported power data are reviewed in Section 11.6.4.

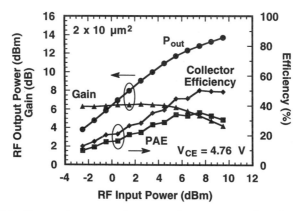

FIGURE 11.22 Measured power performance of a common-emitter 2×10 μm^2 power InP/InGaAs/InP DHBT at 30 GHz.

11.6 CIRCUIT APPLICATIONS

Because of the simplicity in design and fabrication technology, InP SHBTs were first developed and used in the majority of the InP HBT circuits reported in the late 1980s and early 1990s. The poor output conductance and low breakdown voltage of these transistors, however, limit their usefulness to low-voltage or low-power-dissipation applications. A few DHBT circuits were reported as the InP DHBT technology becomes more mature.

11.6.1 Digital Applications

A critical requirement for digital applications is low power dissipation, which requires low operation voltage and current. InP-based HBTs are very attractive in this area because of their superior electron transport properties, low base–emitter turn-on voltage, large current gain, and high cutoff frequency at low current levels compared to GaAs HBTs. Since the lower limit of power supply voltage is determined by the base–emitter turn-on voltage in a bipolar transistor, considerable savings in power consumption (about 40%) can be obtained in abrupt InP, abrupt InGaAlAs, or graded InAlAs emitter InP-based HBTs compared with GaAs-based HBTs. Operation at low current levels is facilitated in InP-based HBTs by their low surface recombination velocity, which allows device scaling without sacrificing current gain. In addition, it is also desirable to attain high speed and high cutoff frequency at such low current levels so as to lower the power–delay product. Yamahata et al. recently reported record 267-GHz f_{max} and 144-GHz f_T values at a collector current of 4 mA in a 0.8×5 μm^2 InP/InGaAs/InP DHBT [4]. Even more impressive is a 1×2 μm^2 InP/InGaAs SHBT exhibiting an f_T value over 100 GHz at a submilliampere collector current of only 0.6 mA [36]. Scaling devices with an emitter area from 2×20 μm^2 to 1×2 μm^2 reduced the f_T value only slightly, from 176 GHz to 163 GHz and the dc current gain from 140 to 60, and improved the f_{max} value from 99 GHz to 112 GHz [36]. The smallest InP SHBT fabricated had an emitter area of only 0.3 μm^2 and exhibited an f_{max} value of 99 GHz [37]. These encouraging results demonstrate that the InP-based HBT technology with small emitter sizes is expected to provide higher speed and a lower power–delay product than those of any other bipolar technology presently under development. Furthermore, it is important to get nearly ideal current characteristics because poor ideality factors reduce the device transconductance and cause the current requirements to increase [97]. This requirement is easily met with the abrupt InP and InAlGaAs emitter HBTs described in Section 11.5.1.

A variety of digital integrated circuits have been fabricated in InP SHBT technology. Frequency dividers were among the first digital circuits demonstrated in the late 1980s and early 1990s [98–100]. The power consumption of such InP HBT divider was only about a fifth of that of a GaAs HBT circuit and a twelfth of that of a silicon bipolar divider [99]. Since then, significant improvements have been made in transistor performance. The record performance to date has been improved to 39.5 GHz for a current mode logic (CML) divide-by-4 circuit, which contained 37 SHBTs and dissipated 425 mW for a 3-V power supply [101]. The most complex digital circuit to date is the 12-bit pipelined accumulator, which consisted of 928 transistors and was fully functional up to 7 GHz [102]. At such speed it required 1 W of power. In addition, a 4-bit MSI programmable divide-by-N counter was demonstrated at a clock frequency of 3.5 GHz. The chip consisted of more than 300 transistors and dissipated 750-mW power [103].

Other digital circuits fabricated include an 11-GHz divide-by-4/5 dual-modulus prescaler implemented using 124 transistors that dissipated a total of 500 mW power [103], a 9-GHz dual-modulus 8/9 divider dissipating 900 mW for a 5-V power supply [100], a 2-GHz phase detector that used 80 InP transistors [103], a 2:1 multiplexer that produced good-looking eye diagrams up to a 24-GHz clock frequency (corresponds to a 48-Gbit/s data output) [104], a 1:2 demultiplexer that operated up to at least a 25-GHz clock frequency (corresponds to demultiplexing a 50-Gbit/s data stream) [104], a 10-Gbit/s master–slave D flip-flop [105], and a 25-GHz emitter-coupled divide-by-2 circuit with a power consumption of 380 mW [106]. A number of these digital components are for eventual use in signal processing and telecommunication systems (Section 11.6.5). The high f_T value of InP-based SHBTs has led to short switching times in logic circuits, with a 14.7-ps/stage propagation delay obtained in a 17-stage NTL ring oscillator [107]. An effort is being made to realize a multi-Gbit/s 4:1 multiplexer [108]. The first DHBT digital circuits demonstrated include a 25-stage ring oscillator implemented with InGaAlAs collector DHBTs with a ECL gate propagation delay of 26 ps [57], and a 11.9-GHz divide-by-4 ECL frequency divider implemented with InGaAlAs collector DHBTs with a total power consumption of 255 mW [58].

11.6.2 Analog Applications

InP HBTs offer numerous advantages to analog circuit applications, including very high f_T, high transconductance, high output impedance, good device matching, and monolithically integrated Schottky diodes. These properties are important for the development of very high-performance analog circuits, including broadband amplifiers, logarithmic amplifiers, and receiver and transmitter circuits for fiber-optic and personal communication ICs. Some circuits, such as differential amplifiers, logarithmic amplifiers, and direct-coupled broadband amplifiers, are more suitable for bipolar device implementation than FET device technologies because of the exponential transconductance characteristics of bipolar transistors. Amplifiers with frequency response from dc to several gigahertz have many applications, including fiber-optics communications, radar signal processing, and instrumentation. FET-based amplifiers face a difficulty in this regime stemming from frequency-dependent output resistance due to the influence of traps and substrate conduction on the substrate–channel interface. HBTs are free of these trapping effects, since the device is well shielded from traps in the bulk and surface regions. Broadband amplifiers are used mostly in communication, instrumentation, and EW systems. Oscillators are needed in many communication systems to provide reference signals and clocks of high spectral purity. For multipliers, oscillators, and broadband communication amplifiers, the $1/f$ noise is quite important. Limited data have shown that InP-based HBTs have lower low-frequency noise than that in GaAs-based transistors, as expected from their low surface recombination velocity [34,109,110] and high electron injection energy from the emitter [32]. However, care must be taken when interpreting these results since they depend greatly on the technologies used.

Impressive circuit results have been demonstrated in feedback amplifiers, including a dc-to-20 GHz monolithic wideband feedback amplifier with a gain of 30 dB [88], a dc-to-33 GHz monolithic cascode feedback amplifier with a 8.6-dB gain [111], and a 15-GHz dc-coupled wideband amplifier with constant group delay and a 9-dB gain [104]. Other analog components reported include a 40-GHz positive inductive transimpedance–transadmittance amplifier with a 10-dB gain and a 600-MHz bandwidth [104]; single-stage and two-stage differential cascode amplifiers with dc gains of 21 and 33 dB, respectively, and unity-gain bandwidths over 15 GHz [112]; a 20-to-40-GHz balanced high-intercept amplifier with a 5-dB gain, and a 20-dBm IP3 at 35 GHz [113]; a dc-to-16 GHz double-balanced active mixer that had over 45 dB of LO rejection without trimming [114]; 18-, 30-, 40-, 46-, 62-, and 94-GHz low-phase-noise voltage-control oscillators (VCOs) [113,115–117]; a 95-GHz frequency-source module using a 23.8-GHz InP HBT MMIC dielectric resonator oscillator (DRO) with an external DR in conjunction with a GaAs-based HEMT MMIC frequency quadrupler and W-band output amplifiers [116]; a 5-to-12-GHz double-balanced mixer with LO-RF and IF-RF isolations greater than 30 dB [118]; and a 2-to-50-GHz InAlAs/InGaAs SHBT distributed amplifier with a peak gain of 6.3 dB, consuming 89 mW from a 4-V supply [84]. Many of these circuits are intended for eventual use in wide-bandwidth communication systems (Section 11.6.5). As the DHBT design and technology become more mature, more microwave circuits based on InP DHBTs are expected. For example, a 24-GHz DHBT-based transimpedance preamplifier was recently reported [119].

11.6.3 Mixed-Signal Applications

As digital signal processing and data processing speeds advance, increasing pressure is placed on analog-to-digital (A/D) converters to interface these systems with the real, analog world. A few A/D conversion functions were demonstrated. Among them, a second-order $\Delta\Sigma$ modulator was fabricated with 12-bit dynamic range for an oversampling ratio (OSR) of 32 and a sampling rate of 3.2 GHz [120]. The chip consisted of 240 SHBTs and dissipated 1 W from ± 5-V power supplies. More impressively, a 4-bit A/D converter was realized using 1138 InAlAs/InGaAs SHBTs and was fully functional at an input frequency of 1 GHz and a clock frequency of 2.4 GHz [121]. More recently, a 3-bit, 8-Gsample/s flash A/D converter was also demonstrated [122]. The chip consisted of approximately 900 InP DHBTs and dissipated 3.5 W power. For direct digital synthesis (DDS) applications, good spurious performance for all fractional frequency cases from dc to Nyquist is required. A 1-GHz, 12-bit digital-to-analog (D/A) converter was demonstrated using 1200 InAlAs/InGaAs HBTs [123]. When synthesizing near one-third the clock frequency, the carrier-adjacent spurious performance exceeded –58 dBc at the 1-GHz clock frequency. The D/A converter dissipated 2.8 W.

11.6.4 Microwave Applications

Microwave circuits typically require high power gain, high output power, and/or low noise. The high values of power gain obtainable with InP HBTs make them usable for amplification over a frequency range that extends into the millimeter-wave regime. For microwave power applications, the base–collector reverse saturation current needs to be minimized for the HBTs to operate at high collector–emitter biases, either by reducing the InGaAs thickness in the collector or other means without increasing the knee voltage in the I_C–V_{CE} characteristics [42]. The InGaAs subcollector layer has to be kept thin to take full advantage of the good thermal conductivity of InP substrates. InP power DHBTs fabricated at Texas Instruments typically have a BV_{CBO} value of 29 V defined at 1 μA leakage current, an f_T value of 69 GHz, an f_{max} value of 166 GHz, and a gain greater than 19 dB at 10 GHz when they are biased at V_{CE} = 3 V [42]. This kind of gain is considerably higher than that in conventional X-band GaAs power devices, which is about 16 dB at the same frequency.

Although no InP power amplifier has been reported to date, impressive microwave power results have been demonstrated in properly designed DHBTs with measured output power densities exceeding 4.3, 6.9, 5.0, and 1.9 W/mm at 4.5, 9, 10, and 30 GHz, respectively [54,55,93,124]. Since InP HBTs have low turn-on voltages, their power capability under low-voltage operation has also been demonstrated for wireless applications at 1.9 GHz [125]. Table 11.5 compares the representative power performance of InP HBTs reported to date at different frequencies [54,55,93,96,124,125]. Low dc current gain is needed for high BV_{CEO}. High output power, high power gain, and high power-added efficiency are expected for these power DHBTs. Since properly designed DHBTs operate at lower junction tempera-

TABLE 11.5 Recently Reported Power Results of InP-Based HBTs

Organization	Technology	Total Emitter Area (μm^2)	Frequency (GHz)	P_{out} (W)	P_{out} Density (W/mm)	PAE (%)	G_a (dB)	V_{CE} Bias (V)	Year
University of Michigan	SHBT	$4 \times (2 \times 10)$	10	0.110	2.74				1996
		$4 \times (2 \times 20)$	10			43			1996
Fujitsu	DHBT	$8 \times (1.5 \times 20)$	1.9	0.049	0.20	49.5	12.9	1.5	1995
Hughes	DHBT	$24 \times (2 \times 20)$	4.5	2.05	4.3	51	7.9		1994
	DHBT	$12 \times (2 \times 30)$	9	2.0	5.6	70	10.0	14	1995
	DHBT	$6 \times (2 \times 20)$	10	0.52	2.2	42	7.2	10	1994
TI	DHBT	2×10	30	0.019	1.91	35.5	5.3	4.76	1996

tures than GaAs HBTs of the same power dissipation [42], one can increase the f_T and f_{max} values simply by increasing the operating current density before thermal limitations take place.

As for the linearity or intermodulation, InP SHBTs are expected to be worse than GaAs HBTs because of their poor output conductance. However, properly designed InP DHBTs should show linearity comparable to that in GaAs HBTs. For low-voltage RF applications, InP SHBTs are still useful because of their simple structures and lower turn-on voltages than GaAs HBTs. Thin collector InP DHBTs with high f_T and f_{max} values are also very attractive because they are easier to design than thick collector InP DHBTs. The 1.5-V operation of InP power DHBTs clearly demonstrated their capability for 1.9-GHz applications [125].

11.6.5 Optoelectronic and Telecommunication Applications

Optical fiber communication offers ultrawide bandwidth and security for audio, video, and data transmission. Extensive research efforts on transmission components and networks at data rates well above 20 Gbit/s are being carried out. A significant amount of work has been based on the relatively mature GaAs technology. However, because of the high transparency and low dispersion characteristics of optical fibers at wavelengths of 1.3 and 1.55 μm, InP-based alloys are the material of choice for application to fiber-optic communications systems. For example, an InP HBT chip set intended for a 40-Gbit/s fiber-optic demonstrator system was designed and fabricated [104]. The chip set consisted of 14 different digital and analog ICs similar to those described in Sections 11.6.1 and 11.6.2.

A typical fiber-optic communication system consists of the following key components: (1) a front-end amplifier for the conversion of the photoelectric current produced by the photodetector into a voltage output, (2) a driver IC capable of driving a laser diode or an external modulator with a large output current or voltage, and (3) a main amplifier having a large dynamic range from dc to at least 15 GHz in the case of 20-Gbit/s systems. Two types of circuits are capable of satisfying this requirement: a differential amplifier and a feedback amplifier (Section 11.6.2). To complete the system, a timing circuit and a decision circuit are also needed.

Since the main function of the front-end amplifier is current–voltage conversion, it can be realized with a transimpedance circuit. The use of InP HBTs has attracted much interest because of their high performance, relative ease in lithographic processing, lack of sensitivity to surface conditions, and compatibility with detectors. For example, an InP/InGaAs SHBT preamplifier showed a S_{21} gain of 19.4 dB with a bandwidth of 18.9 GHz and transimpedance of 52.5 dB·Ω [106], which can then be monolithically integrated with PIN photodiodes to build photoreceivers [94, 119,126–128]. The conventional way to make such optoelectronic integrated circuits (OEICs) is to prepare PIN photodiode layers and HBT layers separately on the same wafer and use different fabrication processes for the two kinds of devices. An alternative and easier method of integrating PIN photodiode and HBTs is to use the base, collector, and subcollector layers of the HBT as PIN layers, or use the HBT as heterojunction phototransistor (HPT), and fabricate the two kinds of devices simultaneously [94,119,127]. In addition to PINs, metal–semiconductor–metal (MSM) photodiodes are also commonly used in photoreceivers. Some recently reported InP HBT OEIC receiver sensitivity data are presented as a function of bit rate in Fig. 11.23 [119,126–132]. Most of the high performance of OEIC receivers is associated with a PIN–HBT combination, although PIN–HEMT and PIN–JFET receivers are also competitive [133]. The best OEICs are now competitive with the best hybrid receivers. Wavelength-division multiplexing (WDM) offers high-data-density transmission. Efficient and compact WDM systems require that arrays of optical and/or electronic devices be used to conserve space and packaging cost. A few 1 × 8 OEIC receiver arrays have been demonstrated for 2.5-Gbit/s per channel operation using InP PIN-HBT structures [134–136] (see Chapter 13, Section 13.4.1 for OEIC receivers).

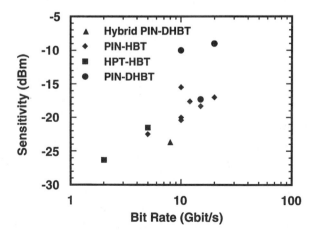

FIGURE 11.23 Recently reported InP HBT–based photoreceiver sensitivity plotted as a function of bit rate.

InP-based HBTs are particularly attractive for laser drivers because of their high switching speed and large current driving capability necessary for laser modulation. Several InP HBT driver ICs have been reported so far. Examples are a 12-Gbit/s differential-output-stage digital driver IC for lasers and optical modulators implemented using InP DHBTs with a maximum output voltage of 3 V across 25 Ω and a current swing of 120 mA [137], a high-voltage InP DHBT driver for optical waveguide modulators and switches with a differential output voltage of 11.4 V [138], and a 10-Gbit/s InP SHBT laser driver delivering 100 mA of modulation current and 50 mA of dc current with less than 1 W of power dissipation [139]. Compared to integrated receivers, however, the integration of high-performance HBTs and lasers has proved to be difficult. Demanding processing and multiple layer growth are required for state-of-the-art lasers. The formation of laser facets, which will ultimately be required for integration with large circuits, also remains a challenge, although a number of demonstrations have been made. Lack of thermal isolation between the laser and the heat generated by the driver circuit for an integrated transmitter has remained a problem, particularly in WDM applications, where precise thermal regulation of the laser is required for wavelength stability. A recent example of an integrated transmitter is the 1.5-μm multiple-quantum-well ridge laser integrated with an InP SHBT driver circuit [140]. This circuit used the stacked single growth technique, and the laser facets were formed using cleaving. Operation speed was 5 Gbit/s, with laser threshold currents in the range 18 to 25 mA.

11.7 SUMMARY AND FUTURE CHALLENGES

InP HBTs have shown significant improvement in both transistor performance and circuit complexity over the past few years, due to better understanding in device design and rapid advances in process technologies. Ultrahigh-frequency performance HBTs with greater than 200-GHz cutoff frequencies have already been demonstrated. This exciting capability has led to numerous very high-speed, large-bandwidth, high-performance integrated circuits not achievable with GaAs HBT technology. Because of the superior intrinsic material properties of InP-based alloys compared with GaAs-based materials, InP HBT technology will play a very important role in next-generation digital, analog, microwave, mixed-signal, and optoelectronic circuit applications. Improved breakdown voltage and better understanding in DHBT design will soon give rise to more DHBT circuits as well as power amplifiers with higher gain and efficiency than their GaAs counterparts. Because of their material compatibility with long-wavelength lightwave devices, InP HBTs will play a key role in future >20-Gbit/s telecommunication systems. The improved HBT results and low turn-on voltages, along with the relative ease in processing, makes InP-based HBTs a strong contender in future low-voltage, high-speed applications.

With over 200-GHz cutoff frequency already demonstrated in both InP SHBTs and DHBTs, the devices are believed to come close to their theoretical speed limit. Future challenges will focus on the implementation of this device technology in the deep-millimeter-wave regime, and in >40-Gbit/s signal processing and telecommunication systems. Cointegration of InP HBTs and resonant tunneling diodes (RTDs) gives rise to high-functionality digital ICs with fewer transistors than those made with conventional technology [141,142]. Compressed-function binary and multivalued logic circuits can be realized with improved capability and performance than those with other resonant tunneling transistor types. In addition, there has been some interest in monolithic integration of InP HBTs and HEMTs by selective MBE [143]. The capability of integrating HEMT and HBT devices on the same substrate offers MMIC designers the flexibility to take advantage of the unique performance characteristics of FET and bipolar devices, which may result in increased circuit functionality per unit weight, size, and volume, at the expense of increased process complexity. For example, a high-complexity S-band receiver was demonstrated using a monolithically integrated HEMT-HBT IC technology [144].

ACKNOWLEDGMENTS

The authors would like to gratefully acknowledge the support, contributions, and technical assistance of E. A. Beam III, D. Chasse, D. Costa, S. Duncan, S. F. Goodman, T. Henderson, P. Ikalainen, M. Jones, Y.-C. Kao, M. A. Khatibzadeh, D. McQuiddy, T. Session, R. Smith, F. H. Stovall, A. H. Taddiken, J. R. Thomason, J. Wilson, and W. R. Wisseman.

REFERENCES

1. J.-I. Song, K. B. Chough, C. J. Palmstrøm, B. P. Van der Gaag, and W.-P. Hong, "Carbon-doped base InP/InGaAs HBTs with f_T = 200 GHz," *Abstr. IEEE Device Research Conf.*, 1994, paper IVB-5.
2. T. Oka, T. Tanoue, H. Masuda, K. Ouchi, and T. Mozume, "InP/InGaAs heterojunction bipolar transistor with extremely high f_T over 200 GHz," *Electron. Lett.*, **31**, 2044–2045 (1995).
3. H.-F. Chau and Y. C. Kao, "High f_{max} InAlAs/InGaAs heterojunction bipolar transistors," *Tech. Dig. IEEE IEDM,* 1993, pp. 783–786.
4. S. Yamahata, K. Kurishima, H. Nakajima, T. Kobayashi, and Y. Matsuoka, "Ultra-high f_{max} and f_T InP/InGaAs double-heterojunction bipolar transistors with step-graded InGaAsP collector," *Tech. Dig. IEEE GaAs IC Symp.*, 1994, pp. 345–348.
5. K. Kurishima, H. Nakajima, S. Yamahata, T. Kobayashi, and Y. Matsuoka, "Growth, design and performance of InP-based heterostructure bipolar transistors," *IEICE Trans. Electron.*, **E78-C,** 1171–1181 (1995).
6. S. Yamahata, K. Kurishima, H. Ito, and Y. Matsuoka, "Over-220-GHz-f_T-and-f_{max} InP/InGaAs double-heterojunction bipolar transistors with a new hexagonal-shaped emitter," *Tech. Dig. IEEE GaAs IC Symp.*, 1995, pp. 163–166.
7. W. Shockley, U.S. patent 2,569,347 (1951).
8. H. Kroemer, "Theory of a wide-gap emitter for transistors," *Proc. IRE,* **45,** 1535–1537 (1957).

9. H. Kroemer, "Heterostructure bipolar transistors and integrated circuits," *Proc. IEEE,* **70,** 13–25 (1982).
10. H. Kroemer, "Heterostructure bipolar transistors: what should we build?" *J. Vac. Sci. Technol. B,* **1,** 126–130 (1983).
11. See, for example, S. Tiwari, *Compound Semiconductor Device Physics,* Academic Press, San Diego, Calif., 1992, pp. 113–135.
12. J. Hu, K. Tomizawa, and D. Pavlidis, "Transient Monte Carlo analysis and application to heterojunction bipolar transistor switching," *IEEE Trans. Electron Devices,* **36,** 2138–2145 (1989).
13. K. Tomizawa, *Numerical Simulation of Submicron Semiconductor Devices,* Artech House, Norwood, Mass., 1993.
14. D. Costa, W. U. Liu, and J. S. Harris, Jr., "Direct extraction of the AlGaAs/GaAs heterojunction bipolar transistor small-signal equivalent circuit," *IEEE Trans. Electron Devices,* **38,** 2018–2024 (1991).
15. A. P. Laser and D. L. Pulfrey, "Reconciliation of methods for estimating f_{max} for microwave heterojunction transistors," *IEEE Trans. Electron Devices,* **38,** 1685–1692 (1991).
16. H. F. Cooke, "Microwave transistors: theory and design," *Proc. IEEE,* **59,** 1163–1181 (1971).
17. P. M. Asbeck, M.-C. F. Chang, and K. C. Wang, "Heterojunction bipolar transistor technology," in *Introduction to Semiconductor Technology: GaAs and Related Compounds,* ed. C. T. Wang, Wiley, New York, 1990, pp. 170–230.
18. J.-L. Pelouard and M. A. Littlejohn, "Indium phosphide-based heterojunction bipolar transistors," in *SPIE Proc.,* Vol. 1144, *Indium Phosphide and Related Materials for Advanced Electronic and Optical Devices,* 1989, pp. 582–601.
19. C. M. Maziar, M. E. Klausmeier-Brown, and M. S. Lundstrom, "A proposed structure for collector transit-time reduction in AlGaAs/GaAs bipolar transistors," *IEEE Electron Device Lett.,* **7,** 483–485 (1986).
20. T. Ishibashi and Y. Yamauchi, "A possible near-ballistic collector in an AlGaAs/GaAs HBT with a modified collector structure," *IEEE Trans. Electron Devices,* **35,** 401–404 (1988).
21. K. Morizuka, R. Katoh, M. Asaka, N. Iizuka, K. Tsuda, and M. Obara, "Transit-time reduction in AlGaAs/GaAs HBT's utilizing velocity overshoot in the p-type collector region," *IEEE Electron Device Lett.,* **9,** 585–587 (1988).
22. T. Ishibashi, H. Nakajima, H. Ito, S. Yamahata, and Y. Matsuoka, "Suppressed base-widening in AlGaAs/GaAs ballistic collection transistors," *Abstr. IEEE Device Research Conf.,* 1990, paper VIIB-3.
23. H.-F. Chau, J. Hu, D. Pavlidis, and K. Tomizawa, "Breakdown-speed considerations in AlGaAs/GaAs heterojunction bipolar transistors with special collector designs," *IEEE Trans. Electron Devices,* **39,** 2711–2719 (1992).
24. H.-F. Chau, D. Pavlidis, J. Hu, and K. Tomizawa, "Breakdown-speed considerations in InP/InGaAs single- and double-heterostructure bipolar transistors," *IEEE Trans. Electron Devices,* **40,** 2–8 (1993).
25. S. Adachi, *Physical Properties of III–V Semiconductor Compounds: InP, InAs, GaAs, GaP, InGaAs, and InGaAsP,* Wiley, New York, 1992.

26. INSPEC, *Properties of Indium Phosphide,* EMIS Data Rev. Ser., No. 6, INSPEC/IEE, London, 1991.
27. J. Böhrer, A. Krost, and D. B. Bimberg, "Composition dependence of band gap and type of lineup in $In_{1-x-y}Ga_xAl_yAs/InP$ heterostructures," *Appl. Phys. Lett.,* **63**, 1918–1920 (1993).
28. V. Swaminathan and A. T. Macrander, *Materials Aspects of GaAs and InP Based Structures,* Prentice Hall, Upper Saddle River, N.J., 1991.
29. M. Toivonen, A. Salokatve, K. Tappura, M. Jalonen, P. Savolainen, J. Näppi, M. Pessa, and H. Asonen, "Solid source MBE for phosphide-based devices," *Proc. 8th Int. Conf. Indium Phosphide and Related Materials,* Schwäbisch-Gmünd, Germany, 1996, pp. 79–82.
30. R. N. Nottenburg, Y. K. Chen, M. B. Panish, D. A. Humphrey, and R. Hamm, "Hot-electron InGaAs/InP heterostructure bipolar transistors with f_T of 110 GHz," *IEEE Electron Device Lett.,* **10**, 30–32 (1989).
31. S. Wang, *Fundamentals of Semiconductor Theory and Device Physics,* Prentice Hall, Upper Saddle River, N.J., 1989, p. 479.
32. R. N. Nottenburg, Y. K. Chen, M. B. Panish, R. Hamm, and D. A. Humphrey, "High-current-gain submicrometer InGaAs/InP heterostructure bipolar transistors," *IEEE Electron Device Lett.,* **9**, 524–526 (1988).
33. H. Fukano, Y. Kawamura, H. Asai, Y. Takanashi, and M. Fujimoto, "Effect of hot-electron injection energy on characteristics of abrupt $In_{0.52}(Ga_{1-x}Al_x)_{0.48}As/InGaAs$ HBT's," *Abstr. IEEE Device Research Conf.,* 1990, paper IIIA-2.
34. Y. K. Chen, L. Fan, D. A. Humphrey, A. Tate, D. Sivco, and A. Y. Cho, "Reduction of $1/f$ noise current with non-equilibrium electron transport in AlInAs/InGaAs heterojunction bipolar transistors," *Tech. Dig. IEEE IEDM,* 1993, pp. 803–806.
35. B. Jalali, R. N. Nottenburg, Y.-K. Chen, D. Sivco, D. A. Humphrey, and A. Y. Cho, "High-frequency submicrometer $Al_{0.48}In_{0.52}As/In_{0.53}Ga_{0.47}As$ heterostructure bipolar transistors," *IEEE Electron Device Lett.,* **10**, 391–393 (1989).
36. H. Nakajima, K. Kurishima, S. Yamahata, T. Kobayashi, and Y. Matsuoka, "High-speed InP/InGaAs HBTs operated at submilliampere collector currents," *Electron. Lett.,* **29**, 1887–1888 (1993).
37. M. Hafizi, W. E. Stanchina, and H. C. Sun, "Submicron fully self-aligned AlInAs/GaInAs HBTs for low-power applications," *Abstr. IEEE Device Research Conf.,* 1995, pp. 80–81.
38. J. Hu, D. Pavlidis, and K. Tomizawa, "Monte Carlo studies of the effect of emitter junction grading on the electron transport in InAlAs/InGaAs heterojunction bipolar transistors," *IEEE Trans. Electron Devices,* **39**, 1273–1281 (1992).
39. R. N. Nottenburg, A. F. J. Levi, B. Jalali, D. Sivco, D. A. Humphrey, and A. Y. Cho, "Nonequilibrium electron transport in heterostructure bipolar transistors probed by magnetic field," *Appl. Phys. Lett.,* **56**, 2660–2662 (1990).
40. A. Miura, T. Yakihara, S. Kobayashi, S. Oka, A. Nonoyama, and T. Fujita, "InAlGaAs/InGaAs HBT," *Tech. Dig. IEEE IEDM,* 1992, pp. 79–82.
41. H. Fukano, H. Nakajima, T. Ishibashi, Y. Takanashi, and M. Fujimoto, "Effect of hot-electron injection on high-frequency characteristics of abrupt $In_{0.52}(Ga_{1-x}Al_x)_{0.48}As/$InGaAs HBT's, "*IEEE Trans. Electron Devices,* **39**, 500–506 (1992).

42. H.-F. Chau, W. Liu, and E. A. Beam III, "InP-based HBTs and their perspective for microwave applications," *Proc. 7th Int. Conf. Indium Phosphide and Related Materials,* Sapporo, Japan, 1995, pp. 640–643.
43. R. J. Malik, N. Chand, J. Nagle, R. W. Ryan, K. Alavi, and A. Y. Cho, "Temperature dependence of common-emitter I-V and collector breakdown voltage characteristics in AlGaAs/GaAs and AlInAs/GaInAs HBT's grown by MBE, "*IEEE Electron Device Lett.,* **13,** 557–559 (1992).
44. M. G. Holland, "Thermal conductivity," in *Semiconductors and Semimetals,* ed. R. K. Willardson and A. C. Beer, Vol. 2, Academic Press, San Diego, Calif., 1967, pp. 20–21.
45. S. Adachi, *Properties of Aluminium Gallium Arsenide,* EMIS Data Rev. Ser. INSPEC/IEE, London, 1993.
46. P. D. Maycock, "Thermal conductivity of silicon, germanium, III-V compounds and III-V alloys," *Solid-State Electron.,* **10,** 161–168 (1967).
47. A. Feygenson, R. A. Hamm, P. R. Smith, M. R. Pinto, R. K. Montgomery, R. D. Yadvish, and H. Temkin, "A 144 GHz InP/InGaAs composite collector heterostructure bipolar transistor," *Tech. Dig. IEEE IEDM,* 1992, pp. 75–78.
48. H.-F. Chau and E. A. Beam III, "High-speed, high-breakdown voltage InP/InGaAs double-heterojunction bipolar transistors grown by MOMBE," *Abstr. IEEE Device Research Conf.,* 1993, paper IVA-1.
49. O. Sugiura, A. G. Dentai, C. H. Joyner, S. Chandrasekhar, and J. C. Campbell, "High-current-gain InGaAs/InP double-heterojunction bipolar transistors grown by metal organic vapor phase epitaxy," *IEEE Electron Device Lett.,* **9,** 253–255 (1988).
50. E. Tokumitsu, A. G. Dentai, C. H. Joyner, and S. Chandrasekhar, "InP/InGaAs double heterojunction bipolar transistors grown by metalorganic vapor phase epitaxy with sulfur delta doping in the collector region," *Appl. Phys. Lett.,* **57,** 2841–2843 (1990).
51. K. Kurishima, H. Nakajima, T. Kobayashi, Y. Matsuoka, and T. Ishibashi, "InP/InGaAs double-heterojunction bipolar transistor with step-graded InGaAsP collector," *Electron. Lett.,* **29,** 258–260 (1993).
52. A. Feygenson, R. K. Montgomery, P. R. Smith, R. A. Hamm, M. Haner, R. D. Yadvish, M. B. Panish, H. Temkin, and D. Ritter, "InP/InGaAs composite collector heterostructure bipolar transistors and circuits," *Proc. 5th Int. Conf. Indium Phosphide and Related Materials,* Paris, 1993, pp. 572–575.
53. P. M. Asbeck, C. W. Farley, M. F. Chang, K. C. Wang, and W. J. Ho, "InP-based heterojunction bipolar transistors: performance status and circuit applications," *Proc. 2nd Int. Conf. Indium Phosphide and Related Materials,* Denver, Colo., 1990, pp. 2–5.
54. M. Hafizi, T. Liu, P. A. Macdonald, M. Liu, P. Chu, D. B. Rensch, W. E. Stanchina, and C. S. Wu, "High-performance microwave power AlInAs/GaInAs/InP double heterojunction bipolar transistors with compositionally graded base-collector junction," *Tech. Dig. IEEE IEDM,* 1993, pp. 791–794.
55. C. Nguyen, T. Liu, M. Chen, H.-C. Sun, and D. Rensch, "AlInAs/GaInAs/InP double heterojunction bipolar transistor with a novel base-collector design for power applications," *Tech. Dig. IEEE IEDM,* 1995, pp. 799–802.
56. C. W. Farley, J. A. Higgins, W.-J. Ho, B. T. McDermott, and M. F. Chang, "Performance tradeoffs in AlInAs/GaInAs single- and double-heterojunction NpN heterojunction bipolar transistors," *J. Vac. Sci. Technol. B,* **10,** 1023–1025 (1992).
57. H. Yamada, T. Futatsugi, Y. Yamaguchi, K. Ishii, Y. Bamba, T. Fujii, and N. Yokoyama,

"Emitter-coupled logic circuits implemented using InAlAs/InGaAs HBTs with improved emitter-collector breakdown voltage," *Abstr. IEEE Device Research Conf.,* 1990, paper IIIA-1.

58. H. Yamada, T. Futatsugi, H. Shigematsu, T. Tomioka, T. Fujii, and N. Yokoyama, "InAlAs/InGaAs double heterojunction bipolar transistors with a collector launcher structure for high-speed ECL applications," *Tech. Dig. IEEE IEDM,* 1991, pp. 964–966.

59. J. C. Vlcek and C. G. Fonstad, "Multiply-graded InGaAlAs heterojunction bipolar transistors," *Electron, Lett.,* **27,** 1213 (1991).

60. S. J. Pearton, C. R. Abernathy, M. B. Panish, R. A. Hamm, and L. M. Lunardi, "Implant-induced high-resistivity regions in InP and InGaAs," *J. Appl. Phys.,* **66,** 656–662 (1989).

61. U. Niggebrügge, M. Klug, and G. Garus, "A novel process for reactive ion etching on InP using CH_4/H_2," *Inst. Phys. Conf. Ser.,* Vol. 79, 1985, pp. 367–372.

62. H.-F. Chau, D. Pavlidis, and T. Brock, "Reactive ion etching-induced damage studies and application to self-aligned InP/InGaAs heterojunction bipolar transistor technology," *J. Vac. Sci. Technol. B,* **11,** 187–194 (1993).

63. E. F. Chor, R. J. Malik, R. A. Hamm, and R. Ryan, "Metallurgical stability of ohmic contacts on thin base InP/InGaAs/InP HBT's," *IEEE Electron Device Lett.,* **17,** 62–64 (1996).

64. Y. Miyamoto, J. M. M. Rios, A. G. Dentai, and S. Chandrasekhar, "Reduction of base-collector capacitance by undercutting the collector and subcollecor in GaInAs/InP DHBT's," *IEEE Electron Device Lett.,* **17,** 97–99 (1996).

65. J.-I. Song, M. R. Frei, J. R. Hayes, R. Bhat, and H. M. Cox, "Self-aligned InAlAs/InGaAs heterojunction bipolar transistor with a buried subcollector grown by selective epitaxy," *IEEE Electron Device Lett.,* **15,** 123–125 (1994).

66. N. Matine, J. L. Pelouard, F. Pardo, R. Teissier, and M. Pessa, "Novel approach for InP-based ultrafast HBTs," *Proc. 8th Int. Conf. Indium Phosphide and Related Materials,* Schwäbisch-Gmünd, Germany, 1996, pp. 137–140.

67. U. Bhattacharya, M. J. Mondry, G. Hurtz, J. Guthrie, M. J. W. Rodwell, T. Liu, C. Nguyen, and D. Rensch, "100 GHz transferred-substrate Schottky-collector heterojunction bipolar transistor," *Proc. 8th Int. Conf. Indium Phosphide and Related Materials,* Schwäbisch-Gmünd, Germany, 1996, pp. 145–148.

68. H. Shigematsu, T. Iwai, Y. Matsumiya, H. Ohnishi, O. Ueda, and T. Fujii, "Ultrahigh f_T and f_{max} new self-alignment InP/InGaAs HBT's with a highly Be-doped base layer grown by ALE/MOCVD," *IEEE Electron Device Lett.,* **16,** 55–57 (1995).

69. M. Ida, S. Yamahata, K. Kurishima, H. Ito, T. Kobayashi, and Y. Matsuoka, "Enhancement of f_{max} in InP/InGaAs HBT's by selective MOCVD growth of heavily-doped extrinsic base regions," *IEEE Trans. Electron Devices,* **43,** 1812–1818 (1996).

70. H.-F. Chau, E. A. Beam III, Y.-C. Kao, and W. Liu, "InP-based heterojunction bipolar transistors," in *Current Trends in Heterojunction Bipolar Transistors,* ed. M. F. Chang, World Scientific, Singapore, 1996, pp. 303–349.

71. W. Liu, S.-K. Fan, T. Henderson, and D. Davito, "Temperature dependences of current gains in GaInP/GaAs and AlGaAs/GaAs heterojunction bipolar transistors," *IEEE Trans. Electron Devices,* **40,** 1351–1353 (1993).

72. H.-F. Chau, W. Liu, and E. A. Beam III, "InP-based heterojunction bipolar transistors:

recent advances and thermal properties," *Microwave Opt. Technol. Lett.,* **11,** 114–120 (1996).
73. R. H. Winkler, "Thermal properties of high-power transistors," *IEEE Trans. Electron Devices,* **14,** 260–263 (1967).
74. D. Ritter, R. A. Hamm, A. Feygenson, and M. B. Panish, "Anomalous electric field and temperature dependence of collector multiplication in InP/Ga$_{0.47}$In$_{0.53}$As heterojunction bipolar transistors," *Appl. Phys. Lett.,* **60,** 3150–3152 (1992).
75. W. Liu, "Failure mechanisms in AlGaAs/GaAs power heterojunction bipolar transistors," *IEEE Trans. Electron Devices,* **43,** 220–227 (1996).
76. W. Liu, "The temperature and current profiles in an emitter finger as a function of the finger length," *Solid-State Electron.,* **36,** 1787–1789 (1993).
77. W. Liu, "Thermal coupling in 2-finger heterojunction bipolar transistors," *IEEE Trans. Electron Devices,* **42,** 1033–1038 (1995).
78. W. Liu, S. Nelson, D. Hill, and A. Khatibzadeh, "Current gain collapse in microwave multi-finger heterojunction bipolar transistors operated at very high power density," *IEEE Trans. Electron Devices,* **40,** 1917–1927 (1993).
79. W. Liu and A. Khatibzadeh, "The collapse of current gain in multi-finger heterojunction bipolar transistor: its substrate temperature dependence, instability criteria and modeling," *IEEE Trans. Electron Devices,* **41,** 1698–1707 (1994).
80. J. Laskar, R. N. Nottenburg, and A. F. J. Levi, "Forward transit delay in In$_{0.53}$Ga$_{0.47}$As heterojunction bipolar transistors with nonequilibrium electron transport," *Abstr. IEEE Device Research Conf.,* 1993, paper IVA-3.
81. A. F. J. Levi, R. N. Nottenburg, B. Jalali, A. Y. Cho, and M. B. Panish, "Physics and high speed devices," *Proc. 2nd Int. Conf. Indium Phosphide and Related Materials,* Denver, Colo., 1990, pp. 6–12.
82. D. Ritter, R. A. Hamm, A. Feygenson, and P. R. Smith, "Role of hot electron base transport in abrupt emitter InP/InGaAs heterojunction bipolar transistors," *Appl. Phys. Lett.,* **64,** 2988–2990 (1994).
83. H. Shimawaki, Y. Amamiya, N. Furuhata, and K. Honjo, "High-f_{max} AlGaAs/InGaAs and AlGaAs/GaAs HBTs fabricated with MOMBE selective growth in extrinsic base regions," *Abstr. IEEE Device Research Conf.,* 1993. paper IVA-6.
84. K. W. Kobayashi, J. Cowles, L. T. Tran, T. R. Block, A. K. Oki, and D. C. Streit, "A 2–50 GHz InAlAs/InGaAs-InP HBT distributed amplifier," *Tech. Dig. IEEE GaAs IC Symp.,* 1996, pp. 207–210.
85. J.-I. Song, B. W.-P. Hong, C. J. Palmstrøm, and K. B. Chough, "InP based carbon-doped base HBT technology: its recent advances and circuit applications," *Proc. 6th Int. Conf. Indium Phosphide and Related Materials,* Santa Barbara, California, 1994, pp. 523–526.
86. H.-F. Chau and E. A. Beam III, "High-speed InP/InGaAs heterojunction bipolar transistors," *IEEE Electron Device Lett.,* **14,** 388–390 (1993).
87. T. R. Fullowan, S. J. Pearton, R. F. Kopf, F. Ren, and J. Lothian, "High yield scalable dry etch process for indium based heterojunction bipolar transistors," *Proc. 4th Int. Conf. Indium Phosphide and Related Materials,* Newport, R.I., 1992, pp. 343–346.
88. R. K. Montgomery, D. A. Humphrey, P. R. Smith, B. Jalali, R. N. Notterburg, R. A. Hamm, and M. B. Panish, "A dc to 20 GHz high gain monolithic InP/InGaAs HBT feedback amplifier," *Tech. Dig. IEEE IEDM,* 1991, pp. 935–938.

89. J. F. Jensen, W. E. Stanchina, R. A. Metzger, D. B. Rensch, Y. K. Allen, M. W. Pierce, and T. V. Kargodorian, "High speed dual modulus dividers using AlInAs–GaInAs HBT IC technology," *Tech. Dig. IEEE GaAs IC Symp.,* 1990, pp. 41–44.

90. C. Nguyen, T. Liu, M. Chen, H.-C. Sun, and D. Rensch, "AlInAs/GaInAs/InP double heterojunction bipolar transistor with a novel base-collector design for power applications," *IEEE Electron Device Lett.,* **17,** 133–135 (1996).

91. E. A. Bcam III, A. C. Seabaugh, H. F. Chau, W. Liu, and T. P. E. Broekaert, "Gas-source molecular beam epitaxy of electronic devices," *Abstr. Material Research Society Spring Meeting,* 1996, paper C1.1.

92. J. Cowles, L. Tran, T. Block, D. Streit, and A. Oki, "A 140 GHz f_{max} InAlAs/InGaAs pulse-doped InGaAlAs quaternary collector HBT with a 20 V Bvceo," *Abstr. IEEE Device Research Conf.,* 1995, pp. 84–85.

93. M. Hafizi, P. A. Macdonald, T. Liu, D. B. Rensch, and T. C. Cisco, "Microwave power performance of InP-based double heterojunction bipolar transistors for C- and X-band applications," *Tech. Dig. IEEE MTT-S,* 1994, pp. 671–674.

94. Y. Matsuoka, H. Nakajima, K. Kurishima, T. Kobayashi, M. Yoneyama, and E. Sano, "Novel InP/InGaAs double-heterojunction bipolar transistors suitable for high-speed IC's and OEIC's," *Proc. 6th Int. Conf. Indium Phosphide and Related Materials,* Santa Barbara, Calif. 1994, pp. 555–558.

95. W. E. Stanchina, T. Liu, D. B. Rensch, P. MacDonald, M. Hafizi, W. W. Hooper, M. Lui, Y. K. Allen, T. V. Kargodorian, R. Wong-Quen, and F. Williams, "Performance of In-AlAs/GaInAs/InP microwave DHBTs," *Proc. 5th Int. Conf. Indium Phosphide and Related Materials,* Paris, 1993, pp. 17–20.

96. D. Sawdai, J.-O. Plouchart, D. Pavlidis, A. Samelis, and K. Hong, "Power performance of InGaAs/InP single HBTs," *Proc. 8th Int. Conf. Indium Phosphide and Related Materials,* Schwäbisch-Gmünd, Germany, 1996, pp. 133–136.

97. J. F. Jensen, W. E. Stanchina, R. A. Metzger, M. E. Hafizi, T. Liu, and D. B. Rensch, "High speed InP-based HBT integrated circuits," *SPIE Proc.,* Vol. 1680, *High-Speed Electronics and Optoelectronics,* 1992, pp. 2–11.

98. P. J. Topham, J. Thompson, I. Griffith, B. A. Hollis, N. A. Hiams, J. G. Parton, and R. C. Goodfellow, "Digital integrated circuit using GaInAs/InP heterojunction bipolar transistors," *Electron. Lett.,* **25,** 1116–1117 (1989).

99. C. W. Farley, K. C. Wang, M. F. Chang, P. M. Asbeck, R. B. Nubling, N. H. Sheng, R. Pierson, and G. J. Sullivan, "A high-speed, low-power divide-by-4 frequency divider implemented with AlInAs/GaInAs HBT's," *IEEE Electron Device Lett.,* **10,** 377–379 (1989).

100. J. F. Jensen, W. E. Stanchina, R. A. Metzger, D. B. Rensch, R. J. Ferro, P. F. Lou, M. W. Pierce, T. V. Kargodorian, and Y. K. Allen, "Improved AlInAs/GaInAs HBTs for high speed circuits," *SPIE Proc.,* Vol. 1288, *High-Speed Electronics and Device Scaling,* 1990, pp. 57–68.

101. J. F. Jensen, M. Hafizi, W. E. Stanchina, R. A. Metzger, and D. B. Rensch, "39.5-GHz static frequency divider implemented in AlInAs/GaInAs HBT technology," *Tech. Dig. IEEE GaAs IC Symp.,* 1992, pp. 101–104.

102. W. E. Stanchina, J. F. Jensen, R. H. Walden, M. Hafizi, H.-C. Sun, T. Liu, G. Raghavan, K. E. Elliott, M. Kardos, A. E. Schmitz, Y. K. Brown, M. E. Montes, and M. Yung, "An InP-based HBT fab for high-speed digital, analog, mixed-signal and optoelectronic ICs," *Tech. Dig. IEEE GaAs IC Symp.,* 1995, pp. 31–34.

103. R. K. Montgomery and J. F. Jensen, "Design trade-offs in InP based HBT ICs," *Proc. 5th Int. Conf. Indium Phosphide and Related Materials,* Paris, 1993, pp. 557–560.

104. T. Swahn, T. Lewin, M. Mokhtari, H. Tenhunen, R. Walden, and W. E. Stanchina, "40 Gb/s, 3 volt InP HBT ICs for a fiber optic demonstrator system," *Tech. Dig. IEEE GaAs IC Symp.,* 1996, pp. 125–128.

105. B. Jalali, P. R. Smith, R. N. Notternburg, M. Banu, D. A. Humphrey, R. K. Montgomery, D. Sivco, and A. Y. Cho, "10 Gbit/s D flipflop using AlInAs/InGaAs HBTs," *Electron, Lett.,* **27,** 1314–1315 (1991).

106. H. Nakajima, "Design and fabrication of high-speed InP-based heterojunction bipolar transistors," *Proc. 5th Int. Conf. Indium Phosphide and Related Materials,* Paris, 1993, pp. 13–16.

107. Y.-K. Chen, R. N. Nottenburg, M. B. Panish, R. A. Hamm, and D. A. Humphrey, "Subpicosecond InP/InGaAs heterostructure bipolar transistors," *IEEE Electron Device Lett.,* **10,** 267–269 (1989).

108. G. Schuppener, B. Willen, M. Mokhtari, and H. Tenhunen, "Application of III-V semiconductor based heterojunction bipolar transistors towards multi-Gbit/s 4:1 multiplexer," *Phys. Scr.,* **T54,** 46–50 (1994).

109. N. Hayama, S.-I. Tanaka, and K. Honjo, "1/f noise reduction for microwave self-aligned AlGaAs/GaAs HBTs with AlGaAs surface passivation layer," *Proc. 3rd Asia-Pacific Microwave Conf.,* 1990, pp. 1039–1042.

110. A. K. Kirtania, M. B. Das, S. Chandrasekhar, L. M. Lunardi, R. A. Hamm, and L. W. Yang, "A comparison of low-frequency noise characteristics of silicon homojunction and III–V heterojunction bipolar transistors," *Proc. 6th Int. Conf. Indium Phosphide and Related Materials,* Santa Barbara, Calif., 1994, pp. 535–538.

111. M. Rodwell, J. F. Jensen, W. E. Stanchina, R. A. Metzger, D. B. Rensch, M. W. Pierce, T. V. Kargodorian, and Y. K. Allen, "33 GHz monolithic cascode AlInAs/GaInAs," *Tech. Dig. IEEE Bipolar Circuits and Technology Meeting,* 1990, pp. 252–255.

112. M. Banu, B. Jalali, D. A. Humphrey, R. K. Montgomery, R. N. Nottenburg, R. A. Hamm, and M. B. Panish, "Wideband HBT circuits for operation above 10 GHz and power supply voltages below 5 V," *Electron. Lett.,* **28,** 354–355 (1992).

113. K. W. Kobayashi, L. T. Tran, S. Bui, J. Velebir, D. Nguyen, A. K. Oki, and D. C. Streit, "InP based HBT millimeter-wave technology and circuit performance to 40 GHz," *Tech. Dig. IEEE Microwave and Millimeter-Wave Monolithic Circuits Symp.,* 1993, pp. 85–88.

114. L. M. Burns, J. F. Jensen, W. E. Stanchina, R. A. Metzger, and Y. K. Allen, "DC-to-Ku band MMIC InP HBT double-balanced active mixer," *Tech. Dig. IEEE Int. Solid-State Circuits Conf.,* 1991, pp. 124–125.

115. K. W. Kobayashi, L. T. Tran, A. K. Oki, T. Block, and D. C. Streit, "A coplanar waveguide InAlAs/InGaAs HBT monolithic Ku-band VCO," *IEEE Microwave Guided Wave Lett.,* **5,** 311–312 (1995).

116. H. Wang, K. W. Chang, L. Tran, J. Cowles, T. Block, D. C. W. Lo, G. S. Dow, A. Oki, D. Streit, and B. R. Allen, "Low phase noise millimeter-wave frequency sources using InP based HBT technology," *Tech. Dig. IEEE GaAs IC Symp.,* 1995, pp. 263–266.

117. J. Cowles, L. Tran, H. Wang, E. Lin, T. Block, D. Streit, and A. Oki, "InP-based HBT technology for millimeter-wave MMIC VCOs," *Tech. Dig. IEEE IEDM,* 1996, pp. 199–202.

118. K. W. Kobayashi, L. T. Tran, S. Bui, A. K. Oki, D. C. Streit, and M. Rosen, "InAlAs/InGaAs HBT X-band double-balanced upconverter," *IEEE J. Solid-State Circuits,* **29,** 1238–1243 (1994).

119. E. Sano, M. Yoneyama, S. Yamahata, and Y. Matsuoka, "23 GHz bandwidth monolithic photoreceiver compatible with InP/InGaAs double-heterojunction bipolar transistor fabrication process," *Electron. Lett.,* **30,** 2064–2065 (1994).

120. J. F. Jensen, G. Raghavan, A. E. Cosand, and R. H. Walden, "A 3.2-GHz second-order delta–sigma modulator implemented in InP HBT technology," *IEEE J. Solid-State Circuits,* **30,** 1119–1127 (1995).

121. L. Tran, S. Southwell, J. Velebir, A. Oki, D. Streit, and B. Oyama, "Fully functional high speed 4-bit A/D converters using InAlAs/InGaAs HBTs," *Tech. Dig. IEEE GaAs IC Symp.,* 1993, pp. 159–162.

122. C. Baringer, J. Jensen, L. Burns, and B. Walden, "3-bit, 8 Gsps flash ADC," *Proc. 8th Int. Conf. Indium Phosphide and Related Materials,* Schwäbisch-Gmünd, Germany, 1996, pp. 64–67.

123. T. A. Schaffer, H. P. Warren, M. J. Bustamante, and K. W. Kong, "A 2 GHz 12-bit digital-to-analog converter for direct digital synthesis applications," *Tech. Dig. IEEE GaAs IC Symp.,* 1996, pp. 61–64.

124. H.-F. Chau, H.-Q. Tserng, and E. A. Beam III, "Ka-band power performance of InP/InGaAs/InP double heterojunction bipolar transistors," *IEEE Microwave Guided Wave Lett.,* **6,** 129–131 (1996).

125. T. Iwai, H. Shigematsu, H. Yamada, T. Tomioka, S. Sasa, K. Joshin, and T. Fujii, "Microwave power InAlAs/InGaAs double heterojunction bipolar transistors with 1.5 V-low voltage operation," *Abstr. IEEE Device Research Conf.,* 1995, pp. 88–89.

126. S. Chandrasekhar, L. M. Lunardi, A. H. Gnauck, D. Ritter, R. A. Hamm, M. B. Panish, and G. J. Qua, "A 10 Gbit/s OEIC photoreceiver using InP/InGaAs heterojunction bipolar transistors," *Electron. Lett.,* **28,** 466–468 (1992).

127. S. Chandrasekhar, L. M. Lunardi, A. H. Gnauck, R. A. Hamm, and G. J. Qua, "High-speed monolithic p-i-n/HBT and HPT/HBT photoreceivers implemented with simple phototransistor structure," *IEEE Photon. Technol. Lett.,* **5,** 1316–1318 (1993).

128. L. M. Lunardi, S. Chandrasekhar, A. H. Gnauck, C. A. Burrus, R. A. Hamm, J. W. Sulhoff, and J. L. Zyskind, "A 12-Gb/s high-performance, high-sensitivity monolithic p-i-n/HBT photoreceiver module for long-wavelength transmission systems," *IEEE Photon. Technol. Lett.,* **7,** 182–184 (1995).

129. E. Sano, M. Yoneyama, S. Yamahata, and Y. Matsuoka, "InP/InGaAs double-heterojunction bipolar transistors for high-speed optical receivers," *IEEE Trans. Electron Devices,* **43,** 1826–1832 (1996).

130. L. M. Lunardi, S. Chandrasekhar, A. H. Gnauck, C. A. Burns, and R. A. Hamm, "20-Gb/s monolithic p-i-n/HBT photoreceiver module for 1.55-μm applications," *IEEE Photon. Technol. Lett.,* **7,** 1201–1203 (1995).

131. L. M. Lunardi, S. Chandrasekhar, A. H. Gnauck, C. A. Burns, A. G. Dentai, and R. A. Hamm, "15 Gbit/s pin/HBT optoelectronic integrated photoreceiver module realised using MOVPE material," *Electron. Lett.,* **31,** 1185–1186 (1995).

132. S. Chandrasekhar, A. H. Gnauck, R. A. Hamm, and G. J. Qua, "The phototransistor revisted: all-bipolar monolithic photoreceiver at 2 Gb/s with high sensitivity," *IEEE Trans. Electron Devices,* **39,** 2677–2678 (1992).

133. R. H. Walden, "A review of recent progress in InP-based optoelectronic integrated circuit receiver front-ends," *Tech. Dig. IEEE GaAs IC Symp.*, 1996, pp. 255–257.
134. S. Chandrasekhar, L. M. Lunardi, R. A. Hamm, and G. J. Qua, "Eight-channel p-i-n/HBT monolithic receiver array at 2.5 Gb/s per channel for WDM applications," *Proc. 6th Int. Conf. Indium Phosphide and Related Materials*, Santa Barbara, Calif., 1994, pp. 243–246.
135. S. Chandrasekhar et al., "Investigation of crosstalk performance of eight-channel p-i-n/HBT OEIC photoreceiver array modules," *IEEE Photon. Technol. Lett.*, 8, 682–684 (1996).
136. R. H. Walden, C. Dreze, K. Warbrick, and C. Chew, "An OEIC-based, 8-channel, optical receiver submodule for a 2.5 Gb/s WDM network access module," presented at *Engineering Foundation High Speed Opto-Electronics for Communications II*, Snowbird, Utah, 1996.
137. R. Bauknecht, H. P. Schneibel, J. Schmid, and H. Melchior, "A 12 Gb/s laser and optical modulator driver circuit with InGaAs/InP double heterostructure bipolar transistors," *Proc. 8th Int. Conf. Indium Phosphide and Related Materials*, Schwäbisch-Gmünd, Germany, 1996, pp. 61–63.
138. R. Bauknecht, H. Duran, M. Schmatz, and H. Melchior, "InGaAs/InP double heterostructure bipolar transistors for high speed and high voltage driver circuit applications," *Proc. 5th Int. Conf. Indium Phosphide and Related Materials*, Paris, 1993, pp. 565–568.
139. M. Banu, B. Jalali, R. Nottenburg, D. A. Humphrey, R. K. Montgomery, R. A. Hamm, and M. B. Panish, "10 Gbit/s bipolar laser driver," *Electron. Lett.*, 27, 278–280 (1991).
140. K. Y. Liou, S. Chandrasekhar, A. G. Dentai, E. C. Burrows, G. J. Qua, C. H. Joyner, and C. A. Burrus, "A 5 Gb/s monolithically integrated lightwave transmitter with 1.5 μm multiple quantum well laser and HBT driver circuit," *IEEE Photon. Technol. Lett.*, 3, 928–930 (1991).
141. A. C. Seabaugh, E. A. Beam III, A. H. Taddiken, J. N. Randall, and Y.-C. Kao, "Co-integration of resonant tunneling and double heterojunction bipolar transistors on InP," *IEEE Electron Device Lett.*, 14, 472–474 (1993).
142. G. I. Haddad, "Resonant tunneling heterojunction bipolar transistors and their applications in high functionality/speed digital circuits," *Proc. 8th Int. Conf. Indium Phosphide and Related Materials*, Schwäbisch-Gmünd, Germany, 1996, pp. 129–132.
143. L. Tran, J. Cowles, R. Lai, T. Block, P. Liu, A. Oki, and D. Streit, "Monolithic integration of InP HBT and HEMT by selective molecular beam epitaxy," *Proc. 8th Int. Conf. Indium Phosphide and Related Materials*, Schwäbisch-Gmünd, Germany, 1996, pp. 76–78.
144. K. W. Kobayashi, A. K. Oki, D. K. Umemoto, T. R. Block, and D. C. Streit, "A monolithic integrated HEMT-HBT S-band receiver," *Tech. Dig. IEEE GaAs IC Symp.*, 1996, pp. 197–200.

CHAPTER TWELVE

Lasers, Amplifiers, and Modulators Based on InP-Based Materials

NILOY K. DUTTA
University of Connecticut

12.1	Introduction	450
12.2	Semiconductor Lasers	450
	12.2.1 Laser Designs	451
	12.2.2 Quantum-Well Lasers	455
	12.2.3 Distributed Feedback Lasers	465
	12.2.4 Surface Emitting Lasers	470
	12.2.5 Laser Reliability	473
12.3	Optical Amplifiers	477
	12.3.1 Semiconductor Optical Amplifier	478
12.4	Optical Modulator	484
	12.4.1 Waveguide Electroabsorption Modulator	485
	12.4.2 Mach–Zehnder Modulator	486
12.5	Monolithically Integrated Lasers and Photonic Integrated Circuits	487
	12.5.1 Laser Arrays	487
	12.5.2 Integrated Laser Detector	488
	12.5.3 Integrated Laser Modulator	488
	12.5.4 Multichannel WDM Sources	490
	12.5.5 Spot-Size Converter Integrated Laser	492

InP-Based Materials and Devices: Physics and Technology, Edited by Osamu Wada and Hideki Hasegawa.
ISBN 0-471-18191-9   © 1999 John Wiley & Sons, Inc.

12.5.6 Heterodyne Receiver 492
12.6 Summary and Future Challenges 494
References 494

12.1 INTRODUCTION

Significant advances in research results, development, and applications of semiconductor lasers, amplifiers, and modulators have occurred over the last decade. The fiber-optic revolution in telecommunication which provided improvement in transmission capacity of several orders of magnitude at low cost would not have been possible without the development of reliable semiconductor lasers. Today, semiconductor lasers are used not only for fiber-optic transmission but also in optical reading and recording (e.g., CD players), printers, fax machines, and in numerous applications as a high-power laser source. Semiconductor injection lasers continue to be the laser of choice for various system applications, primarily because of their small size, simplicity of operation, and reliable performance. For the same set of reasons, optical amplifiers are being investigated by numerous researchers all over the world. Optical amplifiers can be used in place of regenerators in lightwave transmission systems, the latter being significantly more complex than optical amplifiers, as described in detail in Section 12.3. For many transmission system applications the laser output is encoded with data by modulating the current. However, for many high-data-rate applications, which require long-distance transmission, external modulators are needed.

In this chapter we describe the fabrication, performance characteristics, current state of the art, and research directions for semiconductor lasers, optical amplifiers, and optical modulators. The focus is on lasers, amplifiers, and modulators needed for fiber-optic transmission systems. These devices are fabricated using the InP material system. For early work and a thorough discussion of semiconductor lasers, see Refs. 1–4.

12.2 SEMICONDUCTOR LASERS

The semiconductor injection laser was invented in 1962 [5–7]. With the development of epitaxial growth techniques and the subsequent fabrication of double heterojunction, the laser technology advanced rapidly in the 1970s and 1980s [1–4].The demonstration of continuous-wave (CW) operation of the semiconductor laser in the early 1970s [8] was followed by an increase in development activity in several industrial laboratories. This intense development activity in the 1970s was aimed at improving the performance characteristics and reliability of lasers fabricated using the AlGaAs material system [1]. These lasers emit near an early fiber-optic transmission wavelength of 0.8 μm and were deployed in the transmission systems in the late 1970s and early 1980s.

The optical fiber has zero dispersion near 1.3-μm wavelength and has lowest loss near 1.55-μm wavelength. Thus semiconductor lasers emitting near 1.3 and 1.55 μm are of interest for fiber-optic transmission application. Lasers emitting at these wavelengths are fabricated using InGaAsP/InP materials system and were first fabricated in 1976 [9]. Much fiber-optic transmission systems around the world that are in use or are currently being deployed utilize lasers emitting near 1.3 or 1.55 μm. Initially, these lasers were fabricated using a liquid-phase epitaxy (LPE) growth technique. With the development of metal-organic chemical vapor deposition (MOCVD) and gas-source molecular beam epitaxy (GSMBE) growth techniques in the 1980s, not only has the reproducibility of the fabrication process improved but it has also led to advances in laser designs, such as quantum-well lasers and very high speed lasers using semi-insulating Fe-doped InP current blocking layers [10].

12.2.1 Laser Designs

A schematic of a typical double heterostructure used for laser fabrication is shown in Fig. 12.1. It consists of n-InP, undoped $In_{1-x}Ga_xP_yAs_{1-y}$, p-InP and p-InGaAsP grown over (100)-oriented n-InP substrate. The undoped $In_{1-x}Ga_xP_yAs_{1-y}$ layer is the light-emitting layer (active layer). It is lattice matched to InP for $x \sim 0.45y$. The band gap of the $In_{1-x}Ga_xP_yAs_{1-y}$ material (lattice matched to InP) which determines the laser wavelength is given by [11], $E_g(eV) = 1.35 - 0.72y + 0.12y^2$. For lasers emit-

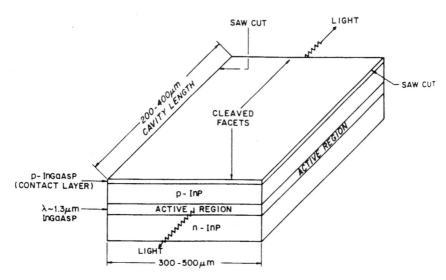

FIGURE 12.1 Schematic of a double-heterostructure laser.

ting near 1.3-μm $y \sim 0.6$. The double-heterostructure material can be grown by a LPE, GSMBE, or MOCVD growth technique.

The double-heterostructure material can be processed to produce lasers in several ways. Perhaps the simplest is the broad-area laser (Fig. 12.1), which involves putting contacts on the p- and n-sides and then cleaving. Such lasers do not have transverse mode confinement or current confinement, which leads to high threshold and nonlinearities in light versus current characteristics. Several laser designs have been developed to address these problems. Among them are the gain-guided laser, weakly index guided laser, and buried heterostructure (strongly index guided) laser. A typical version of these laser structures is shown in Fig. 12.2. The gain-guided structure uses a dielectric layer for current confinement. The current is injected in the opening in the dielectric (typically, 6 to 12 μm wide), which produces gain in that region, and hence the lasing mode is confined to that region. The weakly index guided structure has a ridge etched on the wafer, and a dielectric layer surrounds the ridge. The current is injected in the region of the ridge, and the optical mode overlaps the dielectric (which has a low index) in the ridge. This results in weak index guiding.

The buried heterostructure design shown in Fig. 12.2 has the active region surrounded (buried) by lower-index layers. The fabrication process of a DCPBH (double-channel planar buried heterostructure) laser involves growing a double het-

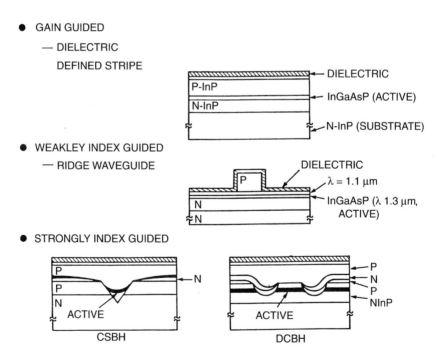

FIGURE 12.2 Schematic of a gain-guided, weakly index-guided, and strongly index-guided buried heterostructure laser.

erostructure, etching a mesa using a dielectric mask, and then regrowing the layer surrounding the active region using a second epitaxial growth step. The second growth can be a single Fe-doped InP (Fe:InP) semi-insulating layer or a combination p-InP, n-InP, Fe:InP layer. Generally, a MOCVD growth process is used for the growth of the regrown layer. Researchers have often given different names to the particular buried heterostructure laser design that they discovered. These are described in detail in Ref. 12. For the structure of Fig. 12.2, the Fe-doped InP layer provides both optical confinement to the lasing mode and current confinement to the active region. Buried heterostructure lasers are generally used in communication system applications because a properly designed strongly index guided buried heterostructure design has superior mode stability, higher bandwidth, and superior linearity in light versus current (L versus I) characteristics compared to the gain-guided and weakly index guided designs. Early recognition of these important requirements of communication grade lasers led to intensive research on InGaAsP BH laser designs all over the world in the 1980s. It is worth mentioning that BH lasers are more complex and difficult to fabricate compared to gain-guided and weakly index guided lasers. A scanning electron micrograph of a buried heterostructure laser is shown in Fig. 12.3.

The light versus current characteristics at different temperatures of an InGaAsP BH laser emitting at 1.3 μm is shown in Fig. 12.4. The typical threshold current of a buried-heterostructure (BH) laser at room temperature is in the range 5 to 10 mA. For gain-guided and weakly index guided lasers, typical room-temperature threshold currents are in the range 25 to 50 mA and 50 to 100 mA, respectively. The external differential quantum efficiency defined as the derivative of the L versus I characteristics above threshold is about 0.25 mW/mA per facet for a cleaved uncoated laser emitting near 1.3 μm.

An important characteristic of the semiconductor laser is that its output can be

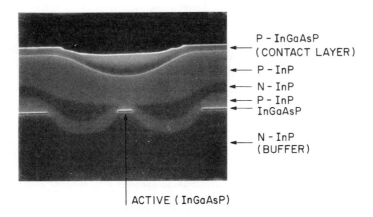

FIGURE 12.3 Scanning electron photomicrograph of a buried heterostructure laser.

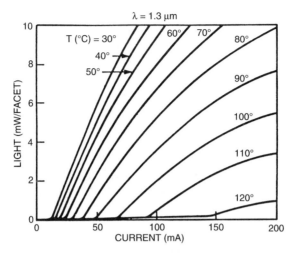

FIGURE 12.4 Light versus current characteristics of an InGaAsP buried heterostructure laser emitting at 1.3 μm.

modulated easily and simply by modulating the injection current. The relative magnitude of the modulated light output is plotted as a function of the modulation frequency of the current in Fig. 12.5 at different optical output powers. The laser is of the BH type (shown in Fig. 12.2), has a cavity length of 250 μm, and the modulation current amplitude was 5 mA. Note that the 3-dB frequency to which the laser can be modulated increases with increasing output power, and the modulation response is maximum at a certain frequency (ω_r). The resonance frequency ω_r is proportional to the square root of the optical power. The modulation response determines the data transmission rate capability of the laser, for example, for 10-Gb/s data transmission, the 3-dB bandwidth of the laser must exceed 10 GHz. However,

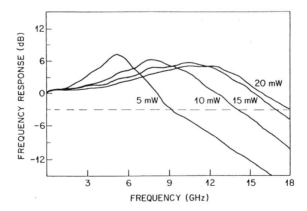

FIGURE 12.5 Modulation response of a laser at different optical output powers.

other system-level considerations, such as allowable error-rate penalty, often introduces much more stringent requirements on the exact modulation response of the laser.

A semiconductor laser with cleaved facets generally emits in a few longitudinal modes of the cavity. Typical spectrum of a laser with cleaved facets is shown in Fig. 12.6. The discrete emission wavelengths are separated by the longitudinal cavity mode spacing, which is about 10 Å for a laser ($\lambda \sim 1.3$ μm) with a 250-μm cavity length. Lasers can be made to emit in a single frequency using frequency-selective feedback, for example, using a grating internal to the laser cavity, as described in Section 12.2.3.

12.2.2 Quantum-Well Lasers

So far we have described the fabrication and performance characteristics of a regular double-heterostructure (DH) laser, which has an active region about 0.1 to 0.2

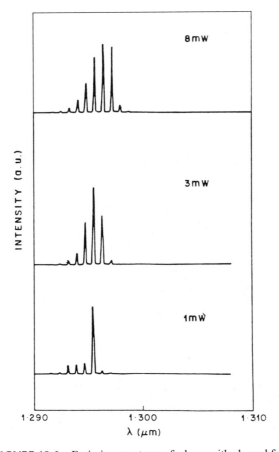

FIGURE 12.6 Emission spectrum of a laser with cleaved facets.

μm thick. Beginning with the 80 s, lasers with very thin active regions, quantum-well lasers, were being developed in many research laboratories [13–22]. Quantum-well (QW) lasers have active regions about 100 Å thick, which restricts the motion of the carriers (electrons and holes) in a direction normal to the well. This results in a set of discrete energy levels, and the density of states is modified to a two-dimensional-like density of states. This modification of the density of states results in several improvements in laser characteristics, such as lower threshold current, higher efficiency, higher modulation bandwidth, and lower CW and dynamic spectral width. All of these improvements were first predicted theoretically and then demonstrated experimentally [23–32].

The development of InGaAsP QW lasers were made possible by the development of MOCVD and GSMBE growth techniques. The transmission electron micrograph (TEM) of a multiple-QW laser structure is shown in Fig. 12.7. Shown are four InGaAs quantum wells grown over n-InP substrate. The well thickness is 70 Å and they are separated by barrier layers of InGaAsP ($\lambda \sim 1.1$ μm). Multi-quantum-well (MQW) lasers with threshold current densities of 600 A/cm^2 have been fabricated [33]. The schematic of a MQW BH laser is shown in Fig. 12.8. The composition of the InGaAsP material from the barrier layers to the cladding layer (InP) is gradually varied in this structure over a thickness of about 0.1 μm. This produces a graded variation in index (GRIN structure) which results in a higher optical confinement of the fundamental mode than that for an abrupt interface design. A larger mode confinement factor results in a lower threshold current. The laser has a MQW active region and utilizes Fe-doped semi-insulating (SI) InP

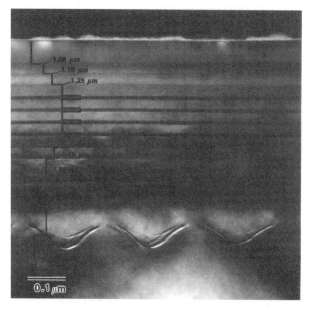

FIGURE 12.7 Transmission electron micrograph of a multi-quantum-well laser structure.

12.2 SEMICONDUCTOR LASERS   457

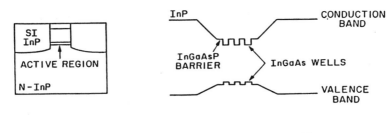

FIGURE 12.8 Schematic of multi-quantum-well buried heterostructure laser.

layers for current confinement and optical confinement. The light versus current characteristics of a MQW BH laser is shown in Fig. 12.9. The laser emits near 1.5 μm. The MQW lasers have lower threshold currents than regular DH lasers. Also, the two-dimensional-like density of states of the QW lasers makes the transparency current density of these lasers significantly lower than that for regular DH

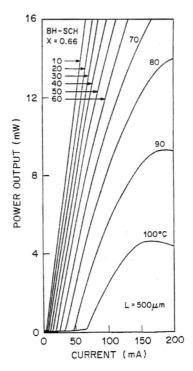

FIGURE 12.9 Light versus current characteristics of a multi-quantum-well buried heterostructure laser at different temperatures.

lasers [30]. This allows the fabrication of very low threshold lasers using high-reflectivity coatings.

The optical gain (g) of a laser at a current density J is given by

$$g = a(J - J_0) \tag{1}$$

where a is the gain constant and J_0 is the transparency current density. Although a logarithmic dependence of gain on current density [23] is often used to account for gain saturation, a linear dependence is used here for simplicity. The cavity loss α is given by

$$\alpha = \alpha_c + \frac{1}{L} \ln \frac{1}{R_1 R_2} \tag{2}$$

where α_c is the free carrier loss, L the length of the optical cavity, and R_1 and R_2 the reflectivity of the two facets. At threshold, gain equals loss, hence it follows from Eqs. (1) and (2) that the threshold current density J_{th} is given by

$$J_{th} = \frac{\alpha_c}{a} + \frac{1}{La} \ln \frac{1}{R_1 R_2} + J_0 \tag{3}$$

Thus for a laser with high-reflectivity facet coatings ($R_1, R_2 \sim 1$) and with low loss ($\alpha_c \sim 0$), $J_{th} \sim J_0$. For a QW laser, $J_0 \sim 50$ A/cm^2 and for a DH laser, $J_0 \sim 700$ A/cm^2; hence it is possible to get much lower threshold current using QW as the active region.

The light versus current characteristics of a QW laser with high-reflectivity coatings on both facets is shown in Fig. 12.10. The threshold current at room tempera-

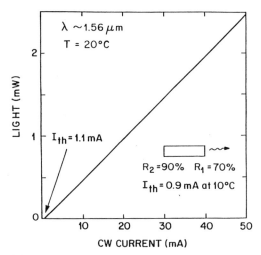

FIGURE 12.10 Light versus current of a quantum-well laser with high reflectivity coatings on both facets. (From Ref. 33.)

ture is about 1.1 mA. The laser is 170 μm long and has 90% and 70% reflective coating at the facets. This laser has a compressively strained MQW active region. For a lattice-matched MQW active region, a threshold current of 2 mA has been reported [34]. Such low-threshold lasers are important for array applications. Recently, QW lasers were fabricated which have higher modulation bandwidth than that of regular DH lasers. The current confinement and optical confinement in this laser is carried out using MOCVD-grown Fe-doped InP lasers similar to that shown in Fig. 12.2. The laser structure is then modified further by using a small contact pad and etching channels around the active-region mesa (Fig. 12.11). These modifications are designed to reduce the capacitance of the laser structure. The modulation response of the laser is shown in Fig. 12.12. A 3-dB bandwidth of 25 GHz is obtained [35].

Strained Quantum-Well Lasers Quantum-well lasers have also been fabricated using an active layer whose lattice constant differs slightly from that of the substrate and cladding layers. Such lasers are known as strained quantum-well lasers. Over the last few years, strained quantum-well lasers have been investigated extensively all over the world [36–43]. They show many desirable properties, such as a very low threshold current density and a lower linewidth than regular MQW lasers under both CW operation and modulation. The origin of the improved device performance lies in the band-structure changes induced by the mismatch-induced strain [44, 45]. Figure 12.13 shows the band structure of a semiconductor under tensile and compressive strains. Strain splits the heavy- and the light-hole valence bands at the Γ point of the Brillouin zone where the bandgap is minimum in direct-bandgap semiconductors.

Two material systems have been widely used for strained quantum-well lasers: InGaAs grown over InP by the MOCVD or the CBE growth technique [36–40] and InGaAs grown over GaAs by the MOCVD or the MBE growth technique [41–43]. The former material system is of importance for low-chirp semiconductor lasers for lightwave system applications, while the latter material system has been used to fabricate high-power lasers emitting near 0.98 μm, a wavelength of interest for pumping erbium-doped fiber amplifiers.

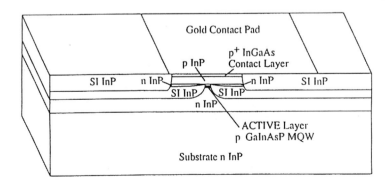

FIGURE 12.11 Schematic of a laser designed for high speed. (From Ref. 35.)

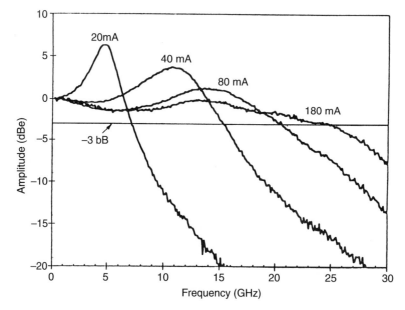

FIGURE 12.12 Modulation response of multiquantum well high speed lasers. (From Ref. 35.)

The alloy $In_{0.53}Ga_{0.47}As$ has the same lattice constant as InP. Semiconductor lasers with an $In_{0.53}Ga_{0.47}As$ active region have been grown on InP by the MOCVD growth technique. Excellent material quality is also obtained for $In_{1-x}Ga_xAs$ alloys grown over InP by MOCVD for nonlattice-matched compositions. In this case the laser structure generally consists of one or many $In_{1-x}Ga_xAs$ quantum-well layers with InGaAsP barrier layers whose composition is lattice matched to that of InP. For $x < 0.53$ the active layer in these lasers is under tensile stress, while for $x > 0.53$ the active layer is under compressive stress.

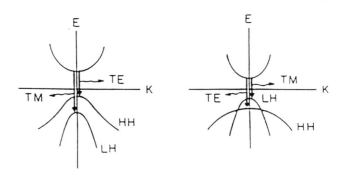

FIGURE 12.13 Band structures under stress. The figures on the left and right represent situations under compressive and tensile strain, respectively.

Superlattice structures of InGaAs/InGaAsP with tensile and compressive stress have been grown by both MOCVD and CBE growth techniques over an n-type InP substrate. Figure 12.14 shows the broad-area threshold current density as a function of cavity length for strained MQW lasers with four $In_{0.65}Ga_{0.35}As$ [39] quantum wells with InGaAsP ($\lambda \sim 1.25$ μm) barrier layers. The active region in this laser is under 0.8% compressive strain. Also shown for comparison is the threshold current density as a function of cavity length of MQW lattice-matched lasers with $In_{0.53}Ga_{0.47}As$ wells. The entire laser structure, apart from the quantum-well composition, is identical for the two cases. The threshold current density is lower for the compressively strained MQW structure than that for the lattice-matched MQW structure.

Buried heterostructure (BH) lasers have been fabricated using compressive- and tensile-strained MQW lasers. The threshold current of these lasers as a function of the In concentration is shown in Fig. 12.15. Lasers with compressive strain have a lower threshold current than that for lasers with tensile strain. This can be explained by splitting of the light- and heavy-hole bands under stress [47,48]. However, more recent studies have shown that it is possible to design tensile strained lasers with lower threshold [37,42].

Strained quantum-well lasers fabricated using $In_{1-x}Ga_xAs$ layers grown over a GaAs substrate have been studied extensively [41,43,49–54]. The lattice constant of InAs is 6.06 Å and that of GaAs is 5.654 Å. The $In_{1-x}Ga_xAs$ alloy has a lattice constant between these two values, and to a first approximation it can be assumed to vary linearly with x. Thus an increase in the In mole fraction x increases the lattice mismatch relative to the GaAs substrate and therefore produces larger compressive strain on the active region.

A typical laser structure grown over the n-type GaAs substrate is shown in Fig.

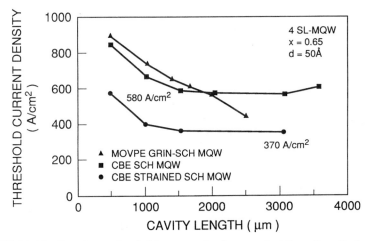

FIGURE 12.14 Broad-area threshold current density as a function of cavity length for strained and lattice-matched InGaAs/InP MQW lasers. (From Ref. 39.)

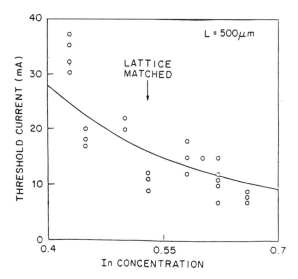

FIGURE 12.15 Threshold current of buried heterostructure $In_xGa_{1-x}As/InP$ MQW lasers plotted as a function of In concentration x. (From Ref. 46.)

12.16 for this material system. It consists of a MQW active region with one to four $In_{1-x}Ga_xAs$ wells separated by GaAs barrier layers. The entire MQW structure is sandwiched between n- and p-type $Al_{0.3}Ga_{0.7}As$ cladding layers, and the P-cladding layer is followed by a p-type GaAs contact layer. Variations of the foregoing structure with different cladding layers or large optical cavity designs have been reported. Emission wavelength depends on the In composition, x. As x increases, the emission wavelength increases, and for x larger than a certain value (typically, about 0.25), the strain is too large to yield high-quality material. For $x \sim 0.2$, the emission wavelength is near 0.98 μm, a wavelength region of interest for pumping fiber amplifiers [49]. Threshold current density as low as 47 A/cm^2 has been reported for

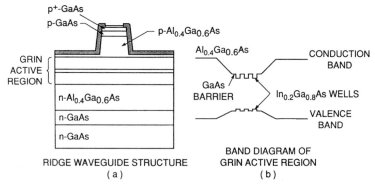

FIGURE 12.16 Typical $In_{1-x}Ga_xAs/GaAs$ MQW laser structure. (From Ref. 41.)

$In_{0.2}Ga_{0.8}As$/GaAs strained MQW lasers [52]. High-power lasers have been fabricated using an $In_{0.2}Ga_{0.8}As$/GaAs MQW active region. Single-mode output powers of greater than 200 mW have been demonstrated using a ridge-waveguide laser structure.

The frequency chirp of strained and unstrained QW lasers has been investigated. Strained QW lasers (InGaAs/GaAs) exhibit the lowest chirp (or dynamic linewidth) under modulation. The lower chirp of strained QW lasers is consistent with a small linewidth enhancement factor (α factor) measured in such devices. The α-factor is the ratio of the real and imaginary parts of the refractive index. A correlation between the measured chirp and linewidth enhancement factor for regular double-heterostructure strained and unstrained QW lasers is shown in Table 12.1. The high efficiency, high power, and low chirp of strained and unstrained QW lasers make these devices attractive candidates for lightwave transmission applications.

Other Material Systems A few other material systems have been reported for lasers in the wavelength range 1.3 μm. These are the AlGaInAs/InP and InAsP/InP materials grown over InP substrates and more recently, the InGaAsN material grown over GaAs substrates. The AlGaInAsP/InP system has been investigated with the aim of producing lasers with better high-temperature performance for uncooled transmitters [55]. This material system has a larger conduction band offset than the InGaAsP/InP material system, which may result is lower electron leakage over the heterobarrier and thus better high-temperature performance. The energy band diagram of a GRINSCH (graded-index separate confinement heterostructure) laser design is shown in the Fig 12.17. The laser has five compressively strained quantum wells in the active region. The 300-μm-long ridge waveguide lasers typically have a threshold current of 20 mA. The measured light versus current characteristics of a laser with 70% high-reflectivity coating at the rear facet is shown in Fig 12.18. These lasers have somewhat better high-temperature performance than that of InGaAsP/InP lasers.

The InAsP/InP material system has also been investigated for 1.3-μm lasers [56]. InAsP with an arsenic composition of 0.55 is under 1.7% compressive strain when grown over InP. Using a MOCVD growth technique, buried heterostructure lasers with an InAsP quantum well, InGaAsP ($\lambda \sim 1.1$ μm) barrier layers, and InP cladding layers have been reported. The schematic of the laser structure is shown in

TABLE 12.1 Linewidth Enhancement Factor and Chirp of Lasers

Laser Type	Linewidth Enhancement Factor	FWHM Chirp at 50 mA (1 Gb/s) (A)
DH laser	5.5	1.2
MQW laser	3.5	0.6
Strained MQW laser	1.0	0.2 (InGaAs/GaAs, $\lambda \sim 1$ μm)
	2.0	0.4 (InGaAs/InP, $\lambda \sim 1.55$ μm)

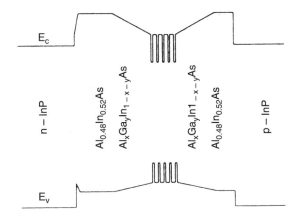

FIGURE 12.17 Band diagram of a AlGaInAs GRINSCH with five quantum wells. (From Ref. 55.)

Fig 12.19. Typical threshold current of the BH laser diodes are about 20 mA for a 300-μm cavity length.

When grown over GaAs, the material InGaNAs can have a very large (~ 300 meV) conduction band offset, which can lead to much better high-temperature performance than for the InGaAsP/InP material system [57]. The threshold temperature dependence is characterized by $I_{th}(T) = I_0 \exp(T/T_0)$, where T_0 is generally called the characteristic temperature. Typical T_0 values for InGaAsP/InP laser is about 60 to 70 K in the temperature range 300 to 350 K. The predicted T_0 values for

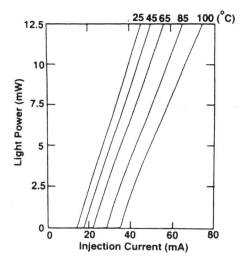

FIGURE 12.18 Light versus current characteristics of a AlGaInAs quantum-well laser with five wells. (From Ref. 55.)

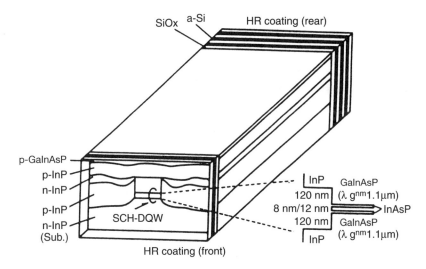

FIGURE 12.19 Schematic of a buried heterostructure InAsP/InGaAsP quantum-well laser. (From Ref. 56.)

the InGaNAs/GaAs system is about 150 K, and $T_0 = 126$ K has been reported for a InGaNAs laser emitting near 1.2 μm [57].

12.2.3 Distributed Feedback Lasers

Semiconductor lasers fabricated using the InGaAsP material system are widely used as sources in many lightwave transmission systems. One measure of the transmission capacity of a system is the data rate. Thus the drive toward higher capacity pushes systems to higher data rates where the chromatic dispersion of the fiber plays an important role in limiting the distance between regenerators. Sources emitting in a single wavelength help reduce the effects of chromatic dispersion and are therefore used in most systems operating at high data rates (>1.5 Gb/s).

The single-wavelength laser source used in most commercial transmission systems is the distributed feedback (DFB) laser, where a diffraction grating etched on the substrate close to the active region provides frequency-selective feedback, which makes the laser emit in a single wavelength. In this section we report the fabrication, performance characteristics, and reliability of DFB lasers [58]. The schematic of our DFB laser structure is shown in Fig. 12.20. The fabrication of the device involves the following steps. First, a grating with a periodicity of 2400 Å is fabricated on a (100)-oriented n-InP substrate using optical holography and wet chemical etching. Four layers are then grown over the substrate. These layers are an n-InGaAsP ($\lambda \sim 1.3$ μm) waveguide layer, an undoped InGaAsP ($\lambda \sim 1.55$ μm) active layer, a p-InP cladding layer, and a p-InGaAsP ($\lambda \sim 1.3$ μm) contact layer. Mesas are then etched on the wafer using a SiO_2 mask and wet chemical etching. Fe-doped InP semi-insulating layers are grown around the mesas using the MOCVD

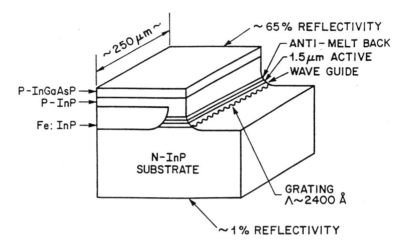

FIGURE 12.20 Schematic of a capped mesa buried heterostructure (CMBH) distributed feedback laser.

growth technique. The semi-insulating layers help confine the current to the active region and also provide index guiding to the optical mode. The SiO_2 stripe is then removed and the p-InP cladding layer and a p-InGaAsP contact layer are grown on the wafer using the vapor-phase epitaxy growth technique. The wafer is then processed to produce 250-μm-long laser chips using standard metallization and cleaving procedures. The final laser chips have an antireflection coating (<1%) at one facet and a high-reflection coating (~65%) at the back facet. The asymmetric facet coatings help remove the degeneracy between the two modes in the stopband.

The CW light versus current characteristics of a laser are shown in Fig. 12.21. Also shown is the measured spectrum at different output powers. The threshold currents of these lasers are in the range 15 to 20 mA. For high fiber coupling efficiency, it is important that the laser emit in the fundamental transverse mode. The measured farfield pattern parallel and perpendicular to the junction plane at different output powers of a device is shown in Fig. 12.22. The figure shows that the laser operates in the fundamental transverse mode in the entire operating power range from threshold to 60 mW. The full width at half maximum of the beam divergences parallel and normal to the junction plane are 40° and 30°, respectively. The dynamic spectrum of the laser under modulation is an important parameter when the laser is used as a source for transmission. The measured 20-dB full width is shown in Fig. 12.23 at two different data rates as a function of bias level. Note that for a laser biased above threshold, the chirp width is nearly independent of the modulation rate.

Tunable Lasers Tunable semiconductor lasers are needed for many applications. Examples of application in lightwave transmission systems are wavelength-division multiplexing, where signals at many distinct wavelengths are simultaneously modulated and transmitted through a fiber, and coherent transmission systems, where the

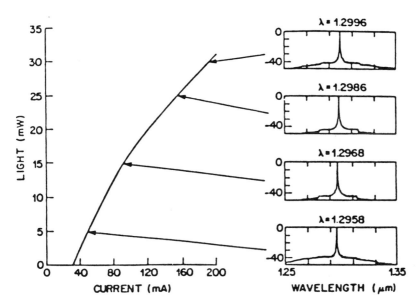

FIGURE 12.21 CW light versus current characteristics and measured spectrum at different output powers. Temperature = 30°C.

wavelength of the transmitted signal must match that of the local oscillator. Several types of tunable laser structures have been reported in the literature [59–64]. Two principal schemes are the multisection DFB laser and multisection distributed Bragg reflector (DBR) laser. Multisection DBR lasers generally exhibit higher tunability than that for multisection DFB lasers. The design of a multisection DBR laser is shown in Fig. 12.24 schematically. The three sections of this device are the active region section, which provides the gain; the grating section, which provides the tunability; and the phase-tuning section, which is needed to access all wavelengths continuously. The current through each of these sections can be varied independently. The tuning mechanism can be understood by noting that the emission wavelength λ of a DBR laser is given by $\lambda = 2n\Lambda$, where Λ is the grating period and n is the effective refractive index of the optical mode in the grating section. The latter can be changed simply by varying the current in the grating section. The extent of wavelength tunability of a three-section DBR laser is shown in Fig. 12.25. Measured wavelengths are plotted as a function of phase-section current for different currents in the tuning section. A tuning range in excess of 6 nm can be obtained by controlling currents in the grating and phase-tuning sections.

An important characteristic of lasers for applications requiring a high degree of coherence is the spectral width (linewidth) under CW operation. The CW linewidth depends on the rate of spontaneous emission in the laser cavity. For coherent transmission applications, the CW linewidth must be quite small. The minimum linewidth allowed depends on the modulation format used. For differential phase-

468 LASERS, AMPLIFIERS, AND MODULATORS BASED ON InP-BASED MATERIALS

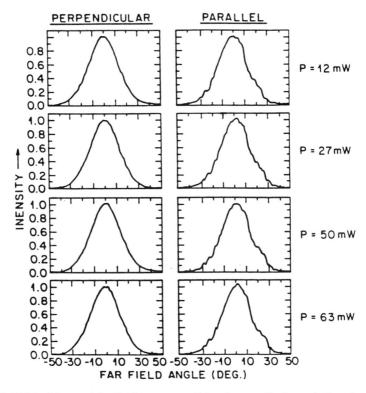

FIGURE 12.22 Measured far-field pattern parallel and perpendicular to the junction plane.

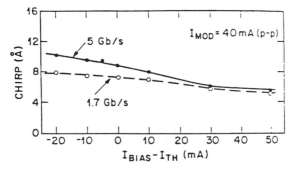

FIGURE 12.23 Measured 20 dB full width of the dynamic linewidth is plotted as a function of bias level for operation at 5 and 1.7 Gb/s.

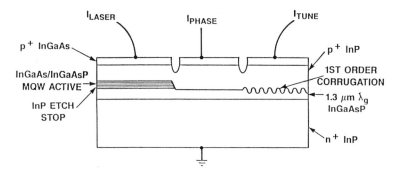

FIGURE 12.24 Schematic of a multisection DBR laser. The laser has a MQW active region. The three sections are optically coupled by the thick waveguide layer below the MQW active region. (From Ref. 59.)

shift keying (DPSK) transmission, the minimum linewidth is given approximately by $B/300$, where B is the bit rate. Thus for a 1-Gb/s transmission rate, the minimum linewidth is 3 MHz. The CW linewidth of a laser decreases with increasing length and varies as α^2, where α is the linewidth enhancement factor. Since α is smaller for a multiquantum-well (MQW) laser, the linewidth of DFB or DBR lasers utilizing an MQW active region is smaller than that for lasers with a regular DH active region. The linewidth varies inversely with the output power at low powers (<10 mW) and shows saturation at high powers. The measured linewidth as a function of the output power of a 850-μm-long DFB laser with an MQW active region is shown in Fig. 12.26. The minimum linewidth of 350 kHz was observed for this device at an oper-

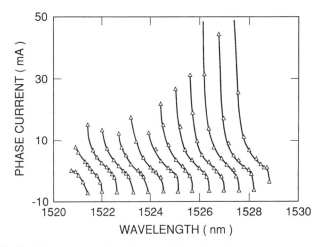

FIGURE 12.25 Frequency tuning characteristics of a three-section MQW DBR laser. (From Ref. 59.)

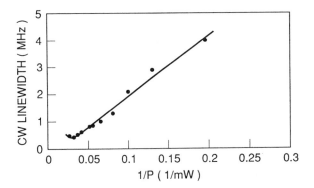

FIGURE 12.26 Measured CW linewidth plotted as a function of the inverse of the output power for a MQW DFB laser with a cavity length of 850 μm.

ating power of 25 mW. For multisection DBR lasers of the type shown in Fig. 12.24, the linewidth varies with changes in currents in the phase-tuning and grating sections. The data measured for a MQW three-section DBR laser are shown in Fig. 12.27. Measured linewidths are plotted as a function of phase-section current for different currents in the tuning section.

12.2.4 Surface Emitting Lasers

Semiconductor lasers described in previous chapters have cleaved facets that form the optical cavity. The facets are perpendicular to the surface of the wafer and light

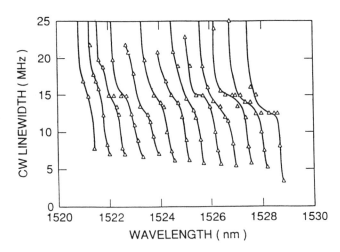

FIGURE 12.27 Measured CW linewidth as a function of wavelength for a three-section MQW DBR laser. (From Ref. 61.)

12.2 SEMICONDUCTOR LASERS

is emitted parallel to the surface of the wafer. For many applications requiring a two-dimensional laser array or monolithic integration of lasers with electronic components (e.g., optical interconnects), it is desirable to have the laser output normal to the surface of the wafer. Such lasers are known as surface-emitting lasers (SELs). A class of surface-emitting lasers also have the optical cavity normal to the surface of the wafer [65–72]. These devices are known as vertical-cavity surface-emitting lasers (VCSEL) to distinguish them from other surface emitters.

A generic SEL structure utilizing multiple semiconductor layers to form a Bragg reflector is shown in Fig. 12.28. The active region is sandwiched between n- and p-type cladding layers, which are themselves sandwiched between two n- and p-type Bragg mirrors. This structure is shown using the AlGaAs/GaAs material system, which has been very successful in the fabrication of SELs. The Bragg mirrors consist of alternating layers of low- and high-index materials. The thicknesses of each layer is one-fourth of the wavelength of light in the medium. Such periodic quarter-wave-thick layers can have very high reflectivity. For normal incidence, the reflectivity is given by [73]

$$R = \frac{[1 - n_4/n_1(n_2/n_3)^{2N}]^2}{[1 + n_4/n_1(n_2/n_3)^{2N}]^2} \qquad (4)$$

where n_2 and n_3 are the refractive indices of the alternating layer pairs, n_4 and n_1 are the refractive indices of the medium on the transmitted and incident sides of the DBR mirror, and N is the number of pairs. As N increases, R increases. Also, for a

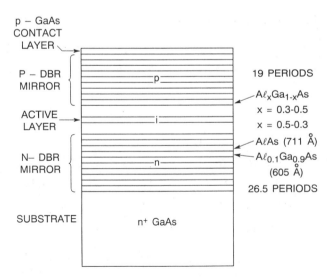

FIGURE 12.28 Schematic illustration of a generic SEL structure utilizing distributed Bragg mirrors formed by using multiple semiconductor layers. DBR pairs consist of AlAs (711 Å thick) and $Al_{0.1}Ga_{0.9}As$ (605 Å thick) alternate layers. Active layer could be either a quantum well or similar to a regular double-heterostructure laser.

given N, R is larger if the ratio of n_2/n_3 is smaller. For a AlAs/Al$_{0.1}$Ga$_{0.9}$As set of quarter-wave layers, typically 20 pairs are needed for a reflectivity of about 99.5%. Various types of AlGaAs/GaAs SELs have been reported [74–82].

For a SEL to have a threshold current density comparable to that of an edge-emitting laser, the threshold gains must be comparable for the two devices. The threshold gain of an edge-emitting laser is about 100 cm^{-1}. For a SEL with an active-layer thickness of 0.1 μm, this value corresponds to a single-pass gain of about 1%. Thus for the SEL device to lase with a threshold current density comparable to that of an edge emitter, the mirror reflectivities must be >99%. Central to the fabrication of low-threshold SELs is the ability to fabricate high-reflectivity mirrors. In the late 1970s, Soda et al. [83] reported on an SEL fabricated using the InP material system. Their device structure is shown in Fig 12.29. The surfaces of the wafer form the Fabry–Perot cavity of the laser. Fabrication of the device involves the growth of a double heterostructure on an n-InP substrate. A circular contact is made on the p-side using an SiO$_2$ mask. The substrate side is polished, making sure that it is parallel to the epitaxial layer, and ring electrodes (using an alloy of Au–Sn) are deposited on the n-side. The laser had a threshold current density of about 11 kA/cm^2 at 77 K and operates at output powers of several milliwatts.

InGaAsP/InP SELs have been investigated by many researchers over the last few years [83–89]. Many of the schemes utilize alternating layers of InP and InGaAsP to produce Bragg mirrors (Fig. 12.30). The refractive index difference between InGaAsP ($\lambda \sim 1.3$ μm) and InP layers is smaller than that in GaAs SELs; hence InGaAsP/InP SELs utilize more pairs (typically, 40 to 50) to produce a high-reflectivity (> 99 %) mirror. Such mirror stacks have been grown by both chemical beam epitaxy (CBE) and MOCVD growth techniques and have been used to fabricate InGaAsP SELs. Room-temperature pulsed operation of InGaAsP/InP SELs using these mirror stacks and emitting near 1.5 μm have been reported [88].

An alternative approach is using the technique of wafer fusion [89,90]. In this technique the Bragg mirrors are formed using an GaAs/AlGaAs system grown by MBE, and the active region of InGaAsP bounded by thin InP layers is formed by

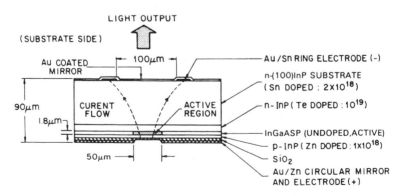

FIGURE 12.29 Schematic of a InGaAsP SEL. (From Ref. 65.)

12.2 SEMICONDUCTOR LASERS

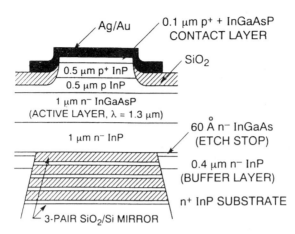

FIGURE 12.30 Schematic of a InGaAsP SEL fabricated using multilayer mirors. (From Ref. 86.)

MOCVD. The post type structure of a 1.5-μm-wavelength SEL formed using this technique is shown in Fig. 12.31. The optical cavity is formed by wafer fusion of the InGaAsP quantum well between the Bragg mirrors. Room-temperature CW threshold currents of 2.3 mA have been reported for the 8-μm-diameter post device [89].

12.2.5 Laser Reliability

The performance characteristics of injection lasers used in lightwave systems can degrade during their operation. The degradation is generally characterized by a

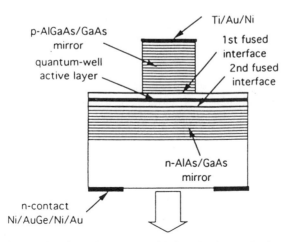

FIGURE 12.31 Schematic of a InGaAsP SEL fabricated using wafer fusion. (From Ref. 89.)

change in the operational characteristics of the lasers and in some cases is associated with the formation and/or multiplication of defects in the active region. The degraded lasers usually exhibit an increase in the threshold current which is often accompanied by a decrease in the external differential quantum efficiency. For single-wavelength lasers the degradation may be a change in the spectral characteristics (i.e., the degraded device may no longer emit in a single wavelength, although the threshold or light output at a given current has changed very little).

The dominant mechanism responsible for the degradation is determined by any or all of several fabrication processes, including epitaxial growth, wafer quality, device processing, and bonding [91–101]. In addition, the degradation rate of devices processed from a given wafer depends on the operating conditions (i.e., the operating temperature and the injection current). Although many of the degradation mechanisms are not fully understood, extensive amounts of empirical observations exist in the literature which have allowed the fabrication of InGaAsP laser diodes with extrapolated median lifetimes in excess of 25 years at an operating temperature of 20°C [93].

The detailed studies of degradation mechanisms of optical components used in lightwave systems have been motivated by the desire to have a reasonably accurate estimate of the operating lifetime before they are used in practical systems. Since for many applications the components are expected to operate reliably over a period in excess of 10 years, an appropriate reliability assurance procedure becomes necessary, especially for applications such as an undersea lightwave transmission system, where the replacement cost is very high. The reliability assurance is usually carried out by operating the devices under high stress (e.g., high temperature), which enhances the degradation rate so that a measurable value can be obtained in an operating time of a few hundred hours. The degradation rate under normal operating conditions can then be obtained from the measured high-temperature degradation rate using the concept of an activation energy [93].

The light output versus current characteristics of a laser change after stress aging. There is generally a small increase in threshold current and a decrease in external differential quantum efficiency following stress aging. Aging data for 1.3-µm InGaAsP lasers used in the first submarine fiber-optic cable are shown in Figure 12.32. Some lasers exhibit an initial rapid degradation, after which the operating characteristics of the lasers are very stable. Given a population of lasers, it is possible to quickly identify stable lasers by a high-stress test (also known as a purge test) [91,92,102,103].The stress test implies that operating the laser under a set of high-stress conditions (e.g., high current, high temperature, high power) would cause the weak lasers to fail and stabilize the possible winners. Observations on the operating current after stress aging have been reported by Nash et al. [91]. It is important to point out that determination of the duration and specific conditions for stress aging are critical to the success of this screening procedure.

The expected operating lifetime of a semiconductor laser is generally determined by accelerated aging at high temperatures, using an activation energy. The lifetime t at a temperature T is found experimentally to vary as $\exp(-E/kT)$, where E is the activation energy and k is the Boltzmann constant [104,105]. The measured injection

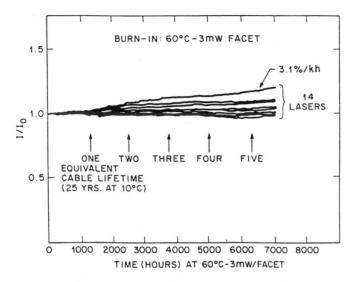

FIGURE 12.32 Operating current for 3 mW output at 60°C as a function of operating time. These data were generated for 1.3 μm InGaAsP lasers used in the first submarine fiber optic cable. (From Ref. 91.)

current for 3-mW output power at 60°C for buried heterostructure DFB lasers as a function of operating (or aging) time is shown in Fig. 12.33. The operating current increases at a rate of less than 1% khr^{-1} of aging time. Assuming a 50% change in operating current as the useful lifetime of the device and an activation energy of 0.7 eV, this aging rate corresponds to a light-emitting lifetime of greater than 100 years at 20°C.

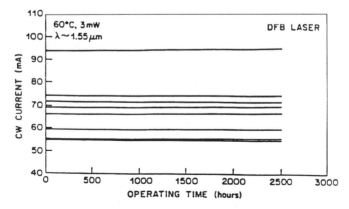

FIGURE 12.33 Operating current as a function of operating time for 3-mW output power at 60°C for lasers emitting near 1.55 μm. (Courtesy of F. R. Nash.)

A parameter that determines the performance of the DFB laser is the side mode suppression ratio (SMSR) (i.e., the ratio of the intensity of the dominant lasing mode to that of the next most intense mode). An example of the spectrum before and after aging is shown in Fig. 12.34. The SMSR for a laser before and after aging as a function of current is plotted in Fig. 12.35. Note that the SMSR does not change significantly after aging, which confirms the spectral stability of the emission. For some applications, such as coherent transmission systems, the absolute wavelength stability of the laser is important. The measured change in emission wavelength at 100 mA before and after aging of several devices is shown in Fig. 12.36. Note that most of the devices do not exhibit any change in wavelength and the standard deviation of the change is less than 2 Å. This suggests that the absolute wavelength stability of the devices is adequate for coherent transmission applications.

Another parameter of interest in certifying the spectral stability of a DFB laser is the change in dynamic linewidth or chirp with aging. Since the chirp depends on data rate and the bias level, a certification of the stability of the value of the chirp is, in general, tied to the requirements of a transmission system application. A measure of the effect of chirp on the performance of a transmission system is the dispersion penalty. We have measured the dispersion penalty of several lasers at 600 Mb/s for a dispersion of 1700 ps/nm before and after aging. The median change in dispersion penalty for 42 devices was less than 0.1 dB and no single device showed a change larger than 0.3 dB. This suggests that the dynamic linewidth or the chirp under modulation is stable.

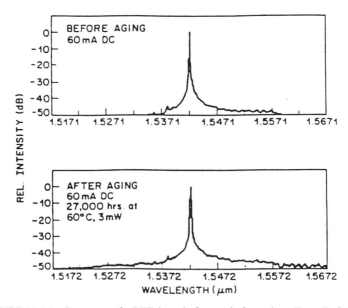

FIGURE 12.34 Spectrum of a DFB laser before and after aging. (From Ref. 105.)

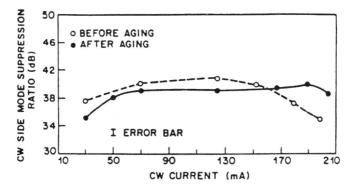

FIGURE 12.35 Side-mode suppression ratio (SMSR) as a function of current before and after aging.

12.3 OPTICAL AMPLIFIERS

An optical amplifier, as the name implies, is a device that amplifies an input optical signal. The amplification factor or gain can be higher than 1000 (>30 dB) in some devices. There are two principal types of optical amplifiers: the semiconductor optical amplifier and the fiber-optic amplifier. In a semiconductor amplifier, amplification of light takes place when it propagates through a semiconductor medium fabricated in a waveguide form. In a fiber amplifier, amplification of light occurs when it travels through a fiber doped with rare earth ions (e.g., Nd^+, Er^+, etc.). Semiconductor laser amplifiers are typically less than 1 mm in length, whereas fiber amplifiers are typically 1 to 100 m in length. The operating principles, design, fabrica-

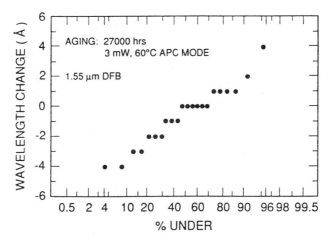

FIGURE 12.36 Change in emission wavelength after aging is plotted in the form of a normal probability distribution.

478 LASERS, AMPLIFIERS, AND MODULATORS BASED ON InP-BASED MATERIALS

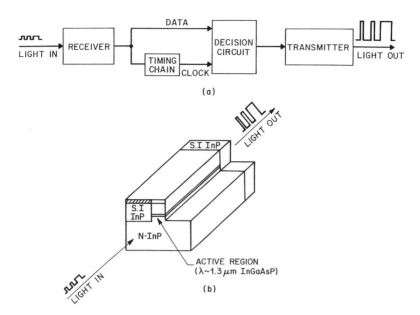

FIGURE 12.37 (a) Block diagram of a lightwave regenerator, (b) schematic of a semiconductor optical amplifier.

tion, and performance characteristics of InP-based semiconductor amplifiers are described in this section.

In a lightwave transmission system, as the optical signal travels through the fiber, it weakens and gets distorted. Regenerators are used to restore the optical pulses to their original form. Figure 12.37a shows the block diagram of a typical lightwave regenerator. Its main components are an optical receiver, an optical transmitter, and electronic timing and decision circuits. Optical amplifiers can nearly restore the original optical pulses and thereby increase the transmission distance without using conventional regenerators. An example of a semiconductor amplifier that function as a regenerator is shown schematically in Fig. 12.37b. The semiconductor amplifiers need external current to produce gain, and fiber amplifiers need pump lasers for the same purpose. Because of its simplicity, an optical amplifier is an attractive alternative for a new lightwave system.

12.3.1 Semiconductor Optical Amplifier

A semiconductor optical amplifier [106–112] is a device very similar to a semiconductor laser. Hence their operating principles, fabrication, and design are also similar. When the injection current is below threshold, the laser acts as an optical amplifier for incident light waves and above threshold it undergoes oscillation. Semiconductor optical amplifiers can be classified into two categories: the Fabry–Perot (FP) amplifier and the traveling-wave (TW) amplifier. A FP amplifier has

12.3 OPTICAL AMPLIFIERS

considerable reflectivity at the input and output ends, which results in resonant amplification between the end mirrors. Thus an FP amplifier exhibits very large gain at wavelengths corresponding to the longitudinal modes of the cavity. The TW amplifier, by contrast, has negligible reflectivity at each end, which results in signal amplification during a single pass. The optical gain spectrum of a TW amplifier is quite broad and corresponds to that of the semiconductor gain medium. Most practical TW amplifiers exhibit some small ripple in the gain spectrum which arises from residual facet reflectivities. TW amplifiers are more suitable for system applications. Therefore, much effort has been devoted over the last few years to fabricate amplifiers with very low effective facet reflectivities. Such amplifier structures either utilize special low-effective-reflectivity dielectric coatings, or have tilted or buried facets. Fabrication and performance of these devices are described later.

Extensive work on optical amplifiers were carried out in the 1980s using the AlGaAs material system. Much of the recent experimental work on semiconductor optical amplifiers has been carried out using the InGaAsP material system, with the optical gain centered around 1.3 or 1.55 μm. The amplifiers used in lightwave system applications, either as preamplifiers in front of a receiver or as in-line amplifiers as a replacement of regenerators, must also exhibit equal optical gain for all polarizations of the input light. In general, the optical gain in a waveguide is polarization, dependent although the material gain is independent of polarization for bulk semiconductors. This arises from unequal mode confinement factors for the light polarized parallel to the junction plane (TE mode) and that for light polarized perpendicular to the junction plane (TM mode). For thick active regions, the confinement factors of the TE and TMs mode are nearly equal. Hence the gain difference between the TE and TM modes is smaller for amplifiers with a thick active region (Fig. 12.38).

Low-Reflectivity Coatings A key factor for good performance characteristics (low gain ripple and low polarization selectivity) for TW amplifiers is very low facet reflectivity [112]. The reflectivity of cleaved facets can be reduced by dielectric coating. For plane waves incident on an air interface from a medium of refractive index n, the reflectivity can be reduced to zero by coating the interface with a dielectric whose refractive index equals $n^{1/2}$ and whose thickness equals $\lambda/4$. However, the fundamental mode propagating in a waveguide is not a plane wave, and therefore the $n^{1/2}$ above law provides only a guideline for achieving very low (ca. 10^{-4}) facet reflectivity by dielectric coatings. In practice, very low facet reflectivities are obtained by monitoring the amplifier performance during the coating process. The effective reflectivity can then be estimated from the ripple at the Fabry–Perot mode spacings, caused by residual reflectivity, in the spontaneous emission spectrum. The result of such an experiment is shown in Fig. 12.39. The reflectivity is very low ($<10^{-4}$) only in a small range of wavelengths. Although laboratory experiments have been carried out using amplifiers that rely only on low-reflectivity coatings for good performance, the critical nature of the thickness requirement and a limited wavelength range of good antireflection (AR) coating led to the investigation of alternative schemes, as discussed below.

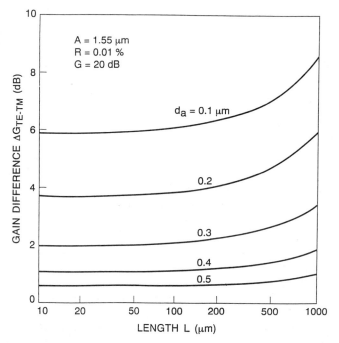

FIGURE 12.38 Optical gain difference between the TE and TM modes of a semiconductor amplifier plotted as a function of device length for various active layer thicknesses. (From Ref. 112.)

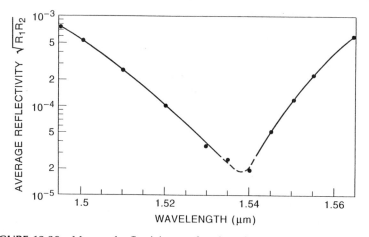

FIGURE 12.39 Measured reflectivity as a function of wavelength. (From Ref. 112.)

Buried Facet Amplifiers The principal feature of buried facet (also known as window structure) optical amplifiers relative to AR-coated cleaved facet devices is a polarization-independent reduction in mode reflectivity due to the buried facet, resulting in better control in achieving polarization independent gain. A schematic cross section of a buried facet optical amplifier is shown in Fig. 12.40. Current confinement in this structure is provided by semi-insulating Fe-doped InP layers grown by the MOCVD growth technique. Fabrication of this device involves a procedure similar to that used for lasers. The first four layers are grown on a (100)-oriented n-InP substrate by MOCVD. These layers are an n-InP buffer layer, an undoped InGaAsP ($\lambda \sim 1.55$ μm) active layer, a p-InP cladding layer, and a p-InGaAsP ($\lambda \sim 1.3$ μm) layer. Mesas are then etched on the wafer along the [110] direction with 15-μm-wide channels normal to the mesa direction using a SiO_2 mask. The latter is needed for buried facet formation. Semi-insulating Fe-doped InP layers are then grown around the mesas by MOCVD with the oxide mask in place. The oxide mask and p-InGaAsP layers are removed, and p-InP and p-InGaAsP ($\lambda \sim 1.3$ μm) contact layers are then grown over the entire wafer by the MOCVD growth technique. The wafer is processed using standard methods and cleaved to produce 500-μm-long buried facet chips with about 7-μm-long buried facets at each end. Chip facets are then AR-coated using a single-layer film of ZrO_2. Fabrication of cleaved facet devices follows the same procedure as described above, except that the mesas are continuous with no channels separating them. The latter is needed for defining the buried facet regions. In both types of devices, the semi-insulating layer provides current confinement and lateral index guiding. For buried facet devices it also provides the buried facet region.

The effective reflectivity of a buried facet decreases with increasing separation between the facet and the end of the active region. The effective reflectivity R_{eff} of such a facet can be calculated by using a Gaussian beam approximation for the propagating optical mode. It is given by [113]

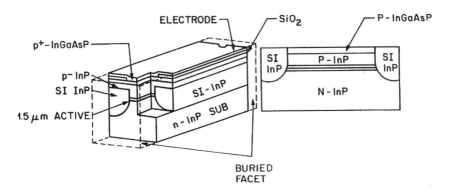

FIGURE 12.40 Schematic of a buried facet optical amplifier.

$$R_{\text{eff}} = \frac{R}{1 + (2S/kw^2)^2} \tag{5}$$

where R is the reflectivity of the cleaved facet; S is the length of the buried facet region; $k \sim 2\pi/\lambda$, where λ is the optical wavelength in the medium; and w is the spot size at the facet. The calculated reflectivity using $w = 0.7$ μm and $R = 0.3$ for an amplifier operating near 1.55 μm is less than 10^{-2} for buried facet lengths larger than about 15 μm.

Although increasing the length of the buried facet region decreases the reflectivity, if the length is too long, the beam emerging from the active region will strike the top metallized surface, producing multiple peaks in the far-field pattern, a feature not desirable for coupling into a single-mode fiber. The beam waist w of a Gaussian beam after traveling a distance z is given by the equation [114]

$$w^2(z) = w_0^2 \left[1 + \left(\frac{\lambda z}{\pi w_0^2} \right)^2 \right] \tag{6}$$

where w_0 is the spot size at the beam waist and λ is the wavelength in the medium. Since the active region is about 4 μm from the top surface of the chip, it follows from the equations above that the length of the buried facet region must be less than 12 μm for single-lobed far-field operation.

The optical gain is determined by injecting light into the amplifier and measuring the output. The internal gain of an amplifier chip as a function of current at two different temperatures is shown in Fig. 12.41. Open circles and squares represent the gain for a linearly polarized incident light with the electric field parallel to the p-n junction in the amplifier chip (TE mode). Solid circles represent the measured

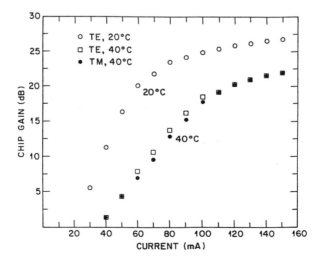

FIGURE 12.41 Measured chip gain as a function of amplifier current. (From Ref. 115.)

gain for the TM mode at 40°C. Measurements were done for low input power (−40 dBm), so that the observed saturation is not due to gain saturation in the amplifier but rather, to carrier loss caused by Auger recombination. Note that the optical gain for the TE and TM input polarizations are nearly equal. Figure 12.42 shows the measured gain as a function of input wavelength for TE-polarized incident light. The modulation in the gain (gain ripple) with a periodicity of 0.7 nm is due to residual facet reflectivities. The measured gain ripple for this device is less than 1 dB. The estimated facet reflectivity from the measured gain ripple of 0.6 dB at 26 dB internal gain is 9×10^{-5}. The 3-dB bandwidth of the optical gain spectrum is 45 nm for this device. It has been shown that the gain ripple and polarization dependence of gain correlate well with the ripple and polarization dependence of the amplified spontaneous emission spectrum. Measurements of amplified spontaneous emission are much simpler to make than gain measurements and provide a good estimate of the amplifier performance [113,115].

Tilted Facet Amplifiers Another way to suppress the resonant modes of the Fabry–Perot cavity is to slant the waveguide (gain region) from the cleaved facet so that the light incident on it internally does not couple back into the waveguide [116]. The process essentially decreases the effective reflectivity of the tilted facet relative to a normally cleaved facet. The reduction in reflectivity as a function of the tilt angle is shown in Fig. 12.43 for the fundamental mode of the waveguide. The schematic of a tilted facet optical amplifier is shown in Fig. 12.44. Waveguiding along the junction plane is weaker in this device than that for the strongly index-guided buried heterostructure device. Weak index guiding for the structure of Fig. 12.44 is provided by a dielectric defined ridge. The fabrication of the device follows a procedure similar to that described previously.

The gain measured as a function of injection current for TM and TE polarized

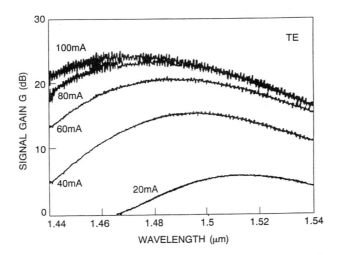

FIGURE 12.42 Measured gain as a function of wavelength for the TE mode.

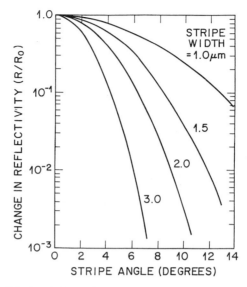

FIGURE 12.43 Calculated change in reflectivity as a function of tilt angle of the facet.

light for a tilted facet amplifier is shown in Fig. 12.45. Optical gains as high as 30 dB have been obtained using titled facet amplifiers. Although the effective reflectivity of the fundamental mode decreases with increasing tilt of the waveguide, the effective reflectivity of the higher-order modes increases. This may cause appearance of higher-order modes at the output (which may reduce fiber-coupled power significantly), especially for large ridge widths.

12.4 OPTICAL MODULATOR

Most commercial fiber-optic transmission systems use a directly modulated laser diode as the source of information. Under direct modulation, the 3-dB spectral width of a single-wavelength DFB laser is typically about 0.1 nm. This finite spectral width results in a dispersion penalty due to chromatic dispersion of the fiber

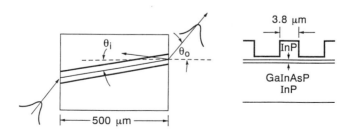

FIGURE 12.44 Schematic of a tilted facet amplifier. (From Ref. 116.)

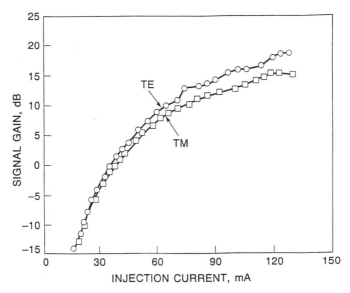

FIGURE 12.45 Measured gain plotted as a function of injection current. (From Ref. 116.)

and limits the transmission distance, particularly at high data rates. This problem can be partially solved if an external modulator is used to modulate the output of a continuously operating laser. For long-distance transmission at high data rates, externally modulated lasers are needed. Such an externally modulated source is a CW DFB laser whose output is coupled to a $LiNbO_3$- or InP-based external optical modulator. The two types of InP-based modulators that have been studied are the Mach–Zehnder (M-Z)-based electrorefraction modulator and the regular waveguide electroabsorption modulator. The M-Z-based modulator is used extensively for $LiNbO_3$ devices, but for InP-based devices the MQW electroabsorption modulator has as good a performance and is simpler to fabricate than M-Z-based modulators.

12.4.1 Waveguide Electroabsorption Modulator

For the electroabsorption modulator, a change in absorption is produced by a change in electric field. The change in absorption at a given wavelength (close to the band gap) occurs due to the Franz–Keldysh effect, according to which the bandgap decreases with increasing electric field. The simplest modulator structure that works quite well has a ridge waveguide design and consists of InGaAsP quantum wells sandwiched between higher-gap InGaAsP barrier layers; 20-Gb/s operation has been demonstrated for such modulators [117]. The waveguide optical modulators are generally polarization sensitive, due both to different optical confinement factors of the TE and TM modes and to polarization dependence of absorption in a MQW structure. Recently, attention has been focused on the fabrication of polarization-independent optical modulators [118,119]. Such modulators have been reported using a strain-compensated structure. In this structure the wells and barriers have opposite strains, and the thicknesses are chosen so that the strain of the total MQW structure is canceled. One such structure is shown in Fig. 12.46. The MQW region consists of ten 12-nm-thick $In_{0.47}Ga_{0.53}As$ quantum wells (tensile strained) and eleven 5-nm-thick $In_{0.60}Al_{0.40}As$ barriers (compressively strained). The addition of compressive strain in the barriers produces strain compensation. The 3-μm-wide mesa structure is produced using reactive ion etching. The modulator length is about 200 μm. The transmission characteristics of the modulator is shown in Fig. 12.47 for both TE and TM polarized incident light. Note that with about 2 to 3 V reverse-bias extinction ratio of >20 dB can be obtained with good polarization independence of transmission. The modulator bandwidth is about 20 GHz.

12.4.2 Mach–Zehnder Modulator

The Mach–Zehnder modulator relies for its operation on the electrooptic effect (i.e., in the presence of an electric field the refractive index is changed). The schematic of a Mach–Zehnder modulator is shown in Fig. 12.48. The incident light (represented by E_0) is split into two parts of equal intensity, and then they recombine at the output. The light traveling through the two arms undergoes different phase changes, depending on the voltage applied, and thus the final output is determined by two interfering signals. By modulating the voltage a differential phase shift of π between the on and off states can be obtained. In principle, Mach–Zehnder modulators can have zero chirp [120] (i.e., the spectral width of the modulated signal would be transform limited). Most InP-based Mach–Zehnder modulators have been fabricated using the ridge waveguide configuration for lateral guiding; 10-Gb/s operation has been demonstrated using these types of modulators [121].

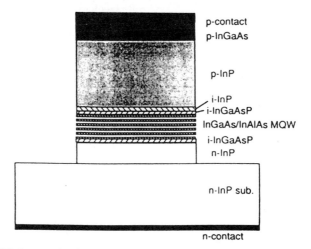

FIGURE 12.46 Strain-compensated InGaAs/InAlAs modulator structure.

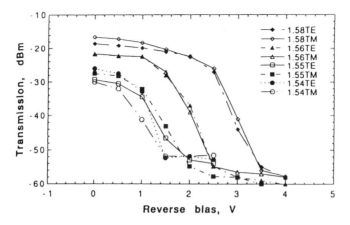

FIGURE 12.47 Transmission of TE and TM modes plotted as a function of reverse bias.

12.5 MONOLITHICALLY INTEGRATED LASERS AND PHOTONIC INTEGRATED CIRCUITS

There has been a significant number of developments in the technology of optical and electronic integration of semiconductor lasers and related devices on the same chip. These chips allow higher levels of functionality than that achieved using single devices. For example, laser and electronic drive circuits have been integrated, serving as simple monolithic transmitters. The name *photonic integrated circuits* (PICs) is generally used when all the integrated components are photonic devices (e.g., lasers, detectors, amplifiers, modulators, and couplers).

12.5.1 Laser Arrays

The simplest of all photonic integrated circuits are one-dimensional arrays of lasers, LEDs, or photodetectors. These devices are fabricated exactly the same way as individual devices except that the wafers are not scribed to make single-device chips but

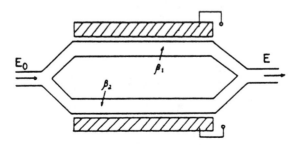

FIGURE 12.48 Schematic of a Mach–Zehnder modulator.

are left in the form of a bar. The main characteristics required for a laser array are low threshold current and good electrical isolation between the individual elements of the array. The schematic of two adjacent devices in a 10-channel low-threshold laser array is shown in Fig. 12.49. These lasers emit near 1.3 μm and are grown by MOCVD on p-InP substrate. The lasers have a multiquantum-well active region with five 7-nm-thick wells, as shown in the insert of Fig. 12.49. The light versus current characteristics of all the lasers in a 10-element array is shown in Fig. 12.50. The average threshold current and quantum efficiency are 3.2 mA and 0.27 W/A, respectively. The cavity length was 200 μm, and the facets of the lasers were coated with dielectric to produce 65% and 90% reflectivity, respectively [122].

The vertical cavity surface-emitting laser (VCSEL) design is more suitable than the edge-emitting laser design for the fabrication of two-dimensional arrays. Several researchers have reported two-dimensional arrays of VCSELs. Among the individual laser design used are the proton-implanted design and the oxide-confined design. These VCSELs and VCSEL arrays have been fabricated so far using the GaAs/AlGaAs material system for an emission near 0.85 μm.

12.5.2 Integrated Laser Detector

Considerable attention has been paid to developing integrated laser-detector structures. In many cases, the purpose of the detector is to serve as a monitor for the output power of the laser. The output of the monitor can then be used to stabilize the power of the laser if it drifts due to aging or temperature change, using a feedback circuit. A typical integrated laser and detector circuit is shown in Fig. 12.51. The emitting region of the laser and the absorbing region of the photodiode are composed of the same material. The detector monitors the back facet output of the laser. In this structure the laser has one cleaved and one etched facet.

12.5.3 Integrated Laser Modulator

The two types of integrated laser modulator structures that have been investigated are the integrated electroabsorption-modulated laser (EML) and the integrated electrorefraction-modulated laser. The electrorefraction property is used in a Mach–Zehnder configuration to fabricate a low-chirp-modulated light source. For some applications it is desirable to have the laser and the modulator integrated on

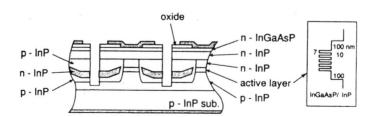

FIGURE 12.49 Schematic of two adjacent devices in a laser array. (From Ref. 122.)

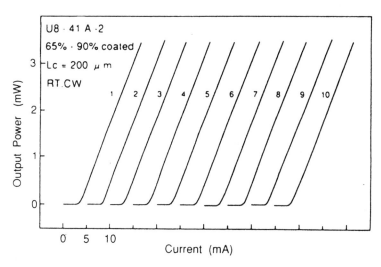

FIGURE 12.50 Light versus current characteristics of all the lasers in a 10-element array.

the same chip. Such devices, known as electroabsorption-modulated lasers (EMLs), are used for high-data-rate transmission systems with large regenerator spacing. The schematic of an EML is shown in Fig. 12.52. In this device, the light from the DFB laser is coupled directly to the modulator. The modulator region has a slightly higher bandgap than that of the laser region, which results in very low absorption of the laser light in the absence of bias. However with reverse bias, the effective bandgap decreases, which results in reduced transmission through the modulator. For very high speed operation, the modulator region capacitance must be sufficient-

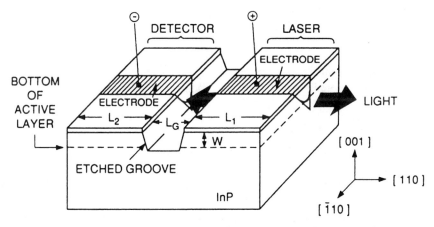

FIGURE 12.51 Schematic of a integrated laser and monitoring photodiode. (From Ref. 123.)

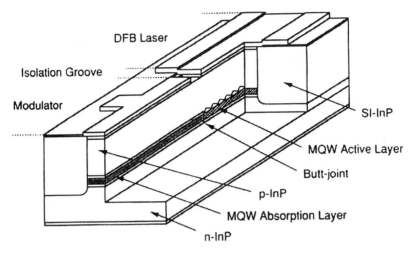

FIGURE 12.52 Schematic of a electroabsorption-modulated laser structure. (From Ref. 124.)

ly small, which makes the modulator length small, resulting in a low on/off ratio. Very high speed EMLs have been reported using a growth technique where the laser and modulator active regions are fabricated using two separate growths, thus allowing independent optimization of the modulator bandgap and length for a high on/off ratio and speed; 40-Gb/s operation has been demonstrated using this device [124].

EML devices have also been fabricated using the selective-area epitaxy growth process [125]. In this process the laser and the modulator active region are grown simultaneously over a patterned substrate. The patterning allows the materials grown to have slightly different bandgaps, resulting in separate laser and modulator regions. EMLs have been fabricated with bandwidths of 15 GHz and have operated error free over 600 km at a 2.5-Gb/s data rate. An integrated laser Mach–Zehnder device is shown in Fig. 12.53. This device has a ridge waveguide DFB laser integrated with a Mach–Zehnder modulator, which also has a lateral guiding provided by a ridge structure. The Mach–Zehnder traveling-wave phase modulator is designed so that the microwave and optical velocities in the structure are identical. This allows good coupling of the electrical and optical signal.

12.5.4 Multichannel WDM Sources

An alternative to single-channel very high speed (> 20 Gb/s) data transmission for increasing transmission capacity is multichannel transmission using wavelength division multiplexing (WDM) technology. In WDM system many (4, 8, 16, or 32) wavelengths carrying data are optically multiplexed and simultaneously transmitted through a single fiber. The received signal with many wavelengths are optically demultiplexed into separate channels, which are then processed electronically in a conventional form. Such a WDM system needs transmitters with many lasers at

12.5 MONOLITHICALLY INTEGRATED LASERS

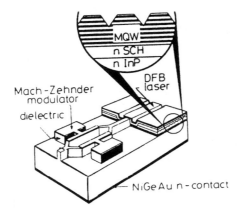

FIGURE 12.53 Schematic of an integrated gain coupled DFB laser and Mach–Zehnder modulator.

specific wavelengths. It is desirable to have all of these laser sources on a single chip for compactness and ease of fabrication, like electronic integrated circuits.

Figure 12.54 shows the schematic of a photonic integrated circuit with multiple lasers for a WDM source. This chip has eight individually addressable DFB lasers, the output of which are combined using a waveguide-based multiplexer. Since the waveguide multiplexer has an optical loss of about 8 dB, the output of the chip is

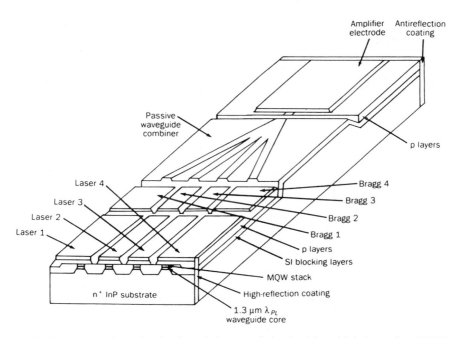

FIGURE 12.54 Schematic of a photonic integrated circuit with multiple lasers for a WDM source. (From Ref. 126.)

amplified further using a semiconductor amplifier. The laser output in the waveguide is TE polarized, and hence an amplifier with a multiquantum-well absorption region that has a high saturation power is integrated in this chip.

12.5.5 Spot-Size Converter Integrated Laser

A typical laser diode has too wide (ca. 30° × 40°) an output beam pattern for good mode matching to a single-mode fiber. This results in a loss of power coupled to the fiber. Thus a laser whose output spot size is expanded to match an optical fiber is an attractive device for low-loss coupling to the fiber without a lens and for wide alignment tolerances. Several researchers have reported such devices [127,128]. Generally, they involve producing a vertically and laterally tapered waveguide near the output facet of the laser. The tapering needs to be done in an adiabatic fashion so as to reduce the scattering losses. The schematic of a spot-size converter (SSC) laser is shown in Fig. 12.55. The laser is fabricated using two MOCVD growth steps. The SSC section is about 200 μm long. The waveguide thickness is narrowed along the cavity from 300 nm in the active region to about 100 nm in the region over the length of the SSC section. The laser emits near 1.3 μm, has a multiquantum-well active region, and a laser section length of 300 μm. The light versus current characteristics of a SSC laser [128] at various temperatures is shown in Fig. 12.56. A beam divergence of 13° was obtained for this device. Beam divergences of 9° and 10° in the horizontal and vertical directions have been reported for similar SSC devices [128].

12.5.6 Heterodyne Receiver

One of the complicated PICs reported so far is the heterodyne receiver for coherent transmission [129,130]. The essential elements of such a receiver are the local oscillator laser, the optics for mixing local oscillator light with the incident signal, and the balanced receiver. The receiver PIC is shown in Fig. 12.57. The device has a tun-

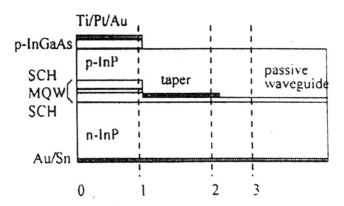

FIGURE 12.55 Schematic of a SSC laser. (From Ref. 127.)

12.5 MONOLITHICALLY INTEGRATED LASERS   493

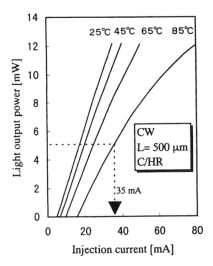

FIGURE 12.56 Light versus current characteristics of a SSC laser at different temperatures. (From Ref. 128.)

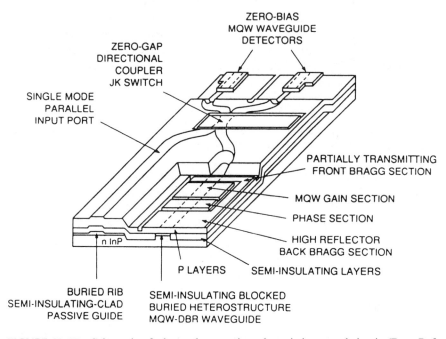

FIGURE 12.57 Schematic of a heterodyne receiver photonic integrated circuit. (From Ref. 130.)

able DBR laser that serves as the local oscillator, several branching waveguides for mixing and splitting the signal, and two photodiodes that serve as balanced receivers. Details of a variety of PICs so far developed can be found in the literature devoted exclusively to this topic [131,132].

12.6 SUMMARY AND FUTURE CHALLENGES

Tremendous advances in InP-based semiconductor lasers have occurred over the last decade. The advances in research and many technological innovations have led to the worldwide deployment of fiber-optic communication systems that operate near 1.3 and 1.55-μm wavelengths. Although most of these systems are based on digital transmission, lasers have also been deployed for carrying high-quality analog cable TV transmission systems. However, many challenges remain.

The need for higher capacity is pushing the deployment of WDM-based transmission which needs tunable or frequency-settable lasers. An important research area will continue to be the development of lasers with very stable and settable frequency. Integration of many such lasers on a single substrate would provide the ideal source for WDM systems. The laser-to-fiber coupling is also an important area of research. Recent development in spot-size converter integrated lasers are quite impressive, but more work perhaps needs to be done to make them easy to manufacture. This may require more process developments.

Lasers with better high-temperature performance is an important area of investigation. The InGaAsN-based system is a promising candidate for making lasers that emit near 1.3 μm with good high-temperature performance. New materials investigation is important. Although WDM technology is currently being considered for increasing the transmission capacity, the need for sources with very high modulation capability will remain. Hence research on new mechanisms for very high speed modulation is important. The surface-emitting laser is very attractive for two-dimensional arrays and for single-wavelength operation. Several important advances in this technology have occurred over the last few years. An important challenge is the fabrication of a device with characteristics superior to that of an edge emitter.

Finally, many of the advances in laser development would not have been possible without the advances in materials and processing technology. The challenges of much of the current laser research is intimately linked with the challenges in materials growth, which include not only the investigation of new material systems but also improvements in existing technologies to make them more reproducible and predictable.

REFERENCES

1. H. Kressel and J. K. Butler, *Semiconductor Lasers and Heterojunction LEDs,* Academic Press, San Diego, Calif., 1977.

REFERENCES   **495**

2. H. C. Casey, Jr. and M. B. Panish, *Heterostructure Lasers,* Academic Press, San Diego, Calif., 1978.
3. G. H. B. Thompson, *Physics of Semiconductor Lasers,* Wiley, New York, 1980.
4. G. P. Agrawal and N. K. Dutta, *Long Wavelength Semiconductor Lasers,* 2nd ed., Van Nostrand Reinhold, New York, 1993.
5. N. Holonyuk, Jr. and S. F. Bevacqua, *Appl. Phys. Lett.,* **1,** 82 (1962).
6. M. I. Nathan, W. P. Dumke, G. Burns, F. H. Dill, Jr., and G. Lasher, *Appl. Phys. Lett.,* **1,** 63 (1962).
7. T. M. Quist, R. H. Retiker, R. J. Keyes, W. E. Krag, B. Lax, A. L. McWhorter and H. J. Ziegler, *Appl. Phys. Lett.,* **1,** 91 (1962); R. N. Hall, G. E. Fenner, J. D. Kingsley, T. J. Soltys, and R. O. Carlson, *Phys. Rev. Lett.,* **9,** (1962).
8. I. Hayashi, M. B. Panish, P. W. Foy, and S. Sumski, *Appl. Phys. Lett.,* **47,** 109 (1970).
9. J. J. Hsieh, *Appl. Phys. Lett.,* **28,** 283 (1976).
10. G. P. Agrawal and N. K. Dutta, *Long Wavelength Semiconductor Lasers,* 2nd ed., Van Nostrand Reinhold, New York, 1993, Chap.
11. R. E. Nahory, M. A. Pollack, W. D. Johnston, Jr., and R. L. Barns, *Appl. Phys. Lett.,* **33,** 659 (1978).
12. G. P. Agrawal and N. K. Dutta, *Long Wavelength Semiconductor Lasers,* 2nd ed., Van Nostrand Reinhold, New York, 1993, Chap. 5.
13. N. Holonyak, Jr., R. M. Kolbas, R. D. Dupuis, and P. D. Dapkus, *IEEE J. Quantum Electron.,* **16,** 170 (1980).
14. N. Holonyak, Jr., R. M. Kolbas, W. D. Laidig, B. A. Vojak, K. Hess, R. D. Dupuis, and P. D. Dapkus, *J. Appl. Phys.,* **51,** 1328 (1980).
15. W. T. Tsang, *Appl. Phys. Lett.,* **39,** 786 (1981).
16. W. T. Tsang, *IEEE J. Quantum Electron.,* **20,** 1119 (1986).
17. S. D. Hersee, B. DeCremoux, and J. P. Duchemin, *Appl. Phys. Lett.,* **44,** 476 (1984).
18. F. Capasso and G. Margaritondo, eds., *Heterojunction Band Discontinuation: Physics and Applications,* North-Holland, Amsterdam, 1987.
19. N. K. Dutta, S. G. Napholtz, R. Yen, R. L. Brown, T. M. Shen, N. A. Olsson, and D. C. Craft, *Appl. Phys. Lett.,* **46,** 19 (1985).
20. N. K. Dutta, S. G. Napholtz, R. Yen, T. Wessel, and N. A. Olsson, *Appl. Phys. Lett.,* **46,** 1036 (1985).
21. N. K. Dutta, T. Wessel, N. A. Olsson, R. A. Logan, R. Yen, and P. J. Anthony, *Electron. Lett.,* **21,** 571 (1985).
22. N. K. Dutta, T. Wessel, N. A. Olsson, R. A. Logan, and R. Yen, *Appl. Phys. Lett.,* **46,** 525 (1985).
23. Y. Arakawa and A. Yariv, *IEEE J. Quantum Electron.,* **21,** 1666 (1985).
24. Y. Arakawa and A. Yariv, *IEEE J. Quantum Electron.,* **22,** 1887 (1986).
25. A. Yariv, C. Lindsey, and V. Sivan, *J. Appl. Phys.,* **58,** 3669 (1985).
26. A. Sugimura, *IEEE J. Quantum Electron.,* **20,** 336 (1984).
27. A. Sugimura, *Appl. Phys. Lett.,* **43,** 728 (1983).
28. N. K. Dutta and R. J. Nelson, *J. Appl. Phys.,* **53,** 74 (1982).
29. L. C. Chiu and A. Yariv, *IEEE J. Quantum Electron.,* **18,** 1406 (1982).
30. N. K. Dutta, *J. Appl. Phys.,* **54,** 1236 (1983).

31. A. Sugimura, *IEEE J. Quantum Electron.*, **19,** 923 (1983).
32. C. Smith, R. A. Abram, and M. G. Burt, *J. Phys. C,* **16,** L171 (1983).
33. H. Temkin, N. K. Dutta, T. Tanbun-Ek, R. A. Logan, and A. M. Sergent, *Appl. Phys. Lett.,* **57,** 1610 (1990).
34. C. Kazmierski, A. Ougazzaden, M. Blez, D. Robien, J. Landreau, B. Sermage, J. C. Bouley and A. Mirca, *IEEE J. Quantum Electron.,* **27,** 1794–1797 (1991).
35. P. Morton, R. A. Logan, T. Tanbun-Ek, P. F. Sciortino, Jr., A. M. Sergent, R. K. Montgomery, and B. T. Lee, *Electron. Lett.*
36. P. J. A. Thijs, L. F. Tiemeijer, P. I. Kuindersma, J. J. M. Binsma, and T. van Dongen, *IEEE J. Quantum Electron.,* **27,** 1426 (1991).
37. P. J. A. Thijs, L. F. Tiemeijer, J. J. M. Binsma, and T. van Dongen *IEEE J. Quantum Electron.,* **QE-30,** 477–499 (1994).
38. H. Temkin, T. Tanbun-Ek, and R. A. Logan, *Appl. Phys. Lett.,* **56,** 1210 (1990).
39. W. T. Tsang, L. Yang, M. C. Wu, Y. K. Chen, and A. M. Sergent, *Electron. Lett.,* **2033,** (1990).
40. W. D. Laidig, Y. F. Lin, and P. J. Caldwell, *J. Appl. Phys.,* **57,** 33 (1985).
41. S. E. Fischer, D. Fekete, G. B. Feak, and J. M. Ballantyne, *Appl. Phys. Lett.,* **50,** 714 (1987).
42. N. Yokouchi, N. Yamanaka, N. Iwai, Y. Nakahira, and A. Kasukawa, *IEEE J. Quantum Electron.,* **QE-32,** 2148–2155 (1996).
43. K. J. Beernik, P. K. York, and J. J. Coleman, *Appl. Phys. Lett.,* **25,** 2582 (1989).
44. J. P. Loehr and J. Singh, *IEEE J. Quantum Electron.,* **27,** 708 (1991).
45. S. W. Corzine, R. Yan, and L. A. Coldren, "Optical gain in III–V bulk and quantum well semiconductors," in *Quantum Well Lasers,* ed. P. Zory, Academic Press, San Diego, Calif., to be published.
46. H. Temkin, T. Tanbun-Ek, R. A. Logan, D. A. Coblentz, and A. M. Sergent, *IEEE Photon. Technol. Lett.,* **3,** 100 (1991).
47. A. R. Adams, *Electron. Lett.,* **22,** 249 (1986).
48. E. Yablonovitch and E.O. Kane, *J. Lightwave Technol.,* **4,** 50 (1986).
49. M. C. Wu, Y. K. Chen, M. Hong, J. P. Mannaerts, M. A. Chin, and A. M. Sergent, *Appl. Phys. Lett.,* **59,** 1046 (1991).
50. N. K. Dutta, J. Lopata, P. R. Berger, D. L. Sivco, and A. Y. Cho, *Electron. Lett.,* **27,** 680 (1991).
51. J. M. Kuo, M. C. Wu, Y. K. Chen, and M. A. Chin, *Appl. Phys. Lett.,* **59,** 2781 (1991).
52. N. Chand, E. E. Becker, J. P. van der Ziel, S. N. G. Chu, and N. K. Dutta, *Appl. Phys. Lett.,* **58,** 1704 (1991).
53. H. K. Choi and C. A. Wang, *Appl. Phys. Lett.,* **57,** 321 (1990).
54. N. K. Dutta, J. D. Wynn, J. Lopata, D. L. Sivco, and A. Y. Cho, *Electron. Lett.,* **26,** 1816 (1990).
55. C. E. Zah, R. Bhat, B. N. Pathak, F. Favire, W. Lin, N. C. Andreadakis, D. M. Hwang, T. P. Lee, Z. Wang, D. Darby, D. Flanders, and J. J. Hsieh, *IEEE J. Quantum Electron.,* **30,** 511–523 (1994).
56. A. Kusukawa, T. Namegaya, T. Fukushima, N. Iwai, and T. Kikuta, *IEEE J. Quantum Electron.,* **29,** 1528–1535 (1993).

57. M. Kondow, T. Kitatani, S. Nakatsuka, Y. Yazawa, and M. Okai, *Proc. OECC'97*, July 1997, Seoul, Korea, pp. 168–169.
58. G. P. Agrawal and N. K. Dutta, *Long Wavelength Semiconductor Lasers,* 2nd ed., Van Nostrand Reinhold, New York, 1993, Chap. 4.
59. T. L. Koch, U. Koren, R. P. Gnall, C. A. Burrus, and B. I. Miller, *Electron. Lett.,* **24,** 1431 (1988).
60. Y. Suematsu, S. Arai, and K. Kishino, *J. Lightwave Technol.,* **1,** 161 (1983).
61. T. L. Koch and U. Koren, *IEEE J. Quantum Electron.,* **27,** 641 (1991).
62. N. K. Dutta, A. B. Piccirilli, T. Cella, and R. L. Brown, *Appl. Phys. Lett.,* **48,** 1501 (1986).
63. K. Y. Liou, N. K. Dutta, and C. A. Burrus, *Appl. Phys. Lett.,* **50,** 489 (1987).
64. T. Tanbun-Ek, R. A. Logan, S. N. G. Chu, and A. M. Sergent, *Appl. Phys. Lett.,* **57,** 2184 (1990).
65. H. Soda, K. Iga, C. Kitahara, and Y. Suematsu, *Jpn. J. Appl. Phys.,* **18,** 2329 (1979).
66. K. Iga, F. Koyama, and S. Kinoshita, *IEEE J. Quantum Electron.,* **24,** 1845 (1988).
67. J. L. Jewell, J. P. Harbison, A. Scherer, Y. H. Lee, and L. T. Florez, *IEEE J. Quantum Electron.,* **27,** 1332 (1991).
68. C. J. Chang-Hasnain, M. W. Maeda, N. G. Stoffel, J. P. Harbison, and L. T. Florez, *Electron. Lett.,* **26,** 940 (1990).
69. R. S. Geels, S. W. Corzine, and L. A. Coldren, *IEEE J. Quantum Electron.,* **27,** 1359 (1991).
70. R. S. Geels and L. A. Coldren, *Appl. Phys. Lett.,* **5,** 1605 (1990).
71. K. Tai, G. Hasnain, J. D. Wynn, R. J. Fischer, Y. H. Wang, B. Weir, J. Gamelin, and A. Y. Cho, *Electron. Lett.,* **26,** 1628 (1990).
72. K. Tai, L. Yang, Y. H. Wang, J. D. Wynn, and A. Y. Cho, *Appl. Phy. Lett.,* **56,** 2496 (1990).
73. M. Born and E. Wolf, *Principles of Optics,* Pergamon Press, Elmsford, N.Y., 1977, Sec. 1.6.5, p. 69.
74. J. L. Jewell, A. Scherer, S. L. McCall, Y. H. Lee, S. J. Walker, J. P. Harbison, and L. T. Florez, *Electron. Lett.,* **25,** 1123 (1989).
75. Y. H. Lee, B. Tell, K. F. Brown-Goebeler, J. L. Jewell, R. E. Leibenguth, M. T. Asom, G. Livescu, L. Luther, and V. D. Mattera, *Electron. Lett.,* **26,** 1308 (1990).
76. K. Tai, R. J. Fischer, C. W. Seabury, N. A. Olsson, D. T. C. Huo, Y. Ota, and A. Y. Cho, *Appl. Phys. Lett.,* **55,** 2473 (1989).
77. A. Ibaraki, K. Kawashima, K. Furusawa, T. Ishikawa, T. Yamayachi, and T. Niina, *Jpn. J. Appl. Phys.,* **28,** L667 (1989).
78. E. F. Schubert, L. W. Tu, R. F. Kopf, G. J. Zydzik, and D. G. Deppe, *Appl. Phys. Lett.,* **57,** 117 (1990).
79. R. S. Geels, S. W. Corzine, J. W. Scott, D. B. Young, and L. A. Coldren, *IEEE Photon. Tech. Lett.,* **2,** 234 (1990).
80. G. Hasnain, K. Tai, L. Yang, Y. H. Wang, R. J. Fischer, J. D. Wynn, B. E. Weir, N. K. Dutta, and A. Y. Cho, *IEEE J. Quantum Electron.,* **27,** 1377 (1991).
81. P. L. Gourley, S. K. Lyo, T. M. Brennan, B. E. Hammons, C. P. Schaus, and S. Sun, *Appl. Phys. Lett.,* **51,** 2698 (1989).

82. N. K. Dutta, L. W. Tu, G. J. Zydzik, G. Hasnain, Y. H. Wang, and A. Y. Cho, *Electron. Lett.*, **27**, 208 (1991).
83. H. Soda, K. Iga, C. Kitahara, and Y. Suematsu, *Jpn. J. Appl. Phys.*, **18**, 2329 (1979).
84. K. Iga, F. Koyama, and S. Kinoshita, *IEEE J. Quantum Electron.*, **24**, 1845 (1988).
85. K. Tai, F. S. Choa, W. T. Tsang, S. N. G. Chu, J. D. Wynn, and A. M. Sergent, *Electron. Lett.*, **27**, 1514 (1991).
86. L. Yang, M. C. Wu, K. Tai, T. Tanbun-Ek, and R. A. Logan, *Appl. Phys. Lett.*, **56**, 889 (1990).
87. T. Baba, Y. Yogo, K. Suzuki, F. Koyama, and K. Iga *Electron. Lett.*, **29**, 913 (1993).
88. Y. Imajo, A. Kasukawa, S. Kashiwa, and H. Okamoto, *Jpn. J. Appl. Phys. Lett.*, **29**, p L1130–1132 (1990).
89. D. I. Babic, K. Streubel, R. Mirin, N. M. Margalit, J. E. Bowers, E. L. Hu, D. E. Mars, L. Yang, and K. Carey, *IEEE Photon. Tech. Lett.*, **7**, 1225 (1995).
90. Z. L. Liau and D. E. Mull, *Appl. Phys. Lett.*, **56**, 737 (1990).
91. F. R. Nash, W. J. Sundburg, R. L. Hartman, J. R. Pawlik, D. A. Ackerman, N. K. Dutta, and R. W. Dixon, *AT&T Tech. J.*, **64**, 809 (1985).
92. The reliability requirements of a submarine lightwave transmission system are discussed in a special issue of *AT&T Tech. J.*, **64**, 3 (1985).
93. B. C. DeLoach, Jr., B. W. Hakki, R. L. Hartman, and L. A. D'Asaro, *Proc. IEEE,* **61**, 1042 (1973).
94. P. M. Petroff and R. L. Hartman, *Appl. Phys. Lett.*, **2**, 469 (1973).
95. W. D. Johnston and B. I. Miller, *Appl. Phys. Lett.*, **23**, 1972 (1973).
96. P. M. Petroff, W. D. Johnston, Jr., and R. L. Hartman, *Appl. Phys. Lett.*, **25**, 226 (1974).
97. J. Matsui, R. Ishida, and Y. Nannichi, *Jpn. J. Appl. Phys.*, **14**, 1555 (1975).
98. P. M. Petroff and D. V. Lang, *Appl. Phys. Lett.*, **31**, 60 (1977).
99. O. Ueda, I. Umebu, S. Yamakoshi, and T. Kotani, *J. Appl. Phys.*, **53**, 2991 (1982).
100. O. Ueda, S. Yamakoshi, S. Komiya, K. Akita, and T. Yamaoka, *Appl. Phys. Lett.*, **36**, 300 (1980).
101. S. Yamakoshi, M. Abe, O. Wada, S. Komiya, and T. Sakurai, *IEEE J. Quantum Electron.*, **17**, 167 (1981).
102. K. Mizuishi, M. Sawai, S. Todoroki, S. Tsuji, M. Hirao, and M. Nakamura, *IEEE J. Quantum Electron.*, **19**, 1294 (1983).
103. E. I. Gordon, F. R. Nash, and R. L. Hartman, *IEEE Electron Device Lett.*, **4**, 465 (1983).
104. R. L. Hartman and R. W. Dixon, *Appl. Phys. Lett.*, **26**, 239 (1975).
105. W. B. Joyce, K. Y. Liou, F. R. Nash, P. R. Bossard, and R. L. Hartman, *AT&T Tech. J.*, **64**, 717 (1985).
106. M. J. Coupland, K. G. Mambleton, and C. Hilsum, *Phys. Lett.*, **7**, 231 (1963).
107. J. W. Crowe and R. M. Graig, Jr., *Appl. Phys. Lett.*, **4**, 57 (1964).
108. W. F. Kosnocky and R. H. Cornely, *IEEE J. Quantum Electron.*, **4**, 225 (1968).
109. T. Saitoh and T. Mukai, in *Coherence, Amplification and Quantum Effects in Semiconductor Lasers*, ed. by Y. Yamamoto, Wiley, New York, 1991, Chap. 7.
110. M. Nakamura and S. Tsuji, *IEEE J. Quantum Electron.*, **17**, 994 (1981).
111. T. Saitoh and T. Mukai, *J. Lightwave Technol.*, **6**, 1656 (1988).

112. M. O'Mahony, *J. Lightwave Technol.,* **5,** 531 (1988).
113. N. K. Dutta, M. S. Lin, A. B. Piccirilli, and R. L. Brown, *J. Appl. Phys.,* **67,** 3943 (1990).
114. H. Kogelnik and T. Li, *Proc. IEEE,* **54,** 1312 (1966).
115. M. S. Lin, A. B. Piccirilli, Y. Twu, and N. K. Dutta, *Electron. Lett.,* **25,** 1378 (1989).
116. C. E. Zah, J. S. Osinski, C. Caneau, S. G. Menocal, L. A. Reith, J. Salzman, F. K. Shokoohi, and T. P. Lee, *Electron. Lett.,* **23,** 990 (1987).
117. F. Devaux, F. Dorgeuille, A. Ougazzaden, F. Huet, M. Carre, A. Carenco, M. Henry, Y. Sorel, J. F. Kerdiles, and E. Jeanney, *IEEE Photon. Tech Lett.,* **5,** 1288 (1993).
118. I. Kotaka, K. Wakita, K. Kawano, H. Asai, and M. Naganuma *Electron. Lett.,* **27,** 2162 (1991).
119. K. Wakita, I. Kotaka, K. Yoshino, S. Kondo, and Y. Noguchi, *IEEE Photon. Tech Lett.,* **7,** 1418 (1995).
120. F. Koyama and K. Iga, *J. Lightwave Technol.,* **6,** 87–93 (1988)
121. J. C. Cartledge, H. Debregeas, and C. Rolland, *IEEE Photon. Tech. Lett.,* **7,** 224–226 (1995).
122. S. Yamashita, A. Oka, T. Kawano, T. Tsuchiya, K. Saitoh, K. Uomi, and Y. Ono *IEEE Photon. Tech. Lett.,* **4,** 954–957 (1992).
123. L. A. Koszi, A. K. Hin, B. P. Segner, T. M. Shen, and N. K. Dutta *Electron. Lett.,* **21,** 1209 (1985).
124. H. Takeuchi, K. Tsuzuki, K. Sato, M. Yamamoto, Y. Itaya, A. Sano, M. Yoneyama, and T. Otsuji *IEEE Photon. Tech. Lett.,* **9,** 572–574 (1997).
125. M. Aoki, M. Takashi, M. Suzuki, H. Sano, K. Uomi, T. Kawano, and A. Takai, *IEEE Photon. Tech Lett.,* **4,** 580 (1992).
126. T. L. Koch and U. Koren, *AT&T Tech. Journal,* **70,** 63 (1992).
127. R. Y. Fang, D. Bertone, M. Meliga, I. Montrosset, G. Oliveti, and R. Paoletti, *IEEE Photon. Tech. Lett.,* **9,** 1084–1086 (1997).
128. H. Yamazaki, K. Kudo, T. Sasaki, and M. Yamaguchi, *Proc. OECC'97* Seoul, Korea, paper 10C1-3, 440–441.
129. H. Takenchi, K. Kasaya, Y. Kondo, H. Yasaka, K. Oe, and Y. Imamura, *IEEE Photon. Tech. Lett.,* **1,** 398 (1990).
130. T. L. Koch, U. Koren, R. P. Gnall, F. S. Choa, F. Hernandez-Gil, C. A. Burrus, M. G. Young, M. Oron, and B. I. Miller, *Electron. Lett.,* **25,** 1621 (1989).
131. M. Dagenais, ed., *Integrated Optoelectronics,* John Wiley, 1995.
132. O. Wada, ed., *Optoelectronic Integration: Physics Technology and Applications,* Kluwer Academic, 1994.

CHAPTER THIRTEEN

Photodiodes and Receivers Based on InP Materials

KENKO TAGUCHI
NEC Corporation

13.1	Introduction	502
13.2	Basic Photodiode Design and Requirements for Use in Optical Fiber Communications	503
	13.2.1 Basic Photodiode Operation	503
	13.2.2 Receiver Sensitivity	504
	13.2.3 Response Speed	506
13.3	Photodiodes	508
	13.3.1 Basic InGaAs PIN Photodiodes	508
	13.3.2 Waveguide PIN Photodiodes	513
	13.3.3 Other Types of Photodiodes	516
13.4	Avalanche Photodiodes	517
	13.4.1 InGaAs/InP Avalanche Photodiodes	517
	13.4.2 Superlattice Avalanche Photodiodes	521
13.5	Integrated Photoreceivers	526
	13.5.1 Receiver OEICs	526
	13.5.2 Photonic Integrated Circuits Including Photodiodes	527

InP-Based Materials and Devices: Physics and Technology, Edited by Osamu Wada and Hideki Hasegawa.
ISBN 0-471-18191-9   © 1999 John Wiley & Sons, Inc.

13.6 Summary and Future Challenges   530
References   531

13.1 INTRODUCTION

Semiconductor photodetectors based on InP materials are the ones most often used in state-of-the-art long-wavelength optical fiber communication systems. Mixed compounds such as InGaAs(P) and In(Al)GaAs lattice matched to InP are the materials responsible for detecting long-wavelength light, especially the nondispersion wavelength (1.3 μm) and loss-minimum wavelength (1.55 μm) of silica optical fibers. The characteristics of these InP-based photodetectors are superior to those of conventional photodiodes composed of elemental Ge, which was the only material applicable for wavelengths below 1.5 μm. By using a heterostructure, which had not been expected in group IV elemental semiconductors such as Si and Ge, new concepts and new device designs for high-performance photodetectors have been developed. For example, the absorption region can be confined to a limited layer, and the InP wide-bandgap layer can serve as a transparent layer for specific communication wavelengths. Recently, InGaAs/InP avalanche photodiodes (APDs) with a SAM (separation of absorption and multiplication) configuration have become commercially available. The SAM configuration is thought to be necessary for high-performance APDs utilizing long wavelengths.

Because photodiodes operate under reverse bias, high-quality semiconductor layers need to be produced. To obtain photodiodes that operate at a low bias and have a low dark current, it is necessary to produce epitaxial layers that are pure and that have few defects (such as dislocations, point defects, and impurity precipitates). To get stable and uniform gain in APDs, in which internal gain is achieved through the carrier avalanche process, the layers in the avalanche region must be uniform and free of dislocations. Furthermore, a planar device structure requires that a guard ring be used to keep the electric field around the photoreceptive area from increasing too much. Fabrication and processing technologies such as impurity diffusion, ion implantation, and passivation will also play important roles in the production of reliable photodetectors.

This chapter mainly deals with three types of photodetectors: PIN photodiodes (PDs) with no internal gain, avalanche photodiodes (APDs), and integrated devices incorporating PDs. In the next section we briefly discuss the device parameters needed to obtain a large signal-to-noise (S/N) ratio in receiver circuits, and then the concept, design, and performance of each of the three types of photodetectors are discussed. In each section, mature basic devices such as InGaAs PIN-PDs and InGaAs/InP heterostructure APDs are first overviewed. Then topical devices, which are expected to be important in the future, are introduced: Two examples of such topical devices are highly efficient high-speed waveguide PIN-PDs and superlattice APDs in which the impact ionization rate is increased by exploiting the band-edge discontinuity in multiple-quantum-well band structures. Integrated devices including photodiodes are also introduced. Finally, a summary is provided and future prospects are briefly described.

13.2 BASIC PHOTODIODE DESIGN AND REQUIREMENTS FOR USE IN OPTICAL FIBER COMMUNICATIONS

The photodiodes used in receiver circuits are required to translate optical signals into electrical signals faithfully and efficiently. Therefore, in this section, basic photodiode operation is introduced, and then photodiode design and the device parameters required are discussed briefly on the basis of a simplified theoretical analysis of receiver sensitivity and photodiode response speed.

13.2.1 Basic Photodiode Operation

Photodiodes operate under reverse bias to create a depleted region in which photogenerated electron–hole pairs are separated. Figure 13.1a shows a schematic

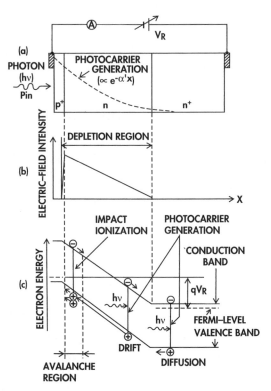

FIGURE 13.1 Basic operation of photodiode: (a) cross-sectional sketch of p^+-n-n^+ diode under reverse bias; (b) electric field distribution; (c) energy band diagram.

cross section of a photodiode (or avalanche photodiode) with a p$^+$-n-n$^+$ structure. It also shows the optical absorption (or photocarriers generation), which is subject to the absorption coefficient α' of the material for the incident light and decreases exponentially with increasing distance from the diode front p$^+$. Figures 13.1b and c show the electric field distribution and the energy band diagram, respectively. Most photocarriers are designed for use in the fully depleted n-region, so that they have a high-speed response: Electrons and holes generated within the depletion region are instantaneously separated by the electric field and drift in the opposite direction, inducing a photocurrent in the external circuit. On the other hand, minority-carrier holes excited within an average diffusion length in the undepleted n$^+$(or n) region adjacent to the depleted region diffuse into the edge of the depleted junction with some recombination and are collected across the high-field region, resulting in a diffusion photocurrent in the external circuit. Diffusion photocurrent is generally characterized by its slow response to the optical signal, since the speed of the response depends on the time it takes for the photogenerated minority carriers to diffuse from where they are generated in the neutral undepleted region into the edge of the depletion region. The photodiode should therefore be designed in such a way that there is no optical absorption in the undepleted neutral region. For the same reason, as well as to reduce the recombination loss of photocarriers generated in the p$^+$-region on the front side of the diode, the p$^+$-region must be as thin as possible.

When the electric field of the diode is elevated to several hundreds of kilovolts per centimeter by increasing reverse bias, an internal gain for a primary photocurrent can be obtained. This gain is a result of the electron–hole pair creation (avalanche process) initiated by the photogenerated carriers, which creation is in turn governed by the relation between the strength of the electric field and the electron and hole impact ionization rates of the material itself.

13.2.2 Receiver Sensitivity

Receiver sensitivity has been analyzed numerically for a variety of signal waveforms by Personick [1] and Smith and Personik [2], but the present discussion of sinusoidal optical signal detection with an APD is simplified by considering a receiver circuit combined with a load resistance R_L followed by a preamplifier with an equivalent input noise F_{amp}. In the case of a sinusoidal signal P_{in} with full modulation depth, the mean-square signal current is described as

$$<i_p^2> = \frac{(I_p M)^2}{2} \qquad (1)$$

where I_p is the primary photocurrent transferred from the optical signal to the electrical signal and M is the multiplication factor of APD. A photodiode with no internal gain is one for which $M = 1$. The relation between the input optical signal and the photocurrent is defined as

13.2 BASIC PHOTODIODE DESIGN AND REQUIREMENTS

$$I_p = \frac{\eta q P_{in}}{h\nu} \quad (2)$$

where q is the electronic charge, h is Planck's constant, ν is the photon frequency, and η is the ratio of electron–hole pairs generated for each incident photon. η is usually called the quantum efficiency. When the absorption layer is fully depleted, the quantum efficiency can be approximated by

$$\eta = (1-R)[1-\exp(-\alpha' W)] \quad (3)$$

where R is the reflection rate of incident light on the photodiode surface, α' the absorption coefficient for the light, and W the thickness of the absorption layer.

The noises in the circuit are shot noise and circuit noise (including the following preamplifier noise). The shot noise is due to the diode dark current and to the photocurrent, including the excess noise caused by the avalanche process. The dark current I_d in an idealized APD can be expressed as the sum of two components: the multiplied dark current I_{dM} and the unmultiplied dark current I_{d0}. That is,

$$I_d = M I_{dM} + I_{d0} \quad (4)$$

The excess noise caused by the avalanche process was analyzed theoretically by McIntyre [3]. According to his theory, when photogenerated carriers from outside the avalanche region are injected into the inside and these injected carriers act as minority carriers and have a larger ionization rate than that of the majority carriers, the excess noise factor F for the minimum noise condition is given by

$$F = M\left[1 - (1-k_i)\left(\frac{M-1}{M}\right)^2\right] \quad (5)$$

Here k_i is assumed to be a constant ionization rate ratio that is less than 1 and that does not depend on the electric field strength. It is defined as α (electron ionization rate)/β (hole ionization rate) if α is smaller than β (or β/α if α is larger than β). This equation indicates that an APD whose avalanche layer has a larger ionization rate ratio has a lower excess noise factor. This is because the feedback impact ionization initiated by the impact-ionized majority carriers causes an additional noise. Equation (5) can often be simplified to

$$F = M^x \quad (6)$$

where x is a parameter depending on the ionization rate ratio and is usually a positive number greater than 0 but not greater than 1. The total mean-square shot noise current can thus be deduced to be

$$\langle i_s^2 \rangle = 2q(I_p + I_{dM})M^{2+x}B + 2qI_{d0}B \quad (7)$$

where B is the objective bandwidth. Receiver circuit noise is simplified to circuit thermal noise, including the following preamplifier noise F_{amp}:

$$\langle i_{sc}^2 \rangle = \frac{4kTF_{amp}B}{R_{eq}} \tag{8}$$

where R_{eq} is the equivalent-circuit resistance usually represented by the load resistance R_L, k the Boltzmann constant, and T the absolute temperature.

The signal-to-noise ratio (S/N) in the circuit can then be expressed by

$$\frac{S}{N} = \frac{\langle i_p^2 \rangle}{\langle i_s^2 \rangle + \langle i_{sc}^2 \rangle}$$

$$= \frac{(I_p M)^2}{2[2q(I_p + I_{dM})M^{2+x}B + 2qI_{d0}B + 4kTF_{amp}B/R_L]} \tag{9}$$

The optimum multiplication factor M_{opt}, for which S/N is maximum, is obtained from $d(S/N)/dM = 0$ as

$$M_{opt} = \left[\frac{4kTF_{amp}}{xq(I_p + I_{dM})R_L}\right]^{1/(2+x)} \tag{10}$$

The unmultiplied dark current I_{d0} was neglected in this calculation because its contribution to noise is the same as that of $I_{dM}M^{2+x}$. Equation (9) can also be solved for the minimum optical power P_{min} required for a given S/N ratio. The resulting expression is

$$P_{min} = \frac{h\nu}{q\eta}\left(\frac{S}{N}\right)^{1/2}\left(\frac{8kF_{amp}B}{R_L}\right)^{1/2}\left(1 + \frac{2}{x}\right)^{1/2}M_{opt}^{-1} \tag{11}$$

The improvement factor obtained by using an APD instead of a PD can thus be estimated to be $(1 + 2/x)^{1/2}M_{opt}^{-1}$.

13.2.3 Response Speed

The main factors limiting the photodiode response speed are the carrier diffusion and drift (transit) times of photogenerated carriers, the diode capacitance (RC time constant in a circuit), and the gain–bandwidth product for an APD. By using a heterostructure, it is relatively easy to be free from the diffusion problem. This is because the light absorption region can be confined in a fully depleted narrow-bandgap layer. Therefore, high-purity epitaxial layers are required so that the depletion layer can be formed under a low bias. In some cases, however, a carrier pile-up problem occurs at heterointerfaces. This is described in Sections 13.3 and 13.4.

Every photodiode has a characteristic photogenerated carrier transit time, which

13.2 BASIC PHOTODIODE DESIGN AND REQUIREMENTS

is the time required for the carriers to drift to the edge of the depletion region. Analytically, the drift current is given by solving the continuity equation and the current density equation [4,5]. The resulting expression for a diode with a depleted absorption layer W and light penetration from the p-side is

$$F(\omega) = \frac{1}{1 + j(\omega/\alpha' v_p)} \left\{ \frac{\exp(-\alpha' W)\{\exp[-j(\omega W/v_p)] - 1\}}{j(\omega W/v_p) + 1 -} + \frac{1 - \exp(-\alpha' W)}{\alpha' W} \right\}$$

$$+ \frac{1}{1 - j(\omega/\alpha' v_n)} \left\{ \frac{1 - \exp[-j(\omega W/v_n)]}{j(\omega W/v_n)} - \frac{1 - \exp(-\alpha' W)}{\alpha' W} \right\} \qquad (12)$$

where ω is the angular modulation frequency, and v_n and v_p are the electron and hole velocities, respectively. From Eq. (12) the 3-dB-down cutoff frequency (bandwidth) for a 1-μm full-width drift at a velocity of 1×10^7 cm/s is calculated to be 44 GHz.

The diode capacitance is also an important factor governing the receiver speed. Reducing the pn-junction area is indispensable for obtaining high speed, and a back-illumination structure with a flip-chip configuration is often used to eliminate the capacitance in the region of the contact-pad metal. In the case of an APD, there is a response speed limit due to the avalanche process. This is called the gain(multiplication)–bandwidth (GB) product, which was theoretically analyzed by Emmons [6]. Theoretically,

$$GB \propto \frac{1}{\tau k_i} \qquad (13)$$

where τ is W_a/v_s (W_a is the avalanche region thickness and v_s is the saturation drift velocity). According to the theory, a larger GB product can be expected for an avalanche layer with a larger ionization rate ratio and a smaller thickness. This is because a feedback impact ionization caused by the impact-ionized majority carriers (carriers with a lower ionization rate) in the avalanche region lengthens the avalanche build-up time. Therefore, the study of materials and structures with a large ionization rate ratio is extremely important to the development of high-performance APDs.

The issues basic to the production of high-performance photodiodes can be summarized as follows:

1. *Dark current reduction.* The diode dark current needs to be as low as possible, and in an APD the reduction of multiplied dark current is more important than the reduction of unmultiplied dark current [see Eqs. (7) and (10)]. The current dependence of the sensitivity is greater at lower signal bit rates. When bit rates are low, the sensitivity of receivers using high-impedance FET preamplifiers and low-dark-current InGaAs PIN-PDs with low capacitance is higher than that of receivers using Ge-APDs [7]. This higher sensitivity is due to the noise reduction due to the low dark current and to the large load resistance allowed by the low diode capacitance [see Eqs. (9) and (11)].

2. *High-speed and high-quantum efficiency.* The minimum received signal level [see Eq. (11)] is inversely proportional to the quantum efficiency. Optimizing the absorption layer thickness is therefore especially important when high-speed signals must be detected [see Eqs. (3), (11), and (12)].
3. *Reduction of diode capacitance.* Diode capacitance should be low basically so that the diode can operate at a high speed, and it also allows sensitivity to be high because it enables the use of a large load resistance [see Eq. (11)].
4. *Large-ionization-rate-ratio materials for the avalanche layer.* Basically excess-noise, gain–bandwidth, and overall receiver sensitivity depend strongly on this ratio [see Eqs. (5), (6), (11), and (13)]. This is why InGaAs/InP-APDs with InP avalanche layers are preferable to Ge-APDs for high-sensitivity detection.

Books introduced in Refs. 8 and 9 are useful for more detail and for a theoretical understanding of photodetectors.

13.3 PHOTODIODES

The photodiodes (PDs) treated in this section are mainly the PIN type with no internal gain. PIN designates a layer structure in which an unintentionally doped high-purity layer is sandwiched between the p$^+$ and n$^+$ layers. The light absorption layer is usually an InGaAs layer lattice matched to InP. This is because the InGaAs is responsive to every wavelength of light sources based on InGaAsP/InP and InAlGaAs/InP material systems.

13.3.1 Basic InGaAs PIN Photodiodes

In making a simple planar structure PD, a double heterostructure consisting of InGaAs/InP with an InP capping layer is grown, and this growth is followed by selective impurity diffusion to form the p$^+$n-junction. A window allowing light to pass through is formed on the front (the grown-layer surface) or back surface of the InP substrate. The back-illumination type is of a structure often used to obtain the low capacitance desirable for high-speed operation.

A cross-sectional view of a front-illuminated planar-structure InGaAs PIN-PD is shown in Fig. 13.2. The front of the pn-junction is formed in the InGaAs absorption layer close to the InGaAs/InP interface by using Zn or Cd thermal diffusion. The dark-current characteristics of a diode with an effective junction diameter of 104 μm and a 4-μm-thick InGaAs absorption layer with a carrier concentration of 2×10^{15} cm^{-3} are shown in Fig. 13.3. The exponential dark-current increase with increasing bias, evident when the reverse bias is large, is due to the InGaAs band-to-band tunneling. Forrest et al. [10] and Takanashi et al. [11] demonstrated that the dark-current characteristics showing an exponential increase in InGaAs PDs (and APDs) are unavoidable due to the low mass of electrons in the narrow-bandgap ma-

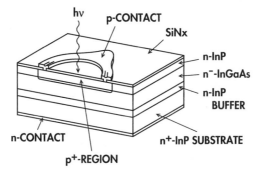

FIGURE 13.2 Schematic cross section of a planar-structure InGaAs photodiode.

terial. The values obtained fit the results well by integrating the following differential equation, which is based on Kane's theory [12]:

$$\frac{dJ}{dx} = \frac{q^3 m^{*1/2} E^2(x)}{2\sqrt{2}\,\pi^3 \hbar^2 E_g^{1/2}} \exp\left[-\frac{\pi m^* E_g^{3/2}}{2\sqrt{2}\,q\hbar E(x)}\right] \tag{14}$$

where m^* is the effective tunneling mass, $\hbar$ the reduced Planck's constant, E_g the energy gap, and $E(x)$ the electric field as a function of position. The diode must be operated under a moderate reverse bias, where it is not influenced by the tunneling current. At a low bias of less than 10 V, the dark current of well-fabricated diodes is less than 1 nA and it changes by about one order of magnitude when the diode tem-

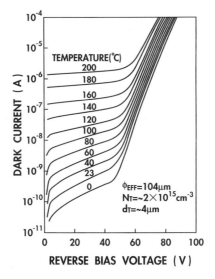

FIGURE 13.3 Dark current versus voltage characteristics at various temperatures.

perature changes by 40°C. These low dark-current characteristics are obtained with a planar structure that terminates the pn-junction in the wide-bandgap InP capping layer. The dark current in a mesa structure, which is controlled by surface leakage current at the mesa wall, is commonly two or more orders of magnitude larger than that in a planar heterostructure.

Figure 13.4 shows the external spectral quantum efficiency for a diode under a 5-V bias. Quantum efficiency is relatively constant and over 80% at wavelengths between 1.0 and 1.55 μm. This wide-range high efficiency is due not only to the antireflection coating on the surface but also to the thin p^+-InGaAs region, which acts as a recombination region for photocarriers generated in it. The high efficiency is also due to the reduction in the interface recombination velocity for photogenerated carriers at the p^+-InP/p^+-InGaAs hererojunction, the configuration of which cannot be expected in conventional homojunction photodiodes. The cutoff at the short wavelength (≈ 0.95 μm) is controlled by the bandgap of the InP ($E_g = 1.35$ eV) capping layer. The bandwidth (3-dB-down cutoff frequency) is about 5 GHz with a 50-Ω load resistance.

We pointed out earlier that the reduction of dark current is a basic issue in producing reliable and high-performance photodiodes. To evaluate the level of the dark current obtained, it might be useful to estimate and analyze the effective carrier lifetime [13] for each diode. In general, the dark current consists of three components: the generation–recombination (g-r) current in the bulk depletion region, the g-r current in the surface depletion region, and the diffusion current in the neutral undepleted region. The surface current in well-fabricated devices with planar structures with wide-bandgap window layers may be neglected. The g-r current in the bulk depletion region is defined for a depletion layer thickness W as [13]

$$J_{\text{g-r}} = \frac{qn_iW}{\tau_{\text{eff}}} \tag{15}$$

where n_i is the intrinsic carrier concentration and τ_{eff} is the effective lifetime. The effective lifetime can be deduced by fitting a line to a plot of the dark current versus

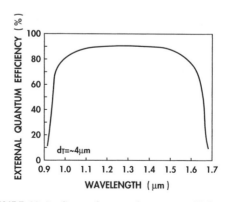

FIGURE 13.4 Spectral external quantum efficiency.

the square root of the reverse-bias voltage. The values obtained for planar structure diodes with InP cap layers grown mainly by hydride vapor-phase epitaxy are plotted against donor concentration in Fig. 13.5, which shows that there is an inverse relation between the lifetime and donor concentration. The data obtained, which are represented by the relation $\tau_{eff} = 10^{11}/N_T$ in the figure, are in the extrapolation from the doping dependence of the lifetime, which was confirmed by PL measurement of doped high-concentration samples [14]. These results indicate that high-purity epitaxial layers are required for both low-bias and low-dark-current operation.

To get a high-speed response, it is necessary to shorten the carrier transit time by reducing the thickness of the absorption layer. But this usually results in a lower quantum efficiency. Figure 13.6 shows the calculated cutoff frequency characteristics against the InGaAs absorption layer thickness evaluated on the basis of Eq. (12), in which v_e and v_h have been exchanged to calculate for the back-illuminated configuration. The absorption coefficient used in these calculations was 6800 cm^{-1}, which corresponds to the 1.55-μm wavelength [15]; the electron and hole saturation velocities were 6.5×10^6 and 4.5×10^6 cm/s, respectively [16]; and the pn-junction diameter was treated as a parameter. The load resistance (equivalent-circuit resistance) was assumed to be 50 Ω. For a given pn-junction diameter there is a maximum cutoff frequency. This is due to the effect of the diode capacitance, which is inversely proportional to the thickness of the depleted absorption layer. The cutoff frequencies limited solely by the transit time, free from the capacitance effect, for back and front illumination, are also shown in Fig. 13.6. At a given thickness the speed is somewhat higher for front illumination than for back illumination. This is because the saturation velocity of electrons is higher than that of holes. It is evident that a 100-μm-diameter diode cannot be expected to operate at 10 GHz or more. Optimization of the absorption layer thickness and junction diameter is needed for a given transmission speed. As can be seen from Fig. 13.6, one of the best ways to obtain an ultrafast response speed is by reducing the junction capacitance.

Cutoff frequencies over 100 GHz have been attained in a thin, small-area diode employing a graded-bandgap layer to reduce the carrier trapping at the InGaAs/InP

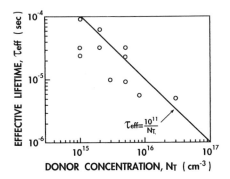

FIGURE 13.5 Effective lifetime as a function of donor concentration in InGaAs p$^+$n-junction.

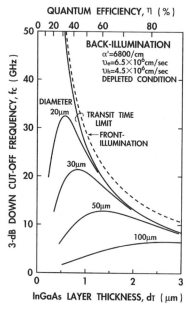

FIGURE 13.6 Bandwidth versus absorption layer thickness characteristics calculated for various pn-junction diameters.

heterointerface [17,18]. Figure 13.7 shows mushroom mesa geometry PIN-PDs with an air-bridge contact metal that was developed for ultrahigh-speed operation by Tan et al. [19]. The mushroom mesa was used to reduce the junction capacitance while maintaining a large contact area for low series resistance, and the air-bridge contact was used to minimize the parasitic capacitance. A 2-μm-diameter photodi-

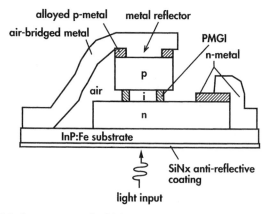

FIGURE 13.7 Mushroom-mesa ultrahigh-speed PIN-PD with an air-bridge metal. (After Ref. 19, © 1995 IEEE, reprinted with permission.)

ode with 0.18-μm-thick InGaAs showed a response time of 2.7 ps (full width at half maximum) for Ti-sapphire laser pulses whose wavelength was 0.98 μm, and a cut-off frequency of 120 GHz was confirmed by the fast Fourier transform of the measured pulse responses. The external quantum efficiency at the wavelength of 1.3 μm was estimated to be 28%.

Some other techniques have also been developed to improve the problems due to a small junction area and the associated low efficiency in high-speed PIN-PDs; for example, the use of signal light reflection at the contact metal deposited on the rear surface of the InP substrate and a Bragg reflector to enhance efficiency [20,21], as well as monolithic lens integration [22] on the substrate to magnify the effective receiving area efficiently, have been reported.

13.3.2 Waveguide PIN Photodiodes

There is a trade-off between the speed and efficiency in conventional vertical-type photodiodes, but waveguide(WG)-structure photodiodes are basically free from this trade-off problem because of the parallel penetration of signal light along the absorption layer. In this structure, internal quantum efficiency is a function of the length, propagation mode, and mode confinement factor. To improve the coupling, several structures have been proposed and developed.

WG-PDs having a bandwidth of 50 GHz and a quantum efficiency of 40% at a wavelength of 1.53 μm were reported by Wake et al. [23], who made PDs with an asymmetric InGaAsP waveguide structure. A thin (0.13-μm) InGaAs absorption layer was sandwiched between 3-μm-thick n-doped InGaAsP (bandgap wavelength $\lambda_g = 1.3$ μm) and 0.1-μm-thick undoped InGaAsP. The absorptive InGaAs and thick InGaAsP layers were designed to give a high external efficiency: The thick InGaAsP layer largely determines the transverse waveguiding properties of the diode, and this ensures a large mode size, comparable with the mode size of the lensed fiber. The InGaAs absorption layer should be thin, so as to attain a large mode size. The diodes were 5 μm wide and 10 μm long, and their capacitance was less than 0.1 pF.

Kato et al. developed a multimode waveguide structure with symmetric InGaAsP intermediate layers inserted between the InGaAs light-absorption and InP cladding layers [24]. They showed numerically that higher-order mode lights in the structure enabled the efficiency of coupling between the waveguide PD and the fiber to be increased. They reported that experiments with a structure having a 0.6-μm-thick InGaAs absorption layer sandwiched between 0.6-μm-thick InGaAsP ($\lambda_g = 1.3$ μm) layers yielded a bandwidth of 50 GHz with an external quantum efficiency of 68%. Their calculations indicated that the coupling efficiency depends on the total thickness of the InGaAs and the two InGaAsP intermediate-bandgap layers, and their results are summarized in Fig. 13.8. These results mean that the InGaAs absorption layer thickness can be designed not only for the coupling but also the objective speed (transit time).

Figure 13.9 illustrates a multimode waveguide PIN-PD with a mushroom-mesa structure [25]. The mushroom-mesa configuration was used to reduce the diode ca-

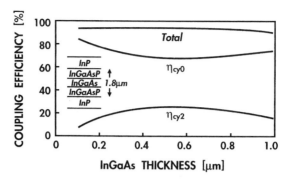

FIGURE 13.8 Calculated coupling efficiency (Total) as a function of InGaAs thickness for multimode WG PDs in which the total thickness of the InGaAs and InGaAsP layers is kept constant at 1.8 μm. η_{cy0} and η_{cy2} show calculated fundamental- and second-order mode contributions to the coupling efficiency, respectively. (After Ref. 24, © 1992 IEEE, reprinted with permission.)

pacitance, and this configuration left a wide area for metal contact, to minimize series resistance. This is important for reaching a response speed over 50 GHz. The layer structure consists of 0.8-μm-thick p-doped and n-doped InGaAsP layers with a 0.2-μm-thick unintentionally doped InGaAs-core layer. The mushroom mesa was made by forming 6μm-wide cladding layers that were then selectively wet etched to decrease the junction capacitance. The external quantum efficiency of structures

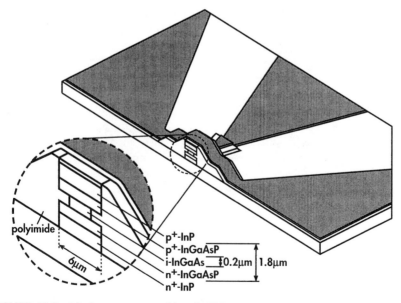

FIGURE 13.9 Mushroom-mesa multimode WG PIN-PD. (After Ref. 25, © 1994 IEEE, reprinted with permission.)

with various InGaAs core-layer widths was measured, and a diode with a 1.5-μm-width core showed an efficiency of 50% at a wavelength of 1.55 μm. For core layers more than 3 μm wide, the efficiency was about 60%. This dependence on core-layer width is due to the coupling loss in the horizontal direction, which loss is caused by the optical field mismatch to the fiber with a 1.3-μm spot size. The 1.5-μm-wide core mesa had a capacitance of 15 fF and a series resistance of 10 Ω. The frequency response for a circuit with an impedance of 50 Ω was measured and is shown in Fig. 13.10. The response is almost flat over the frequency range 0 to 75 GHz. The Fourier transform of the measured short pulse responses indicates a bandwidth of 110 GHz. The bandwidth-efficiency product of 55 GHz obtained is 1.6 times larger than that of the 120-GHz-bandwidth vertical PIN-PD in Section 13.3.1.

Multimode waveguide structures have also attracted much attention because of their potential ability to couple easily to fibers and to planar lightwave circuits (PLCs) without requiring the use of a focusing lens, thereby lowering the cost of receiver modules. The operation speed of the receiver modules in access networks is not expected to be more than a few gigahertz, so a key issue for this application is the alignment tolerance to fibers and optical waveguides. Figure 13.11 shows the tolerance curves obtained experimentally for the coupling of WG-PDs to dispersion-shifted fibers (DSFs) and PLCs [26]. The 1-dB tolerance in the vertical direction is ±2 μm, which is comparable to spot-size-converted LD tolerance. The WG-PD has an asymmetric multimode structure consisting of a 1-μm-thick p-InP cladding layer, a 3-μm-thick InGaAsP ($\lambda_g = 1.4$ μm) core layer, and a 1.5-μm-thick n-InGaAsP ($\lambda_g = 1.2$ μm) layer on the InP substrate. The responsivity obtained at a wavelength of 1.31 μm is 0.95 A/W, and the average coupling loss between WG-PDs and cleaved DSFs is 0.44 dB without coupling lenses. Recently, greatly simplified structures that achieve easy coupling with low bias have been reported [27]. The layer structure consists of a 3-μm-thick InGaAsP ($\lambda_g = 1.4$ μm) photoabsorbing core layer and two 2-μm-thick InGaAsP ($\lambda_g = 1.2$ μm) intermediate layers. A selective impurity diffusion was used to form the pn-junction and to form a slab waveguide, and the pn-junction front was designed to be deep in the light-absorption core layer. A tolerance of 5.5 μm in the vertical direction, a bandwidth of 500

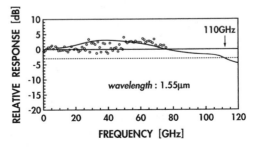

FIGURE 13.10 Frequency response measured by a spectrum analyzer (circles) and deduced from the Fourier transform. (After Ref. 25, © 1994 IEEE, reprinted with permisson.)

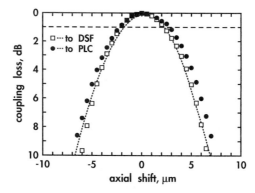

FIGURE 13.11 Coupling tolerance characteristics of a WG PIN-PD to DSF and PLC. (After Ref. 26, © 1995 IEE, reprinted with permission.)

MHz, and a responsivity of 0.87 A/W at a wavelength of 1.31 μm were achieved under a 1-V bias.

13.3.3 Other Types of Photodiodes

A metal–semiconductor–metal (MSM) structure with an interdigitated contact has also been used in photodetectors because this structure is easy to fabricate and because the simplicity of the planar contact makes the structure suitable for integration with electrical circuits. MSM photodiodes have therefore often been used in receiver OEICs. They also have a low capacitance per unit area, which is advantageous and useful in photoreceivers having a large photoreceptive area. The basic problems with these photodiodes are the dark current and quantum efficiency. Dark current degradation in the InGaAs/InP material system is due basically to the low Schottky barrier heights of InP and InGaAs(P). This problem has been solved by using an InAlAs capping layer [28], but the degradation in quantum efficiency caused by electrode masking cannot be eliminated. There have been reports on transparent electrodes [29,30] and back illumination [31,32] for the substrate side of MSM-PDs to improve the efficiency. However, these new structures result in response-speed degradation due to the low electric field underneath the electrodes. Electron beam lithography has also been used to reduce the size of the contact metal region. For high-speed applications, an MSM structure having a submicrometer line-and-space layout has also been studied using electron-beam lithography [33].

As another approach to high-speed photodetectors, a traveling-wave photodiode has been proposed and fabricated [34,35]. This is a waveguide photodiode with an electrode structure designed to support traveling electromagnetic waves with a characteristic impedance matched to that of the external circuit. A bandwidth of 172 GHz has been reported for a GaAs/AlGaAs diode 1 μm wide and 7 μm long [35]. On the other hand, to improve the quantum efficiency of photodiodes with a thin absorption layer, the use of mirrors for multiple optical passes through the absorption layer (similar to the resonances of Fabry–Perot microcavities) has been reported [20,21]. An InP-based resonant cavity photodiode was reported by Dentai et al. [36], and its quantum efficiency at a wavelength of 1.48 μm was 82%. This diode had a 0.2-μm-thick InGaAs absorption layer, which was embedded in a cavity consisting of InP spacer layers, a bottom mirror of an InP/InGaAsP quarter-wave stack, and a top mirror of a single ZnSe/CaF$_2$ pair.

13.4 AVALANCHE PHOTODIODES

Although the highest overall receiver sensitivity in the wavelength range 1.55 μm has been achieved in systems using PIN photodiodes with Er-doped fiber amplifiers [37], avalanche photodiodes (APDs) are still the most sensitive semiconductor photodetectors, serving as a simple, low-power-consumption receiver module in wavelength regions 1.3 and 1.55 μm. In this section we review InGaAs/InP heterostructure APDs in which the InGaAs light-absorptive layer is separated from the InP avalanche layer. We then introduce new superlattice APDs in which the carrier ionization rate is increased, exploiting the bandgap discontinuities between the well and barrier layers in superlattice structures.

13.4.1 InGaAs/InP Avalanche Photodiodes

The SAM (separation of absorption and multiplication) configuration was developed for heterostructure APDs used at long wavelengths. The early InP-based APD device structures had homojunctions in which the pn junction was within the InGaAs(P) absorption layer. The homojunction-type APDs, however, showed an exponential rise of the dark current with increasing reverse bias and also had poor multiplication factors (less than 10). It was shown that this nonsaturation dark current is due to the band-to-band tunneling current caused by the physical properties of InGaAs(P), mainly the low effective mass and the narrow bandgap [10,11]. Taguchi et al. [38,39] reported the first demonstration of a high-gain, low-dark-current InGaAsP/InP-APD breaking through the limitations of homojunction-type APDs. In this new photodiode the p$^+$-n junction was located in the InP wide-bandgap layer and was separated from the light-absorbing layer. Since these reports appeared, almost all studies and development of heterostructure APDs for long-wavelength applications have been based on the SAM concept.

Figure 13.12 shows schematically the layer structure of a planar InGaAs/InP-APD with the SAM configuration. In the early stages of development, the mesa structure was often used to demonstrate the feasibility of the device [41,42]. Although there are other types of planar configuration [43,44], they all have vertical layer structures that are almost the same. Incident light at wavelengths between 1.0 and 1.6 μm is absorbed in the InGaAs absorption layer. Photogenerated holes in the InGaAs are injected into the InP avalanche layer through the thin intermediate InGaAsP layers inserted to eliminate the hole pileup that occurs at the valence band offset between the InP and InGaAs when the bias is low [45–47]. The maximum

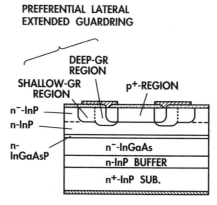

FIGURE 13.12 Schematic structure of a planar InGaAs/InP APD with a SAM (separation of absorption and multiplication) configuration. (From Ref. 40.)

electric field in the InGaAs absorption layer must be kept below 200 kV/cm to suppress the tunneling current [40]. A zinc source (Zn_3P_2) or cadmium source (Cd_3P_2) is used to form the p^+-region, whose depth is controlled by adjusting the annealing time to tailor the InGaAsP/InGaAs heterojunction electric field to about 150 kV/cm under breakdown conditions. The conductivity of each layer is chosen to allow photo-excited holes to be injected into the n-InP avalanche layer, because in InP the hole ionization rate is larger than the electron ionization rate. This enables the excess noise in the avalanche process to be minimized [see Eq. (5)]. The guard ring (GR) suppressing the breakdown at the p^+-n junction edge is formed around the n^-/n-InP interface region by two-step Be ion implantation.

Plots of dark current and photocurrent versus reverse bias for a typical InGaAs/InP APD are shown in Fig. 13.13. The dark current is about 15 nA when the bias voltage is 90% of the breakdown voltage, and a gain of over 60 is observed at a primary photocurrent of 1 µA. A distinctive feature of the planar structure is that it results in two photocurrent kinks. The first knee at a low bias (≈ 18 V) corresponds to the occurrence of depletion-layer punch-through to the InGaAs layer within the GR. The flat region next to this knee well indicates unity gain (multiplication factor $M = 1$), because the photocarriers generated in the small photoreceptive area under the p^+-n junction almost always diffuse into the GR without recombination. The other kink at a higher bias (≈ 35 V) corresponds to the occurrence of depletion-layer punch-through into the InGaAs within the p^+-n photoreceiving area. The relation between the dark current and multiplication factor is shown in Fig. 13.14, from which the multiplied dark current can be extrapolated to 2 nA at $M = 1$, and the unmultiplied dark current can be seen to be negligibly small [see Eq. (4)]. The dependence of dark current characteristics on temperature is shown in Fig. 13.15. When the bias voltage is high, the activation energy is about 0.38 eV (about half the InGaAs bandgap), which means that the high-bias dark current is dominated by the InGaAs generation–recombination current. The spectral quantum efficiency is al-

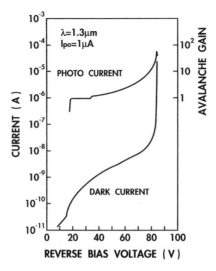

FIGURE 13.13 Reverse-bias characteristics in an InGaAs/InP APD.

most the same as that in InGaAs PIN-PDs (see Fig. 13.4) and depends on the InGaAs layer thickness, especially at wavelengths near 1.5 μm. Figure 13.16 shows the excess noise factor versus multiplication characteristics measured at 1.3- and 1.55-μm wavelengths. It also shows the theoretical curves, based on Eq. (5), for various ionization rate ratios. From these curves, the effective ionization rate ratio ($k_i = \alpha/\beta$) in this APD is inferred to be about 0.4 (the hole/electron ratio in the figure is 2.5) and the x-value in Eq. (6) to be about 0.7. This ratio is much larger than that in Ge (where $k_i \approx 1$ and $x \approx 0.9$). The InGaAs/InP-APD receiver is therefore more sensitive than the Ge-APD receiver [see Eq. (11)].

Figure 13.17 shows the bandwidth characteristics of five samples with different

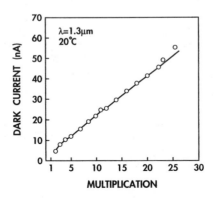

FIGURE 13.14 Relation between dark current and multiplication in an InGaAs/InP APD.

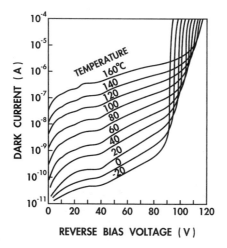

FIGURE 13.15 Dark current characteristics at temperatures from –20 to 160°C.

values of maximum electric field in the InGaAs layer. These samples were fabricated from the same wafer, in which the concentration of the n-InP avalanche layer was about 2.5×10^{16} cm^{-3}. Gain–bandwidth products of 40 to 90 GHz have been obtained. The maximum bandwidth, about 4 GHz, has been limited by the RC time constants. The GB product theoretically decreases with increasing thickness of the avalanche layer [see Eq. (13)], but the experimental data indicate that an APD with a thicker n-InP avalanche layer and lower maximum electric field in the InGaAs has a

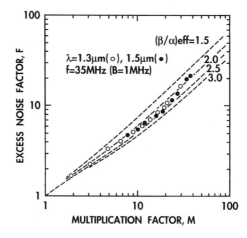

FIGURE 13.16 Excess-noise-factor characteristics for an InGaAs/InP APD.

13.4 AVALANCHE PHOTODIODES

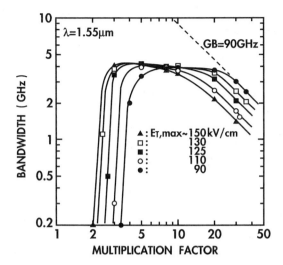

FIGURE 13.17 High-frequency characteristics as a function of multiplication factor for InGaAs/InP APDs with different maximum electric fields ($E_{T,\text{max}}$) in the InGaAs layer.

larger GB product. This can be explained by the effect of the avalanche process in the InGaAs layer [48,49], where the gain factor is indeed much less than 2. This effect is exaggerated in the structure because the carriers whose ionization rate is high in InGaAs [50] are electrons and are holes in InP. Thereby tight feedback of impact ionization between the InP and InGaAs layers occurs, and this process results in deterioration of the effective ionization rate ratio and avalanche buildup time [see in Eq. (13)].

To improve GB products, avalanche layer reformation by δ-doping was proposed [51,52]. A thin charge sheet layer between the InP and InGaAs layers was introduced by Taroff et al. [49], who measured the largest GB product yet obtained in the InGaAs/InP system: 122 GHz. As seen in Fig. 13.17, bandwidth degradation appears in the low-multiplication region and depends on the electric field in the InGaAs layer. An APD with a higher maximum electric field in the InGaAs can be responsive with high speed at a lower multiplication factor. This is because the photogenerated holes pile up at the valence band offset between the InP and InGaAs layers. To be free of this pileup at the lowest bias (at low multiplication), multiple-step InGaAsP intermediate layers [52] or graded quaternary layers [53] were inserted between the InP and InGaAs layers.

13.4.2 Superlattice Avalanche Photodiodes

As discussed in Section 13.2, APD performance is governed basically by the ionization rates of electrons and holes within the avalanche region, and these rates usually depend on the composition of the material. In the early 1980s, Chin et al. [54] and Capasso et al. [55] indicated that the ionization rates could be increased by using the band offsets in superlattice (multiple-quantum-well) structures. This concept has since been studied and used in various material systems. The report of Kagawa et al. [56] on the ionization rate elevation due to the use of an InGaAs/InAlAs superlattice system accelerated the development of these InP-based APDs for use in optical fiber communications. Their results are summarized in Fig. 13.18, in which the ionization rates of bulk InGaAs reported [50] are also depicted. The figure indicates the enhancement of the electron ionization rate and the suppression of the hole rate in the InGaAs/InAlAs superlattice. As a result, superlattice APDs based on InGaAs/InAlAs are expected to break through the limitations of conventional InGaAs/InP APDs with a bulk InP avalanche layer.

In the initial stages of the development, InGaAs/InAlAs superlattice (SL) APDs with a SAM configuration were widely studied. It became clear, however, that 20-nm-order wells/barriers result in a large dark current (exceeding 10 μA) because of the tunneling in InGaAs wells [57]. Since then, research has focused on obtaining a higher GB product and lower dark current in practical applications, especially applications in the 10-Gb/s range. Several material systems, such as InGaAsP/InAlAs [58], InAlGaAs/InAlAs [59], and narrow-well InGaAs/InAlAs [60], have been studied and a GB product over 110 GHz has been reported for each system.

Figure 13.19 shows the mesa structure of an InAlGaAs/InAlAs flip-chip SL-APD. The flip-chip back-illuminated configuration has been used to reduce the diode capacitance and enable operation beyond 10 GHz. The thin InP field-buffer layer has been introduced to create the SAM structure. The InGaAs light-absorption layer is lightly doped with Be and is p-type to allow the photogenerated electrons to be injected into the SL avalanche layer. Injection of photo-excited electrons into the SL avalanche layer is required to minimize the excess noise caused by the avalanche process [see Eq. (5)]. Figure 13.20 shows the reverse-bias characteristics for an APD with a 0.23-μm InAlGaAs/InAlAs SL layer, a 1-μm InGaAs absorption layer, and a 30-μm-diameter photoreceptive area. The SL layer consists of 11 periods of an 8-nm-thick InAlGaAs (E_g = 1 eV) well and a 13-nm-thick InAlAs barrier. The

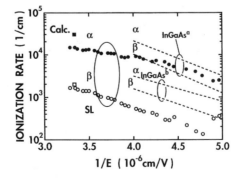

FIGURE 13.18 Ionization rates for an InGaAs/InAlAs superlattice: the solid circles denote electron ionization rates and the open circles denote hole ionization rates. (After Ref. 56, © 1989 AIP, reprinted with permission.)

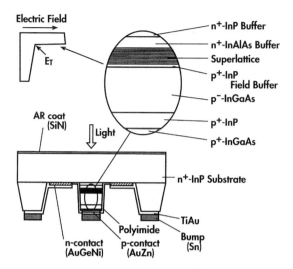

FIGURE 13.19 InAlGaAs/InAlAs mesa-structure flip-chip superlattice APD.

dark current has been decreased drastically by using InAlGaAs wells instead of conventional InGaAs wells: Dark current is less than 1 µA at multiplication factors over 20 [61]. The relation between the dark current and multiplication yields a multiplied dark current between 10 and 20 nA, which is sufficiently low for applications in the 10-Gb/s range. Figure 13.21 shows cutoff frequency (bandwidth) versus multiplication factor for different SL APDs which have different maximum electric fields in the InGaAs absorption layer at breakdown. The maximum cutoff frequency, which is limited by the RC time constant, is more than 15 GHz and the GB prod-

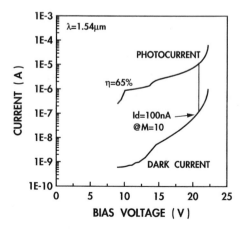

FIGURE 13.20 Dark current and photocurrent versus reverse-bias voltage in an InAlGaAs/InAlAs SL-APD.

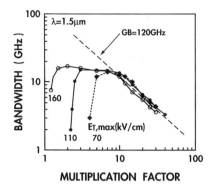

FIGURE 13.21 Bandwidth versus multiplication factor for InAlGaAs/InAlAs SL-APDs with different maximum electric fields in the InGaAs layer.

ucts can be deduced from the inversely proportional relation to be about 120 GHz. These data show the same tendency as the results shown in Fig. 13.17 for InGaAs/InP APDs. However, the GB products obtained do not depend as strongly on the electric field of the InGaAs absorption layer. This is because the carriers with larger ionization rates are electrons in both the SL and the InGaAs. Simulation also predicted the same tendency [60]. On the other hand, optimization of the field-buffer layer introduced to create a SAM structure is also important for obtaining a larger GB product. A thick InP buffer layer lowers the GB product because of the avalanche process caused by the large ionization rate carrier holes within the InP. The largest GB product, 150 GHz, was reported for a thin (33 nm) InP field-buffer layer [62]. An InAlAs field buffer, in which the ionization rate of electrons is larger than that of holes, is preferable for obtaining a higher GB value. The sensitivities of receivers (with non-return-to-zero sequence and bit error rate of 10^{-9}) using these SL-APDs were measured to be in a range from -27 to -29 dBm at 10 Gb/s [63–65].

Figure 13.22 shows the relationship between GB product and multiplication lay-

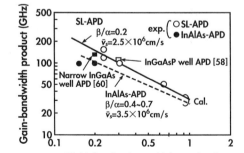

FIGURE 13.22 Relation between gain–bandwidth product and multiplication layer thickness for a variety of APD structures.

er thickness [66]. A larger GB product is obtained in a thinner SL layer. Data from APDs of different material systems and from InAlAs bulk avalanche layer APDs fabricated with the same device structure and dimensions as those of InAlGaAs/InAlAs-APDs is also depicted. It is evident that GB products for different SL systems show the same tendency, and the improvement due to superlattice structures is by a factor of at least 20%.

Device reliability is also important in practical system applications, and a lifetime at 50°C of about 10^5 h was reported for simple polyimide-coated InAlGaAs/InAlAs mesa APDs [67]. To extend this lifetime, a new planar structure with a new type of guard ring (GR) has been developed [68]. Figure 13.23 shows the planar structure layout. The layer structure is inverted from that of conventional mesa structure to make the GR formed on the wafer surface. Ti-ion implantation was performed to decrease the p-concentration of the InP field-buffer layer outside the photoreceptive area, thereby decreasing the maximum electric field in that region and implementing the GR effect.

Staircase APDs, in which the avalanche layer is composed of a sawtooth bandgap layer, were proposed by Capasso et al. [69]. There is no barrier for electrons at a reverse-bias in the staircase APD, which in Fig. 13.24 is compared with a rectangular superlattice. Simulation predicted the superiority of staircase APDs over rectangular-well superlattice APDs [70], and SAM-configuration staircase APDs composed of an avalanche layer with 10 periods of a 20-nm linearly graded layer from InAlAs to InAlGaAs ($E_g = 1$ eV) have been reported to exhibit a GB product of 100 GHz [71]. Ionization rates reported for various different layer structures [72,73] are summarized in Fig. 13.25. The ionization rate ratio of an InAlGaAs/InAlAs SL is about 0.33 when the electric field is about 500 kV/cm. For the staircases, on the other hand, the ratio at the same electric field strength is 0.20 to 0.28. Ionization rate ratios tend to be higher in staircase structures than in superlattice structures. Staircase APDs have mainly been grown by metal-organic vapor-phase epitaxy [71–73], which can relatively easily be used to make good-quality sawtooth-bandgap InAl-

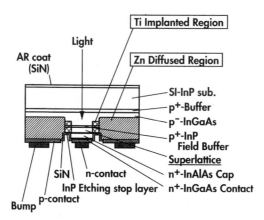

FIGURE 13.23 InAlGaAs/InAlAs planar-structure SL-APD with a Ti-implanted guard ring.

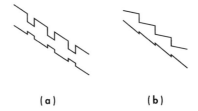

FIGURE 13.24 Band diagram of (*a*) rectangular superlattice and (*b*) sawtooth (staircase) under reverse bias.

GaAs layers. However, the thickness ($\approx$ 0.2 μm) of the Zn-doped InP field-buffer layer, which was introduced to create a SAM configuration, was almost the same as that of the sawtooth avalanche layer. High-speed characteristics better than those of conventional SL-APDs have therefore not yet been obtained [71].

13.5 INTEGRATED PHOTORECEIVERS

Monolithic integration of optical and electronic devices on a substrate is expected to break through the limitations in performance, production costs, integration scale, and reliability that restrict the use of their hybrid counterparts. Such advantages of monolithic integration have already been seen in Si LSI development. Receiver OE-ICs and integrated receivers with multiple functions are reviewed briefly here.

13.5.1 Receiver OEICs

Since Leheny et al. [74] first reported InP-based OEICs with simple pin/FETs, their integration scale and operational speed have been increased steadily. Their use

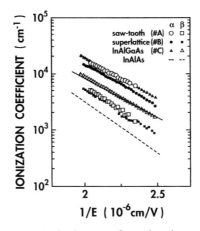

FIGURE 13.25 Ionization rates for various layer structures.

rapidly expanded in the latter half of the 1980s, after many fabrication difficulties due to the basic structural differences between optical and electronic devices were overcome. Although speeds of several gigahertz were obtained and up to 100 elements were integrated, a few serious disadvantages became apparent: low production yield, which would result in high cost, and some penalty in performance (compared with that of hybrid OEIC systems). These disadvantages are of course due to the InP-based materials and processing technology being less mature than the Si- and GaAs-based materials and processing technology. The low processing yield, which probably exists at the present, is due basically to the different device structures of optical devices and electronic circuits and to insufficiently planarized regrowth wafers. Many studies, however, have addressed the difficulties that must be overcome to get higher speed, better sensitivity, and more signal channels in PIN/HEMT [75,76], PIN/HBT [77,78], and MIS/MODFET [79] combinations.

Figure 13.26 shows the reported receiver sensitivity plotted against transmission speed. OEIC photoreceiver technology has improved recently so much that the performance of OEIC photoreceivers is comparable to that of the best hybrid systems, especially at speeds over 10 GHz [78]. In turn, in hybrid systems the deterioration in performance due to the hybrid layout and due to the wiring used between photodiodes and electric circuits is exaggerated and serious at high speeds over 10 GHz. OEIC performance now depends more on the characteristics of the particular transistors than on the integration technology.

13.5.2 Photonic Integrated Circuits Including Photodiodes

Higher transmission capacity in both trunklines and access networks is being called for, and a number of new transmission systems are being developed. These applications require new devices with new functions and more suitable circuit configura-

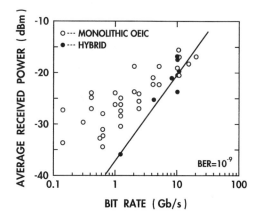

FIGURE 13.26 Receiver sensitivity against signal bit rate for monolithic OEICs and hybrid receivers. The straight line shown in the figure represents about the best signal level for hybrid receiver systems.

tions. One key element in a wavelength-division-multiplexed (WDM) system is a wavelength-demultiplexing receiver that can resolve the wavelength channels and receive the signals. The integration of these functions on a single chip has been studied [80]. Figure 13.27 shows the layout of a monolithic eight-wavelength demultiplexing receiver [81]. The waveguide grating router (WGR) consists of eight input waveguides, eight output waveguides, two star couplers, and the array grating waveguide section. When eight wavelengths are launched simultaneously into any one input waveguide, the WGR spectrally resolves the eight signals, sending one into each of the eight output waveguides. Each signal is then coupled into a PIN-PD and the photocurrent is amplified by an integrated amplifier. Figure 13.28 shows a cross section of the devices in the demultiplexing receiver. The WGR has a buried rib waveguide which consists of an n^--InP lower cladding layer, a 0.3-μm n^+-InGaAsP (λ_g = 1.3 μm) waveguide, a 12-nm n^--InP stop-etch layer, a 40-nm n^--InGaAsP rib layer, and a 1.5-μm undoped InP layer burying the rib waveguides. The chip can demultiplex eight wavelengths spaced 0.81 nm apart with a nearest-neighbor crosstalk of less than −15 dB. The receiver must be polarization independent because input signal polarization is unknown. As a result, polarization-independent WDM receivers have also been studied [82,83]. Steenbergen et al. [84] recently reported an eight-channel high-speed (10 GHz) polarization-independent phase-array demultiplexer monolithically integrated with photodiodes. Polarization-independent operation was achieved using polarization dispersion compensation, with two different sections for the array waveguide. The on-chip losses were 3 dB for TE polarization and 5 dB for TM polarization. The two transmission spectra for the TE and TM modes overlapped almost completely.

With regard to access networks, bidirectional links for broadband integrated services digital networks and fiber-to-the-home architectures are also receiving a great deal of attention, and optical receivers with many circuit configurations have there-

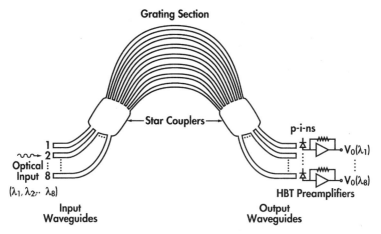

FIGURE 13.27 Layout of a monolithic eight-wavelength demultiplexing receiver. (After Ref. 81, © 1995 IEEE, reprinted with permission.)

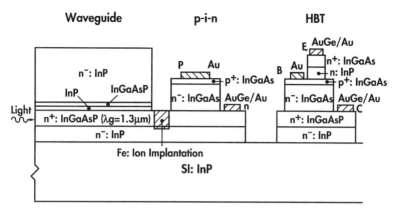

FIGURE 13.28 Cross section of devices used in the demultiplexing receiver. (After Ref. 81, © 1995 IEEE, reprinted with permission.)

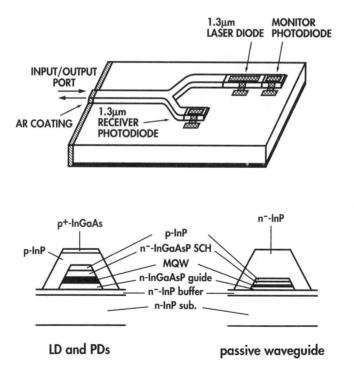

FIGURE 13.29 Conceptual structure of a WDM receiver PIC.

fore been studied and developed. For such a subscriber-use device, cost reduction as well as high performance is essential. There have accordingly been several investigations of transceiver devices, both the semiconductor photonic integrated-circuit (PIC) type [85,86] and the hybrid type using silica waveguides [87]. All the optical components in PICs are monolithically integrated on a semiconductor substrate. Thus the number of optical alignments can be reduced and assembly costs can be lower than they are for hybrid transceivers. The most common fabrication process for PICs, however, requires repetitive etching and regrowth, which in turn reduce uniformity and lower the fabrication yield. A new technology allowing selective-area metal-organic vapor-phase epitaxy is expected to break through the limitations of conventional fabrication techniques because in-plane bandgap control can be attained by using this technology [88], which is introduced in more detail in Chapter 6. The bandgap energy of InGaAs(P) layers can be controlled by varying the mask width. Figure 13.29 shows a conceptual diagram of the WDM transceiver PIC [89]. The active layer of the LDs, the absorption layer of PDs, and the core layers of passive waveguides can be grown simultaneously on a mask-patterned substrate, and this growth is followed by InP regrowth over layers. Active and passive components can be fabricated without complicated etching and partial regrowth and without forming waveguide discontinuities. Thus high device yield, resulting in low device cost, can be expected.

13.6 SUMMARY AND FUTURE CHALLENGES

This chapter has dealt with the design and performance of photodetectors for use in optical fiber communications. New-concept photodiodes using a heterostructure, which was not expected for conventional elemental materials such as Si and Ge, have been proposed and developed. Furthermore, planar InGaAs PIN-PDs with an InP capping layer and InGaAs/InP APDs with a separation of absorption and multiplication configuration are both commercially available.

It is not easy to predict the future of photoreceivers, but high-speed and high-sensitivity receivers will continue to be studied. Multimode waveguide photodiodes have already reached a bandwidth of 110 GHz with a quantum efficiency of 50%, and devices based on new technologies such as the traveling-wave structure and the Fabry–Perot microcavity have the potential to provide even higher speed and higher efficiency. Research to improve the effective ionization rate ratio, resulting in high-speed and low-noise characteristics has been ongoing. The superlattice structure is the most likely candidate to improve the rate artificially. InAlGaAs/InAlAs superlattice APDs have demonstrated gain–bandwidths as high as 150 GHz and maximum bandwidths of over 15 GHz in the low-multiplication region. These new APDs are expected to be used in moderate-span trunkline receivers for 10-Gb/s range transmission.

Cost performance, easy "coupling to the fiber," and multifunctionality are expected to be key words in new device development. Multimode waveguide photodiodes for access networks are expected to have the potential for application to passive alignment. Photonic integrated circuits that handle many wavelengths and many channels will play an important role in the next generation of intelligent systems. There will be a need for continuous evolution in the InP-based materials and their fabrication processing, especially to achieve high uniformity and high production yield.

REFERENCES

1. S. D. Personick, "Receiver design for digital fiber optic communication systems, Parts I and II," *Bell Syst. Tech. J.*, **52**, 843–886 (1973).
2. R. G. Smith and S. D. Personick, "Receiver design for optical communication systems," in *Semiconductor Devices for Optical Communication*, 2nd ed., ed., H. Kressel, Springer-Verlag, New York, 1982, pp. 89–160.
3. R. J. McIntyre, "Multiplication noise in uniform avalanche diodes," *IEEE Trans. Electron Devices*, **13**, 164–168 (1966).
4. G. Lucovsky, R. F. Schwarz, and R. B. Emmons, "Transit-time consideration in p-i-n diodes," *J. Appl. Phys.*, **35**, 622–628 (1964).
5. J. E. Bowers, C. A. Burrus, and R. J. McCoy, "InGaAs PIN photodetectors with modulation response to millimeter wavelengths," *Electron. Lett.*, **21**, 812–814 (1985).
6. R. B. Emmons, "Avalanche-photodiode frequency response," *J. Appl. Phys.*, **38**, 3705–3714 (1967).
7. D. R. Smith, R. C. Hooper, P. P. Smyth, and D. Wake, "Experimental comparison of a germanium avalanche photodiode and InGaAs PINFET receiver for long wavelength optical communication systems," *Electron. Lett.*, **18**, 453–454 (1982).
8. G. E. Stillman and C. M. Wolfe, "Avalanche photodiodes," in *Semiconductor and Semimetals*, ed., R. K. Willardson and A. C. Beer, Vol. 5, Academic Press, San Diego, Calif., 1977, Chap. 5, pp. 291–393.
9. R. K. Willardson and A. C. Beer, *Semiconductors and Semimetals*, Vol. 22, *Lightwave Communication Technologies*, Part D, *Photodiodes*, vol. ed., W. T. Tsang, Academic Press, San Diego, Calif., 1985.
10. S. R. Forrest, R. F. Leheny, R. E. Nahory, and M. A. Pollack, "InGaAs photodiodes with dark current limited by generation-recombination and tunneling," *Appl. Phys. Lett.*, **37**, 322–325 (1980).
11. Y. Takanashi, M. Kawashima, and Y. Horikoshi, "Required donor concentration of epitaxial layers for efficient InGaAsP avalanche photodiodes," *Jpn. J. Appl. Phys.*, **19**, 693–701 (1980).
12. E. O. Kane, "Theory of tunneling," *J. Appl. Phys.*, **32**, 83–91 (1961).
13. A. S. Grove, *Physics and Technology of Semiconductor Devices*, Wiley, New York, 1967, p. 174.
14. C. H. Henry, R. A. Logan, F. R. Merritt, and C. G. Bethea, "Radiative and nonradiative lifetimes in n-type and p-type 1.6 μm InGaAs," *Electron. Lett.*, **20**, 358–359 (1984).
15. D. A. Humphreys, R. J. King, D. Jenkins, and A. J. Moseley, "Measurement of absorption coefficients of $Ga_{0.47}In_{0.53}As$ over the wavelength range 1.0–1.7 μm," *Electron. Lett.*, **21**, 1187–1189 (1985).

16. P. Hill, J. Schlafer, W. Powazinik, M. Urban, E. Eichen, and R. Olshansky, "Measurement of hole velocity in n-type InGaAs," *Appl. Phys. Lett.,* **50,** 1260–1262 (1987).
17. J. E. Bowers and C. A. Burrus, "Ultrawide-band long-wavelength p-i-n photodetectors," *J. Lightwave Technol.,* **5,** 1339–1350 (1987).
18. Y. G. Wey, D. L. Crawford, K. Giboney, J. E. Bowers, M. J. Rodwell, P. Silvestre, M. J. Hafich, and G. Y. Robinson, "Ultrafast graded double-heterostructure GaInAs/InP photodiode," *Appl. Phys. Lett.,* **58,** 2156–2158 (1991).
19. I.-H. Tan, C.-K. Sun, K. S. Giboney, J. E. Bowers, E. L. Hu, B. I. Miller, and R. J. Capik, "120-GHz long-wavelength low-capacitance photodetector with an air-bridge coplanar metal waveguide," *IEEE Photon. Technol. Lett.,* **7,** 1477–1479 (1995).
20. A. Chin and T. Y. Chang, "Enhancement of quantum efficiency in thin photodiodes through absorptive resonance," *J. Lightwave Technol.,* **9,** 321–328 (1991).
21. K. Kishino, M. S. Ünlü, J-I. Chyi, J. Reed, L. Arsenault, and H. Morkoç, "Resonant cavity-enhanced (RCE) photodetectors," *IEEE J. Quantum Electron.,* **27,** 2025–2033 (1991).
22. O. Wada, T. Kumai, H. Hamaguchi, M. Makiuchi, A. Kuramata, and T. Mikawa, "High-reliability flip-chip GaInAs/InP photodiode," *Electron. Lett.,* **26,** 1484–1486 (1990).
23. D. Wake, T. P. Spooner, S. D. Perrin, and I. D. Henning, "50GHz InGaAs edge-coupled pin photodetector," *Electron. Lett.,* **27,** 1073–1075 (1991).
24. K. Kato, S. Hata, K. Kawano, J. Yoshida, and A. Kozen, "A high-efficiency 50 GHz InGaAs multimode waveguide photodetector," *IEEE J. Quantum Electron.,* **28,** 2728–2735 (1992).
25. K. Kato, A. Kozen, Y. Muramoto, Y. Itaya, T. Nagatsuma, and M. Yaita, "110-GHz,50%-efficiency mushroom mesa waveguide p-i-n photodiode for a 1.55 m wavelength," *IEEE Photon. Technol. Lett.,* **6,** 719–721 (1994).
26. Y. Akatsu, Y. Muramoto, K. Kato, M. Ikeda, M. Ueki, A. Kozen, T. Kurosaki, K. Kawano, and J. Yoshida, "Long-wavelength multimode waveguide photodiodes suitable for hybrid optical module integrated with planar lightwave circuit," *Electron. Lett.,* **31,** 2098–2100 (1995).
27. K. Kato, M. Yuda, A. Kozen, Y. Muramoto, K. Noguchi, and O. Nakajima, "Selective-area impurity-doped planar edge-coupled waveguide photodiode (SIMPLE-WGPD) for low-cost, low-power-consumption optical hybrid modules," *Electron. Lett.,* **32,** 2078–2079 (1996).
28. J. B. D. Soole and H. Schumacher, "InGaAs metal–semiconductor–metal photodetectors for long wavelength optical communications," *IEEE J. Quantum. Electron.,* **27,** 737–752 (1991).
29. J. W. Soe, C. Caneau, R. Bhat, and I. Adesida, "Application of indium-tin-oxide with improved transmittance at 1.3 μm for MSM photodetectors," *IEEE Photon. Technol. Lett.,* **5,** 1313–1315 (1993).
30. R.-H. Yuang, J.-I. Chyi, Y.-J. Chan, W. Lin, and Y.-K. Tu, "High-responsivity InGaAs MSM photodetectors with semi-transparent Schottky contacts," *IEEE Photon. Technol. Lett.,* **7,** 1333–1335 (1995).
31. J. H. Kim, H. T. Griem, R. A. Friedman, E. Y. Chan, and S. Ray, "High-performance back-illuminated InGaAs/InAlAs MSM photodetector with a record responsivity of 0.98 A/W," *IEEE Photon. Technol. Lett.,* **4,** 1241–1244 (1992).
32. O. Vendier, N. M. Jokerst, and R. P. Leavitt, "Thin-film inverted MSM photodetectors," *IEEE Photon. Technol. Lett.,* **8,** 266–268 (1996).

33. D. Kuhl, E. H. Böttcher, F. Hieronymi, E. Dröge, and D. Bimberg, "Inductive bandwidth enhancement of sub-m InAlAs-InGaAs MSM photodetectors," *IEEE Photon. Technol. Lett.,* **7,** 421–423 (1995).
34. K. S. Giboney, M. J. W. Rodwell, and J. E. Bowers, "Traveling-wave photodetectors," *IEEE Photon. Technol. Lett.,* **4,** 1363–1365 (1992).
35. K. S. Giboney, R. L. Nagarajan, T. E. Reynolds, S. T. Allen, R. P. Mirin, M. J. W. Rodwell, and J. E. Bowers, "Travelling-wave photodetectors with 172-GHz bandwidth and 76-GHz bandwidth–efficiency product," *IEEE Photon. Technol. Lett.,* **7,** 412–414 (1995).
36. A. G. Dentai, R. Kuchibhotla, J. C. Campbell, C. Tsai, and C. Lei, "High quantum efficiency, long wavelength InP/InGaAs microcavity photodiode," *Electron. Lett.,* **27,** 2125–2127 (1991).
37. K. Hagimoto, K. Iwatsuki, A. Takeda, M. Nakazawa, M. Saruwatari, K. Aida, K. Nakagawa, and M. Horiguchi, "A 212 km non-repeated transmission experiment at 1.8 Gb/s using LD pumped Er^{3+}-doped fiber amplifiers in an IM/direct-detection repeater system," presented at *OFC '89,* Houston, Texas, 1989, postdeadline paper PD15.
38. K. Taguchi, Y. Matsumoto, and K. Nishida, "InP–InGaAsP planar avalanche photodiodes with self-guard-ring effect," *Electron. Lett.,* **15,** 453–455 (1979).
39. K. Nishida, K. Taguchi, and Y. Matsumoto, "InGaAsP heterostructure avalanche photodiodes with high avalanche gain," *Appl. Phys. Lett.,* **35,** 251–253 (1979).
40. K. Taguchi, T. Torikai, Y. Sugimoto, K. Makita, and H. Ishihara, "Planar-structure InP/InGaAsP/InGaAs avalanche photodiodes with preferential lateral extended guard ring for 1.0–1.6 μm wavelength optical communication use," *J. Lightwave Technol.,* **6,** 1643–1655 (1988).
41. S. R. Forrest, G. F. Williams, O. K. Kim, and R. G. Smith, "Excess-noise and receiver sensitivity measurements of $In_{0.53}Ga_{0.47}As$/InP avalanche photodiodes," *Electron. Lett.,* **17,** 917–919 (1981).
42. J. C. Campbell, W. T. Tsang, G. J. Qua, and J. E. Bowers, "InP/InGaAsP/InGaAs avalanche photodiodes with 70 GHz gain–bandwidth product," *Appl. Phys. Lett.,* **51,** 1454–1456 (1987).
43. M. Kobayashi, S. Yamazaki, and T. Kaneda, "Planar InP/InGaAsP/InGaAs buried-structure avalanche photodiode," *Appl. Phys. Lett.,* **45,** 759–761 (1984).
44. H. Ando, Y. Yamauchi, and N. Susa, "Reach-through type planar InGaAs/InP avalanche photodiode fabricated by continuous vapor phase epitaxy," *IEEE J. Quantum Electron.,* **20,** 256–264 (1984).
45. R. Yeats and K. V. Dessonneck, "Detailed performance characteristics of hybrid InP-InGaAsP APDs," *IEEE Trans. Electron Device Lett.,* **2,** 268–271 (1981).
46. S. R. Forrest, O. K. Kim, and R. G. Smith, "Optical response time of $In_{0.53}Ga_{0.47}As$/InP avalanche photodiodes," *Appl. Phys. Lett.,* **41,** 95–98 (1982).
47. Y. Matsushima, S. Akiba, K. Sakai, Y. Kushiro, Y. Noda, and K. Utaka, "High-speed-response InGaAs/InP heterostructure avalanche photodiode with InGaAsP buffer layers," *Electron. Lett.,* **18,** 945–946 (1982).
48. J. N. Hollenhorst, "Fabrication and performance of high speed InGaAs APDs," *Tech. Dig. OFC '90,* San Francisco, 1990, p. 148.
49. L. E. Tarof, J. Yu, R. Bruce, D. G. Knight, T. Baird, and B. Oosterbrink, "High-frequency performance of separate absorption grading, charge, and multiplication InP/InGaAs avalanche photodiodes," *IEEE Photon. Technol. Lett.,* **5,** 672–674 (1993).

50. F. Osaka, T. Mikawa, and T. Kaneda, "Impact ionization coefficients of electrons and holes in (100)-oriented $Ga_{1-x}In_xAs_yP_{1-y}$," *IEEE J. Quantum Electron.*, **21**, 1326–1338 (1985).
51. M. Ito, T. Mikawa, and O. Wada, "Optimization of -doped InGaAs avalanche photodiode by using quasi-ionization rates," *J. Lightwave Technol.*, **8**, 1046–1050 (1990).
52. R. Kuchibhotla and J. C. Campbell, "Delta-doped avalanche photodiodes for high bit-rate lightwave receivers," *J. Lightwave Technol.*, **9**, 900–905 (1991).
53. H. Kuwatsuka, Y. Kito, T. Uchida, and T. Mikawa, "High-speed InP/InGaAs avalanche photodiodes with a compositionally graded quaternary layer," *IEEE Photon. Technol. Lett.*, **3**, 1113–1114 (1991).
54. R. Chin, N. Holonyak Jr., G. E. Stillman, J. Y. Tang, and K. Hess, "Impact ionization in multilayered heterojunction structures," *Electron. Lett.*, **16**, 467–469 (1980).
55. F. Capasso, W. T. Tsang, A. L. Hutchinson, and G. F. Williams, "Enhancement of electron impact ionization in a superlattice: a new avalanche photodiode with a large ionization rate ratio," *Appl. Phys. Lett.*, **40**, 38–40 (1982).
56. T. Kagawa, Y. Kawamura, H. Asai, M. Naganuma, and O. Mikami, "Impact ionization rates in an InGaAs/InAlAs superlattice," *Appl. Phys. Lett.*, **55**, 993–995 (1989).
57. T. Kagawa, H. Asai, and Y. Kawamura, "An InGaAs/InAlAs superlattice avalanche photodiode with a gain bandwidth product of 90 GHz," *IEEE Photon. Technol. Lett.*, **3**, 815–817 (1991).
58. T. Kagawa, Y. Kawamura, and H. Iwamura, "InGaAsP–InAlAs superlattice avalanche photodiode," *IEEE J. Quantum Electron.*, **28**, 1419–1423 (1992).
59. I. Watanabe, K. Makita, M. Tsuji, T. Torikai, and K. Taguchi, "Extremely low dark current InAlAs/InGaAlAs quaternary well superlattice APD," *Proc. 4th Int. Conf. InP and Related Materials,* Newport, R.I., 1992, pp. 246–249.
60. S. Hanatani, H. Nakamura, S. Tanaka, C. Notsu, H. Sano, and K. Ishida, "Superlattice avalanche photodiode with a gain-bandwidth product larger than 100 GHz for very-high-speed systems," *Tech. Dig. OFC/IOOC'93,* San Jose, Calif., 1993, pp. 187–188.
61. K. Makita, I. Watanabe, M. Tsuji, and K. Taguchi, "Dark current and breakdown analysis in In(Al)GaAs/InAlAs superlattice avalanche photodiodes," *Jpn. J. Appl. Phys.*, **35**, 3440–3444 (1996).
62. I. Watanabe, M. Tsuji, K. Makita, and K. Taguchi, "Gain–bandwidth product analysis of InAlGaAs–InAlAs superlattice avalanche photodiodes," *IEEE Photon. Technol. Lett.*, **8**, 269–271 (1996).
63. Y. Miyamoto, K. Hagimoto, M. Ohhata, T. Kagawa, N. Tsuzuki, H. Tsunetsugu, and I. Nishi, "10 Gb/s strained MWQ DFB-LD transmitter module and superlattice APD receiver module using GaAs MESFET IC's," *J. Lightwave Technol.*, **12**, 332–341 (1994).
64. L. D. Tzeng, O. Mizuhara, T. V. Nguyen, K. Ogawa, I. Watanabe, K. Makita, M. Tsuji, and K. Taguchi, "A high-sensitivity APD receiver for 10-Gb/s system applications," *IEEE Photon. Technol. Lett.*, **8**, 1229–1231 (1996).
65. T. Y. Yun, M. S. Park, J. H. Han, I. Watanabe, and K. Makita, "10-gigabit-per-second high-sensitivity and wide-dynamic-range APD-HEMT optical receiver," *IEEE Photon. Technol. Lett.*, **8**, 1232–1234 (1996).
66. K. Taguchi, K. Makita, I. Watanabe, M. Tsuji, and S. Sugou, "InAlGaAs quaternary well superlattice avalanche photodiodes with large gain–bandwidth and low dark current,"

Optoelectron. Devices Technol., **10,** 97–108 (1995).
67. I. Watanabe, M. Tsuji, M. Hayashi, K. Makita, and K. Taguchi, "Reliability of mesa-structure InAlGaAs–InAlAs superlattice avalanche photodiodes," *IEEE Photon. Technol. Lett.*, **8,** 824–826 (1996).
68. I. Watanabe, M. Tsuji, K. Makita, and K. Taguchi, "A new planar-structure InAlGaAs–InAlAs superlattice avalanche photodiode with a Ti-implanted guardring," *IEEE Photon. Technol. Lett.*, **8,** 827–829 (1996).
69. F. Capasso, W. T. Tsang, and G. F. Williams, "Staircase solid-state photomultipliers and avalanche photodiodes with enhanced ionization rates ratio," *IEEE Trans. Electron Devices*, **30,** 381–390 (1983).
70. K. Brennan, "Theoretical study of multiquantum well avalanche photodiodes made from the GaInAs/AlInAs material system," *IEEE Trans. Electron Devices*, **33,** 1502–1510 (1986).
71. M. Tsuji, I. Watanabe, K. Makita, and K. Taguchi, "InAlGaAs staircase avalanche photodiodes," *Jpn. J. Appl. Phys.*, **33,** L32–L34 (1994).
72. M. Tsuji, K. Makita, I. Watanabe, and K. Taguchi, "InAlGaAs impact ionization rates in bulk, superlattice, and sawtooth band structures," *Appl. Phys. Lett.*, **65,** 3248–3250 (1994).
73. M. Tsuji, K. Makita, I. Watanabe, and K. Taguchi, "Band offset dependence on impact ionization rates in InAlGaAs staircase avalanche photodiodes," *Jpn. J. Appl. Phys.*, **34,** L1048-L1050 (1995).
74. R. F. Lehney, R. E. Nahory, M. A. Pollack, A. A. Ballman, E. D. Beebe, J. C. DeWinter, and R. J. Martin, "Integrated $In_{0.53}Ga_{0.47}As$ p-i-n F.E.T. photoreceiver," *Electron. Lett.*, **16,** 353–355 (1980).
75. Y. Akahori, M. Ikeda, A. Kohzen, and Y. Akatsu, "11 GHz ultrawide-bandwidth monolithic photoreceiver using InGaAs pin PD and InAlAs/InGaAs HEMTs," *Electron. Lett.*, **30,** 267–268 (1994).
76. K. Takahata, Y. Muramoto, Y. Akatsu, Y. Akahori, A. Kozen, and Y. Itaya, "10-Gb/s two-channel monolithic photoreceiver array using waveguide p-i-n PD's and HEMT's," *IEEE Photon. Technol. Lett.*, **8,** 563–565 (1996).
77. S. Chandrasekhar, L. M. Lunardi, R. A. Hamm, and G. J. Qua, "Eight-channel p-i-n/HBT monolithic receiver array at 2.5 Gb/s per channel for WDM applications," *IEEE Photon. Technol. Lett.*, **6,** 1216–1218 (1994).
78. L. Lunardi, S. Chandrasekhar, A. H. Gnauck, C. A. Burrus, and R. A. Hamm, "20-Gb/s monolithic p-i-n/HBT photoreceiver module for 1.55-μm applications," *IEEE Photon. Technol. Lett.*, **7,** 1201–1203 (1995).
79. P. Fey, W. Wohlmuth, C. Caneau, and I. Adesida, "15GHz monolithic MODFET-MSM integrated photoreceiver operating at 1.55 μm wavelength," *Electron. Lett.*, **31,** 755–756 (1995).
80. H. Takahashi, Y. Hibino, Y. Ohmori, and M. Kawachi, "Polarization-insensitive arrayed-waveguide wavelength multiplexer with birefringence compensating film," *IEEE Photon. Technol. Lett.*, **5,** 707–709 (1993).
81. S. Chandrasekhar, M. Zirngibl, A. G. Dentai, C. H. Joyner, F. Storz, C. A. Burrus, and L. M. Lunardi, "Monolithic eight-wavelength demultiplexed receiver for dense WDM applications," *IEEE Photon. Technol. Lett.*, **7,** 1342–1344 (1995).
82. J. B. D. Soole, M. R. Amersfoort, H. P. LeBlanc, N. C. Andreadakis, A. Rajhel, and C.

Caneau, "Polarisation-independent monolithic eight-channel 2 nm spacing WDM detector based on compact arrayed waveguide demultiplexer," *Electron. Lett.,* **31,** 1289–1291 (1995).

83. M. Zirngibl, C. H. Joyner, and P. C. Chou, "Polarisation compensated waveguide grating router on InP," *Electron. Lett.,* **31,** 1662–1664 (1995).
84. C. A. M. Steenbergen, C. Dam, A. Looijenl, C. G. P. Herben, M. Kok, M. K. Smith, J. W. Pedersen, I. Moerman, R. G. Baets, and B. H. Verbeek, "Compact low loss 8 × 10 GHz polarisation independent WDM receiver," *Proc. 22th European Conf. Optical Communication,* Oslo, 1996, Vol. 1, pp. 129–132.
85. R. Matz, J. G. Bauer, P. Clemens, G. Heise, H. F. Mahlein, W. Metzger, H. Michel, and G. Schulte-Roth, "Development of a photonic integrated transceiver chip for WDM transmission," *IEEE Photon. Technol. Lett.,* **6,** 1327–1329 (1994).
86. G. M. Foster, J. R. Rawsthorne, J. P. Hall, M. Q. Kearley, and P. J. Williams, "OEIC WDM transceiver modules for local access networks," *Electron. Lett.,* **31,** 132–133 (1995).
87. Y. Yamada, S. Suzuki, K. Moriwaki, Y. Hibino, Y. Tohmori, Y. Akatsu, Y. Nakasuga, T. Hashimoto, H. Terui, M. Yanagisawa, Y. Inoue, Y. Akahori, and R. Nagase, "Application of planar lightwave circuits platform to hybrid integrated optical WDM transmitter/receiver module," *Electron. Lett.,* **31,** 1366–1367 (1995).
88. T. Sasaki, M. Kitamura, and I. Mito, "Selective metalorganic vapor phase epitaxial growth of InGaAsP/InP layers with bandgap energy control in InGaAs/InGaAsP multiple-quantum well structures," *J. Cryst. Growth,* **132,** 435–443 (1993).
89. T. Takeuchi, T. Sasaki, M. Hayashi, K. Hamamoto, K. Makita, K. Taguchi, and K. Komatsu, "A transceiver PIC for bidirectional optical communication fabricated by bandgap energy controlled selective MOVPE," *IEEE Photon. Technol. Lett.,* **8,** 361–363 (1996).

CHAPTER FOURTEEN

Hybrid Integration and Packaging of InP-Based Optoelectronic Devices

WERNER HUNZIKER
Swiss Federal Institute of Technology

14.1. Introduction	538
14.2. Functions of a Package	540
14.2.1 Optical Connection	540
14.2.2 Electrical Connection	541
14.2.3 Thermal Management	543
14.2.4 Protection and Reliability	545
14.3. Optical Connection	547
14.3.1 Coupling Efficiency	547
14.3.2 Coupling Losses	549
14.3.3 Optical Coupling and Low-Cost Packaging	551
14.4. Integrated Waveguide Tapers	554
14.4.1 Lateral Tapers	556
14.4.2 Vertical Tapers	557
14.5. Hybridization on Motherboards	560
14.5.1 Motherboard Materials	562
14.5.2 Silica-on-Silicon Hybridization Boards	563
14.6. Self-Aligned Array Connection Techniques	564
14.6.1 Fiber Array Alignment	565
14.6.2 Alignment with Optical Marks	566

InP-Based Materials and Devices: Physics and Technology, Edited by Osamu Wada and Hideki Hasegawa.
ISBN 0-471-18191-9 © 1999 John Wiley & Sons, Inc.

14.6.3 Solder Bump Alignment   568
14.6.4 Alignment by Mechanical Features on Device
        and Board   569
14.6.5 Device Itself Acting as Mounting Board   572
14.7. Conclusions   573
References   574

14.1 INTRODUCTION

Optical communication systems with their ability to provide very high data transmission have become the driving force for the development of optoelectronic devices. These components are based mainly on III-V semiconductor materials and are combined with passive optical devices. For the latter, the transmission medium itself (i.e., the optical fiber) or planar silica-on-silicon waveguides are most often used. A wide introduction of optical communication systems beyond high-speed long-distance trunk networks into access networks and parallel optical interconnects makes high performance, reliable, low-cost optoelectronic modules mandatory. The optical transport medium, the optical fiber, influences all the packaging levels, from the network down to the component, mounting platform and module level. Analysis shows that the costs of optoelectronic module reside to a dominant extent (i.e., 60 to 90%) in packaging. The remainder are composed 5 to 20% of component processing and 5 to 20% of testing. For low-cost modules it is therefore necessary to reduce the dominant packaging part by choosing appropriate packaging technologies and special device features that diminish packaging complexity and time. Both hybrid and monolithic integration increase functionality within the module without much additional packaging effort.

Monolithic integration reduces the number of interfaces and allows for more complex and advanced functions with higher compactness, factors that made the Si integrated-circuit history so successful. Transfer from the Si to the III-V semiconductor devices is not as straightforward, for several reasons [1]. The Si integration is based primarily on a combination of very few basic cells (transistor, resistor, capacitance) which are compatible in the fabrication process. High-functionality optoelectronic integrated circuits (OEICs) instead include active planar waveguide devices (lasers, optical amplifiers, switches, modulators), passive waveguide devices (splitters, couplers, wavelength multiplexers, interconnections, crossings, tapers), optical surface components [vertical cavity surface-emitting lasers (VCSELs), photodiodes] and electronics (drivers, electrical amplifiers). A large variety of technological processes makes it possible to achieve desired functions through different material combinations and growth and structuring processes. This flexibility is one of the advantages of III-V semiconductor devices that makes it possible to implement almost all functionalities, but also greatly complicates the monolithic integration process. The establishment of standardized basic components, their fabrication processes, and corresponding design tools will make a large step toward a wider application of monolithic integration.

Up to now, monolithic integration has been most successful for devices that are based on very similar material structures and have no or only a few optical on-chip interconnections. Integration of photodiodes with amplifying electronics has shown that similar receiver performances can be achieved as with the hybrid version [2] and that parallel interconnect modules profit from array devices [3]. The realization of different waveguide devices on one chip, sometimes referred to as photonic integrated circuits (PICs), is complicated on the one hand by the combination of different materials or structures (material with gain for amplifiers or lasers, transparency for waveguides, absorption for detectors) and on the other hand by an efficient on-chip optical coupling between them. A wide variety of monolithic integrated devices has been demonstrated. They include coherent receivers [4], wavelength-division-multiplexing (WDM) devices with lasers and/or amplifiers [5], transceivers with photodiodes, and laser and WDM coupler [6], and a few examples are described in Chapters 12 and 13. However, fully monolithic integrated devices are rarely found on the market. Cost studies favor partly monolithic, partly hybrid integration, taking performance and yield into account. The amount of monolithic integration will increase with process developments and component introduction into the mass market.

Hybrid integration together with packaging remains therefore a very challenging topic for the years to come. A large part of the optoelectronic modules will be based on hybridization, also taking silica-on-silicon devices into account, for example. Even for fully monolithic integrated components, solutions for low-cost optical and electrical interconnections are needed. Both hybridization and packaging have many common issues on the lower levels of the packaging hierarchy and will profit from the same developments. A lot of effort has been (and still is) being undertaken to reduce the large packaging part of optoelectronic module costs.

This chapter therefore covers specific problems of optoelectronic module packaging and summarizes the main technical approaches toward cost-effective hybridization and packaging: the use of waveguide mode-shape adapters and self-alignment mounting techniques. First, the primary functions of a module package are given in Section 14.2, focusing on the special problems of optoelectronic devices compared to the electronic devices, to illustrate the requirements of the packaging process. Section 14.3 is dedicated to the specific interconnection problem arising with the optical coupling to fibers. Considerations of coupling efficiencies and needed alignment precision for waveguide-to-fiber coupling give evidence of the important technological developments, which are described next. Section 14.4 deals with the monolithic integration of waveguide tapers that adapt the typically small semiconductor waveguide mode to that of the optical fiber, which has a large impact on the packaging process. They not only improve the coupling efficiency and alignment tolerances but also avoid microoptic components. This reduces the number of components as well as alignment and fixation steps.

In the following two sections we discuss the use of self-alignment technologies for the fiber-to-chip coupling that reduces packaging time and cost, because active driving of the device, its connection to feedback, and control tools can be omitted, and precision requirements of the pick and placing tools are reduced. As a mounting platform for those techniques, processed silicon wafers, also called motherboards or waferboards, have a wide application range through compatibility in processing with the devices themselves and the physical properties of Si. These advantageous features can be used for direct hybridization with silica-on-silicon devices combining a passive waveguide device and mounting platform. In Section 14.5 we describe hybridization using motherboard techniques. The packaging of array components with fiber arrays illustrates the most challenging connection problem, due to the large number of devices and degrees of freedom for the alignment. Solutions using self-alignment principles are described in Section 14.6. These techniques give fundamental packaging approaches also valuable for single-fiber connections for packaging and hybridization. Section 14.7 concludes the chapter.

14.2 FUNCTIONS OF A PACKAGE

The package of an optoelectronic module has to provide defined and stable interfaces between the OEIC and the outside world. This includes the optical connection from the chip to strain-relieved, protected, and eventually connectorized optical fibers, the electrical connection from the chip to case pins or connectors, and a thermal interface allowing heat dissipation and/or temperature stabilization. Furthermore, the module package has to protect the components from different environmental influences to maintain the device properties in a long term. The packaging of OE modules has been based in the early stage primarily on the developments made for electronic ICs. Even though these technologies still cover an important part of the packaging issue, only a few specific remarks can be given in the frame of this book. Additional information can be found in Refs. 7 to 10. The main topic of this chapter is dedicated to the specific problem of the optical interconnection that drives the packaging technology developments for the OE modules and opens the way to new electronic component packaging concepts.

14.2.1 Optical Connection

The optical coupling provides the optical interconnection between the optoelectronic device and the optical fibers to give optical access to the chip from the outside. The involved coupling partners are on one side the fibers, either of multimode (MMF), single-mode (SMF), or polarization-maintaining (PMF) type, as single fiber or as fiber ribbon. The standard geometrical dimensions are 125 μm for the outer glass diameter and 250 μm for the primary plastic coating, which is then protected further by various layers, depending on application. The optical core diameter of the MMFs is usually 50 or 62.5 μm and 9 to 10 μm for the telecommunication SMF for the 1300 and 1550 nm wavelength window. Special SMF, such as dispersion shifted, doped fibers, those for shorter wavelengths or PMF have core diameters in the range 4 to 8 μm. The latter also often exhibit an elliptic mode distribution. The coupling partners on the other side include planar waveguide devices (e.g., lasers, amplifiers, modulators, switches, wavelength selective elements) or surface oriented components (e.g. photodiodes, photoreceivers, VCSELs) as single devices or as one-dimensional or two-dimensional arrays. Figure 14.1 shows two typical planar waveguide components and different optical mode distributions that are normally guided in these devices. The size and shape of these modes represent the important factor of the optical connection. The parameters that characterize the optical interface are the coupling efficiency, alignment tolerances, number of coupling partners, and the sensitivity to optical, thermal, chemical, and mechanical influences. As this interface is specific for OE modules and dominates the needed packaging efforts, it is treated in the next section.

14.2.2 Electrical Connection

The electrical connection provides the electrical interface from the chip electrodes to the module pins or connectors. Each interface within the module affects the transmission properties and is a potential site for failure. A single component between package and chip therefore reduces the number of interfaces, eases the packaging process, and improves reliability. Usually, metallizations on printed circuit boards (PCBs) or ceramic substrates provide the pin-to-chip connection. Special technologies such as wire bonding or soldering make the final step to the chip electrodes. Important parameters characterizing the electrical connection are electrical

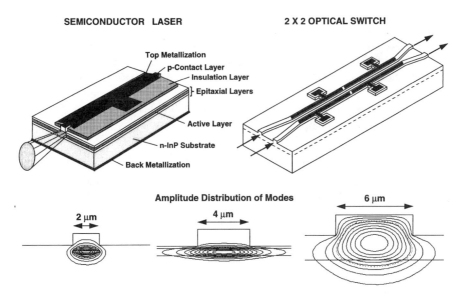

FIGURE 14.1 Two typical planar waveguide devices in OE modules and various amplitude-mode distributions representing lasers, electrooptic switches, or modulators and diluted or tapered waveguide devices used to investigate optical coupling.

impedance, needed number and space of connections, electrical crosstalk, and physical properties of the carrier when used as a chip mounting board. In the case of OE modules especially, the influence of the electrical connection process on the optical coupling also has to be considered.

As optoelectronic devices are dedicated in many applications to broadband information transmission, high-speed electrical connections play an important role in OE packaging. Transmission lines have to provide low-loss, impedance-matched (low electrical reflection) connections from the module case to the chip. Techniques used for microwave component packaging are therefore used in high-speed OE modules. Electrical transmission lines are realized with single- or multilevel metallizations on dielectric substrates. Planar technologies for the transmission lines allow photolithographic structuring for optimal adaptation to impedance and geometry. From the various types of transmission lines [11], microstrip and coplanar lines are employed most often. Table 14.1 provides a comparison of the two. The coplanar type is used primarily because lines and ground are on the same surface, allowing easy shunt configuration without through-plated holes and geometrical adaptation of the linewidth from connector to chip side while maintaining a constant impedance.

For connection from the transmission line to the chip, three main technologies are used: wire bonding, beam leads, or solder bumps. Table 14.2 summarizes important parameters of this techniques. Wire bonding is used in many cases because of the high flexibility and because no specific processes are needed on the chip and transmission-line side. Its main drawbacks for application in OE modules are the limited bandwidth, crosstalk, and the fact that a lot of OE devices are mounted upside down for optical alignment reasons. The beam lead technology has been invented for electronic IC packaging [12] but has not been widely used, as a special technology is needed for their fabrication and chip handling becomes critical. In OE module packaging a special feature of these projecting electrodes, their flexibility, diminishes drastically the influence of the electrical connection process on the optical coupling [13]. Flip-chip solder bump technologies are increasing in electronic packaging due to the access to all chip positions, the multiconnection process, the large bandwidth, and because no extra space besides the chip is needed [14]. For OE applications the solder bumps are smaller in diameter, about 20 μm instead of about

TABLE 14.1 Comparison of Microstrip and Coplanar Transmission Lines

Property	Microstrip	Coplanar
Power-handling capability	High	Medium
Radiation loss	Low	Medium
Area needed	Small	Large
Component mounting in series configuration	Easy	Easy
Component mounting in shunt configuration	Difficult	Easy
Crosstalk	Medium	Small
Geometrical dimensions for given substrate and impedance	Fixed	Variable

TABLE 14.2 Comparison of Transmission Line-to-Chip Connections

Wire Bonding	Beam Leads	Solder Bumps
Upside-up mounting	Upside-down mounting	Upside down (flip-chip)
Au, Al, Pd, Cu, and alloy wires 17–50 μm in diameter	Thickened electrodes projecting over chip edge	Solder balls from deposited and reflowed solder
Thermocompression, ultrasonic or thermosonic bonding	Thermocompression, ultrasonic bonding or soldering	Solder flow control by wettable and nonwettable areas
No special chip processing needed	Needs beam lead fabrication process	Needs solder-wettable chip pads
High flexibility in horizontal and vertical direction	All connections guided to chip edges	Midchip connections possible
Limited high-speed capability	Good high-speed capability	Good high-speed capability
Single-connection technique	Multi- or single-connection technique	Multiconnection technique
Large extra space needed	Small extra space needed	No extra space needed

100 μm, to achieve higher placement precision and also to allow higher bandwidth [15].

Other electrical interconnection technologies combine transmission lines and chip electrode connection. Tape automated bonding (TAB) is a combination of the beam lead technique with the transmission line. TAB uses a thin polymer tape containing a metallic circuitry for chip-to-card connections. The connections from the tape to the chip are called *inner lead bonds*. These bonds are made by dissolving some of the polymer away so that the electrical connections are left as small cantilever beams. They are connected directly to the chip pads by thermal compression or soldering. Connections from the tape to the pins are called *outer lead bonds* and are also carried out by soldering or by thermal compression bonding [16]. The film carrier interconnection technique combines the TAB process with microstrip broadband transmission lines to connect a driving board with the OEIC. It enhances the high-speed performance and reduces the crosstalk [17]. Also, flexible microstrip lines can be used for connector-to-chip interconnections, avoiding stress to the critical optical alignment [18], [19].

14.2.3 Thermal Management

The thermal management plays an important role in the design and realization of OE modules in various ways. The functions of the package include heat dissipation, temperature stabilization for sensitive elements, and maintaining of optical coupling under thermal stress. Important parameters that have to be considered in this packaging aspect are the type and number of components to be packaged, their temperature sensitivity and power consumption, and the type of materials in the package with their thermal conductivity and coefficient of thermal expansion (CTE), not to mention the thermal stress of the packaging process itself.

There are two main properties of InP-based optoelectronic devices that enforce temperature stabilization: the strong temperature dependence of threshold current and quantum efficiency of light-emitting devices and the large temperature sensitivity of the refractive index. The first causes large output intensity changes with temperature. The second produces wavelength shifts of about 0.1 nm/deg for wavelength-selective elements such as DFB and DBR lasers or wavelength demultiplexing devices (WDMs) [20]. In the later case this dependence can be advantageous for some applications, as it allows wavelength tuning over a relatively wide range. The packaging itself becomes more complicated and much more costly because Peltier cooler and temperature sensors have to be incorporated, the number of electrical connections increases, and external control equipment is needed. Hence developments toward less-temperature-sensitive components have a strong impact on low-cost OE modules.

Packages providing good heat dissipation are important for high-power laser applications as MOPAs (master oscillator power amplifiers) and pump lasers for fiber amplifiers and especially for pumping of solid-state lasers where several hundred W/cm^2 of thermal energy have to be dissipated. Submounts with microchannels for liquid cooling are used for these applications [21]. Other OE modules needing heat dissipation are those for broadband applications, including high-speed driving electronics close to the OEIC (e.g., in optical switches, modulators, or transmitters), where the thermal influence from the drivers on the OEIC has to be minimized by the package design. Efficient cooling is important to increase the lifetime of the device and to improve the reliability of the modules.

The reliability aspect in connection with temperature influences is of special concern for OE modules because of the critical optical interfaces. Precise alignment has to be achieved and maintained first at the packaging process itself, where several high temperature processes, such as welding, soldering, and/or curing, are often used. The coupling properties achieved have then to be kept over the operating time of the module for a large temperature range for storage and operation. As the package of an OE module includes a wide variety of materials, including metals, semiconductors, glass, ceramic, and junction and sealing materials, the thermal properties play an important role in material and process selection. Figure 14.2 summarizes the thermal conductivity and CTE for many different materials used in packaging, collected from various sources. The conductivity values from different sources can cover a wide range, especially for compound materials, as they depend on fabrication process and specific structure. Materials of special interest are those with similar CTE values as the III/V semiconductor material, around 5 ppm/deg or below, and with high thermal conductivity for the materials situated close to the active device. This makes Si, AlN, BeO, and CuW alloy the preferred submount materials, although BeO is used reluctantly, as it is hazardous, especially in machining. A parameter for selection is the large electrical resistivity difference between these materials. A low thermal expansion mismatch between materials is important to avoid stress and position shifts for various temperatures and this becomes increas-

14.2 FUNCTIONS OF A PACKAGE   545

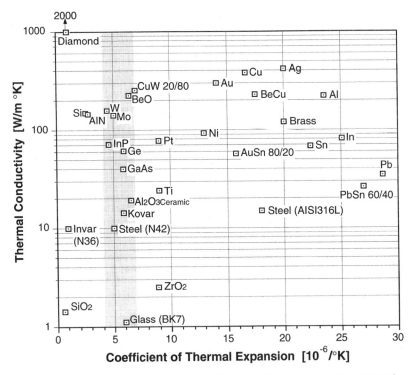

FIGURE 14.2 Thermal conductivity and coefficient of thermal expansion (CTE) for various materials used in packaging. (Courtesy of B. Valk and O. Anthamatten, Ascom Tech, Berne.)

ingly important for growing packaging size. Figure 14.3 gives an example of a hermetic, extended temperature range package of a laser diode using an appropriate selection of materials and arrangement.

14.2.4 Protection and Reliability

A further function of the package is to protect the components from various environmental influences to assure the lifetime of the components and the stability of operation. The influences are of chemical, radiational, and mechanical type. The chemical influences depend a lot on the environment of operation and include different types of gases and humidity. The presence of optical surfaces within packages makes this protection for OE modules particularly important, as deposition on the facets can shorten the lifetime. For full protection, not only hermeticity of the package is required, but also the materials and atmosphere inside the package have to be chosen appropriately. This limits the use of organic compounds, due to their tendency to outgas and adsorb. For high-performance modules with long-lifetime requirements, care has to be taken not to limit the module lifetime compared to the

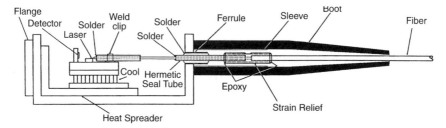

FIGURE 14.3 Layout of a hermetic, extended-temperature laser diode package. (From Ref. 22.)

component itself through packaging-induced failures [23]. Hermetic packaging is very demanding and limits the choice of materials and packaging processes drastically. It will therefore generally be used for high-performance packages and for application in special environments. On the other hand, low-cost modules not providing full hermeticity will also have a large range of applications.

The second type of environmental influences includes the various wavelength ranges of electromagnetic radiation. High-energy or short-wavelength radiation protection is needed against particles in space and nuclear environment or against x-rays in medical applications. Radiation in the optical wavelength range can produce noise and crosstalk, infrared radiation raises the temperature, and microwaves disturb the electrical characteristics. Depending on the specific application, different packaging frames are chosen, of which the metal type provides good protection for most of these interferences. They are also very useful against mechanical perturbations such as shock, impact, vibration, or pressure. As these influences affect primarily the properties of the optical interface, the stability is determined mainly by the fiber-to-chip connection and fixation technique and is sustained by the material type of the frame and layout of the package.

Three main types of fixation of the optical interconnection can be distinguished: gluing, soldering, and welding techniques. The use of organic adhesives and filler materials to connect and protect optical interfaces is promising for low-cost, high-volume packaging. The main concerns are long-term stability and reliability. The coupling to silica-on-silicon passive waveguide devices using ultraviolet-curable glue has shown good reliability results in 5000 h testing under various conditions [24]. This technology takes advantage of different facts: Both waveguides are from the same inert material; the optical modes are large, which results in larger alignment tolerances and avoids optical imaging elements; and adhesive with the same refractive index as that of the waveguides improves the coupling properties by reducing reflections and the effects of surface roughness. Similar results are now achieved using this technique for packaging of $LiNbO_3$ devices [25], and first attempts have been made to mold 1.3 μm lasers entirely with the fiber into a package [26].

Solder is an important junction material for packaging. It replaces organic materials in hermetic packages and provides much lower thermal resistivity, where heat dissipation and thermal stabilization are desired. Solder is used in OE component mounting in various ways. It can be used just to hold the device and fibers in place on separate submounts that are aligned afterward, or it can be used in the form of solder bumps for optical alignment (see Subsection 14.6.3) or the metallized fiber is soldered in place directly using a small amount of solder on the device submount locally heated by an integrated resistor [27].

Laser welding techniques to fix two parts in place have become widely used in OE modules because they provide the high stability that is needed for the optical interface. Pulsed Nd:YAG lasers that are fiber coupled to distribute the light are commonly used for this purpose. OEIC and fibers are normally mounted on separate steel holders which are actively aligned and then welded together, but first attempts have been made to weld a semiconductor device and fiber directly onto a single mounting board [28]. The main problem is the high accuracy in the micron-to-submicron range needed, together with the fact that the welding pulse tends to shift the aligned parts. Improvements are achieved by applying different welding spots symmetrically at the same time, to control precisely the energies of these pulses, or to use correction welding. In this way, the shift during the welding step is measured and further spots are applied to correct the shift, resulting in 0.5 μm alignment precision [29].

14.3 OPTICAL CONNECTION

The optical connection from the OEIC to the optical fibers makes the main difference and represents a great challenge in OE modules compared to electronic IC packaging. On one hand, this interface has very critical characteristics regarding alignment precision and cleanness. On the other hand, packaging becomes further complicated by the large variety of devices, as described in Section 14.2. They are very different in size (tenths of mm^2 to several cm^2) and have different numbers of optical connections on one or two chip sides or even perpendicular to the chip surface. This makes it nearly impossible to define a device-independent interface feature as represented by the bonding pads of the electronic IC. These make it possible to use the same packages and packaging processes for a lot of different devices, as the electrical interface is almost fully decoupled from the device properties and situated at the chip border.

The main parameter used to characterize the optical connection is coupling efficiency, defining the amount of light guided in the coupled waveguide relative to that in the input waveguide. After some theoretical considerations of the coupling efficiency, the main reasons for the loss of waveguide to SMF coupling are discussed below. Based on these results, some conclusions are drawn regarding approaches to low-cost packaging.

14.3.1 Coupling Efficiency

High coupling efficiencies between optical fibers and optoelectronic devices are highly desirable for most applications, as the losses reduce the performances not only of the devices but especially of the systems. Less fiber-coupled power reduces the repeater distance or the performance of diode-pumped devices. It reduces the amplifier gain and the signal-to-noise ratio, or the sensitivity of photodiodes and receivers. The coupling characteristics depend strongly on the types of fibers and devices, and they are influenced by a lot of parameters, as described below.

Coupling into multimode fibers and from fibers to nonguiding components as photodiodes can be described by the numerical aperture and the optical active area, circles in most cases. The numerical aperture, $NA = \sin \theta$, with θ being half the acceptance angle, is given for step-index and graded-index fibers as

$$NA = \sqrt{n_C^2 - n_{Cl}^2} \quad \text{and} \quad NA = \sqrt{n_C^2(r) - n_{Cl}^2} \tag{1}$$

where n_C and n_{Cl} are the core and cladding refractive index, respectively. If NA and the diameter of the optical active area of the coupled device are both larger than those guiding the incident light, there is no coupling loss. In the other cases the NA and/or diameter mismatch loss becomes

$$\text{loss(dB)} = 10 \log\left(\frac{NA_c}{NA_i}\right)^2 \quad \text{or} \quad 10 \log\left(\frac{\text{diameter}_c}{\text{diameter}_i}\right)^2 \tag{2}$$

The influence of misalignment, gaps, reflections, and so on, on the optical coupling efficiency are discussed in the frame of the single-mode coupling. This is the most important connection for OE modules, as most OEICs use monomode waveguides and middle-to-long-distance optical communication systems are based on SMF. Furthermore, due to the much smaller mode sizes, the effects of misalignment are close to an order of magnitude bigger, making single-mode coupling the real challenge in OEIC packaging. We therefore concentrate on the characteristics and packaging technologies for single-mode optical connections.

The coupling between guided optical modes was studied in detail by Kogelnik [30] for coupling light between different laser resonators with different geometries and misalignment. The modes of light guiding structures are regarded as transversal electromagnetic (TEM) waves. They are represented by their transverse field distributions $\Psi(x,y)$ or $\Psi(r,\varphi)$. Coupling between incident and coupled modes is then described by the overlap of their field distributions Ψ_i and Ψ_c. The coupled power is given by the power transmission coefficient T (sometimes also referred to as overlap integral η):

$$T = \frac{|\int\int \Psi_i(x, y)\overline{\Psi_c(x, y)}\, dx\, dy|^2}{\int\int |\Psi_i(x, y)|^2\, dx\, dy \int\int |\Psi_c(x, y)|^2\, dx\, dy} \tag{3}$$

If single-mode coupling is considered, optimum coupling is achieved for identical—amplitude and phase!—field distributions Ψ_i and Ψ_c. It is also remarkable that there is no difference between light coupling from the device to the fiber, or vice versa, due to the symmetry of the integral. So the same amount of light is coupled from a small to a large mode as in the opposite direction.

For an investigation of coupling properties the Gaussian beam approximation [31] is a very useful tool, as analytical solutions exist for the overlap integral. Modes of step-index fibers can be approximated very well by Gaussian beams, which allows us to describe nicely the coupling properties between SMFs [32]. Actual mode forms of semiconductor waveguides show not that close Gaussian profiles, but good predictions of coupling characteristics are still possible. The mathematical model is also very useful for coupling arrangements that include optical imaging elements, such as lenses for mode matching. The power transmission coefficients or coupling efficiencies are summarized in Fig. 14.4 for the coupling of two Gaussian beams characterized by their beam waist radii for different waist sizes and the various misalignment possibilities. These equations show that the alignment tolerances become tighter for smaller modes in the case of longitudinal and lateral misalignment, but larger for the tilt. As the differences in mode size and lateral tolerances contribute much more to the coupling losses, larger modes are preferred for coupling (see also below), but more attention then has to be paid to the angular alignment.

14.3.2 Coupling Losses

The optical throughput at the interface between two waveguides is affected by various parameters (see also the summary in Fig. 14.6). They are given partly by the inherent properties of the guided modes in the two waveguides, partly due to the rela-

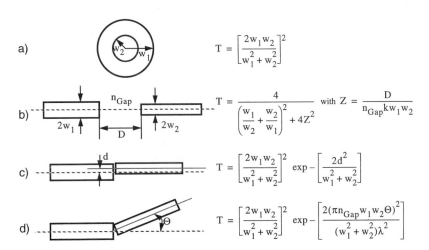

FIGURE 14.4 Power transmission coefficients (or coupling efficiencies) for coupling of two Gaussian beams with (*a*) different beam waist radii w_i as well as for (*b*) longitudinal, (*c*) lateral, and (*d*) angular misalignment. (Based on Ref. 32.)

tive alignment between the modes and partly caused by the interface itself (i.e., the transition between media with different refractive indices). The first two types of properties are strongly related mainly through the mode size, while the interface-related parameters are almost mode-shape independent.

The *difference in mode size and mode symmetry* of the two modes to be coupled causes a large part of the losses. This is illustrated in Fig. 14.5, where the power transmission coefficient is calculated between three different waveguide modes and rotational symmetric Gaussian beams plotted as a function of the beam radius w. The latter stands for the fiber mode either without and with optical imaging element. A beam radius of about 5 μm represents the guided fiber mode in the standard telecom fiber and therefore gives values for butt fiber coupling. It clearly demonstrates the large effect of mode-size mismatch, resulting in about 10 dB coupling loss from lasers [mode distribution and curve (c)] to cleaved fibers and the large improvements that are made possible by larger waveguide structures [part (a)]. Coupling losses can also be strongly reduced by using optical imaging elements, say lenses, to reduce the fiber mode and adapt it to the waveguide mode, reaching values below 1 dB. The effect of difference in mode symmetry is illustrated by the elliptic waveguide mode [part (b)], where the overlap with the rotational mode never results in very low coupling losses. This could be corrected only by using anamorphotical optics to adapt the mode shapes. Another important point is the value of the beam radius w, where the maximum coupling efficiency occurs. The value of about 1 μm for lasers needs an optical imaging system with a very high numerical aperture that can hardly be achieved with glass lenses, whereas values of 2 μm are reached with most types of microlenses. This includes small ball lenses, GRIN

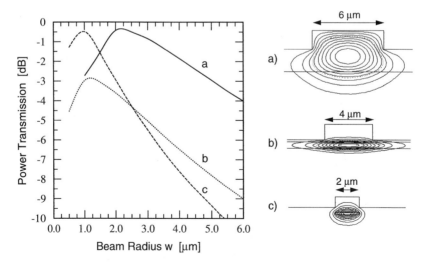

FIGURE 14.5 Calculated power transmission coefficients (or overlap integrals) between different waveguide modes (amplitude mode distribution given on the right) and rotational symmetric Gaussian beams as functions of their beam waist radius w.

(graded index) lenses [33], and lensed fibers, where the fiber end itself is processed to act as a lens [34–38]. The reduction in fiber mode size to very small values for laser coupling drastically increases the efficiency but makes the alignment very critical. The difference in position between the two modes then becomes the most crucial parameter (see Fig. 14.7), which is again relaxed for larger waveguide structures. The *difference in angular alignment,* on the other hand, becomes more critical for larger modes, as described before with the Gaussian beams. Based on this approximation, about 1 dB excess loss results from 9° misalignment for two 3-μm-diameter modes and from about 2° for cleaved fiber coupling with 10 μm diameter.

The *difference in radial profile* in most cases makes a minor contribution to the coupling losses, because the single-mode waveguides guide similar mode profiles that are often approximated by Gaussian beams. Correction would require complicated, mode-specific, aspherical optics.

Optical *reflections* play an important role at the semiconductor–fiber interface, for two reasons. First, semiconductor materials have high refractive indices, close to 3.5. This results in a Fresnel reflection at each chip-to-air transition of about 30%. Second, optical active devices such as lasers and especially amplifiers are very sensitive to back reflections for values above 10^{-5}. Antireflection coatings are therefore applied to the chip facets. To reduce just the transmission loss, single quarter-wave dielectric coatings can be used. For especially sensitive devices, multilayer coatings and direct control during coating process are employed [39]. These layers can also improve the lifetime of high-power devices by preventing facet oxidation and corrosion [40]. An additional way to reduce drastically the back reflections from the facets into the device itself is to tilt the waveguide away from the perpendicular direction to the facet [41]. Angles of 6 to 10° are widely used to ease the reflection problem. But this creates many new packaging problems, as the light is refracted at the facets to angles of 18 to 35°.

Losses due to *surface roughness* can be avoided for III-V semiconductor to waveguide coupling by proper cleaving of chip facets and fibers. They become important if on-chip facets are desirable, for example, for full wafer processing and testing [42] or for motherboards with integrated waveguide devices, mainly silica-on-silicon. Again, due to the smaller modes and the higher refractive index of semiconductor waveguides, this parameter is more crucial than for the silica types.

Figure 14.6 summarizes the various origins of optical coupling loss and shows some techniques used to reduce them. Regarding mode- and alignment-related losses, large, rotational symmetric modes are clearly preferred, as only the angular alignment sensitivity is increased. Interface-related losses do not depend on the alignment but are sensitive to environmental influences during the packaging process and operation.

14.3.3 Optical Coupling and Low-Cost Packaging

In previous sections we have shown clearly that the packaging expenses for optoelectronic devices are dominated by the optical connection. The coupling efficiency between chip waveguides and single-mode fibers plays an important role during the

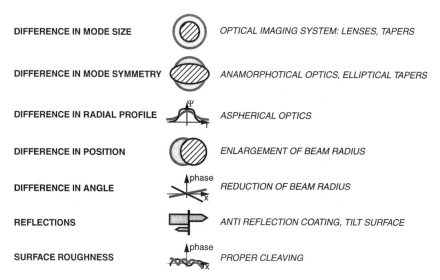

FIGURE 14.6 Summary of origins of optical coupling losses with countermeasures.

packaging process itself but also during module operation to keep the value long term. For typical InP waveguide–SMF connections the difference in mode size largely dominates the losses if no optics are used. As illustrated in Fig. 14.5, only about 10% of the light is coupled, which can be improved using optics to adapt the size of the two modes. But, by doing this, the alignment tolerances, mainly in the lateral directions, become much tighter. Figure 14.7 illustrates this effect. It gives the calculated overlap integrals between the mode distributions shown on top and Gaussian beams with different beam radii w_2 as a function of the lateral displacement between the two. In the case of the typical small mode on the left, a large improvement in coupling efficiency is achieved by reducing the fiber mode of 5 μm radius with optics, say to 2 μm (or magnification of the waveguide mode in the opposite direction by a factor of 2.5). But this results in two negative effects on the alignment tolerances. First, the alignment tolerances for 0.5- and 1 dB excess loss become very small (0.6 and 0.8 μm instead of 1.2 and 1.8 μm), and second, the tolerance curves become much steeper outside the maximum. If the components are fixed in a nonperfectly aligned position, for example, minor shifts then cause large coupling differences of several decibels for less than 0.5 μm displacement. This effect strongly influences the stability and reliability properties of a module, making the trade-off between coupling efficiency and alignment precision very important in view of low-cost packaging and stability. The curves on the right side, illustrating coupling to a larger waveguide mode, clearly demonstrate a double advantage: much better coupling efficiency and flatter alignment tolerance curves.

Optics between typical III-V waveguides and fiber can be used to improve the coupling efficiency. There are different possibilities for realization of optical mode imaging. The use of discrete lenses can keep the tolerances during packaging higher

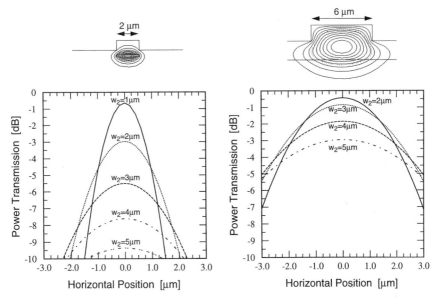

FIGURE 14.7 Calculated coupling efficiencies and lateral alignment tolerances between the waveguide modes given on top and Gaussian beams with different beam radii w_2 as functions of the horizontal displacement.

by placing the fiber at the magnified image from the waveguide mode. The stability problem between chip and lens remains the same, as this shift is also magnified. This approach is very useful if other optical elements, such as isolators, are included in the package or to achieve easier hermeticity, as no fiber feeds through the package are required. Drawbacks of these lens coupling techniques are that at least three or more components per coupling have to be adjusted and fixed in place. Furthermore, it is very difficult to realize array connections. This is eased if lensed fibers are used, because lens and fiber are already aligned and stably connected.

Much easier coupling schemes are possible if the mode size of the waveguide itself is better adapted to the fiber mode. As large-core waveguide structures cause the properties of active devices to deteriorate, the use of mode-shape adapters, also called waveguide tapers, is an important factor in reducing packaging problems. Tapers enlarge the mode on the chip close to the facet and are therefore self-aligned to the waveguide. Improvement in the coupling properties by an enlarged waveguide mode is shown at the right of Fig. 14.7. The coupling efficiencies increase from −9.5 to −3 dB and from −3 to −0.5 dB for cleaved and lensed fibers with a 5- and 2-μm beam radius, respectively. Not only is the efficiency improved, but the alignment tolerances for 0.5 and 1 dB excess loss are also relaxed to values of 0.8 and 1.2 μm for lensed and 1.4 and 2.0 μm for cleaved fibers, respectively. As mentioned earlier, the tolerance curves are flatter, causing fewer power fluctuations for alignment variations.

To overcome cost barriers in OE module packaging, the optical connection process has to be optimized to reduce packaging time and the necessary precision of the alignment and fixation process, by (1) reducing the number of components to be aligned, (2) using coupling schemes with large alignment tolerances, (3) reducing the number and precision of device placing and alignment steps, and (4) avoiding active driving of devices during packaging.

To achieve low-loss optical coupling, the optical modes of the two partners have to be matched and a high-precision alignment technique is needed. Two primarily technological contributions open the way to low-loss, low-cost optical coupling. First, *integrated mode-shape adapters or waveguide tapers* will drastically reduce packaging expenses by (1) increasing the coupling efficiency, (2) increasing the alignment tolerances, (3) reducing the number of parts to be aligned through monolithic integration and by avoiding discrete optical elements, and (4) allowing batch processing to integrate this mode-size adapter. Second, the use of a *mounting board with alignment features* allowing passive self-alignment techniques eases the packaging and reduces the packaging time by (1) reducing the number of alignment steps, (2) reducing the needed precision of the placing tools and processes, (3) avoiding equipment and time-intensive active driving during packaging, and (4) providing stable fixation of coupling parts on a single mounting board.

The next sections are therefore dedicated to the considerable contributions that have been made in these fields within the last years. As already mentioned, progress in other areas of packaging will also help to reduce the cost of OE module packaging (e.g., the developments of less temperature-dependent devices, which would make it possible to omit temperature stabilization). A very important point, especially for optoelectronic devices, is that *packaging aspects have to be taken into consideration in the design, development, and processing of optoelectronic components.*

14.4 INTEGRATED WAVEGUIDE TAPERS

Optical coupling plays a dominant role in optoelectronic packaging, as described earlier. The coupling between single-mode fibers and semiconductor waveguide devices is complicated by the fact that the optical guiding structures in semiconductor devices are based on higher-refractive-index differences between core and cladding than in the fiber. Mode sizes in device waveguides are therefore much smaller than those of the single-mode fibers, say about 2 and 10 μm, respectively. Furthermore, the semiconductor waveguide modes normally exhibit strong asymmetry between vertical and horizontal mode profiles due to the layer structure, the processing techniques used, and the enhanced device properties. The resulting mode mismatch between the device waveguide and the fiber strongly affects the coupling. A mode-shape adaptation must be performed to achieve good coupling efficiencies, and large alignment tolerances should be envisaged for low-cost, stable packaging. Both features can be achieved if the device mode is enlarged at the fiber interface. In such a case, not only are coupling and tolerances improved, but an otherwise necessary optical imaging element is replaced by one that is self-aligned and fixed to the active waveguide, and can be fabricated in a batch process.

Various structures and techniques have been developed within the last few years to enlarge the mode of semiconductor waveguides at the fiber interface. An excellent overview, including results for passive waveguide and laser tapering, has been worked out by Moerman et al. [43]. The mode size is changed along a waveguide normally by changing the thickness of the core, because changes in the refractive indices are technically hardly possible. The effect of the thickness variation of the core on the mode size is illustrated in Fig. 14.8 for a symmetrical slab waveguide structure with typical refractive indices for III-V heterojunction devices. The mode can be enlarged either by increasing the core size (up-tapering), but also by decreasing the thickness to very small values (down-tapering). The latter is very efficient to increase the mode size by small thickness variations but is also very critical for proper dimensioning. The option used depends on the device-specific waveguide structure and the available process technology. There are also very different geometrical suppositions in the vertical and horizontal directions. The first are determined primarily by epitaxial growth, where high resolution on a nanometer scale are easily achieved, but thicknesses of several microns drastically increase processing time and costs. In the second case, the lateral direction (in the taper terminology *lateral* always means horizontal), the dimensions are defined by photolithography. Structures in the micron range are achieved with common technology, whereas well-defined structures in the submicron range need special techniques. So down-tapering is preferred for the vertical direction and up-tapering for the lateral direction. For equivalent reasons of geometry and processing, tapering in the lateral direction

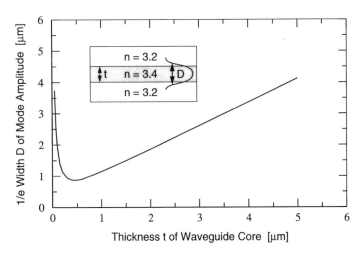

FIGURE 14.8 $1/e$ width of the optical amplitude-mode distribution as a function of the core thickness for a typical InGaAsP/InP symmetric slab waveguide structure at a wavelength of 1550 nm.

is less difficult but becomes really efficient only together with vertical tapering, as otherwise the mode asymmetry increases even more.

The improvement in coupling efficiency using tapered waveguide structures should not be achieved at the expense of much higher waveguide propagation losses. An adiabatic change in the mode size allows for a loss-free mode transition. The minimum length of the taper is depending on the refractive index difference between core and cladding, as well as on the geometrical form of the taper. For weakly guiding waveguide structures, the transition angle has to be very small [44], resulting in a taper length of several hundred microns. A reduction in length can be achieved if more complicated taper forms than linear tapering, such as parabolic or exponential forms, are used [45]. Besides the proper design of the taper, the optical properties of the taper are determined primarily by the limitations of the technology and fabrication process used.

14.4.1 Lateral Tapers

The most advantageous feature of lateral (i.e., horizontal) tapers is that their definition by photolithographic mask and standard etch techniques to realize the desired taper form is straightforward. On the other hand, efficient coupling to symmetrical modes becomes difficult. Pure lateral up-tapering just enlarges the mode in the horizontal direction, complicating efficient coupling by further increasing the mode asymmetry. Therefore, if pure lateral tapers are used, down-tapering is preferred. Today's principal approaches to lateral tapers are summarized in Fig. 14.9. To achieve an enlargement in the mode by down-tapering, the waveguide width has to be reduced to values below 0.8 μm [46] or even below 0.5 μm [47], depending on the device structure. Electron beam lithography or double processing is needed for precise waveguide definition. Down-tapering to very small waveguides or even tapering off to an end also allows an increase in the vertical mode size. But through the slow loss of mode confinement, the mode size at the interface becomes dependent on the cleaving position of the device facet. To guide the enlarged modes, dou-

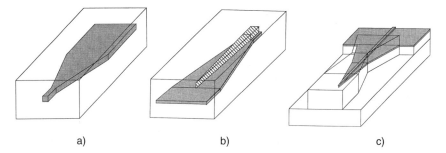

FIGURE 14.9 Main principles of lateral taper structures: (*a*) lateral down-tapered buried waveguide; (*b*) double-layer overlapping buried waveguide taper; (*c*) joint waveguide transition from a ridge waveguide to a fiber-matched waveguide. (From Ref. 43 with permission from Elsevier Science Ltd.)

ble-core waveguides (Fig. 14.9*b*) or joint waveguides with different structures (Fig. 14.9*c*) [48] have been developed.

In double-core devices a second weakly guiding core layer is structured underneath the device waveguide. By tapering off the latter, the increased mode is guided to the fiber interface by the underlying core [49]. This change of the waveguide in the taper region is especially attractive for optically active devices, where for the lower waveguide a higher-bandgap, transparent waveguide material can be used. It allows for short active devices and avoids inefficient pumping of the taper region to avoid absorption for lasers [50] and amplifiers [51]. The passive waveguide section also allows for integration of grating structures for wavelength selection or stabilization [52]. Coupling efficiencies to cleaved single-mode fibers down to 2 to 3 dB are reported with alignment tolerances of 1.5 to 2.5 μm for a 1 dB excess loss. Drawbacks of these double-core structures are the special process technologies needed to achieve very pointed active core taper ends to avoid back reflections, multiprocessing, and regrowth on nonplanar wafers. Furthermore, the design of the device layer structure has to take clearly into account the underlying core layer, so as not to cause the main device properties to deteriorate too much.

14.4.2 Vertical Tapers

The development of vertical tapers at the input and/or output interface of III-V semiconductor devices has become one of the challenges in device developments in the last few years in overcoming the strong vertical mode confinement and related optical coupling problems. From a geometrical point of view, the main vertical taper approaches are very similar to the lateral ones, as illustrated in Fig. 14.10. As described earlier, down-tapering is clearly preferred in this direction. For guiding of the increased mode, the same restrictions apply as for lateral tapers. Larger guiding structures around the device core (Fig. 14.10*c*) or underneath the core [53] have been developed, and combinations with lateral tapering have been realized. But

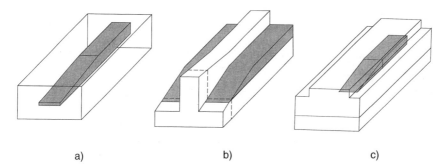

FIGURE 14.10 Main principles of pure vertical taper structures: (*a*) vertical down-tapered buried waveguide; (*b*) vertical down-tapered ridge waveguide; (*c*) overlapping waveguide transition from a buried waveguide to a fiber-matched waveguide. (From Ref. 43 with permission from Elsevier Science Ltd.)

first, the main task is the development of technologies allowing a thickness variation in the waveguide core layer over a long distance on the wafer to achieve adiabatic mode transformation. The two possibilities are special etch processes on the planar core layer and the production of the taper during the growth process.

Figure 14.11 shows some of the special etch techniques investigated for vertical taper fabrication. The dip-etch technique (Fig. 14.11a) makes it possible to control taper shape through the pull-out speed as a function of the etch rate. As this technology is limited to the chip edge and hardly allows additional chip processing, technologies are needed that permit full wafer processing. Attempts have been made using a dynamic mask etch technique where the lower of two masks is attacked much faster by the etchant than the core layer, resulting in large underetching to achieve a slow thickness variation. Realization of a high length/thickness taper ratio and reproducibility are most difficult to achieve. Multistep etching of a core layer with intermediate etch-stop layers (Fig. 14.11b) allows the use of common photolithographic structuring techniques. But for almost adiabatic tapers with low back reflections, a lot of processing steps are required. Another wet etch technique is based on diffusion-limited etching between masked areas on the wafer, resulting in position-dependent etch rates. The complete process for waveguide taper fabrication based on this concept [54] is given in more detail below. A technical approach using dry etching is shown in Fig. 14.11c. A shadow mask with a silicon spacer is fixed on the substrate and placed under an angle in a nitrogen ion milling machine. The taper shape is dependent on the spacer thickness, incidence angle, and shape of the ion beam. Further taper shaping is possible during fabrication by variation of the beam incidence angle [55]. The use of dry etching techniques makes it possible to overcome some of the reproducibility problems of wet etching techniques for high length/thickness aspect ratios.

Figure 14.12 illustrates the two main technologies used to realize vertical tapers during the growth process using MOVPE. Masking of the wafer prior to growth results in position-dependent growth rates during epitaxy. In the shadow masked technique (Fig. 14.12a) [56] the size of the opening restricts the deposition rate. Smaller

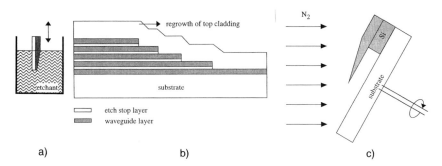

FIGURE 14.11 Etch techniques for vertical taper structures: (a) dip etching; (b) step etching; (c) ion-beam shadow etching. (From Ref. 43 with permission from Elsevier Science Ltd.)

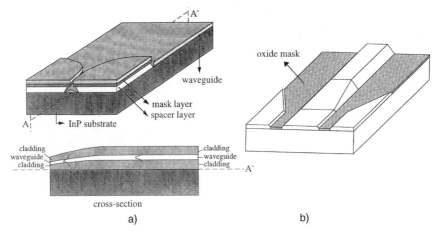

FIGURE 14.12 Growth techniques for vertical taper structures: (*a*) shadow masked growth; (*b*) selective-area growth. (From Ref. 43 with permission from Elsevier Science Ltd.)

mask openings and higher reactor pressure result in higher rate reduction relative to the nominal growth rate on the nonmasked substrate. In this first approach epitaxially grown spacer and shadow mask layers are used. This allows precise thickness control but requires a great effort for mask fabrication and removal. A newer approach uses a silicon shadow mask to fabricate beam expander integrated laser diodes [57]. The selective-area growth technique (Fig. 14.12*b*) uses an in-plane oxide mask that enhances the growth rate in the openings in between [58]. Smaller mask openings, larger oxide areas next to it, and higher reactor pressures result in higher growth enhancement relative to the nonmasked area.

The shadow masked growth and selective-area growth processes are very similar but have an inverse influence on the growth rate, which is reduced in the first and enhanced in the second case in the masked region. As not only the growth rate but also the layer composition are influenced, shadow masked growth is advantageous in the sense that the device area lays in the nonmasked, undisturbed growth part of the wafer. In the selective-area growth process the device waveguides are fabricated in the enhanced growth region, making thickness control and precise layer composition more delicate. Especially for optically active devices where the taper should exhibit a higher-bandgap, transparent core material, the taper core is sometimes grown in a separate step. It also offers the possibility of realizing different types of guiding structures for the increased mode in the taper [59]. Although the fabrication process is complicated by an additional regrowth step to achieve a low-loss, low-reflection interface between the gain and taper region, uniform fabrication of lasers over a full 2-in. wafer has been reported [60]. For multi-quantum-well waveguides the slower-growing thinner part becomes automatically transparent because the bandgap increases. Tapered lasers can be realized with these structures without a waveguide core interface [61]. For good laser properties the appropriate gain region length, which includes a part of the taper, has to be chosen.

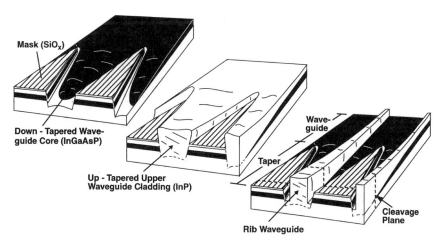

FIGURE 14.13 Fabrication sequence of optical waveguide mode-shape transformer using the same oxide mask for vertical down-tapering of the core with diffusion-limited wet etching and for selective-area regrowth for up-tapering of the upper cladding layer before etching of the laterally guiding rib structure. (From Ref. 54.)

Another taper fabrication process combines a special etching technique with selective-area growth using only a single mask, as illustrated in Fig. 14.13. The oxide mask window tapered down toward the fiber interface enhances the etch rate for smaller openings using a diffusion-limited wet chemical etching process that produces down-tapering of the core layer. The same mask is used afterward for selective growth of the upper cladding. This results in an increased cladding thickness in the taper region, where the mode size is also increased. The rib waveguide structure in the device and taper area is then defined together in a final step, including lateral up-tapering toward the cleaving position.

Optoelectronic devices with good vertical tapers show approximately the same optical coupling properties as those based on lateral tapers: coupling efficiencies of 2 to 3 dB to cleaved fibers and alignment tolerances of 1.5 to 2.5 μm for a 1 dB excess loss. Although some of the processes make it possible to enhance the coupling properties even more by increasing the mode size further, care has to be taken not to degrade the device properties themselves by restrictions caused by the growth and structuring processes (e.g., very thick cladding layers and nonplanar wafer surfaces during device processing). The choice of the taper technology used is a trade-off between a lot of parameters, such as type of device, available process tools and technology, difficulty and number of process steps, optimization of application-specific device properties, yield, device-to-packaging costs, and so on.

14.5 HYBRIDIZATION ON MOTHERBOARDS

In the packaging hierarchy the motherboard or mounting platform acts as a subcarrier for the basic functional devices of a module. For optoelectronic modules this includes passive waveguide components and optoelectronic and purely electronic integrated circuits building an optoelectronic multichip module (MCM). The optoelectronic and electronic chips are mounted directly on the platform together with optical components such as fibers, lenses, and waveguides. The use of monolithic integrated devices reduces the size, number of devices, and number of electrical and optical interconnections and has a large impact on assembly costs. Special array devices are therefore important for parallel optical interconnects. Several technical solutions for the most critical optical waveguide or fiber-to-chip connection with arrays are described in the next section. Besides the high-precision requirements for alignment of the optical components, the motherboard has to provide several other functions, as described next.

Mechanical Function The motherboard acts as a physical support for the devices and the electrical and optical connections. Device materials are Si, GaAs, and InP. Electrical connections are based on structured metal layers; the optical connections include fibers, optical elements such as lenses or mirrors, and even dielectric layers processed to build waveguide structures. The platform has to provide mechanical stability for all the connections on the board against thermal (see below) and mechanical influences (vibrations, shock) during the packaging process and under operation.

Processing Function The platform material should allow the processing of additional features, such as fine electrical interconnections or dielectric layers, and has to withstand the related chemical process influences. Furthermore, an easy and precise machining possibility for the board material itself is very advantageous for optoelectronic platforms to allow the fabrication of alignment features for precise placing of the components.

Thermal Function The thermal properties are very important parameters of a board material. This includes primarily the coefficient of thermal expansion (CTE) and the thermal conductivity, which are given for many materials found in optoelectronic modules in Fig. 14.2. The CTE of the platform has to match as closely as possible those of the mounted devices, to avoid high mechanical stress and component displacements at different temperatures. High thermal conductivity is important to dissipate the heat efficiently away from components that are either temperature sensitive, such as laser diodes or optical amplifiers, or that produce a large amount of heat, such as high-power optical devices or high-speed electronic circuits.

Electrical Function The platform material should be a good electrical insulator with good dielectric properties for broadband applications. The electrical connections between the components and to the outside are then directly patterned into metal layers on the platform. Otherwise, multilayer structures with dielectric spacer layers can be used. This also makes higher integration possible, by direct fabrication of electrical devices within the platform.

14.5.1 Motherboard Materials

The choice of the platform material is determined primarily by the thermal and mechanical properties. To match the CTE of the mounted devices as Si, GaAs, InP, and SiO_2, boards with a CTE of 4 to 7 ppm/K (shadowed area in the graph of Fig. 14.2) are preferred. Ceramics, especially alumina (Al_2O_3), are widely used for electronic MCMs, due to the low cost, high mechanical strength, and good dielectric properties. New ceramics such as BeO and AlN have been developed, particularly to improve the heat dissipation behavior.

Another very important material is silicon, with a high thermal conductivity and a low CTE, in between those of III-V semiconductors materials and optical waveguides or fibers. The nonideal electrical property, showing a much higher electrical conductivity than for ceramics, is more than compensated for in optoelectronic applications by its processing possibilities. The use of IC photolithographic and structuring techniques not only allows the realization of very fine, dense, broadband electrical connections but also opens the way to the fabrication of high-precision alignment features needed in optoelectronic mounting platforms. These features can be realized by structuring metallic or dielectric layers on the silicon or by structuring the platform itself. The crystal structure and wet chemical etch techniques make it possible to realize precise and deep V-grooves or pyramidal-shaped openings. They are used to position fibers or optical components, or to insert alignment features realized on the chips. Figure 14.14 shows an example where the etched grooves are used not only to align the optical fibers, but their endwalls are used as turning mirrors to illuminate photodiodes. The solder bumps used for electrical connection and fixation can even be used to self-align the photodiode chip on the board (see the next section). As further techniques have been developed to realize stable low-loss waveguides in thick dielectric layers based on SiO_2 (silica) directly on the Si wafer, the latter has become the most widely used motherboard material for optoelectronic modules. The main techniques combining silica-on-Si and mounting plat-

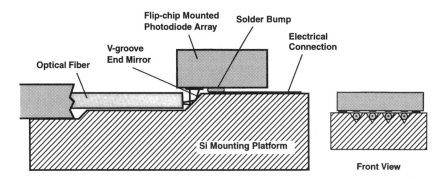

FIGURE 14.14 Flip-chip-mounted photodiode array module on processed Si motherboard with V-grooves for the optical fibers, mirrors on the endwalls, and solder bump electrical connections.

form are summarized next. The use of Si motherboards with alignment features for self-aligned packaging in optoelectronic modules is then described in the following section.

14.5.2 Silica-on-Silicon Hybridization Boards

Various technologies have been developed to realize silica-based single-mode waveguides on a planar substrate with mode sizes similar to those of single-mode fibers (SMF) using specific deposition and structuring techniques. Due to its geometry, surface quality, price, and existing adapted processing equipment, the silicon wafer is very well suited as a substrate for these silica waveguide components. Silica-on-silicon devices are primarily passive optical components such as couplers, power splitters and, with gaining importance for wavelength-division-multiplexing (WDM) applications, wavelength filters, wavelength demultiplexers, and add-drop filters. As they are used mainly directly between SMFs and the fabrication technology makes it possible to realize similar waveguide geometries, the optical modes of silica waveguides and fiber are often well adapted. Good coupling efficiencies are achieved and large alignment tolerances ease the optical coupling. Also using the silicon wafer as a mounting platform offers a possibility of stable connection on a single board and for passive self-alignment. The guiding grooves for the fibers can be defined as part of the waveguide fabrication process, to achieve optimum alignment precision. Main drawbacks of the on-wafer hybridization are the realization of on-chip waveguide facets (no polishing possible) and the coprocessing of waveguides and grooves with large vertical level differences. One realization of on-wafer hybridization is based on etched V-grooves for horizontal and vertical alignment of the fibers in front of the silica waveguide array with a diced facet [62]. Another approach uses etched openings in the silica layer as an etch mask for the silicon U-grooves and the waveguide facets in the same step. The silica opening then defines the horizontal alignment, and the depth of the isotropically etched Si U-groove the vertical position [63].

Even so, most silica devices are passive; they are generally used in systems in combination with active semiconductor components at source, receiver, or in a switching fabric. The hybridization of the silica devices with semiconductor components saves a large amount of packaging effort by reducing the number of components and optical interfaces. Optical connections from silica waveguides to planar active devices such as vertical cavity surface emitting lasers (VCSELs) or photodetectors are less critical in terms of optical alignment, but the light has to change direction by 90°. Light turning mirrors based on metallized etched slopes normally provide the desired deflection, as shown in Refs. 64 and 65 and Fig. 14.14. Edge emitting devices are usually butt coupled to the silica waveguides but need tighter alignment tolerances. If alignment features on the Si board and semiconductor device are combined, the alignment and fixation process is eased. The use of alignment ridge and groove [66] or standoffs and solder bumps [67] for passive alignment are described in more detail in the next section.

Electrical connections and thermal management become important with the hy-

564 HYBRID INTEGRATION AND PACKAGING

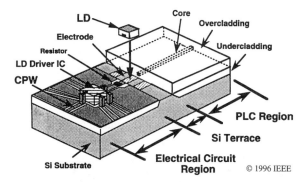

FIGURE 14.15 Silica-on-silicon hybridization platform, including a planar waveguide section (PLC), a silicon terrace for optoelectronic device mounting, and an electrical circuit region with high-speed connections and electronic drivers. (From Ref. 68.)

bridization of silica devices with active components. Whereas the dielectric layer between the silicon and the metallization improves the electrical broadband properties, it drastically reduces the heat dissipation behavior. The use of a terraced Si platform makes it possible to combine the various features [68]. Figure 14.15 shows the hybridization of a high-speed transmitter module with silica waveguide, laser diode, and driving electronics on such a platform. The planar lightwave circuit (PLC) and the electrical circuit are based on a thick dielectric layer providing both optical undercladding and electrical separation from the Si. The laser diode itself is placed on the Si terrace to achieve good heat dissipation.

14.6 SELF-ALIGNED ARRAY CONNECTION TECHNIQUES

Compared to single-fiber pigtailing, the optical coupling of fiber arrays to waveguide arrays introduces additional critical alignment degrees of freedom. These include the distance between the fibers, the alignment of the fibers in a plane, and the adjustment to the same length, as well as the rotation of the array plane relative to the waveguide plane. A large reduction in the number of alignment steps and therefore packaging time and equipment is needed to achieve competitive fiber to waveguide array packaging. In this section we summarize the various types of self-alignment techniques developed in the recent years.

There are different levels of self-alignment, starting from self-alignment of fibers in a V-groove array with subsequent active alignment to the waveguides to almost fully passive self-alignment technologies. Self-alignment means that during the placing or during a further process, a component is guided to a much more precise position than the placing equipment is able to achieve. Active alignment in the common sense means that light guided in the waveguides to be connected is used to optimize the position of the fibers. Light-generating devices such as lasers or amplifiers therefore require electrical connection, driving and thermal management during packaging, and fiber connection to a receiver. For passive waveguide components, source and detector have to be connected to the input and output fibers, and the alignment procedure is further complicated by the fact that both waveguide interfaces have to be taken into account at the same time. Technologies using alignment marks on a chip and mounting platform for optical "active" alignment are also often referred to as passive alignment techniques, as the device waveguide is not used. Passive alignment drastically reduces the number of tools needed and the setup time before packaging. Passive self-alignment requires the lowest preparation effort for the packaging process and no further alignment control during the final positioning step of the components, and is therefore best suited for fully automated packaging.

14.6.1 Fiber Array Alignment

A first step in reducing the high degree of freedom for positions in the fiber array-to-waveguide array coupling are holders for the fiber array. They define precisely the distance between the fibers and their adjustment in one plane. In most cases they consist of silicon with anisotropically etched grooves defined by photolithography. Use of the same technique for the waveguide components and the mounting boards allows for the same geometrical resolution and good control of tolerances. The V-groove-shaped openings provide self-alignment.

Si V-grooves are used extensively, for example, for lithium niobate switch matrices with up to 16 input and outputs [69] and grooved glass substrates for large array connections to silica-on-silicon waveguide devices [70]. In a special application, rectangular U-grooves allow array coupling of special polarization maintaining fibers [71]. A further step in self-alignment uses alignment features for the positioning of fibers along the optical axis in combination with etched stops for butt-ended fibers or with pyramidal holes for lensed fibers [72]. Figure 14.16 shows an example where optical beam-shaping elements, ball lenses in this case, have been self-aligned together with the fiber array on a specially designed Si board [73]. Arrays of fibers aligned on a board and angle polished by about 45° are often used to connect

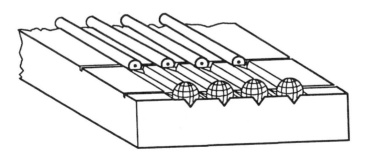

FIGURE 14.16 Silicon fiber-lens array with V grooves for fixing the fibers and pyramidal cavities for holding the ball lenses. (From Ref. 73.)

optically arrays of surface devices (VCSELs, photodetectors). The light deflection makes it possible to fix the fibers and fiber holder in the same plane as the mounting board, with the electrical chip connections easing the module layout and packaging process [74].

14.6.2 Alignment with Optical Marks

Despite the large reduction in the number of alignment steps using fiber array holders, a very critical step remains regarding alignment (three angular directions) and fixation of waveguide device and fiber array. Active alignment can be avoided and the number of degrees of freedom needed for alignment can be reduced by using a mounting board with appropriate alignment features to couple the chip and fiber array.

Figure 14.17 illustrates an index alignment method that uses a transparent auxiliary mounting plate with alignment marks and vacuum manifolds for laser array packaging. First, the fiber array holder and the laser array are aligned separately using fiducial marks on the chips and the alignment plate and fixed to the latter by

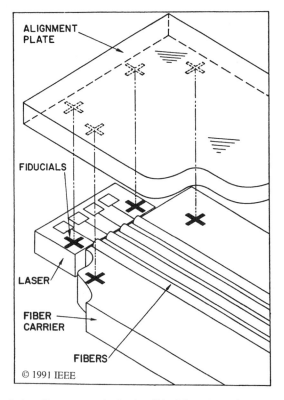

FIGURE 14.17 Index alignment method using fiducial marks on laser array chip, fiber array carrier, as well as on the bottom of the overlying glass alignment plate. (From Ref. 75.)

vacuum. With further marks the mounting plate with chips is aligned to a substrate. Laser array and fiber holder are then fixed in place by solder on the substrate metallization [75]. Critical points of the process are vertical alignment, component holder, alignment procedure, and fixing in the desired position on another board.

Further reduction in alignment and fixing steps, together with improvement in stability and reliability, is achieved using the fiber array holder itself also as a base for the waveguide component. As in most cases, Si V-grooves are used to align and hold the fibers, the optical axis lies close to the Si surface, which leads to the consequence that the waveguide device has to be turned over (flip-chip mounting). This makes necessary the use of other electrical connection techniques, as the chip electrode pads are no longer accessible for wire bonding. Soldering technologies make is possibile to obtain multielectrical connections and fixation in one step. Together with the possibility of realizing midchip connections, good heat dissipation, and electrical high-speed properties, solder connections are widely used for flip-chip mounting of components.

Alignment with marks becomes more difficult, as the processed surfaces of substrate and optoelectronic device including the precise marks face each other. Figure 14.18 shows the two main alignment principles that have been developed and commercialized as flip-chip aligner bonders. As the Si, GaAs, and InP wafers are transparent in the near infrared-to-infrared wavelength region, such light can be used to visualize metallic marks by light transmission through substrate and device [76,77]. Parallelism of substrate and die, which is important for large devices with numerous solder connections, can be controlled with an autocollimator. Magnified images of substrate and die marks are used for rotational and lateral alignment. Before soldering, only a small movement is needed to close the small gap between die and mounting board (Figure 14.18a). In the case of a system using double magnifying

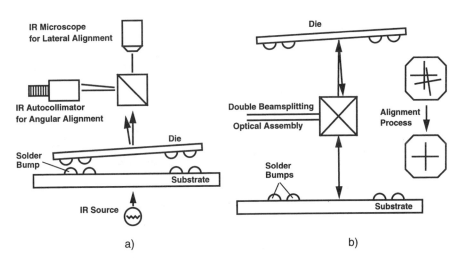

FIGURE 14.18 Principle of flip-chip aligner bonders using IR light in transmission (*a*) or double-side optics between substrate and chip (*b*) for alignment with marks.

optics in between the components to align them, large high-precision translation stages are needed (Figure 14.18b). As both surfaces of the devices are made visible, no restrictions for substrate and die materials apply and "standard" optics and cameras for visible light can be used. Some equipment also makes it possible to use a fluxless thermocompression bonding of the die to specially designed metal pads on the substrate [78]. Solder processes without flux are difficult to achieve, and flux residues are especially critical on the optical surfaces of optoelectronic devices. Both alignment and bonding techniques achieve alignment precision in the range 1 to 2 μm which is not enough for efficient coupling to array devices with small modes but is sufficient for butt coupling and larger waveguide modes.

14.6.3 Solder Bump Alignment

The solder bump technology has been developed for the flip-chip packaging of electronic integrated circuits before the realization of optoelectronic components [79]. Solder bumps feature multiconnections in one step, allow for midchip connections, avoid excess area on the mounting board, have large broadband capability, and fix the chip at the same time. Figure 14.19a shows the typical structure of a solder bump. The nonwettable dielectric layer avoids solder flow and attachment besides the defined pads. Under bump metallurgy (UBM) is important to provide good electrical contact and adhesion to the chip electrode pads, to avoid atomic diffusion into the device, to provide a base for a strong intermetallic connection to the solder, and to define the precise position of the solder bump. With the thickness of

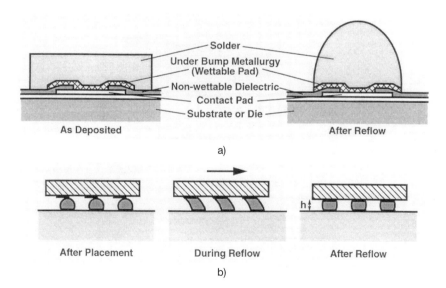

FIGURE 14.19 (a) Typical structure and fabrication of a solder bump for flip-chip mounting with definition of important wettable and nonwettable areas; (b) Self-alignment process between die and substrate by surface tension of liquid solder during reflow.

the deposited solder and the ratio of its area relative to that of the UBM, the solder ball dimensions after reflow can be changed. The surface tension forming the solder ball during the liquid phase can be used to self-align the chip on a mounting board [80]. Figure 14.19b illustrates the principle of self-alignment using solder bumps. A chip is placed on the respective substrate. Large alignment tolerances are possible, because just a section of the wettable pad has to get in contact with the corresponding solder bump. As the solder attaches only to the wettable areas during reflow, they are aligned against each other to minimize the liquid solder surface. During cooling the chip is also pulled down to a certain vertical level. Proper and precise self-alignment depends on various parameters, such as size and number of bumps, type of solder and UBM, and the reflow conditions, including temperature profile, flux, or specific atmosphere [81]. A short overview of soldering technologies for optoelectronic devices is given in Ref. 82.

Primary applications with growing importance are the packaging of photodiodes and photodetector arrays, as the alignment precision requirements are not stringent and the advantages of the technology can be exploited. The use of multiple connections for array devices of parallel optical interconnects [83], the possibility to mount III/V components directly on silicon [84] and GaAs integrated circuits [85], as well as the high-speed properties [86] will assure this technology an important place in optoelectronic device packaging. Furthermore, several developments have been made to self-align waveguide components to SMFs [87] and the hybridization with various components [88] on silicon mounting boards. Alignment accuracies are normally in the range 1 to 2 μm, which limits the single-mode waveguide coupling to the same applications as the one with alignment marks described above.

Although it has been shown that using a larger number (more than 20) of solder bumps with 20 to 30 μm diameter, a lateral alignment precision below 1 μm can be achieved [89], the vertical distance h between chip and substrate remains a critical issue. Spacers on either die or substrate can be used to define the vertical distance [90], as shown in Fig. 14.20a. Nevertheless, the solder bump height remains critical, because these standoffs can affect the lateral alignment process during reflow. The process illustrated in Fig. 14.20b is often referred to also as a solder bump self-alignment process, although lateral and vertical positions are defined by special features on chip and substrate, therefore using an alignment principle, as described in the next subsection. The vertical distance is given by standoffs. Intentionally misaligned wettable pads on die and board together with microstops provide the alignment force and end position [67]. The latter is defined by the alignment features and not the solder bumps themselves.

14.6.4 Alignment by Mechanical Features on Device and Board

The use of mechanical features on the mounting board and the optoelectronic component can overcome the limitations in alignment precision of pick and placement tools, together with the problem of maintaining the alignment during fixation. As these alignment features can be realized with the same fabrication processes as the waveguide devices themselves (e.g., photolithographic definition and dry or wet

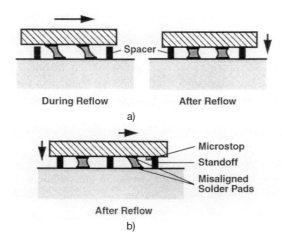

FIGURE 14.20 Combination of solder bump self-alignment process (*a*) with spacers to provide specific vertical distance and (*b*) together with microstops for lateral alignment using intentionally misaligned solder pads to "pull" the chip into position.

etching techniques), the same geometrical precision can be achieved. Minimum deviation on the die is achieved if the waveguide and alignment features are defined by the same mask level. The same holds for the alignment features and fiber V-grooves on the substrate. The passive self-aligned Si V-groove flip-chip packaging technique takes advantages of this and the fact that the slopes of the etched grooves on the Si guide the chip to the precise end position [91,92]. The principle of the self-aligned Si V-groove flip-chip packaging technique is shown in Fig. 14.21 taking the packaging of a space switch matrix as an example [93]. A Si motherboard acts as a mounting platform for the OEIC and the fiber array. The front and back parts include V-grooves for the fibers and the alignment features of the turned-over OEIC; the center part is used for the electrical connections. The optical self-alignment between fibers and device waveguides is illustrated in the inset of Fig. 14.21. On the waveguide device, alignment indentations are etched parallel to the waveguides. The edges of these trenches guide the turned-over chip on the sidewalls of the V-grooves on the Si board into position. Maximum precision is achieved using the same V-groove sidewalls, (111) planes of an (100)-oriented wafer, for the optical alignment of the chip and fibers. Alignment in both the horizontal and vertical directions is thereby defined by a single parameter D, the distance between the waveguide center and alignment edge. The etching of these indentations is not critical regarding the angle of the etched plane as well as smoothness of the surfaces; it is only necessary to ensure that the edge position on top is maintained. The use of the same V-grooves for chip-to-fiber alignment allows large Si etch tolerances while maintaining optical alignment. This is due to the fact that chip and fiber move together vertically, maintaining their relative alignment for different V-groove widths. A further advantage of this alignment technique is the self-positioning effect of the sidewall slope, which guides the fibers and the chip from a relatively loose initial

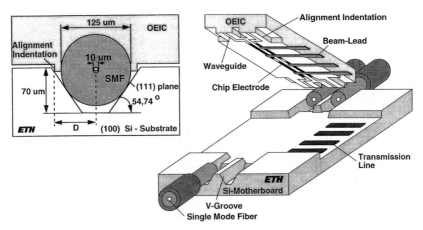

FIGURE 14.21 Principle of passive self-aligned flip-chip OEIC packaging technique on a Si motherboard using V-grooves for precise alignment of single-mode fibers and device waveguides.

position to its precise end position. For a chip insertion depth of 10 µm, the opening in the silicon is 7 µm larger on both sides, which relieves the required placement tool precision. Figure 14.22 shows an SEM picture of a flip-chip-mounted laser array coupled to a fiber ribbon. Alignment trenches on the laser bar fit over alignment mesas on the Si motherboard that are defined and processed in the same steps as the fiber-guiding V-grooves [94].

Submicron alignment precision at the packaging of electrooptical switches and laser arrays has been achieved. A special layout of the alignment features on the chip and the Si motherboard also makes it possible to self-align amplifier arrays with waveguides tilted relative to the facet [95]. This device packaging is complicated by the fact that the light changes its direction strongly at the chip facet, due to the refraction at the large refractive index step. The electrical connections are carried out either by beam leads or by solder areas opposite to the chip electrodes. The beam leads project over the edge of the chip [12] and can be bonded to transmission lines on the motherboard when the device is turned over. Soldering is used for lasers and amplifiers where efficient heat dissipation is needed.

The same alignment technique has been used for the packaging of switch arrays on a silicon board with a through hole in the center to allow wire bonding after turning board and chip back over [96]. Using this technique with prealigned fiber arrays on a holder which then self-aligns on the Si motherboard, eight InP chips with two 1 × 8 switch matrices have been mounted and finally, hybrid assembled to a strictly nonblocking 8 × 8 switch matrix [97]. For the coupling of a laser to a silica-on-silicon waveguide, an alignment ridge on the laser is inserted into a V-groove, providing horizontal alignment. Vertical alignment is done by the solder thickness and laser-to-waveguide distance by a mechanical stop beside the waveguide [98]. The use of mechanical stops for definition of the horizontal alignment and standoffs for

FIGURE 14.22 SEM picture of a passive self-aligned laser-to-fiber array with etched alignment trenches fitting over alignment mesas on the V-grooved Si motherboard.

the vertical alignment is the other common alignment principle, using features on both the chip and mounting platform. The one already described (Fig. 14.20b) uses solder surface-tension forces that have to overcome the friction between the alignment planes to pull the chip into position. Figure 14.23 shows a similar alignment scheme for laser-to-fiber array coupling, where the chip is pushed by the placement tool on the standoffs against the alignment pedestals before soldering [99]. Local heating for the solder process makes it possible to equip the full Si waferboard with optoelectronic devices before dicing, providing an additional step on the way to low-cost automated packaging [100].

14.6.5 Device Itself Acting as Mounting Board

Fiber array-to-waveguide array optical coupling using only a single part in addition to the fibers seems to be the most promising technique considering alignment steps and tolerances as well as stability. This implies that the waveguide chip itself acts as the mounting board. Hence new processing steps have to be introduced, such as photolithographic processing with large vertical level differences and fabrication of on-chip facets. Thus, due to the substrate material, the larger tolerances of the larger modes, and smaller refractive-index difference to air, silica-on-silicon waveguide device packaging has been envisaged first as described in the corresponding sec-

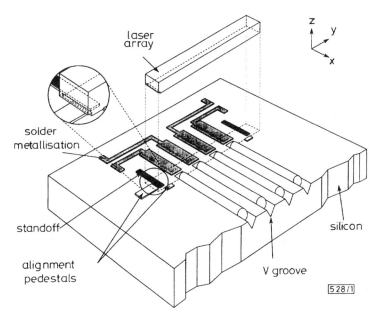

FIGURE 14.23 Waferboard for self-alignment of laser and fiber array with standoffs and pedestals as well as alignment notches on the chip. (From Ref. 99.)

tions. On III-V semiconductor materials, facet fabrication is much more critical, and deep grooves for the fibers are not as easily etched as in silicon. Nevertheless, various approaches have been proposed for laser-to-fiber coupling. One uses a microcleaving process with ultrasonics to prepare the laser facet and active alignment for appropriate positioning of the lensed fiber in the groove [101]. Self-alignment is achieved with a butt-ended fiber in an approach with an etched facet [102]. With mechanical cleaving of the free-standing waveguide structure over the V-groove termination, smooth waveguide facets have been achieved for on-chip waveguide array-to-fiber coupling [103]. In addition to the more complicated fabrication process, direct fiber positioning on the III-V device also suffers from the high material costs as long as the device area is much smaller than the area for fiber alignment and fixing.

14.7 CONCLUSIONS

Hybrid integration and packaging of optoelectronic devices is a major issue for the realization of modules, as they contribute 60 to 90% to the costs. This dominating factor can be reduced by partly increasing the degree of monolithic integration and by choosing appropriate hybridization and packaging processes. Optoelectronic monolithic integration has its limitations by the large variety of specific material combinations, growth, and structuring techniques used to achieve the desired device functionality. Complicated fabrication technologies to integrate electronics with optical waveguides or different types of waveguides with efficient coupling between them affect production costs and yield. Hybrid integration will maintain its importance for optoelectronic modules, also taking the combination with silica-on-silicon waveguide devices into account. The needed technologies are strongly related to the lowest levels of the packaging hierarchy—module realization.

The packaging of optoelectronic modules is based on the one for electronic ICs. Main difference, or more specifically, main addition, is the optical interconnection of devices with optical fibers. Very high alignment requirements (down to the submicron range), coupling losses, cleanliness of the optical facets, and stable fixation with protection against the various environmental influences are the major factors in the optoelectronic packaging challenge. To overcome cost barriers, the optical connection process has to be optimized to reduce packaging time and the needed precision of the alignment and fixation process, by reducing the number of components to be aligned, using coupling schemes with large alignment tolerances, reducing the number and precision of the device placing and alignment steps, and avoiding active driving of devices during packaging.

Two main technological contributions open the way to low-loss, low-cost optical coupling. First, monolithic integrated waveguide tapers adapt the typically small semiconductor waveguide mode to that of the optical fiber. They not only improve the coupling efficiency and alignment tolerances but also avoid microoptic components. This reduces the number of components as well as the alignment and fixation steps. Second, the use of self-alignment technologies for fiber-to-chip coupling reduces packaging time and cost because active driving of the device and its connection to feedback and control tools can be omitted, precision requirements of the pick and placing tools are reduced, and stable fixation on a single mounting platform can be achieved. The Si motherboard concept is the most popular approach with the highest potential for hybridization and packaging. The physical properties and compatibility in processing with the device fabrication itself make it possible to realize mounting platforms with precise alignment features. These can be used for the self-alignment of semiconductor devices to optical fibers or for direct hybridization with silica-on-silicon waveguide components.

Progress in other areas of packaging will also help to reduce the cost of OE module packaging, such as the development of less-temperature-dependent devices that would make it possible to omit temperature stabilization for various components. In essence, for the realization of low-cost optoelectronic modules, packaging aspects have to be taken into consideration at the design, development, and processing stages of optoelectronic components.

REFERENCES

1. M. Erman, "Monolithic vs. hybrid approach for photonic circuits", *Proc. 5th European Conf. Integrated Optics, ECIO'93,* Neuchatel, Switzerland, 1993, pp. 2–1 to 2–3.
2. M. Blaser and H. Melchior, "High-performance monolithically integrated InGaAs/InP p-i-n/JFET optical receiver front-end with adaptive feedback control," *Photon. Technol.*

Lett., **4**(11), 1244–1247 (1992).
3. K.-C. Syao, K. Yang, X. Zhang, G. I. Haddad, and P. Bhattacharya, "Monolithically integrated 16-channel 1.55 μm pin/HBT photoreceiver array with 11.5 GHz bandwidth," *Electron. Lett.*, **33**(1), 82–83 (1997).
4. R. J. Deri, E. C. M. Pennings, A. Scherer, A. S. Gozdz, C. Caneau, N. C. Andreadakis, V. Shah, L. Curtis, R. J. Hawkins, J. B. D. Soole, and J.-I. Song, "Ultracompact monolithic integration of balanced, polarization diversity photodetectors for coherent lightwave receivers," *Photon. Technol. Lett.*, **4**(11), 1238–1240 (1992).
5. C. E. Zah, M. R. Amersfoort, B. Pathak, F. Favire, P. S. D. Lin, A. Rajhel, N. C. Andreadakis, R. Bhat, C. Caneau, and M. A. Koza, "Wavelength accuracy and output power of multiwavelength DFB laser arrays with integrated star couplers and optical amplifiers," *Photon. Technol. Lett.*, **8**(7), 864–866 (1996).
6. P. J. Williams, P. M. Charles, I. Griffiths, N. Carr, D. J. Reid, N. Forbes, and E. Thom, "WDM transceiver OEICs for local access networks," *Electron. Lett.*, **30**(18), 1529–1530 (1994).
7. D. P. Seraphim, R. Lasky, and C.-Y. Li, *Principles of Electronic Packaging,* McGraw-Hill, New York, 1989.
8. R. R. Tummala and E. J. Rymazewski, *Microelectronics Packaging Handbook,* Van Nostrand Reinhold, New York, 1989.
9. C. A. Harper, *Handbook of Microelectronics Packaging,* McGraw-Hill, New York, 1991.
10. M. Pecht, *Integrated Circuit, Hybrid, Multichip Module Package Design Guidelines,* Wiley, New York, 1994.
11. K. C. Gupta, R. Garg, I. Bahl, and P. Bhartia, *Microstrip Lines and Slotlines,* 2nd ed., Artech House, Norwood, Mass., 1996.
12. M. P. Lepselter, "Beam-lead technology," *Bell Syst. Tech. J.*, **45**(2), 233–253 (1966).
13. W. Hunziker, W. Vogt, and H. Melchior, "OEIC packaging with self-aligned fiber to waveguide array coupling by Si V-groove flip-chip technique," *Proc. Integrated Photonics Research, IPR'94,* San Francisco, Calif., February 1994, pp. 335–337.
14. J. H. Lau, *Flip Chip Technologies,* McGraw-Hill, New York, 1996.
15. H. Tsunetsugu, K. Katsura, T. Hayashi, F. Ishitsuka, and S. Hata, "A new packaging technology for high-speed photoreceivers using micro solder bumps," *Proc. 41st Electronic Components and Technology Conf., ECTC'91,* Atlanta, Ga., 1991, pp. 479–482.
16. J. H. Lau, *Handbook of Tape Automated Bonding,* Van Nostrand Reinhold, New York, 1992.
17. M. Usui, K. Katsura, T. Hayashi, M. Hosoya, K. Sato, S. Sekine, and H. Toba, "Novel packaging technique for laser diode arrays using film carrier," *Proc. 43rd Electronic Components and Technology Conf., ECTC'93,* Orlandlo, Fla., 1993, pp. 818–824.
18. J. Schlafer and R. B. Lauer, "Microwave packaging of optoelectronic components," *IEEE Trans. Microwave Theory Tech.*, **38**(5) (1990).
19. C. A. Armiento, A. J. Negri, M. J. Tabasky, R. A. Boudreau, M. A. Rothman, T. W. Fitzgerald, and P. O. Haugsjaa, "Gigabit transmitter array modules on Si waferboard," *IEEE Trans. Components Hybrids Manuf.*, **15**(6), 1072–1080 (1992).
20. E. Gini and H. Melchior, "Thermal dependence of refractive index of InP measured with integrated optical demultiplexer," *J. Appl. Phys.*, **79**(8), 4335–4337 (1996).
21. D. Mundinger, R. Beach, W. Benett, R. Solarz, V. Sperry, and D. Ciarlo, "High average

power edge emitting laser diode arrays on silicon microchannel coolers," *Appl. Phys. Lett.,* **57**(21), 2172–2174 (1990).
22. "Extended temperature package/RWG laser," *Lasertron Commun.,* **5,** March 1990.
23. P. A. Jakobson, J. A. Sharp, and D. W. Hall, "Requirements to avert packaging induced failures (PIF) of high power 980 nm laser diodes," *Proc. IEEE Lasers and Electro-Optics Society Annual Meeting, LEOS'93,* San Jose, Calif., November 1993.
24. Y. Hibino, F. Hanawa, H. Nakagome, M. Ishii, and N. Takato, "High reliability optical splitters composed of silica-based planar lightwave circuits," *J. Lightwave Technol.,* **13**(8), 1728–1735 (1995).
25. A. O'Donnell, "Packaging and reliability of active integrated optical components," *Proc. 7th European Conf. Integrated Optics, ECIO'95,* Delft, The Netherlands, April 1995, pp. 585–590.
26. M. Fukuda, F. Ichikawa, H. Sato, S. Tohno, and T. Sugie, "Pigtail type laser modules entirely moulded in plastic," *Electron. Lett.,* **31**(20), 1745–1747 (1995).
27. O. T. Strand, M. E. Lowry, S. Y. Lu, D. C. Nelson, D. J. Nikkel, M. D. Pocha, and K. D. Young, "Automated fiber pigtailing technology," *Proc. 44th Electronic Components and Technology Conf., ECTC'94,* Washington, D.C., 1994, pp. 1000–1003.
28. M. Becker, R. Güther, R. Staske, R. Olschewsky, H. Gruhl, and H. Richter, "Laser micro-welding and micro-melting for connection of optoelectronic micro-components," *Proc. Int. Congress LASER'93,* Munich, Lasers in Engineering, Springer Verlag, New York, pp. 457–460.
29. B. Valk, R. Bättig, and O. Anthamatten, "Laser welding for fiber pigtailing with long-term stability and submicron accuracy," *Opt. Eng.,* **34**(9), 2675–2682 (1995).
30. H. Kogelnik, "Coupling and conversion coefficients for optical modes," *Microwave Research Institute Symp. Ser.,* Vol. 14, Polytechnic Press, New York, 1964, pp. 333–347.
31. R. D. Guenther, *Modern Optics,* Wiley, New York, 1990.
32. D. Marcuse, "Loss analysis of single-mode fiber splices," *Bell Syst. Tech. J.,* **56**(5), 703–718 (1977).
33. *SELFOC Product Guide,* Nippon Sheet Glass Co., Tokyo.
34. M. C. Farries and W. J. Stewart, "Fibre Fresnel phaseplates with efficient coupling to semiconductor lasers and low reflective feedback," *Proc. European Conf. Optical Communications, ECOC'90,* Amsterdam, The Netherlands, 1990, pp. 291–294.
35. W. Hunziker, E. Bolz, and H. Melchior, "Elliptically lensed polarisation maintaining fibers," *Electron. Lett.,* **28**(17), 1654–1656 (1992).
36. H. M. Presby and C. A. Edwards, "Near 100% efficient fibre microlenses," *Electron. Lett.,* **28**(6), 582–584 (1992).
37. C. A. Edwards, H. M. Presby, and C. Dragone, "Ideal microlenses for laser to fiber coupling," *J. Lightwave Technol.,* **11**(2), 252–257 (1993).
38. K. Shiraishi, N. Oyama, K. Matsumura, I. Ohishi, and S. Suga, "A fiber lens with a long working distance for integrated coupling between laser diodes and single-mode fibers," *J. Lightwave Technol.,* **13**(8), 1736–1744 (1995).
39. M. Serenyi and H.-U. Habermeier, "Directly controlled deposition of antireflection coatings for semiconductor lasers," *Appl. Opt.,* **26**(5), 845–849 (1987).
40. H. Meier, "Recent developments of 980 nm pump lasers for optical fiber amplifiers,"

Proc. European Conf. Optical Communications, ECOC'94, Florence, Italy, 1994, pp. 947–954.
41. P. A. Besse, J. S. Gu, and H. Melchior, "Reflectivity minimization of semiconductor lasers with coated and angled facets considering two dimensional beam profiles," *J. Quantum Electron.,* **27,** 1830–1836 (1991).
42. P. Vettiger, M. K. Benedict, G. Bona, P. Buchmann, E. C. Cahoon, K. Däetwyler, H. P. Dietrich, A. Moser, H. K. Seitz, O. Voegeli, D. J. Webb, and P. Wolf, "Full wafer technology: a new approach to large-scale laser fabrication and integration," *IEEE J. Quantum Electron.,* **27**(6), 1319–1331 (1991).
43. I. Moerman, G. Vermeire, M. D'Hondt, W. Vanderbauwhede, J. Blondelle, C. Coudenys, P. Van Daele, and P. Demeester, "III-V semiconductor waveguide devices using adiabatic tapers," *Microelectron. J.,* **25,** 675–690 (1994).
44. J. D. Love, W. M. Henri, W. J. Stewart, R. J. Black, S. Lacroix, and F. Gonthier, "Tapered single-mode fibers and devices: 1. Adiabaticity criteria," *IEE Proc. J.,* **138**(5), 343–354 (1991).
45. G. R. Hadley, "Design of tapered waveguides for improved output coupling," *Photon. Technol. Lett.,* **5**(9), 1068–1070 (1993).
46. H. Sato, M. Aoki, M. Takahashi, K. Komori, K. Uomi, and S. Tsuji, "1.3 μm beam-expander integrated laser grown by single-step MOVPE," *Electron. Lett.,* **31**(15), 1241–1242 (1995).
47. K. Kasaya, Y. Kondo, M. Okamoto, O. Mitomi, and M. Naganuma, "Monolithically integrated DBR lasers with simple tapered waveguide for low-loss fibre coupling," *Electron. Lett.,* **29**(23), 2067–2068 (1993).
48. Th. Schwander, S. Fischer, A. Krämer, M. Laich, K. Luksic, G. Spatschek, and M. Warth, "Simple and low-loss fibre-to-chip coupling by integrated field-matching waveguide in InP," *Electron. Lett.,* **29**(4), 326–328 (1993).
49. R. Zengerle, B. Jacobs, W. Weiershausen, K. Faltin, and A. Kunz, "Efficient spot-size transformation using spatially separated tapered InP/InGaAsP twin waveguides," *J. Lightwave Technol.,* **14**(3), 448–453 (1996).
50. I. F. Lealman, L. J. Rivers, M. J. Harlow, and S. D. Perrin, " InGaAsP/InP tapered active layer multiquantum well laser with 1.8 dB coupling loss to cleaved singlemode fibre," *Electron. Lett.,* **30**(20), 1685–1687 (1994).
51. B. Mersali, H. J. Brückner, M. Feuillade, S. Sainson, A. Ougazzden, and A. Carenco, "Theoretical and experimental studies of a spot-size tansformer with integrated waveguide for polarization insensitive optical amplifiers," *J. Lightwave Technol.,* **13**(9), 1865–1871 (1995).
52. M. Bachmann, P. Doussiere, J. Y. Emery, R. N'Go, F. Pommereau, L. Goldstein, G. Soulage, and A. Jourdan, "Polarization-insensitive clamped-gain SOA with integrated spot-size convertor and DBR gratings for WDM applications at 1.55 μm wavelength," *Electron. Lett.,* **32**(22), 2076–2077 (1996).
53. P. Albrecht, H. Heidrich, R. Löffler, L. Mörl, F. Reier, and C. M. Weinert, "Integration of polarization independent mode transformers with uncladded InGaAsP/InP rib waveguides," *Electron. Lett.,* **32**(13), 1196–1198 (1996).
54. T. Brenner and H. Melchior, "Integrated optical modeshape adapters in InGaAsP/InP for efficient fiber-to-waveguide coupling," *Photon. Technol. Lett.,* **5**(9), 1053–1056 (1993).

55. G. Wenger, L. Stoll, B. Weiss, M. Schienle, R. Müller-Nawrath, S. Eichinger, J. Müller, B. Acklin, and G. Müller, "Design and fabrication of monolithic optical spot size transformers (MOSTs) for highly efficient fiber-chip coupling," *J. Lightwave Technol.*, **12**(10), 1782–1790 (1994).

56. I. Moerman, M. D'Hondt, W. Vanderbauwhede, G. Coudenys, J. Haes, P. De Dobbelaere, R. Baets, P. Van Daele, and P. Demeester, "Monolithic integration of a spot size transformer with a planar buried heterostructure InGaAsP/InP laser using the shadow masked growth technique," *Photon. Technol. Lett.*, **6**(8), 888–890 (1994).

57. M. Aoki, M. Komori, M. Suzuki, H. Sato, M. Takahashi, T. Ohtoshi, K. Uomi, and S. Tsuji, "Wide-temperature-range operation of 1.3 μm beam expander-integrated laser diodes grown by in-plane thickness control MOVPE using a silicon shadow mask," *Photon. Technol. Lett.*, **8**(4), 479–481 (1996).

58. E. Colas, C. Caneau, M. Frei, E. M. Clausen, Jr., W. E. Quinn, and M. S. Kim, "In situ definition of semiconductor structures by selective area growth and etching," *Appl. Phys. Lett.*, **59**(16), 2019–2021 (1991).

59. O. Mitomi, K. Kasaya, Y. Tohmori, Y. Suzuki, H. Fukano, Y. Sakai, M. Okamoto, and S. Matsumoto, "Optical spot-size converters for low-loss coupling between fibers and optoelectronic semiconductor devices," *J. Lightwave Technol.*, **14**(7), 1714–1719 (1996).

60. H. Okamoto, Y. Suzaki, Y. Tohmori, M. Okamoto Y. Kondo, Y. Kadota, M. Yamamoto, K. Kishi, Y. Sakai, M. Wada, M. Nakao, and Y. Itaya, "1.3 μm laser diodes with butt-jointed selectively grown spot-size converters uniformly fabricated on a 2 inch InP substrate," *Electron. Lett.*, **32**(12), 1099–1101 (1996).

61. T. Yamamoto, H. Kobayashi, T. Ishikawa, T. Takeuchi, T. Watanabe, T. Fujii, S. Ogita, and M. Kobayashi, "Low threshold current operation of 1.3 μm narrow beam divergence tapered-thickness waveguide lasers," *Electron. Lett.*, **33**(1), 55–56 (1997).

62. S. Day, R. Bellerby, G. Cannell, and M. Grant, "Silicon based fibre pigtailed 1 × 16 power splitter," *Electron. Lett.*, **28**(10), 920–922 (1992).

63. G. Grand, H. Denis, and S. Valette, "New method for low cost and efficient optical connections between single mode fibres and silica guides," *Electron. Lett.*, **27**(1), 16–18 (1991).

64. H. Terui, M. Shimokozono, M. Yanagisawa, T. Hashimoto, Y. Yamada, and M. Horiguchi, "Hybrid integration of eight channel PD-array on silica-based PLC using micro-mirror fabrication technique," *Electron. Lett.*, **32**(18), 1662–1664 (1996).

65. M. V. Bazylenko, M. Gross, E. Gauja, and P. L. Chu, "Fabrication of light turning mirrors in buried-channel silica waveguides for monolithic and hybrid integration," *J. Lightwave Technol.*, **15**(1), 148–153 (1997).

66. C. G. Crookes, I. R. Croston, and C. R. Pescod, "Selfaligned integrated silica-on-silicon waveguide–photodiode interface," *Electron. Lett.*, **30**(12), 1002–1003 (1994).

67. K. P. Jackson, E. B. Flint, M. F. Cina, D. Lacey, J. M. Trewhella, T. Caulfield, and S. Sibley, "A compact multichannel transceiver module using planar-processed optical waveguides and flip-chip optoelectronic components," *Proc. 42nd Electronic Components and Technology Conf., ECTC'92,* San Diego, Calif., 1992, pp. 93–97.

68. S. Mino, T. Ohyama, Y. Akahori, T. Hashimoto, Y. Yamada, M. Yanagisawa, and Y. Muramoto, " A 10 Gb/s hybrid-integrated receiver array module using a planar lightwave circuit (PLC) platform including a novel assembly region structure," *J. Lightwave Technol.*, **14**(11), 2475–2482 (1996).

69. P. J. Duthie and M. J. Wale, "16 × 16 single chip optical switch array in lithium niobate," *Electron. Lett.,* **27**(14), 1265–1266 (1991).
70. M. Ishii, Y. Hibino, and F. Hanawa, "Multiple 32-fiber array connection to silica waveguides on Si," *Photon. Technol. Lett.,* **8**(3), 387–389 (1996).
71. G. A. Bogert, E. J. Murphy, and R. T. Ku, "Low crosstalk 4 × 4 Ti $LiNbO_3$ optical switch with permanently attached polarization maintaining fiber array," *J. Lightwave Technol.,* **4**(10), 1542–1545 (1986).
72. K. Mizuishi, T. Kato, H. Inoue, and H. Ishida, "InP-based 4 × 4 optical switch package qualification and reliability," in *Semiconductor Device Reliability,* Kluwer Academic, Norwell, Mass., 1990, pp. 329–342.
73. A. Greil, H. Haltenorth, and F. Taumberger, "Optical 4 × 4 InP switch module with fiber-lens-arrays for coupling," *Proc. European Conf. Optical Communications, ECOC'92,* Berlin, Germany, 1992, pp. 529–532.
74. J. F. Ewen, K. P. Jackson, R. J. S. Bates, and E. B. Flint, "GaAs fiber-optic modules for optical data processing networks," *J. Lightwave Technol.,* **9**(12), 1755–1763(1991).
75. M. S. Cohen, M. F. Cina, E. Bassous, M. M. Oprysko, and J. L. Speidell, "Passive laser-fiber alignment by index method," *Photon. Technol. Lett.,* **3**(11), 985–987 (1991).
76. K. Kurata, K. Yamauchi, A. Kawatani, H. Tanaka, H. Honmou, and S. Ishikawa, "A surface mount type single-mode laser module using passive alignment," *Proc. 45th Electronic Components and Technology Conf., ECTC'95,* Las Vegas, Nev., 1995, pp. 759–765.
77. T. Hashimoto, Y. Nakasuga, Y. Yamada, H. Terui, M. Yanagisawa, K. Moriwaki, Y. Suzaki, Y. Tohmori, Y. Sakai, and H. Okamoto, "Hybrid integration of spot-size converted laser diode on planar lightwave circuit platform by passive alignment technique," *Photon. Technol. Lett.,* **8**(11), 1504–1506 (1996).
78. A. Ambrosy, H. Richter, J. Hehmann, and D. Ferling, "Silicon motherboards for multichannel optical modules," *IEEE Trans. Components Packaging Manuf. Technol. A,* **19**(1), 34–40 (1996).
79. L. F. Miller, "Continued collapse chip joining," *IBM J. Res. Dev.,* **13**(3), 239 (1969).
80. M. J. Wale, C. Edge, F. A. Randle, and D. J. Pedder, "A new self-aligned technique for the assembly of integrated optical devices with optical fibre and electrical interfaces," *Proc. European Conf. Optical Communications, ECOC'89,* Gothenburg, Sweden, 1989, pp. 368–371.
81. J. F. Kuhmann and D. Pech," In situ observation of the self-alignment during FC-bonding under vacuum with and without H_2," *Photon. Technol. Lett.,* **8**(12), 1665–1667 (1996).
82. Q. Tang and Y. C. Lee, "Soldering technology for optoelectronic packaging," *Proc. 46th Electronic Components and Technology Conf., ECTC'96,* Orlando, Fla., 1996, pp. 26–36.
83. J. W. Parker, P. J. Ayliffe, T. V. Clapp, M. C. Geear, P. M. Harrison, and R. G. Peall, "Multifibre bus for rack-to-rack interconnects based on opto-hybrid transmitter/receiver array pair," *Electron. Lett.,* **28**(8), 801–803 (1992).
84. J. Wieland, H. Melchior, M. Q. Kearly, C. R. Morris, A. M. Moseley, M. J. Goodwin, and R. C. Goodfellow, "Optical reciever array in silicon bipolar technology with self-aligned, low parasitic III/V detectors for DC-1Gbit/s parallel links," *Electron. Lett.,* **27**(24), 2211–2213 (1991).

85. O. Wada, M. Makiuchi, H. Hamaguchi, T. Kumai, and T. Mikawa, "High-performance, high-reliability InP/GaInAs p-i-n photodiodes and flip-chip integrated receivers for lightwave communications," *J. Lightwave Technol.,* **9**(9), 1200–1207 (1991).
86. K. Katsura, T. Hayashi, F. Ohira, S. Hata, and K. Iwashita, "A novel flip-chip interconnection technique using solder bumps for high-speed photoreceivers," *J. Lightwave Technol.,* **8**(9), 1323–1325 (1990).
87. J. Sasaki, Y. Kaneyama, H. Honmou, M. Itho, and T. Uji, "Self-aligned assembly technology for optical devices using AuSn solder bumps flip-chip bonding," *Proc. IEEE Lasers and Electro-Optics Society Annual Meeting, LEOS'92,* Boston, Mass., 1992, pp. 260–261.
88. S.-H. Lee, G.-C. Joo, K.-S. Park, H.-M. Kim, D.-G. Kim, and H.-M. Park, "Optical device module packages for subscriber incorporating passive alignment techniques," *Proc. 45th Electronic Components and Technology Conf., ECTC'95,* Las Vegas, Nev., 1995, pp. 841–844.
89. H. Tsunetsugu, T. Hayashi, K. Katsura, M. Hosoya, N. Sato, and N. Kukutsu, "Accurate, stable, high-speed interconnections using 20–30 μm diameter microsolder bumps," *IEEE Trans. Components Packaging Manuf. Technol. A,* **20**(1), 76–82 (1997).
90. T. H. Ju, W. Lin, Y. C. Lee, D. J. McKnight, and K. M. Johnson, " Packaging of a 128 by 128 liquid-crystal-on-silicon spatial light modulator using self-pulling soldering," *Photon. Technol. Lett.,* **7**(9), 1010–1012 (1995).
91. H. Kaufmann, P. Buchmann, R. Hirter, H. Melchior, and G. Guekos, "Self-adjusted permanent attachment of fibres to GaAs waveguide components," *Electron. Lett.,* **22**(12), 642–644 (1986).
92. W. Hunziker, W. Vogt, and H. Melchior, "Self-aligned optical flip-chip OEIC packaging technologies," *Proc. European Conf. Optical Communications, ECOC'93,* Montreux, Switzerland, 1993, Vol. 1, invited papers, pp. 84–91.
93. R. Kraehenbuehl, M. Bachmann, W. Vogt, T. Brenner, H. Duran, R. Bauknecht, W. Hunziker, R. Kyburz, Ch. Holtmann, E. Gini, and H. Melchior, "High-speed low-loss InP space switch matrix for optical communication systems, fully packaged with electronic drivers and single-mode fibers," *Proc. European Conf. Optical Communications, ECOC'94,* Florence, Italy, 1994, pp. 511–514.
94. W. Hunziker, W. Vogt, H. Melchior, R. Germann, and C. Harder, "Low-cost packaging of semiconductor laser arrays using passive self-aligned flip-chip technique on Si motherboard," *Proc. 46th Electronic Components and Technology Conf. ECTC'96,* Orlando, Fla., 1996, pp. 8–12.
95. W. Hunziker, W. Vogt, H. Melchior, D. Leclerc, P. Brosson, F. Pommereau, R. Ngo, P. Doussiere, F. Mallecot, T. Fillion, I. Wamsler, and G. Laube, "Self-aligned flip-chip packaging of tilted semiconductor optical amplifier arrays on Si motherboard," *Electron. Lett.,* **31**(6), 488–490 (1995).
96. B. Acklin, J. Bellermann, M. Schienle, L. Stoll, M. Honsberg, and G. Müller, "Self-aligned packaging of an optical switch array with integrated tapers," *Photon. Technol. Lett.,* **7**(4), 406–408 (1995).
97. G. Wenger, M. Schienle, J. Bellermann, M. Heinbach, S. Eichinger, J. Müller, B. Acklin, L. Stoll, and G. Müller, "A completely packaged strictly nonblocking 8 × 8 optical matrix switch on InP/InGaAsP," *J. Lightwave Technol.,* **14**(10), 2332–2337 (1996).
98. C. A. Jones, K. Cooper, M. W. Nield, J. D. Rush, R. G. Waller, J. V. Collins, and P. J. Fid-

dyment, "Hybrid integration of a laser diode with a planar silica waveguide," *Electron. Lett.,* **30**(3), 215–216 (1994).

99. C. A. Armiento, M. Tabasky, C. Jagannath, T. W. Fitzgerald, C. L. Shieh, V. Barry, M. Rothman, A. Negri, P. O. Haugsjaa, and H. F. Lockwood, "Passive coupling of InGaAsP/InP laser array and singlemode fibres using silicon waferboard," *Electron. Lett.,* **27**(12), 1109–1110 (1991).

100. P. Zhou, R. Boudreau, and T. Bowen, "Wafer scale photonic-die attachment," *Proc. 47th Electronic Components and Technology Conference, ECTC'97,* San Jose, Calif., 1997, pp. 763–767.

101. M. Hamacher, H. Heidrich, and F. Reier, "A novel fibre/chip coupling technique with an integrated strain relief on InP," *Proc. European Conf. Optical Communications, ECOC'92,* Berlin, Germany, pp. 537–540.

102. M. A. Rothman, C. L. Shieh, A. J. Negri, J. A. Thompson, C. A. Armiento, R. P. Holmstrom, and J. Kaur, "Monolithically integrated laser/rear-facet monitor arrays with V-groove for passive optical fiber alignment," *Photon. Technol. Lett.,* **5**(2), 169–171 (1993).

103. A. Bruno, B. Mersali, and L. Menigaux, "Multi-pigtailing using V-grooves and mechanical cleaving on the same InP substrate," *Electron. Lett.,* **33**(12), 1075–1077 (1997).

Index

Abrupt heterojunction, 40, 432
Absorption
  coefficient, 504–505, 511
  microwave/millimeter wave, 39–40
Activation energy, 453, 474
Adaptive cruise control, 39
ADC, see Analog-to-digital converter
Adsorption of reactive species, 292
AES, see Auger electron spectroscopy
a-factor, 463
AFM, see Atomic force microscopy
Air exposure, 327
Air-post structure, 313, 316
Al, easily oxidized, 292
AlAs, 81
AlGaInAs, 93, 96, 100
Alignment, optical
  active, 564
  passive, 563–565
  see also Self-alignment, optical
Alignment tolerances, optical, 539, 541, 549, 552, 554, 569
AlInAs, 93, 96, 99–100
$Al_xIn_{1-x}As$, 88
Alloys
  atom configuration, 90
  characteristics of, generally, 75, 81
  optical properties, 100
AlP, 81
Aluminum-free HFETs, 373
Analog applications, 433–434

Analog circuits, 433–434
Analog-to-digital converters (ADCs), 40, 42, 53, 56
Anisotropic surface diffusion, 210–211
Anisotropy, 74–75
Annealing, see Wafer annealing
APD, see Avalanche photodiode
ArF-excimer-laser, 303
Array connection, 564–565
Arrayed waveguide grating (AWG), 23
Arsine, 188
Aspect ratio, 297, 367
Asynchronous transfer mode (ATM), 16
ATM, see Asynchronous transfer mode
Atmospheric window, 39
Atomic force microscopy (AFM), 168
Auger electron spectroscopy (AES), 303
Autocompensation, 219
Automotive radars, 39–41
Avalanche build-up time, 507, 521
Avalanche photodiode (APD)
  generally, 502, 504, 507, 517–526
  staircase, 525–256
  superlattice, 517, 521–526
Avalanche process, 502, 505, 507

Band alignment, 403
Band discontinuity, 251
Bandgap, generally
  control, 530
  energy, 4–5

583

INDEX

Band lineup
  artificial control of, 260
  broken gap, 252
  charge neutrality level, 255
  generally, 99–100
  lattice matched heterointerfaces, 257–258
  natural, 254–255
  staggered, 258
  strained heterointerfaces, 256, 259–260
  Type I/Type II, 100, 251–252
Band offset, 174, 251–253
Band-to-band tunneling, 508, 518
Bandwidth
  generally, 456
  photodiode, 505–507
Bardeen limit, 262
Barrier height, 86
Base
  current, 400–401, 406
  generally, 393
  resistance, 404, 415
  transit time, 405, 409
Base-collector
  breakdown, 401
  capacitance, 399–401, 414, 427
  heterojunction design, 407–408, 410, 416, 425
  reverse saturation current, 406, 420–422
Base-emitter
  capacitance, 394, 399, 401
  heterojunction, 394–395, 401, 404–405
  turn-on voltage, 395, 423
Beam lead, 542–543, 571
Binary compounds, 73, 81
Boltzmann transport equation, 396
Bond
  length, 73, 90
  radius, 73
  vectors, 77
Boundary layer, 123, 133–134
Bragg mirrors, 471–472
Breakdown
  collector-emitter, 417
  generally, 401, 411–412
  voltage, 401, 407, 412, 417, 432
Broken gap band lineup, 252
Bromium, 292
Buffer layer, 343, 347, 368
Bulk crystals, 110, 134
Bulk growth, 87
Buoyancy forces, 134
Buried facet amplifier, 481–483
Butt coupling, 235

C_2H_6/H_2 gas, 302

Cap layer, 363
Car radar, 4
Carbon contamination, in etching, 303
Carrier, generally
  concentration, 220–222
  diffusion length, 326
  lifetime, 326
  mobility, 308
CBE, see Chemical beam epitaxy
Channel layer, 343
Characteristic temperature, 464
Charge, generally
  buildup, positive, in etching, 321
  control model, 353–354
  neutrality energy level, 255, 271
  transfer, 345–346
Chemical beam epitaxy (CBE), 189
Chirp/chirping, 18, 463, 476
Chirped superlattice, 411, 418
Chlorine, 292
Chlorine-based gases, 292
Circulator, 380
Clean surface, 269
Cleavage surface, 74
Co-doping, 140, 220–221
Coefficient of thermal expansion (CTE), 544, 562
Collector
  charging time, 400
  current, 394, 407, 425, 432
  -emitter offset voltage, 417
  -emitter saturation voltage, 431
  generally, 393
  material, 401–402, 407
  resistance, 393, 398
Collision avoidance radar, 342
Collision warning radar, 39
Common-base current gain, 399
Common-emitter current gain, 394
Communication links, 40–41
Compensation, 140–141
Compensation ratio, 219, 224
Compliant universal (CU) substrate, 251
Composition dependence, 90–91, 93
Conduction band
  discontinuity, 96, 395–396, 403–404, 406–408, 417
  generally, 79, 343, 394
  minimum, 88
  offset, 407, 426
Constitutional supercooling, 134–135
Contamination due to the air-exposure, 330
Continuity equations, 396
Control of interface, 267–268, 273–274

Controllability of the polarization, 316, 318
Coplanar waveguide (CPW), 45, 62, 369, 378–379
Cost-effectiveness, 18–20
Cost reduction, 19, 25–26
Coupling, optical, 19, 513
Coupling efficiency, 539–540, 547–549, 551–552, 554, 561
Covalent bond, 75
CPW, *see* Coplanar waveguide
Cracking efficiency, 192
Cracking pattern, 192
Critical layer thickness, 250, 348–349, 365
Critical resolved shear stress (CRSS), 126, 128–129
Critical thickness, 90, 188
Crystallographic orientation, 131
CuPt ordering, 90
Current, in photodiodes
   dark, 505, 507–508, 510, 523
   diffusion, 510
   drift, 506
   generation-recombination (g-r), 510, 518
   multiplied dark, 505, 507, 517
Current blocking, 408–411, 416–418
Current gain
   compression, 418
   cut-off frequency, *see* f_{max}, f_T
   HBT, 394, 403, 433

DAC, *see* Digital-to-analog converter
Damage reduction, in etching, 310–312
Dangling bonds, 74
DBR laser, *see* Distributed Bragg reflector (DBR) laser
DBS, *see* Direct broadcast satellite
DDS, *see* Direct digital synthesis
Degradation, 86
Delta-doped i-AlGaAs layer, 324
Dense WDM (D-WDM), 14–17
Density of states, 345
Depletion width, 401
Depth monitoring, etching, 309–310
Desorption rate, 173
Device structure, HFET, 343, 345
DEZn, 216
DFB laser, *see* Distributed feedback (DFB) laser
DH, *see* Double heterostructure
DHBT, *see* Double heterojunction bipolar transistor
Dielectric constants, 90–91
Dielectric films, 370
Dielectric midgap energy (DME), 257
Digital applications, 432–433

Digital circuits, 432–433
Digital-to-analog converters (DACs), 40
DIGS (disorder-induced gap state) model, 264–265, 271
Direct broadcast satellite (DBS), 40, 59
Direct digital synthesis (DDS), 42, 61
Direct gap, 100
Direct injection, 118–120
Disilane, 217–218
Dislocation density, 114–115, 126, 133
Disorder-induced gap state model, *see* DIGS (disorder-induced gap state) model
Distributed Bragg reflector (DBR) laser, 18, 467
Distributed feedback (DFB) laser, 17–18, 465–470, 491
Distribution coefficient, 115, 123, 138
Doping, 115, 133–134
Double crucible, 137
Double crystal diffraction mapping (DCDM), 147
Double heterojunction bipolar transistor (DHBT)
   DC characteristics, 416–419
   design, 108, 418, 436
   generally, 53, 58, 392–393, 406–412, 433
   power DHBT, 418–419
   RF characteristics, 427–430
Double heterostructure (DH)
   characteristics of, generally, 88, 451–452, 455, 457
   laser structure, 225–226
Drain and gate I-V characteristics, HEMT, 308
Drift-diffusion approach, 397–398
Drift time, 506
Drift velocity, 93
Dry-etched
   corrugation, 321
   laser, 313–314
Dry etching, 371–372
D-WDM, *see* Dense WDM

EA modulator, *see* Electroabsorption modulator
Eave-shaped lateral growth, 292
EDFA, *see* Erbium doped fiber amplifier
Edge facet, 126
Effective carrier lifetime, 510
Effective mass, 81, 85, 92–93
Effective V/III ratio, 225
Effective workfunction (EWF) model, 264–266
E-k dispersion, 79
Elastic constant, 77
Elastic energy, 76
Electrical connection, 541–543, 557, 563
Electroabsorption, generally, 18, 485–486, 488

Electroabsorption modulated laser (EML), 488–490
Electroabsorption modulator, 18
Electrochemical deposition, 268
Electron beam (EB) lithography, 321
Electronic band structure, 350
Electron mobility, 350, 372
Electron shower, 322
Electron transfer, 85, 93
Electron transport
   characteristics, 404, 430
   generally, 398, 408, 429
   properties, 403
   saturation velocity, 403, 407
Electron velocity, 350, 352
Electroreflection, 485
Embedded SAE, 213–214
Emitter, HBT
   -base charging time, 399
   current, 394
   dynamic resistance, 398
   generally, 394, 412
   injection efficiency, 394, 405
   resistance, 393, 398
Emitter-coupled logic (ECL), 51
Endpoint monitoring, etching, 309–310
Energy band diagram, 393, 395
Energy band structure
   lattice constant vs., 88
   overview, 79–83
Energy gaps, 81, 85, 88, 96
Ensemble Monte Carlo approach, 398
Epitaxial growth, 87, 90, 93, 367–369
Epitaxial layer, 90, 93
Erbium doped fiber amplifier (EDFA), 2, 14
Etched mesa-stripe, 313
Etched mirror laser, 313
Etched surface
   grassy-roughened bottom surface, 297
   smooth, 297
Etching
   anisotropic, 295
   chemically enhanced, 296
   chlorine-based, 295–296
   crystallographic, 297
   hydrocarbon-based, 295, 301–303
   selective, 303–308
   temperature-dependent, 295
Ethane, 292
EXAFS oscillation, 172. *See also* Extended X-ray absorption fine structure
Exciton, 23
Excitonic transitions, 223–224
Ex-situ step, 329

Extended X-ray absorption fine structure (EXAFS), 170, 172
Fabrication process, 369–370
Facet evolution, 226
Fe doping, 237
Fe-InP, 453
Fermi level pinning, 249, 251, 253, 262–265
Fiber
   dispersion compensating fiber, 14
   dispersion shifted fiber, 14
Fiber-to-the-home (FTTH), 3, 13, 19
Field effect transistors (FETs), 43, 59
Flip-chip, 507, 522, 567, 570
Fluorine, 292
f_{max}
   HBT, 400, 426, 428, 436
   HFET, 358
   photodiodes, 507, 510–511
Frequency multipliers, 42, 45
f_T
   HBT, 400, 405, 426, 428, 432, 436
   HFET, 357
FTTH, *see* Fiber-to-the-home
Full-wafer-technology, 322

$Ga_{0.47}In_{0.53}As$, 91–93
GaAs, 74–75, 79, 81, 86, 93
GaAsSb, 96, 100
Gain, optical, 458
Gain-bandwidth (GB) product, 506–507, 520–521, 525
Gain ripple, 483
GaInAs, 81, 96, 99–100
GaInAsP, 93, 96, 98–102
GaInAsP/InP, 188–189
GaInNAs alloy, 5
GaN-based materials, 290
GaP, 79, 81
Gas-etching mode, 297
Gas-surface chemistry, 292
GaSb, 81
Gate leakage, 356, 361–362, 367
Gate-recessed GaAs/AlGaAs wafer, 324
$Ga_xIn_{1-x}As_yP_{1-y}$, 88
GB product, *see* Gain-bandwidth (GB) product
Ge, 74–76
Generation-recombination (G-r) noise, 360
Glass waveguide, 19
Glow discharge mass spectrometry (GDMS), 124–125, 144
Graded index separate confinement heterostructure (GRINSCH), 456, 463
Graded index structure, 456
Gradual channel model (GCM), 354

Grashof number, 120
GRINSCH, 230–232. *See also* Graded-index separate confinement heterostructure
GRIN structure, *see* Graded index structure
Growth interruption, 174, 327
Guard-ring (GR), 502, 517–518, 521, 525
Gummel characteristics, 416
Gunn diodes, 59

Hall mobilities, 222, 224
HBr/F$_2$ gas mixture, 303
HBT, *see* Hetorojunction bipolar transistor
Heat dissipation, 540, 544, 546–547, 562
Heat transfer, 127
Heat transfer plates, 227, 229
Heavy hole, 79, 93
HEMT, *see* High electron mobility transistor
Heterodyne receiver, 492–494
Heteroepitaxy, 96, 98, 103
Heterojunction, 345, 372
Heterojunction bipolar transistor (HBT)
  AlGaAs/GaAs, 53
  analog, 57
  characteristics, 412–415
  circuits, 432–438
  digital, 56–57
  fabrication, 415–416
  InP, 53–58
  input response delay time, 399
  modeling, 392–393
  operation, 392–393
  power, 57–58
  Si/SiGe, 55–56, 61
  turn-on voltage, 395, 432
Heterojunction interface, 170
Heterostructure field effect transistor (HFET)
  generally, 5
  status quo, 374–377
HFET, *see* Heterostructure field effect transistor
High electron mobility transistor (HEMT)
  AlGaAs/GaAs, 44–45
  AlGaAs/GaInAs, 44–45
  digital circuits, 51–52
  drain and gate I-V characteristics, 308
  generally, 5
  InP, 43–52
  low-noise, 46–47
  mixers, 50–51
  power, 48–50
  structure, 308
High frequency noise, 361–362
High-index surface, growth on, 177
Horizontal Bridgeman, 111, 116

Hybridization, 539, 560–561, 563–564
Hydrogen defect, 144–145

ICs
  generally, 61–62
  HBT, 57
Ideality factor, 416
Ideal Schottky limit, 260
II-VI-based materials, 290
III-V compound semiconductors, 72–81, 290
Immiscible region, 96
Impact ionization, 372
IMPATT diode, 59
In, nonvolatile, 292
In-situ characterization, 167, 327–330
In-situ processing, 6, 291
In-situ steps, 329
InAs, 81
Indirect gap, 79
Information transmission volume, 2
InP, 73–74, 79, 81, 85–86, 96, 98–100, 102–103
InP-based materials, 292
Instability of threshold voltage, 311
Insulator deposition, 273
Insulator films, 272–273
Insulator interlayer, 268
Integration
  generally, 561
  hybrid, 538–539, 574
  monolithic, 538–539, 574
Interface, 99
Interface control layer (ICL), 273
Interface dipole, 255
Interface Fermi-level, 86
Interface states, 271
Interfacet diffusion, 208, 210
Interfacial layer theory, 263
Internet, 12
Intervalley scattering, 410
Intervalley separation, 350
Intrinsic lifetime, 326
Ion extraction voltage, 297
Ion extractor, 295
Ionicity, 75
Ionization rate, 502, 505, 507, 518, 521
Ionization rate ratio, 505, 507–508, 521, 525
IR absorption, 145–146
ISDN, 13

Junction temperature, 402, 421–423, 435–436

Kink defects, 270
Kink effect, 355–356
Knee voltage, 417, 426

588 INDEX

LAN, see Local area network
Laser(s), see specific types of lasers
   characteristics of, generally, 450–476
   diode, 438
   driver, 438
   generally, 312, 438
   threshold current, 438
Laser interferometry, 309
Laser-waveguide butt coupling, 235
Lateral DH junctions, 215
Lateral patterning, 177
Lattice constant, 4, 73, 81, 88, 90, 348, 460
Lattice match, 91, 96, 101
Lattice matching, 88, 91, 93, 96, 103, 368, 373
Lattice mismatch, 90, 348
LEC, see Liquid encapsulated Czochralski
Light hole, 79, 93
Linewidth, 467
Linewidth enhancement factor, 463
Liquid encapsulated Czochralski (LEC), 111–112, 120–122, 126–130
LMDS, see Local multipoint distribution services
Local area network (LAN), 3–4, 12, 39, 41, 59
Local multipoint distribution services (LMDSs), 40
Localized grown DH laser, 225
Lorentz force, 123
Low field mobility, 350
Low frequency noise, 359–360, 380
Low noise amplifier (LNA), 41, 45–47, 341, 358–359

Mach-Zehnder modulator, 486, 488
Macrosegregation, 115
Magnetically stabilized liquid-encapsulated Kyropoulos (MLEK), 124–125
Mass transport, 179
Material properties, 402–404
Maximum frequency of oscillation, 358
Maximum oscillation frequency, see f_{max}
MBE, see Molecular beam epitaxy
Mean time to failure (MTTF), 47
Mesa sidewall isolation, 369
Mesa-stripe RIBE, 309
MESFET, GaAs, 43, 51, 59
Metal-induced gap state model, see MIGS (metal-induced gap state) model
Metal organic chemical vapor deposition (MOCVD), 166, 175, 451, 453, 459, 473, 488
Metal-organic molecular beam epitaxy (MOMBE)
   growth system, generally, 189–191
   specific system design criteria, 192–198

   surface-selective growth, 198–222
Metal organic vapor phase epitaxy (MOVPE), 166, 188, 192, 233–234. See also Metal organic chemical vapor deposition (MOCVD)
Metal-semiconductor junction FET, see MESFET
Metamorphic layer, 373
Methane, 292
Methane/ethane-based gases, 292
Microcavity laser, 318
Micro-roughness, 297
Microstrip line (MSL), 378
Microwave
   applications, 435–436
   circuits, 435
   power, 435–436
Microwave-photonics, 4
Midgap energy, 257
MIGS (metal-induced gap state) model, 264–265, 267
Millimeter wave imaging, 39–40
Mirror reflectivity, 458
Missile seekers, 40
Missing dimer, 269–270
Mixers, 42, 45, 50–51, 57, 380
MLEK, see Magnetically stabilized liquid encapsulated Kyropoulos
MMI coupler, see Multi-mode interference (MMI) coupler
MMIC, see Monolithic microwave integrated circuit
Mobility, 92–93, 96, 166, 325
MOCVD, see Metal organic chemical vapor deposition
Mode-lock/mode-locked, 23
Model solid theory, 256–257
Mode shape adapter, 539, 553
Mode size, 548, 550–551, 554
Modulation doping, 345, 352
Molecular beam epitaxy (MBE), 373, 451
Molecular beam geometry, 235
MOMBE, see Metal-organic molecular beam epitaxy
Monolayer step, 75
Monolithic microwave integrated circuits (MMICs)
   characteristics, generally, 4, 43, 50, 57, 61
   status quo, 377–380
Motherboard, 540, 560–563, 570–571
Mott-Schottky limit, 260–261
Mounting platform, 562
MOVPE, see Metal organic vapor phase epitaxy
MQW, see Multiple quantum well
MQW laser, 229–230, 232–233

MTTF, *see* Mean time to failure
Multimode, 540
Multimode interference (MMI) coupler, 21
Multiple quantum well (MQW), 18, 21, 93, 99–100, 102–103, 456–457, 459
Multiplication factor
  generally, 504, 518, 521
  optimum, 506
Multiwafer growth, 227–229
Multi-zone furnace, 131–133

Nanostructure fabrication, 177
Natural band lineup, 254–255
Negative differential resistivity, 85, 93
Noise
  circuit, 505
  circuit, thermal, 506
  equivalent input, 504
  excess, 505, 518
  preamplifier, 505
  shot, 505
  temperature, 363
Nonvolatile reaction product, 292
Nonvolatility indium chlorides, 292
Nonradiative recombination, 311, 326
Nonuniformity, 322
Normal freezing, 134
Numerical simulation, 126

OEIC, *see* Optoelectronic integrated circuit
Ohmic contact, 86, 266–267, 370
ONU, *see* Optical network unit (ONU)
Optical amplifier, 477–484
Optical communication, generally, 88
Optical-communication systems
  technological advances, 2
  trunk-line, 2–3
Optical connection, coupling, 540–541, 543, 547–554
Optical fiber
  butt, 563, 565
  cleaved, 563
  communication, 92
  generally, 538
  lensed, 550–551
  multi mode, 540, 548
  single mode, 540
Optical interconnection, 3–4, 471
Optical loss, 484–485
Optical mode, 540, 549, 552
Optical modulator, 487
Optical network unit (ONU), 17
Optical time-division multiplexing (OTDM), 3, 22–23

Optoelectronic applications, 436–438
Optoelectronic integrated circuits (OEICs), 7, 20–21, 58, 516
Ordering, 90
Organometallic vapor-phase epitaxy (OMVPE), 75
Oscillators, 42, 45, 57, 380
OTDM, *see* Optical time-division multiplexing
Output conductance, 416, 435–436
Overlap integral, 548–549, 552
Oxidation, 272

Packaging
  functions of, 540–547
  generally, 538, 547, 551–554, 566
Passive double star (PDS), 16
PDS, *see* Passive double star
Personal communication systems (PCS), 41
Phase-array, 526–527
Phase diagram, 96
pHEMT, *see* Pseudomorphic HEMT
Phosphine, 188
Phosphorus, 116–117, 119, 122
Photo-assisted gas-surface chemistry, 303
Photocarrier generation, 504
Photochemical mechanism, 305
Photoconducting antenaas, 110
Photocurrent
  diffusion, 505
  gain, 505, 518
  generally, 505
Photodiode (PD)
  generally, 503, 507
  MSM (metal-semiconductor-metal), 516
  PIN, 502, 508–513, 528
  resonant cavity, 517
  traveling-wave, 516
  waveguide PIN, 502, 513–516
Photodetector, 91–92, 392, 436
Photoluminescence (PL)
  generally, 96, 100–101, 178
  scanning, 147–151
  spectra, 311
  time-resolved, 326–327
Photoluminescence excitation (PLE) spectrum, 182–183
Photonic integrated circuit (PIC), 7, 20–21, 235, 455–465, 515
Photonic networks, 14, 20–22
Photorefractive, 110
Physical sputtering, 296
PIC, *see* Photonic integrated circuit
Pile-up, carrier, 506, 517
PL, *see* Photoluminescence

Planar lightwave circuit (PLC), 19
Planar SAE, 206–213
Plasma
  chemistry, 292
  density, 295
  ECR, 294–295
  -induced contamination, 311
  RF, 294–295
  surface treatment, 268
PLC, see Planar lightwave circuit
PLE, see Photoluminescence excitation (PLE) spectrum
Poisson's equation, 345–346, 396, 398
Polarization
  dependence, 23
  insensitive, 23
Polycrystal synthesis, 116–117
Polymerization, 292
Power, HBT
  amplifier, 435, 438
  dissipation, 403–404, 406, 416, 419, 423, 428, 436
  gain, 435
  gain cut-off frequency, see f_{max}
  microwave, 435–436
Power-added efficiency (PAE), HBT, 41, 49, 372
Power amplifiers, 41, 45, 48–50, 57–58
Power transmission coefficient, 549. See also Overlap integral
Precursors, 167, 193
Protection, 545–547
Pseudomorphic HEMT, GaAs, 45–46, 51, 61
Pseudomorphic HFET, 349, 378
p-symmetry, 72
Pulse-doped donor layer, 343

Q-band, 379
Quantum dot
  disk-shaped InGaAs quantum dots, 178–179
  -laser, 177–178
Quantum efficiency, in photodiodes, 505, 508, 510, 513, 516
Quantum well, 473–476
Quantum wire, 177
Quaternary alloy, 81, 88, 90

Radar sensors, 342
Radical-etching mode, 297
Radiometers, 341
Rate limiting process, in etching, 297
Reaction products
  desorption, 292
  formation, 292
Reactive ion beam etching (RIBE), 290–291, 295

Reactive ion etching (RIE), 290, 413
Real surface, 269
Recombination, 504, 510
Recombination current, 394, 413, 417
  nonradiative, 311, 326
Reflections, 546, 551, 557
Refractive index, 23–24, 91, 96
Reliability, 25, 473–476, 525–526, 544–547, 552
Residence time, 174
Residual strain, 147, 152–153
Resonant tunneling diode, 439
Resonant tunneling transistor, 439
Response speed, 506–508
RF operation, 357–364
Rib waveguides, 235
RIBE, see Reactive ion beam etching
RIBE-induced damage, 330
Richardson constant, 262–263
RIE, see Reactive ion etching
RIE-ed facet, 312–313

SAM, see Separation of absorption and multiplication
Saturation (drift) velocity, 354, 507, 511
Sawtooth bandgap, 525
Saw-toothed undulation, 316
Scattering, 350
SCFL, see Source-coupled FET logic
Schottky barrier height (SBH), 86, 96, 99, 261–262, 266, 363, 369
Schottky barrier layer, 343, 363
Schottky diodes, 57
Schottky layer, 43–44, 49
Schroedinger's equation, 345–346
Selective area epitaxy (SAE)
  generally, 6, 206, 490
  planar, 206, 213
Selective area growth, 175–177
Selective etching, 303–308
Selective infill, 213
Self-alignment, optical, 540, 553, 555, 564–565, 570
Self-bias voltage, 294
Self organized growth, 177–183
Semi-conducting InP, 110, 137–138, 142–146
Semiconductor interlayer, 268–269
Semiconductor waveguide, 19–20, 23
Sensitivity, 504–507, 524, 527
Separation of absorption and multiplication (SAM), 502, 517, 526
SHBT, see Single heterojunction bipolar transistor
Sheet carrier concentration, 325
Sheet carrier density, 308

Shot noise
  generally, 505
  surface leakage, 510
  tunneling, 508–509
  unmultiplied dark, 505–507
Side mode suppression ratio (SMSR), 476
Sidewall profile, etched-, 297
Signal processing, 392, 433
Signal-to-noise (S/N) ratio, 40, 502, 506
Silica-fibers, 88
Silica on silicon, 538,5 46, 551, 563–564, 572
Silicon, 558, 568–569, 571
Silicon bipolar transistors, 53
Single-electron transistor, 177
Single-heterojunction bipolar transistor (SHBT), 53
  DC characteristics, 415–416
  design, 413
  generally, 392, 404–406, 438
  RF characteristics, 425–427
Single mode, 540, 548
SIS, *see* Superconductor–insulator– superconductor
S-K growth mode, *see* Stranski-Krastanow (S-K) growth mode
Small signal equivalent circuit (SEC), 357, 398–402
SMSR, *see* Side mode suppression ratio
Solder bump, 542
Solubility, 138
Solute, 123
Stochiometric InP, 119
Source-coupled FET logic (SCFL), 52
Sp^3-hybridized structure, 72–74
SPA, *see* Surface photo-absorption
Space-charge region (SCR), 393, 399, 401, 407, 411
Spacer layer, 343, 364
S-parameters, 398
Split channel concept, 366, 372
Spontaneous emission, 100
Spot-size converter, 18–20, 492
SS-LDs (spot-size-converter-integrated laser diodes), 19
s-symmetry, 72
Step edge, 75
Step flow growth, 170
Stoichiometric InP, 111
Strain
  compensated structure, 251
  generally, 93, 348–349
Strained disk laser, 183
Strained layer quantum-well structures, 5
Strained quantum well, 459–463

Stranski-Krastanow (S-K) growth mode, 177–178
Stress test, 474
Striations, 123–124, 153, 155–156
Subbands, 346
Submerged heater, 134
Subscribers, 16
Substrate, (311)B, 179–180, 182
Substrates, 110, 135
Superconductor–insulator–superconductor (SIS), 46
Superlattice channel, 372
Superlattice grading, 413, 418
Surface, generally
  (111)A, 78
  (001), 77–78
  (110), 77–78
Surface breakdown, 413
Surface depleted cap layer concept, 367
Surface diffusion, 208–211
Surface emitting laser, 470–473
Surface passivation, 269–273
Surface photo-absorption (SPA), 167–168
Surface recombination velocity (SRV)
  defined, 311
  nonradiative, 326
Surface reconstruction, 167–168
Surface selective growth, 198–222
Surface states, 270
Surface step density, 209–210
Surface temperature, 196, 198
Surface treatment
  plasma, 268
  sulphur and selenium, 267
Surface-state density, 86–87
Synthesis, 110–111, 116–117

T_0, *see* Characteristic temperature
TBAs, *see* Tertiarybutylarsine
TBP, *see* Tertiarybutylphosphine
TDCM, *see* Time-dependent charge measurement
TDM systems, 22. *See also* Time-division multiplexing
TE/TM gain, 483
Telecommunication applications, 436–438
TEM, *see* Transmission electron microscopy
Terahertz waves, 110
Ternary alloy, 81, 88, 90–92, 110, 134–135
Ternary bulk crystal, 5
Tertiarybutylarsine (TBAs), 167, 193–194, 224
Tertiarybutylphosphine (TBP), 167, 193–196, 222–224
Thermal, generally
  characteristics, 418

592 INDEX

Thermal, generally *(continued)*
  conductivity, 402, 412, 419–423, 544, 561
  gradients, 121
  instability, 407, 423–425
  stress, 126–127
Three-dimensional, 100
Three-dimensional growth, 373–374
Threshold current, 453, 458
Threshold current density, 225, 235
Tilted facet amplifier, 483–484
Time-dependent charge measurement (TDCM), 147, 151–152
Time-division multiplexing (TDM), 2–3
TMAl, *see* Trimethylaluminum
TMGa, *see* Trimethylgallium
TMIn, *see* Trimethylindium
Transconductance, 355–356, 365, 395
Transit time
  generally, 357
  in HBT, 401, 403
  in photodiode, 506, 511, 513
Transitivity rule, 254–256
Transmission electron microscope (TEM), 90, 170
Transmission line
  coplanar, 542
  generally, 542–543, 571
  microstrip, 542–543
Transmit/receive (T/R) system, 40, 42
Transport equation, 396
  Boltzmann, 396
Travelling wave amplifier, 478–479
Trimethylaluminum (TMAl), 166
Trimethylgallium (TMGa), 166
Trimethylindium (TMIn), 166, 188
T/R system, *see* Transmit/receive system
Tunneling, 408
Turbulent convection, 123
Turn-on delay, 26
Twin lamella, 113
Twinning, 113–114, 133
Two-dimensional, 93, 103
Two-dimensional electron gas (2DEG), 44, 49, 311, 345–347, 371
2D nucleation, 168
2DEG, *see* Twodimensional electron gas
Type-I, 100, 251–252
Type-II, 100–101, 251–252

UHV-based multiple-chamber system, 328
Unified defect model, 265

Unstable region, 96

Valence band
  band offset, 404, 413
  discontinuity, 406, 424
  generally, 79, 397–398
Valence electron, 75
Valence-force-field (VFF) model, 76–79
Valley, 79
Vapor pressure, 116, 190
Vapor pressure controlled Czochralski (VCZ), 112
VCSEL, *see* Vertical cavity surface emitting laser
VCZ, *see* Vapor pressure controlled Czochralski
Vegard's law, 90
Velocity-field (v-E) characteristic, 85–87
Velocity saturation model (VSM), 354
Vertical cavity surface emitting laser (VCSEL), 7, 313, 315, 471, 488
Vertical profile, etched, 297–299
Vertical gradient freeze (VGF), 112, 120, 130–133
VFF, *see* Valence-force-field (VFF) model
VGF, *see* Vertical gradient freeze
V-grooves, 564–565, 570
Volatile reaction production, 292

Wafer annealing, 142–146
Wafer fusion, 472–473
Wafer-fusion technique, 6
Wavefunction, 73
Waveguide, 235
Waveguide filter, 24
Waveguide taper, *see* Mode shape adapter
  generally, 554
  lateral, 555–557
  vertical, 556–560
Waveguide type photodetector, 22
Wavelength-division multiplexing (WDM), 2–3, 16–17, 21, 528
W-band, 380
WDM, *see* Wavelength-division multiplexing
Wet-etched lasers, 313–314
Wire bonding, 542, 571
Wireless LANs, 4

X-ray topography, 128, 156

Zincblende structure, 72–73, 75
ZnSe, 74–76